Salmonid Field Protocols Handbook
Techniques for Assessing Status and Trends in Salmon and Trout Populations

David H. Johnson
Brianna M. Shrier
Jennifer S. O'Neal
John A. Knutzen
Xanthippe Augerot
Thomas A. O'Neil
Todd N. Pearsons

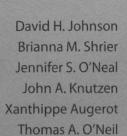

American Fisheries Society
in association with
State of the Salmon

Salmonid Field Protocols Handbook
Techniques for Assessing Status and Trends in Salmon and Trout Populations

David H. Johnson
Brianna M. Shrier
Jennifer S. O'Neal
John A. Knutzen
Xanthippe Augerot
Thomas A. O'Neil
Todd N. Pearsons

American Fisheries Society
Bethesda, Maryland

in association with

State of the Salmon
Portland, Oregon

Suggested citation formats follow.

Entire book
Johnson, D. H., B. M. Shrier, J. S. O'Neal, J. A. Knutzen, X. Augerot, T. A. O'Neil, and T. N. Pearsons, 2007. Salmonid field protocols handbook: techniques for assessing status and trends in salmon and trout populations. American Fisheries Society, Bethesda, Maryland.

Chapter within the book
Wagner, P. G., 2007. Fish counting at large hydroelectric projects. Pages 173–195 in D. H. Johnson, B. M. Shrier, J. S. O'Neal, J. A. Knutzen, X. Augerot, T. A. O'Neil, and T. N. Pearsons. Salmonid field protocols handbook: techniques for assessing status and trends in salmon and trout populations. American Fisheries Society, Bethesda, Maryland.

Front cover: cast netter Sergey Zolotukhin photo by Brian Caouette, electrofishing crew by Gabe M. Temple, rotary screw trap by Mike Ackley (WDFW), snorkeler by Keith Wolf, weir by Christian E. Zimmerman, beach seiners by Richard A. Henderson, and spawning salmon by Todd N. Pearsons.
Back cover: Wild salmon design by Port Gamble S'Klallam artist R. Brian Perry. (Digitized by Andrew Fuller.)

© 2007 by the American Fisheries Society

All rights reserved. Photocopying for internal or personal use, or for the internal or personal use of specific clients, is permitted by AFS provided that the appropriate fee is paid directly to Copyright Clearing Center (CCC), 222 Rosewood Drive, Danvers, Massachusetts, 01923, USA; phone 978-750-8400. Request authorization to make multiple copies for classroom use from CCC. These permissions do not extend to electronic distribution or long-term storage of articles or to copying for resale, promotion, advertising, general distribution, or creation of new collective works. For such uses, permission or license must be obtained from AFS.

Printed in the United States of America on acid-free paper

Library of Congress Control Number: 2007923213
ISBN 978-1-888569-92-6

American Fisheries Society
5410 Grosvenor Lane, Suite 110
Bethesda, Maryland 20814-2199
www.fisheries.org

in association with

State of the Salmon
a joint program of Wild Salmon Center and Ecotrust
721 NW Ninth Avenue, Suite 200
Portland, Oregon 97209
www.stateofthesalmon.org

CONTENTS

Executive Summary ... v
Acknowledgments ... vii

ESSAYS

Introduction ... 1
    Xanthippe Augerot

Evolving Towards a Common Global Language for Salmon Conservation ... 7
    Samantha Chilcote

The Role of Sample Surveys: Why Should Practitioners Consider Using a Statistical Sampling Design? ... 11
    Donald L. Stevens, Jr., David P. Larsen, and Anthony R. Olsen

Data Management: From Field Collection to Regional Sharing ... 25
    Part I: Field Data Collection
    Stewart Toshach, Richard J. O'Connor, Thomas A. O'Neil, Bruce A. Patten, Cedric Cooney, Bill Kinney, Paul Huffman, and Frank Young
    Part II: Regional Data Sharing
    Thomas A. O'Neil, Stewart Toshach, and Wayne Luscombe

Methods ... 55

FIELD PROTOCOLS

Carcass Counts ... 59
    Bruce Crawford, Thaddeus R. Mosey, and David H. Johnson

Cast Nets ... 87
    Kaneaki Edo

Electrofishing: Backpack and Drift Boat ... 95
    Gabriel M. Temple and Todd N. Pearsons

Hydroacoustics: Rivers ... 133
    Suzanne L. Maxwell

Hydroacoustics: Lakes and Reservoirs ... 153
    J. Christopher Taylor and Suzanne L. Maxwell

Fish Counting at Large Hydroelectric Projects ... 173
    Paul G. Wagner

Redd Counts ... 197
    Sean P. Gallagher, Peter K. J. Hahn, and David H. Johnson

Rotary Screw Traps and Inclined Plane Screen Traps ... 235
    Gregory C. Volkhardt, Steven L. Johnson, Bruce A. Miller, Thomas E. Nickelson, and David E. Seiler

Beach Seining ... 267
    Peter K. J. Hahn, Richard E. Bailey, and Annalissa Ritchie

Snorkel Surveys ... 325
    Jennifer S. O'Neal

Tangle Nets ... 341
    Charmane E. Ashbrook, Kyong W. Hassel, James F. Dixon, and Annette Hoffmann

Tower Counts ... 363
    Carol Ann Woody

Weirs ... 385
    Christian E. Zimmerman and Laura M. Zabkar

SUPPLEMENTAL TECHNIQUES

Aerial Counts — 399
Edgar L. Jones III, Steve Heinl, and Keith Pahlke

Fyke Nets (in Lentic Habitats and Estuaries) — 411
Jennifer S. O'Neal

Variable Mesh Gill Nets (in Lakes) — 425
Bruce Crawford

Foot-based Visual Surveys for Spawning Salmon — 435
Bruce Crawford, Thaddeus R. Mosey, and David H. Johnson

Video Methodology — 443
Jennifer S. O'Neal

Glossary — 459

Index — 477

We dedicate this book to salmon and to the systems they reside within.
We recognize their spirit and all they stand for, and we hope that
by preserving their way of life we will in no small way preserve ours.
We honor those who have come before us and our current and
future earth stewards who come with waders, wet suits,
wet hands, coldwater headaches, and sore thumbs from counters,
and live a semiaquatic, slip-resistant, felt-soled lifestyle.

In the spirit of interdependence, we put forth the protocols
in this book as we strive to create a common language
of salmon knowledge for their survival and ours.

Executive Summary

Worldwide, many salmonid populations have been extirpated, many more are declining, and overall distribution throughout their natural range is shrinking. The management, conservation, monitoring, and research regarding salmonids (salmon, trout, and chars) around the world require detailed data that have been reliably and consistently gathered. This book reflects the advancement of scientifically rigorous techniques for assessing salmonid populations in freshwater and estuarine environments. While based largely upon work conducted in the Pacific Northwest of the United States and western Canada, the materials within this book can be used for salmonids anywhere.

Government agencies, fisheries resource managers, and scientists depend on reliable, consistent, high-quality data to make informed management decisions and test hypotheses. State of the Salmon's inventory of long-term assessment, research, and monitoring efforts across the North Pacific reveals that an estimated 8,000 individual activities are underway in the United States and Canada, with an additional but unknown number of activities in Russia and Japan. This considerable geographic scope of activities presents a challenge and an opportunity for the salmonid management community. International treaty organizations such as the Pacific Salmon Commission need to be able to connect and compare numerous independent research efforts across great distances and multiple jurisdictions to characterize status or trends regarding particular salmonid species and stocks.

Our research and salmonid management questions are increasingly complex. For instance, determining key aspects of regional and species-wide population status and distributions and the reasons behind changes in status is critical for effective management. The effects of hatchery fish or habitat changes on wild stocks, the impacts of marine-derived nutrients on ecosystem health, and basic salmonid life history and genetic characteristics are also key considerations for researchers and managers. All these questions require well-structured and well-planned data acquisition programs.

The standardized methods for collecting salmonid data detailed in this book represent an opportunity for scientists and managers to link independent monitoring efforts and improve the quality and consistency of data gathered during fieldwork. Common protocols allow scientists to compare results among projects more reliably and will give managers greater confidence in data collected to adjust harvest levels or prioritize research and conservation efforts. Funding institutions supporting field research will be more inclined to do so if monitoring is grounded in proven techniques that yield the most useful data.

During this three-year project, we reviewed and drew from more than 375 published and unpublished techniques and guidelines. Herein, fisheries managers, scientists, and students will find 13 principle techniques for assessing salmonid populations in the wild. These scientifically rigorous, peer-reviewed methods are carcass counts, cast nets, dam counts, boat and backpack electrofishing, hydroacoustics in rivers and lakes, redd counts, seining, rotary screw and inclined plane traps, snorkeling, tangle nets, counting towers, and weirs. We have included five additional techniques—aerial surveys, fyke nets, gill nets, foot-based spawner counts, and video surveys—that can be used with any of the 13 principle methods to supplement information gathered. The prefatory chapters explain the book's genesis and goals and include instructive essays on the importance of sampling design and data management.

We encourage fisheries agencies, funding organizations supporting salmonid research, and educational programs involved with these important fishes to embrace these protocols in the spirit of the advancement of scientifically based conservation.

Acknowledgments

We are indebted to many institutions and individuals who contributed time, expertise, inspiration and financial support to the making of this book.

In particular, we are grateful to John Piccininni at Bonneville Power Administration for handling the original BPA grant that funded the project's initial work, and to Dr. Tom Karier at the Northwest Power and Conservation Council for his support. We also recognize the peer review contributions from members of the Pacific Northwest Aquatic Monitoring Partnership (PNAMP) and the Collaborative and Systemwide Monitoring and Evaluation Project (CSMEP). PNAMP and CSMEP members represent a solid core of scientific and on-the-ground knowledge regarding salmonid ecology and monitoring methods. Their unyielding enthusiasm and support for this project has made the product richer, broader, and stronger.

We would like to praise Dana Foley, editorial manager, for pulling this project and its many pieces together—a herculean task. Andrew Fuller, senior designer at Ecotrust, has done a fine job of taking a drab document and giving it a much more inviting appearance. We appreciate the efforts of Chris Robbins, policy coordinator for State of the Salmon, for helping bring closure to the project.

Major Funding Sources
The Gordon and Betty Moore Foundation, Wild Salmon Center, Bonneville Power Administration, the Northwest Power and Conservation Council, and Washington Department of Fish and Wildlife.

Funding and Project Partners
Alaska Department of Fish & Game; BioAnalysts, Inc.; Confederated Tribes of the Siletz Indians, Oregon; Fisheries and Oceans Canada, British Columbia; Columbia Basin Fish & Wildlife Authority; Hokkaido Institute of Environmental Sciences, Japan; National Oceanic & Atmospheric Administration, Northwest Fisheries Science Center; Northwest Habitat Institute; Northwest Power & Conservation Council; Oregon Department of Fish & Wildlife; Oregon State University; Public Utility District No. 1 of Chelan County, Washington; Smithsonian Institution, Division of Fishes; StreamNet; Skagit River System Cooperative; Sustainable Fisheries Foundation; Talon Scientific; Tetra Tech EC Inc.; Office of the Interagency Committee, Washington Salmon Recovery Funding Board; Wild Salmon Center; USDA Forest Service; U.S. Department of Energy, Bonneville Power Administration; U.S. Geological Survey; Washington Department of Fish & Wildlife; and Yakama Nation, Washington.

Experts Workshop
A significant milestone reached during this book's development was an experts workshop that State of the Salmon and the Washington Department of Fish and Wildlife jointly convened in Welches, Oregon, in March 2004. Workshop participants contributed their expertise and time to culling, reviewing, and shaping a standardized suite of monitoring protocols that evolved into the final set of protocols presented in this book. The group of international professionals in attendance (listed below) contributed scientific insight and moral support that moved this book forward.

Individual	Affiliation
Carmen Andonaegui	Washington Department of Fish and Wildlife
Richard Bailey	Fisheries and Oceans Canada
James Brady	North Cape Fisheries Consulting
Brian Bue	State of Alaska
Kelly Burnett	USDA Forest Service
Patrick Connolly	United States Geological Survey
Cedric Cooney	Oregon Department of Fish and Wildlife
Bruce Crawford	Washington Interagency for Outdoor Recreation, Salmon Recovery Funding Board
Kaneaki Edo	Hokkaido Institute of Environmental Sciences
Jim Geiselman	Bonneville Power Administration
Peter Hahn	Washington Department of Fish and Wildlife

Individual	Affiliation
Dave Heller	USDA Forest Service
Tracy Hillman	BioAnalysts, Inc.
Paul Huffman	Yakama Nation
Jim Irvine	Fisheries and Oceans Canada
Bill Kinney	Pacific States Marine Fish Commission, StreamNet
Steve Lanigan	USDA Forest Service
Hiram Li	Oregon State University
Suzanne Maxwell	Alaska Department of Fish and Game
Mitsuhiro Nagata	Hokkaido Fish Hatchery
Dick O'Connor	Washington Department of Fish and Wildlife
Tom O'Neil	Northwest Habitat Institute
Bruce Patten	Fisheries and Oceans Canada
Todd Pearsons	Washington Department of Fish and Wildlife
Ned Pittman	Washington Department of Fish and Wildlife
Derek Poon	U.S. Environmental Protection Agency
Jeff Rodgers	Oregon Department of Fish and Wildlife
Brett Roper	USDA Forest Service
Sam Sharr	Idaho Department of Fish and Game
Sergei Sinyakov	KamchatNIRO
Don Stevens	Oregon State University
Stewart Toshach	National Oceanic and Atmospheric Administration
Stan Van Der Wetering	Confederated Tribes of Siletz Indians
Greg Volkhardt	Washington Department of Fish and Wildlife
Keith Wolf	KWA Ecological Sciences, Inc.
Sergei Zolotukhin	KhabarovskTINRO

Pacific Northwest Aquatic Monitoring Partnership

The authors consulted and collaborated with the Pacific Northwest Aquatic Monitoring Partnership (PNAMP) and relied on PNAMP's Fish Population Monitoring Work Group for technical input. Members of this group that contributed are listed below. PNAMP, formed by leaders of aquatic monitoring programs and encompassing a geographic area from northern California to Canada, provides a forum in which state, federal, and tribal aquatic habitat and salmonid monitoring programs can be better coordinated and integrated. The forum is intended to facilitate interagency agreements and craft recommendations on consistent monitoring approaches and data sharing within the community. For more information, visit <www.pnamp.org>.

Individual	Affiliation
John Arterburn	Colville Tribe
Will Beattie	Northwest Indian Fish Commission
Patrick Connolly	United States Geological Survey
Bruce Crawford	Washington Interagency for Outdoor Recreation, Salmon Recovery Funding Board
Tim Fisher	Fisher Consulting
Kurt Fresh	National Oceanic and Atmospheric Administration
Jill Hardiman	United States Geological Survey
Tracy Hillman	BioAnalysts, Inc.

Individual	Affiliation
Mike Hurley	M and M Consulting
David Jepsen	Oregon State University
Frank McCormick	USDA Forest Service
Kelly Moore	State of Oregon
Tom Nickelson	Oregon Department of Fish and Wildlife
Jennifer S. O'Neal	TetraTech EC Inc.
Dan Rawding	Washington Department of Fish and Wildlife
Jeff Rodgers	Oregon Department of Fish and Wildlife
Dave Seiler	Washington Department of Fish and Wildlife
Stewart Toshach	National Oceanic and Atmospheric Administration
Greg Volkhardt	Washington Department of Fish and Wildlife
Paul Wagner	KWA Ecological Sciences, Inc.

Collaborative and Systemwide Monitoring and Evaluation Project

This book has also benefited from the involvement of scientists affiliated with the Collaborative and Systemwide Monitoring and Evaluation Project (CSMEP). CSMEP's mission is twofold: (1) to coordinate efforts to improve the quality, consistency, and focus of fish population and habitat data; and (2) to answer key monitoring and evaluation questions relevant to major decisions in the Columbia River basin. Initiated in 2003, CSMEP is administered by the Columbia Basin Fish and Wildlife Authority, with more than 30 participating scientists from federal, state, and tribal fish and wildlife agencies and consulting firms.

Other Contributors

We also want to acknowledge a legion of other individuals for contributing their time and talent to this book project.

Individual	Affiliation
Lew Adkins	Washington Department of Fish and Wildlife
Stan Allen	Pacific States Marine Fish Commission
Jim Ames	Washington Department of Fish and Wildlife
Dale Bambrick	National Oceanic and Atmospheric Administration
Mike Bannock	Pacific States Marine Fish Commission
Eric Beamer	Skagit River Cooperative
Todd Bennet	National Oceanic and Atmospheric Administration
Ed Bowles	State of Oregon
Rich Carmichael	Eastern Oregon University
Craig Contor	Confederated Tribes of the Umatilla Indian Reservation
Michelle DeHart	Fish Passage Center, Oregon
S. Downie	California Department of Fish and Game
David Fast	Yakama Indian Nation
Jim Geiselman	Bonneville Power Administration
Correigh Green	National Oceanic and Atmospheric Administration
Stan Gregory	Oregon State University
Jeff Haymes	Washington Department of Fish and Wildlife
Lillian Herger	U.S. Environmental Protection Agency
C. Hunter	State of Montana

Individual	Affiliation
Chris Jordan	National Oceanic and Atmospheric Administration
Tom Karier	Northwest Power and Conservation Council
Steve Katz	National Oceanic and Atmospheric Administration
Phil Kaufmann	U.S. Environmental Protection Agency
Cathy Kellon	State of the Salmon
G. Knott	United States Bureau of Reclamation
Steve Lanigan	USDA Forest Service
Phil Larsen	U.S. Environmental Protection Agency
Steve Leider	Governor's Office, Washington
Bob Leland	Washington Department of Fish and Wildlife
Jerry Marco	Colville Tribe
Vince Moore	Idaho Department of Fish and Game
Phil Mundy	State of Alaska
Mike Newsome	United States Bureau of Reclamation
Tony Nigro	Oregon Department Fish and Wildlife
George Pess	National Oceanic and Atmospheric Administration
Charlie Petrosky	Idaho Fish and Game
Chuck Peven	Chelan County Public Utility District
Todd Reeve	Bonneville Environmental Foundation
Phil Roger	Columbia River Inter-Tribal Fish Commission
Phil Roni	National Oceanic and Atmospheric Administration
Jim Ruzycki	Oregon Department of Fish and Wildlife
Bruce Sandford	Washington Department of Fish and Wildlife
Howard Schaller	United States Fish and Wildlife Service
Bruce Schmidt	Pacific States Marine Fish Commission
Jim Scott	Washington Department of Fish and Wildlife
Kevin Shaffer	California Department of Fish and Game
Sam Sharr	Idaho Department of Fish and Game
Greg Sieglitz	State of Oregon
Michael Sparkman	California Department of Fish and Game
Kate Terrell	United States Fish and Wildlife Service
Chris Toole	National Oceanic and Atmospheric Administration
Christian Torgersen	United States Geological Survey
Bill Tweit	Washington Department of Fish and Wildlife
Ian Waite	United States Geological Survey
Bruce Ward	British Columbia Natural Resources
Steve Waste	Northwest Power and Conservation Council
D. Weigel	United States Bureau of Reclamation
Timothy Whitesel	United States Fish and Wildlife Service
Christian Zimmerman	United States Geological Survey

Introduction
Xanthippe Augerot

As of 2003, the Order *Salmoniformes* (i.e., salmon, trouts, chars) contained 217 valid taxa (Integrated Taxonomic Information System, <www.cbif.gc.ca>); an additional species was described in 2005 (Safronov and Zvezdov 2005). While this book has applications to the full array of salmonid taxa, it is based largely on the science developed on Pacific salmon.

Pacific salmon inhabit streams stretching from California to Japan, encompassing some 31 million square kilometers and spanning hundreds of jurisdictions. They begin their complex life history in freshwater, migrate and mature at sea, and return to their natal streams, where they spawn, die, and deliver vital ocean-derived nutrients to freshwater, riparian, and terrestrial ecosystems. Few fish species are as important to North Pacific ecosystems, native peoples, and coastal economies as Pacific salmon. Their role in nature and value to humankind inspires intensive and frequent study. Whether surveying salmon on the Kamchatka peninsula in Russia or in the Columbia River basin in the United States, scientists need a set of standard monitoring protocols that minimizes methodological errors, maximizes the validity and consistency of data, and allows them to make reliable comparisons and reasonable conclusions across projects and river basins and over time. This book's objective is to describe a standard set of monitoring protocols and best practices that decision makers and funding organizations can adopt and practitioners can use to design study and sampling techniques, conduct field activities, and manage spatial and tabular data.

Background
The Wild Salmon Center (WSC) and the joint WSC–Ecotrust State of the Salmon Program recently concluded its first range-wide assessment of the risk of extinction for Pacific salmon (Augerot 2005). A formidable challenge was trying to piece together fragmented species data across jurisdictional boundaries, understanding that differences in data collection methods and the uneven distribution of data by species and region were largely responsible. Similar difficulties have been encountered in the Pacific Northwest region of the United States. Government decision makers at the Bonneville Power Authority and the Northwest Power and Conservation Council and members of the United States Congress have spent hundreds of thousands of dollars over the years to track salmon status and temporal trends across the Columbia River basin only to realize that individual projects cannot easily be "stitched together" due to inconsistent approaches used to collect and count fish.

To address these fundamental differences in data collection, David Johnson, then at the Washington Department of Fish and Wildlife, proposed bringing together a group of regional experts to determine best practices for a suite of commonly used field techniques and to merge the results into an easy-to-use resource. In 2004 State of the Salmon and the Washington Department of Fish and Wildlife coconvened a workshop of international experts in Welches, Oregon to discuss a monitoring strategy for Pacific Rim salmonids. Workshop participants reviewed 375 documents and took a major step towards producing standardized monitoring protocols that could be used in the field. Following the workshop we

INTRODUCTION

whittled down, refined, and standardized the array of field monitoring techniques to 18 core techniques that were then subject to extensive and intensive review by our peers. This handbook is the culmination of these efforts and represents the most comprehensive set of monitoring protocols ever compiled for Pacific salmon and trout.

Meeting the needs of policy makers

Utilizing best monitoring practices is an essential first step to estimating salmonid population status and prioritizing conservation actions. Decision makers and fishery managers find project-level data most useful if field research and monitoring objectives are clearly defined. One of the most frequently overlooked elements of salmon monitoring is selecting appropriate sample design—how we space samples over habitats and over time—to answer relevant questions. Sample designs need to yield the best possible data in a timely and cost-effective manner to be of maximum value to decision makers. While key research and monitoring questions may seem straightforward on paper, in reality the process is often complicated. This is especially true in regions where salmon populations straddle land ownerships and site accessibility becomes an issue. Competing demands for land, water, fish, and wildlife can also create friction among user groups and put researchers in the potentially awkward position of limiting or altering the scope of studies for political reasons. The Columbia River basin in the Pacific Northwest embodies many of these challenges.

The data collection and enumeration protocols in this book serve two distinct monitoring and management goals: to support sustainable salmon harvest and to ensure the viability of threatened and endangered salmon populations. In Alaska, northern British Columbia, and much of the Russian Far East, the principal salmon monitoring objective is stock forecasting, to allocate fishing effort in a manner that allows for sustainable salmon productivity in perpetuity. In shorthand, we can refer to this monitoring and management approach as monitoring to optimize biomass (Hyatt 1996). As one progresses farther south into British Columbia and the U.S. Pacific Northwest, where river basins have been more heavily altered by people, endangered species mandates such as Canada's Species at Risk Act (SARA, 2003) and the U.S. Endangered Species Act (ESA, 1973) drive another monitoring mandate: to provide status assessments, trend assessments, and assessments of salmon population viability across conservation units.[1] For the sake of convenience we will refer to this mandate as managing for biodiversity conservation. Last, the research community is amassing evidence to support what many long suspected, that the annual return of the salmon to our North Pacific river systems is a vital element of the nutrient supply system, feeding riparian vegetation, insects, fish, and birds and supporting ecosystem health (Gende et al. 2002). Canada's new Policy for Conservation of Wild Pacific Salmon explicitly acknowledges maintenance of ecosystem integrity as one of its three objectives.

As salmon are listed as threatened or endangered after experiencing precipitous declines in distribution and abundance, additional legal protections are triggered under endangered species laws. However, we find time and again that we do not have the basic building blocks of conservation knowledge to support salmon recovery efforts. We lack information about the distribution of the dozens of discrete biological populations across the landscape. Traditionally, the larger, mixed population fishery "stock" units have been used to gather information

[1] The European Union Directive on the Conservation of Natural Habitats and of Wild Fauna and Flora provides a similar biodiversity mandate (Cowx and Fraser 2003).

and infer status, but the information these units offer is too coarse. As a result, we cannot characterize populations in terms of demographic performance (e.g., recruit-per-spawner, fecundity, survival rates, size-at-age, sex ratios) and life history diversity. We rarely know which populations in a given conservation unit, or evolutionarily significant unit (ESU) for salmon in the United States), are the most productive and represent the strongholds for conservation and restoration. To improve our understanding of population productivity and its bottlenecks in freshwater or at sea, we particularly need low-cost, reliable means to estimate smolt out-migration. Such knowledge about relative population productivity will be vital to designing effective networks to protect wild salmon productivity in salmon conservation reserves.

The Columbia Basin conundrum

In the Columbia River basin, where the initial idea for this book arose, policy makers operate under both the biomass and biodiversity mandates, which are not always in alignment. Agencies setting policy face the need to balance fish and wildlife conservation with a major hydropower system, tribal treaty fishing rights, commercial fishing interests, and aquatic habitats affected by a growing human population. For instance, salmon hatcheries reflect the complexities of the biomass-versus-biodiversity debate. Hatcheries, mandated by the federal government as a mitigation strategy for habitat loss, account for much of the salmon caught by commercial and tribal fishermen; however, hatcheries conflict with biodiversity objectives because they introduce large numbers of artificially cultivated fish to the system, creating the perception that wild populations are healthy and masking declines in natural population—the backbone of salmon diversity and resilience.

In addition, the Columbia River basin is a web of jurisdictions and competing sovereign interests and responsibilities. Two countries, multiple tribal nations, dozens of government agencies, and thousands of land ownerships exist within the Columbia River basin, one of the most complex salmon monitoring and management regions in the North Pacific.

A typical menu of field research and monitoring questions for the Columbia River basin may include the following, among other themes:

- Which populations are recovering sufficiently to delist ESUs listed as endangered or threatened under the U.S. Endangered Species Act?
- Which populations are robust enough to tolerate recreational and commercial fishing?
- Are hatchery releases competing with listed wild salmon in the river, in the estuary, and at sea?
- What is the magnitude of hatchery postrelease mortality or sublethal stress associated with passage in-river and trucking or barging smolts?
- What is the incremental benefit in fish survival of retrofitting dams with smolt passage weirs?
- Can we describe the cumulative effect of competition and predation on wild salmon by native predators? By invasive fish species?
- What areas within the basin would be highest priority for habitat restoration, given the presence of highly productive "core" populations?

- Do new mining ventures present a significant threat to salmon abundance, distribution, productivity, and diversity?

To answer any of these questions requires careful consideration of the field research approaches, desired level of certainty for answers, frequency of information needed, and scale of information needed (e.g., ESU, county, whole-basin), among other factors. Research questions evolve into research objectives and methods, the cornerstones of well-defined study plans. When field data and initial study plans are poorly documented or described (e.g., little or no metadata), the resulting analyses are scientifically less credible, leading to questions about how they should be interpreted and used by decision makers. As a community, we should strive to document study objectives, sample design rationale, and all limitations with respect to valid inferences from the data we collect.

We cannot guarantee that management decisions or policies will be crafted using the best data, but we can strive to ensure that the best data are available to decision makers for salmon conservation.

Improving the efficiency of monitoring: A common research template

Last, we must recognize that as resource conditions change and management priorities evolve, we will need to ask a different set of questions and rely on new data gathering and management tools to inform salmon conservation objectives. The salmon research community could collect data, generate new information, and expand its knowledge base more economically if it used the same methods, metrics, and information system for sharing basic but essential data gathered at different sites over varying lengths of time. One approach is a master sample framework, a monitoring template designed for and applied to river basins, such as the Columbia River basin, where the distribution of fish populations is documented. Developed by survey statisticians, a master sample frame is designed to provide a pool of sample locales across an area, ensuring statistical reliability at multiple spatial scales. If a master sample framework were established for Columbia River basin fish monitoring—and maintained and updated by a statistical committee representing the sampling interests of county, state, tribal, and federal biologists and housed in a central data system—it would provide cost efficiencies in survey design and facilitate the integration of data across individual studies (Turner 2003). Larsen et al. (2003) recommend a master sample consisting of a dense, spatially balanced suite of sites on stream networks that could be tailored to meet multiple agency design needs for fish monitoring in Oregon. The State of the Salmon program is compiling an accessible database by cataloging, standardizing, and integrating biological data from thousands of study sites to lay the groundwork for the selection of a master sample at the scale of the North Pacific. This effort will improve our ability to monitor trends in salmon abundance, diversity, population structure, and productivity over the range of Pacific salmon.

We recognize that the goal of creating a useful template for documenting salmon monitoring activities across boundaries and providing a research reference point for future studies is extremely ambitious; however, this is an achievable goal if the salmon research and conservation community can learn to speak the same language. This means that our salmonid management objectives, key research questions, and design of monitoring programs must be consistent, repeatable, and verifiable to yield the best possible information. We believe that this handbook represents a significant step towards developing that common language.

How to use this handbook

The next two chapters address study and sampling design and the implementation of robust data systems. These two chapters provide a framework for the heart of the handbook—the technique-based field protocols. Each adult and juvenile salmon collection and enumeration technique has its own chapter. While each chapter was authored by a different team of field biologists, each technique is presented and described using the same template for usability:

- Background and objectives
- Sampling design
- Field/office methods
- Data handling, analysis, and reporting
- Personnel requirements and training
- Operational requirements (including budget)
- Literature Cited

The background and objectives section provides context for the sampling technique and will help guide researchers to the most appropriate one, depending on the focal species and site conditions. It also describes the evolution of the technique, the most common objectives for deploying the technique, and for which species, age classes, and environments it is best suited.

The sampling design section contains more detailed information about the applicability of the technique to meeting specific research objectives for particular species and environments. This section also describes elements of overall study design and appropriate spacing of samples in space and time. (For example, weirs, smolt traps, and sonar enumeration systems all have specific siting requirements that may determine where and how each is used to accomplish study objectives.)

The field/office methods section provides an overview of field logistics and related office support needed to assist operations at study sites. (In the case of fixed gear such as smolt traps and weirs, field descriptions are very robust and list alternative specifics of gear deployment for a variety of stream conditions.)

The data handling, analysis, and reporting section provides best practices for analyzing field data, for documenting and handling data, and for conducting quality analysis and quality control on resulting data. Some chapters provide detailed discussions about alternative approaches for conducting status and trend analyses.

Sections on personnel requirements and training and other operational requirements further facilitate the comparison across techniques. The roles and responsibilities of personnel and training and safety issues are addressed for every technique. The operational requirements section explains the typical amount of time needed to plan and implement the study and covers field schedule, equipment, facility needs, and budgetary considerations.

INTRODUCTION

Literature Cited

Augerot, X. 2005. Atlas of Pacific salmon. UC Berkeley Press, Berkeley, California.

Gende, S. M., R. T. Edwards, M. F. Willson, and M. S. Wipfli. 2002. Pacific salmon in aquatic and terrestrial ecosystems. Bioscience 52(10):917–928.

Hyatt, K. D. 1996. Stewardship for biomass or biodiversity. Fisheries 21(10):4–5.

Larsen, D., A. R. Olsen, and D. L. Stevens, Jr. 2003. The concept of a master sample: an Oregon case study. Presented at Designs and Models for Aquatic Resource Surveys. Oregon State University, 7–9 September, 2003.

Safronov, S. N., and T. V. Zvezdov. 2005. *Salvelinus vasiljevae sp. nova*. A new species of freshwater chars (*Salmonidae, Salmoniformes*) from northwestern Sakhalin. Journal of Ichthyology 45(9):700-711.

Turner, A. G. 2003. Sampling frames and master samples, draft. Expert Group Meeting to Review the Draft Handbook on Designing of Household Sample Surveys, 3–5 December 2003. United Nations Secretariat Statistics Division, ESA/STAT/AC.93/3.

Evolving towards a Common Global Language for Salmon Conservation

Samantha Chilcote

I recently had the distinct honor to work with Dr. Anatoly Semenchenko, a field biologist at the Russian fisheries agency TINRO Center. Dr. Semenchenko is an esteemed scientist and, for the last several years, has focused much of his efforts on the Samarga River in the southern Russian Far East, largely without institutional support. The Samarga remains relatively untouched, a center of biodiversity in the eastern Sikhote-Alin Mountains and home to one of the world's largest populations of endangered Sakhalin taimen *Hucho perryi*. Dr. Semenchenko was so interested in the Samarga that he gave up his vacation time to join us for an expedition.

As part of my postdoctoral work at the University of Montana and in conjunction with Wild Salmon Center, I was tasked with conducting a rapid assessment of this beautifully complex river and its astounding biotic diversity. It is a 600,000 ha roadless watershed where logging has just begun. Fortunately, the logging company has received Forest Stewardship Council certification and is working with conservation groups to set aside critical habitat and otherwise minimize environmental impacts. Therefore, our scientific research would have direct application on the land and to the fish. I had a lot to learn about this wonderful watershed and had to learn it fast.

Of course, Dr. Semenchenko knows that river better than any fish biologist—but the question nagged at me: what exactly does he know? We studied his earlier rapid watershed assessment; in sum, it is a phenomenal body of information about fish richness, abundance, distribution, and spawning areas. Acknowledging the wealth of information that could be gleaned from a larger, collaborative study, I wanted to tap his experience for the full breadth of what he could tell me.

I asked Dr. Semenchenko about his observations on life history diversity. To me, it was a basic platform: If we do not understand life history diversity, we do not understand what habitats the fish are using. Life history diversity is so often how we frame salmon study. I simply assumed that his observations had the same foundation. But Dr. Semenchenko could not fathom my objectives, and I sensed his annoyance at the possibility that we were planning to impose an entirely new methodology onto his protocols. This went on for about an hour, with a translator as a go-between to try to define what was to me an elemental term. Without clarifying the differences in terminology and the nuances and basic foundations that separate our research, we were stuck. Finally, we determined that what we call life history variations, the Russians call ecological forms.

Of course, my experience in Russia is an extreme example of scientists quite literally using different languages, but we face this problem everywhere at every scale in salmon study—if not in science all around. I am confronting this same lack of clarity here at home right now while working with Wild Salmon Center field biologist John McMillan to refine research and monitoring on the Hoh River in Washington's Olympic Peninsula. We are developing protocols for the Hoh River regarding mapping different floodplain habitats, such as parafluvial and orthofluvial springbrooks. Yet, as John and I were talking, we realized that he and I have very different notions of what constitutes a springbrook habitat. Before we could advance any further, we would need to define our term; otherwise, we would have to stratify our study area and describe the Hoh system very differently.

As scientists, fish managers, and conservationists, we need a common language. Ultimately, our task is to produce precise lab work and accurate fieldwork. Both must be replicable in order to be tested and validated, but without standardized protocols, we will continue to collect data that reflect variations in methodology and cannot be stitched together across project or jurisdictional boundaries. Without common definitions, shared objectives, and standardized protocols, our data will continue to be localized, incomparable between researchers and localities, and, ultimately, fixed in scale.

The ecosystem approach we are moving towards in salmon study is an evolution. Essentially, with the development of ecosystem and North Pacific perspectives on salmon conservation, many methodologies still utilize only site-based studies, capturing only small fragments of larger river basin ecosystems. Unfortunately, although there is increasing collaboration across stakeholders, there remains relatively little collaboration across federal agencies, state agencies, and universities. Seldom are there good ecosystem descriptions with consistent methodology at regional and larger scales.

Certainly we have difficulty speaking across nations, but we also stumble when we attempt to reconcile data at different scales or when comparing data collected in different river basins using similar monitoring techniques. Specialties present yet another level of complexity; perhaps a scientist is well practiced in snorkeling but may not have a thorough grasp of other monitoring techniques conducted in the same watershed for the same purpose.

We are at a crossroads in salmon study. We are advancing concepts by applying an ecosystem approach to our research; at the same time we are making exciting technical advancements through geographic information systems, which enable us to work at multiple scales and provide an opportunity to aggregate data in a spatially explicit manner.

Our objectives as scientists studying a migratory species at multiple scales require us to evolve our techniques as practitioners. We are in a similar place the scientists were in 250 years ago, when our predecessors acknowledged the need for a common language to study salmon. It is not that long ago, after all, that George Wilhelm Steller, in the 1740s, phonetically transcribed the Koryak name for salmon into his German manuscript. Over the next 50 years, his book was translated into French, English, and Russian; it was not until 1792 that Steller's book was Latinized by Dr. Johann Julius Walbaum to conform to scientific naming conventions. The Linnean classification system was needed to identify organisms across the world, but there was no extension of that concept to ecological terminology and methodology.

It is time to develop a shared vision and standardize techniques and to create new methodologies that are scaleable and comparable. As we begin to craft these techniques, we first need to ask some hard questions:

- How do we bring the methodology up to the conceptual level?
- How do we choose the best method to achieve our objective?
- How can we ensure that we can cross-validate findings?
- How can we collect data that can be used at multiple scales?

- How can we ensure that our data can be seamlessly used by other scientists using different methods?
- And conversely, how can we use aggregate data from our colleagues to enhance our own work?

If, for example, our objective is species composition and habitat assessment, then try snorkel surveys. Is the water too turbid? Try electrofishing. This also requires weighing the benefits of accurate size measurements against stress delivered to fish and the food web from electroshocking.

Moving in the direction of a common language and standardized data collection methods, we will be able to make our work more replicable and therefore more credible and useful in the policy process. We will be able to compile the best available information, increasing the precision of field-derived data. This in turn will strengthen the application of theory to real-life circumstances. We will fulfill the needs of multiple stakeholders on a more neutral footing. We will be able to work at multiple spatial scales and compare data from different studies and different rivers around the world, which would open up the types of analysis that could be done on existing data sets (e.g., comparisons from different teams within and between river basins).

These are worthy goals, but here is an even more convincing one: our resources as field biologists and managers are becoming increasingly limited, as is the time remaining to make a difference in the conservation and preservation of our study subjects. We need to optimize our efforts and maximize the use of data that is now available and increase the quality of data that will become available. By giving managers valid information at a broader scale, we can make the leap to using reliable research to inform policy and to link science, management, and conservation.

The protocols in this book have been developed, fine-tuned, tailored, and perfected by field biologists and other practitioners. Let us take these efforts to another level and share this vision. Common language and standardized protocols will improve our work as individual researchers, fishery and habitat managers, and stewards of the river basins we call home.

The Role of Sample Surveys: Why Should Practitioners Consider Using a Statistical Sampling Design?

Donald L. Stevens, Jr., David P. Larsen, and Anthony R. Olsen

Introduction

The primary purpose of this book is to describe the great variety of field sampling protocols for determining the abundance, distribution, and productivity of salmonid populations, especially in stream and river networks. These protocols guide the field practitioner in the selection of appropriate methods to collect fish once the sampling locations have been determined. Equally important is the selection of the locations where fish are to be collected, especially when it is impractical to conduct a census by which all the fish are counted or when information is required for all locations on the stream network. Statisticians sometimes distinguish these two aspects as the sampling or survey design (Where should I collect the fish?) and the response design (How should I collect the fish?) (Stevens and Urquhart 2000).

This chapter provides a small amount of balance across these two critical parts of developing a program to monitor salmonid populations by describing some important components of survey designs relevant to the estimation of the abundance, distribution, and productivity of salmonid populations. A variety of statistical books (e.g., Särndal 1978; Cochran 1987; Thompson 1992; Lohr 1999) cover many of the aspects of survey designs in great detail. Some information on environmental sampling is provided by Gilbert (1987), Olsen et al. (1999), and Stehman and Overton (1994). Stevens and Urquhart (2000) discuss technical issues that arise with response designs when conducting a survey in environmental settings. Two recent books provide some insights useful to fisheries workers; chapter 7 of Thompson et al. (1998) is devoted to statistical sampling of fishes, and Thompson (2004) focuses on the topic of sampling for rare or elusive species. Finally, a forthcoming American Fisheries Society book, Analysis and Interpretation of Freshwater Fisheries Data, edited by Michael Brown and Christopher Guy, will greatly aid fisheries workers in freshwater systems of North America.

We strongly recommend that survey design statisticians be involved as part of the planning team from the beginning in the development of a monitoring project. As members of the team, they have a contribution to make in setting goals that can be objectively evaluated through the development of survey designs and analytical procedures that are consistent with the project goals. More often than not, however, monitoring proceeds without input from a statistician until the analytical phase, when a statistician is often asked to assist in the analysis. By this point, it is often too late; objectives may not have been clearly defined, or the monitoring plan may not match the stated objectives. We liken the circumstances to driving to a new destination without a map: we may get there with the help of a good sense of direction and lucky hunches, but we would certainly save time, expense, and angst if we first invest in a spatial plan before setting out on a trip.

Why use statistical surveys?

Regardless of the field of inquiry—whether it is fisheries or human health, economic vitality or agricultural resources, or human population demographics or labor statistics—an accurate representation of the resource is a necessity if a census cannot be conducted. In all these fields, there is a long history in the development and application of statistical surveys to meet analytical needs. An instructive book that covers the history of election polling—Survey Research in the United States: Roots and Emergence 1890–1960 (Converse 1987)—illustrates the evolution of sampling techniques, from error-prone judgmental selection techniques and targeted sampling to sophisticated statistical surveys currently used.

Some of the important objectives of sampling salmon populations, driven by various agency legislative mandates (e.g., ESA) and management objectives, include the estimation of the number of fish of a particular species within a population, metapopulation, or other demographic unit across that unit's spatial domain and whether these numbers are changing over time. This type of objective is shortened to a "status and trends" estimation. Knowing the spatial distribution or spatial structure of these populations is also important: Are they clustered in one part of their domain or more or less evenly distributed? Is that spatial structure changing over time? Other questions include the following: What proportion of the fish population is of hatchery origin? What proportion is wild? What is the age structure of the population? How large is the breeding population? To what extent do the fish stray from their natal domains? In some instances (e.g., migratory populations), some of these questions can be answered if the fish can be counted accurately (i.e., through a census) as they pass a particular location; however, many of the objectives cannot be achieved in this way, especially if the fish do not migrate. If a census is not feasible, then we must realize our objective by extrapolating from the characteristics of a sample and applying them to the characteristics of the population.

Commonly used techniques for selecting a sample include convenience, representative, model-based, and probabilistic. The first two have been widely used in fisheries and environmental management in general but have some substantial deficiencies. A convenience sample is just that: there is no particular reason for selecting a site other than because it was easy to do so. The relationship between sample data and population characteristics is unknown, and there is no reasonable or defendable basis for extrapolating data from the sample to population characteristics. Such data are often inappropriately analyzed using common statistical tools. A representative sample is most often based on professional judgment founded on an informal synthesis of the investigator's experience. One problematic issue is that a site representative of one variable is not necessarily representative for any other variable; another is that if the sites truly are representative of central tendency, then the extremes are suppressed. A major weakness of this technique is that humans fare poorly when integrating new data due to the existence of prior conceptions; this theory is supported by many experiments in cognitive psychology.

Model-based sample selection uses prior knowledge and theoretical population characteristics to choose sample sites. The model defines the relationship of the sample to the target population and provides a prescription for extrapolation from the sample to the population. A significant advantage of model-based procedures is that they maximize the leverage of data: strong inferences can be made with relatively little information. Although there is a minimal need for data, model-based procedures are not necessarily easy or quick to apply. Construction and calibration of an appropriate model can be very time consuming and expensive. The inference is based on the assumed completeness of knowledge and applicability of the model, and there is often no direct way to verify model assumptions. Without demonstrated reliability based on extensive field data, model results may not be viewed with confidence, and the usefulness of the model in fisheries management, especially in controversial situations, is limited.

The final way to select a sample is by using probability-based methods. Probabilistic sampling has a number of advantages as compared to other sample designs. Survey methodology furnishes a rich array of ready-made inference tools to estimate population characteristics, with known, quantified certainty. Prior knowledge and theoretical understanding can be incorporated, both to focus the design and to sharpen the analysis. Without a census, a statistical survey with the incorporation of probability sampling is the only way to assure the selection of a representative sample from which can be drawn unbiased conclusions about the population as a whole. In fisheries an unbiased estimate of the number of fish is critical; if the estimates are biased, consequences (e.g., species extinction) can be expensive or unacceptable.

An example taken from Oregon Department of Fish and Wildlife's (ODFW) coastal coho salmon *Oncorhynchus kisutch* population monitoring program illustrates the importance of obtaining a representative sample by applying probability site selection methods (Jacobs and Nickelson 1999). Beginning in the late 1940s, salmon spawning runs were monitored with standard spawning surveys at index sites to evaluate escapement—first, past the commercial net fishery; later, from offshore troll and sport fisheries. These index sites were likely selected in better-than-average spawning locations in Oregon's coastal watersheds. By 1980, ODFW realized the need for more exact data to establish population levels and develop harvest regulations. During the 1990s, ODFW incorporated random site selection into the evaluation of population sizes; the agency has continued to use random sample surveys since that time. A side-by-side comparison was made between coho spawner densities at the standard index sites and densities estimated from the random surveys; results indicated that estimates derived from the random surveys were, on average, 27% of the densities derived from the index surveys. Index surveys are used frequently in environmental monitoring, and there is often strong reluctance to shift towards the use of statistical surveys, even in the presence of information that reveals the biases of index surveys. Clearly, as illustrated by this example, results from judgmental or convenience sampling can be severely biased, and unless the bias is corrected, poor management decisions could be implemented.

The key point here is that adoption of statistical surveys allows unbiased, rigorous estimates of many of the important aspects of salmon populations that interest resource managers, policy makers, research scientists, and, perhaps most importantly, the public. Moreover, the estimates have a known, quantified level of uncertainty. Effective, cost-efficient management requires sound data, and appropriately designed statistical surveys can facilitate the integration of data obtained from multiple projects and agencies to provide additive statistical power.

Moreover, surveys can allow information gathered in a particular setting to be combined with data collected under other settings. This capability is especially important because many different agencies have the same objectives and often monitor the same salmonid populations. Furthermore, because salmonid populations have ranges that cross jurisdictional boundaries, comparable data are imperative for sound management. For example, in 2002, the Bonneville Power Authority wanted to knit together discrete salmonid monitoring efforts conducted in the Columbia River basin to create a whole picture of salmonid population status from the sum of its parts; however, in part because statistically reliable sampling designs were rarely used, the data could not be combined to create the bigger picture. Information pertained only to the setting in which it was collected.

Data collected in the field are only as good as the sites selected for monitoring; much like building on a solid foundation, if the frame of our inquiry is correct, our data will stand up. If, however, we build on a shaky foundation—or worse, on no foundation—the information we collect will have little or no meaning.

The importance of establishing survey objectives

A clear statement of objectives is essential to developing a sampling design. Working out the objectives in sufficient detail to guide the development of a sampling design can be a lengthy process. The process begins with conceptual questions (e.g., what are the status and trends of coho salmon on the Oregon coast?). The conceptual question is necessary, but it is neither precise nor detailed enough to design a sample. The final objectives statement must include an operational definition of the target population as well as specifications about which characteristics of the population are to be estimated, what measurements are to be made at the sample sites, and what level of precision is required.

The objectives drive the sampling design because the design is created to satisfy the objectives. Frequently, however, the process is iterative as the objectives are refined subject to sampling feasibility. For example, the nominal objective may be to estimate the size of a native trout population in a basin. If a portion of the basin is too difficult or impossible to sample, we may have to settle for a more modest objective, such as estimating the size of a population in the accessible part of the basin.

Characteristics of survey designs for fisheries

The use of statistically designed sample surveys (or probability surveys), in conjunction with appropriate field sampling protocols, allows for robust estimation of salmonid numbers and for changes in abundance over time, spatial structure, and various other aspects of population structure, as well as for an estimate of sampling precision or uncertainty. In what follows, we will introduce several common approaches for developing survey designs, such as simple random sampling, stratified random sampling, and systematic sampling; we will also describe some of their shortcomings with respect to sampling fish in stream/river networks.

At the same time, we will advocate the use of spatially balanced sampling. In particular, we encourage the use of the generalized random-tessellation stratified (GRTS) design (Stevens 2003; Stevens and Olsen 2004), which overcomes many of the shortcomings of other survey designs. GRTS is described later in this essay. We will also describe the concept of a master sample (Yates 1953) and its potential usefulness to facilitate the integration of multiple monitoring programs across a region such as the Columbia River basin.

The development and implementation of sample surveys applied to stream/river networks presents a variety of practical challenges that must be accommodated by the chosen survey design. GRTS is flexible enough to overcome many of the challenges that the other design approaches cannot. For example, the design should accommodate the following practical issues:

- Spatial relationships among sites. The target population exists in a spatial matrix, and spatial relationships in the population are critical, both to our understanding and to sample design. Sites near one another tend to be similar because they tend to share a number of characteristics such as substrate, climate, topography, and natural and anthropogenic stressors. These spatial relationships lead to patterns in the response, such as gradients, patches, or periodicities. Good survey design takes advantage of the patterns.

- Accurate and relevant frame representation. The representation of the stream/river network (or the frame), such as a digital file, from which to draw the sample is rarely an accurate portrayal of the target population's domain. The frame may include stream segments that are not in the target population (e.g., the segment may be dry). Alternatively, the frame may exclude segments that are in the target population. Sites selected from the frame sometimes cannot be sampled because, for example, access is denied by reluctant landowners or because the site is too dangerous to reach. A good survey design should have some means of addressing unreliable frame material and access difficulty.

- Ability to focus on subpopulations as well as the overall population. In many assessments, subsets of the population will be of particular significance. The significance may arise from ecological considerations, genetics, economic importance, environmental stressor levels, scientific interest, or political pressure. Whatever the source, a survey design must be able to focus on selected subpopulations as well as overall population characteristics.

- Ability to evolve goals and objectives. For monitoring plans that are anticipated to persist for several years, it is almost certain that the goals and objectives of the program will evolve. The survey design must have a substantial amount of flexibility to respond to such changes while maintaining continuity.

- Consideration of time and seasonal constraints. Field crews run out of time and cannot sample all sites designated to be sampled. Some species are distributed in disconnected patches across stream/river networks, for example, and only occupy headwaters during a portion of the year. In some cases, it might be important to sample multiple species simultaneously (as a cost-effective measure), even though those species domains might only partially overlap.

The approach advocated in this chapter is designed to accommodate many of these practical aspects of applying survey designs to stream/river networks.

Survey designs for fisheries

The simplest probability-based or statistical approach to the selection of locations in stream/river networks at which to sample is simple random sampling (SRS; see Figure 1). In SRS, every location in the target network domain is given an equal opportunity to be selected in the sample. This equal opportunity meets the basic statistical criterion that every location has a known, nonzero chance of being included in the sample. No location is selected for convenience or on the basis of expert opinion about a valid reason to select a particular site. Locations selected for specific reasons, such as convenience or because a site has been monitored for a long time or is thought to be representative, do not meet this most basic of statistical requirements.

SRS can be applied to a stream network by splitting the network into segments, arranging the segments in a list, and then selecting segments to sample from the list. The segments can be fixed length, defined by starting at the mouth of the network and working up. They can also be defined by physical features of the network, such as confluences, riffles, or pools. Alternatively, points on the stream network can be picked by viewing the network as a continuum of points. The network can be mapped onto a line, and the SRS sample from the line can be mapped back to the network. In either case, the important characteristic is that each point or segment is chosen independently of any other point or segment. It is then easy to augment the design to account for missing or inaccessible sites or to refocus the sample to account for evolving goals. But SRS makes no attempt to account for spatial pattern in the response and thus will likely be an inefficient design.

At times, knowledge about the distribution of fish can be used to stratify the sampling to devote proportionally greater sampling to some strata. Stratified random sampling does allow greater precision of estimates if the information used to create the strata is correct. For example, we may choose to sample segments with a history of high fish density more intensively than segments with historically low density. If that turns out to be the case, then the estimate of total number of fish will be more precise. In many cases, what we think we know about the distribution and abundance of a species or population across its domain is often erroneous. A consequence of misclassification of the domain into strata is that precision can be lower than it would be if simple random sampling had been used. In any case, the basic rule regarding equal opportunity applies within strata: every location within a stratum is equally likely to be selected. Stratification is also used to ensure that estimates can be made for subpopulations of special interest.

One of the primary disadvantages of SRS and stratified random sampling is that the spatial pattern of sites can be clustered, leaving gaps in parts of the network (see Figure 1). Our visual expectation of a random sample is that the distribution of points will be approximately evenly spread across the relevant domain (or across each stratum within the domain); often, however, this is not the case. Completely random processes such as SRS are much more variable than is commonly thought. Spatial simple random samples exhibit apparent patterns in clusters and voids. That is, the sample exhibits a certain amount of clumping. Spatial stratification with a low number of points per stratum (less than five) can be used to increase spatial regularity and reduce clumping. This comes at a price. Small strata can be difficult to identify or describe, and the effect of missing data is magnified, contradicting the intended increased precision of estimates that could be derived from well-defined small strata.

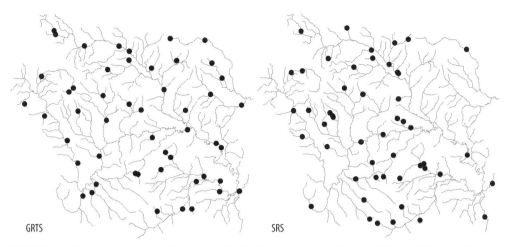

FIGURE 1.—These panels contrast the spatial distribution of sample points established with generalized randon-tessellation stratefied design (GRTS) and a simple random sample (SRS). Notice the relative absence of points in the upper right portion of the network for the SRS sample.

An overriding characteristic of fish populations is that the distribution of fish over their domain often has spatial pattern; parts of the domain might have relatively high densities, and parts low densities. An efficient survey design will utilize this characteristic through spatial balance in the selection of monitoring locations, resulting in increased precision of estimates. Even if we do not know what the pattern is, we can take advantage of its potential existence by ensuring that our sample has spatial balance (i.e., the sample is more or less regularly distributed over the domain). One way of creating an even spatial coverage (or spatial regularity) is to use systematic sampling with a random start. In two dimensions, a systematic sample could be a set of points at the centers or intersections of a square grid or at the intersections of a triangular grid. The known chance of inclusion rule is met by locating the grid with a random start point. For stream networks, a systematic design would look like a set of evenly spaced points across the stream network of interest. But there are disadvantages to systematic sampling. The most severe is that, by accident, the systematic grid might align with the natural feature being investigated. For example, suppose the quality of lakes in Oregon and Washington were being investigated. One set of lakes falls along the north–south Cascade Range. It is conceivable that the random start for the systematic grid aligns to miss this set of high elevation lakes or that the systematic grid preferentially selects these lakes. It is not so easy to identify examples of this type of systematic alignment for stream networks, except for the geomorphic pattern of pools and riffles that roughly occurs in a predictable pattern. Furthermore, poor frame information and missing or inaccessible sites break up the regularity of the initial sample, and it is difficult to add sites while maintaining the regularity.

Adaptive sampling

The distribution of fish sometimes requires adapting the specified design based on what is found at the sites selected. For example, some fish species are distributed in patches across their range. A desirable feature of a sampling design would be to allow increased sampling effort in the vicinity of sites where fish are found when the design selected sites are visited. The concept of adaptive sampling meets this need (S. K. Thompson 1990, 1991, 1992; W. L. Thompson 2004). When a site selected in the original design is visited and no fish are found using the specified sampling protocol, no further sampling occurs in the vicinity of the site; however, if fish are collected at the site, then additional sampling occurs upstream and downstream of the site, following specified rules regarding the length of stream sampled. If no fish are found, then sampling stops. If fish are found, then sampling continues, again following the sampling protocol. Sampling continues until a specified stopping rule is reached. If fish distribution is patchy, the use of adaptive sampling allows for improved estimates of the abundance and spatial structure of target species.

Multistage sampling

Efficient selection of a probability sample requires a good representation of the resource to be sampled. For example, the frame must be a faithful representation of the stream network or estuarine resource to be sampled; however, if the frame is an inadequate representation of the resource, two-stage or multistage sampling can be used. In the first stage, a probability sample of the potential target resource is obtained; then this sample is evaluated to determine whether it is indeed part of the target resource. Ideally, this determination would be reasonably quick and inexpensive, relative to sampling for the relevant indicators. For those first-stage samples that meet the target resource criterion, a second-stage sample, from which the target populations of interest would be measured, could be selected. The concept can be extended to multiple stages if the selection of the final set of sites is efficiently facilitated. Otherwise, the multistage sampling provides no benefit.

The concept of multistage sampling can be extended to include refining a sample even if the frame is an accurate depiction of the resource. Relatively inexpensive measurements could be made on the first-stage sample; then a subset of the first-stage sample could be visited to make more expensive measurements. The first-stage sample could be classified to direct the allocation of the second-stage sample to maximize efficiency with respect to the target estimates. Results from the second stage sample can be extrapolated back to the first stage sample and then to the target resource.

Generalized random-tessellation stratified design

Numerous methods have been developed that adapt and extend the concept of a systematic sample applied to environmental resources to achieve a spatial representation through spatial balance. Stevens and Olsen (2004) reviewed many of these methods, indicating that "... all do reasonably well at getting a spatially balanced sample under favorable circumstances, but have difficulties with some aspect of environmental populations." They proposed a solution, via the GRTS design, that builds on and overcomes the difficulties associated with the available designs and creates a spatially well-balanced, random selection of locations on linear (e.g., stream networks) or areal (e.g., estuaries) resources (see Figure 1). GRTS accommodates the following possibilities:

- The resource of interest might exhibit spatial patterns (parts of the domain might have relatively high abundances and other parts low abundances) or patterns might be regular;
- The representation of the resource used to select the sample can be imperfect (e.g., the digital representation of stream networks contains errors; the map of the estuary strata of interest is not perfect);

- The locations can be selected with variable but known likelihood of inclusion in the sample (to achieve flexible stratification);
- Data might not be collected at some sites for a variety of reasons (e.g., site access might be denied, site is too dangerous to visit, field crews run out of time and do not sample all selected sites);
- Sample points can be added to the initial list to accommodate visit denials or to increase sample sizes (points can be added to sample the entire population or only selected subpopulations or subdomains); and
- Multistage or adaptive sampling might be necessary.

The current version of GRTS creates an ordered list of sites such that each successive site on the list maintains the spatial balance of the full set of sites in the sample. The significance of this property is that an investigator can work down the ordered list to achieve his/her target sample size. If a site is not accessible for some reason, that site is skipped (keeping track of the reason that the site is skipped). This process continues down the list until the requisite number of sites has been sampled. If by some fortuitous circumstance a greater number of sites can be sampled than originally planned (perhaps due to a budget surplus), continuing down the list of ordered sites allows the investigator to increase the sample size, yet maintain the spatial balance of the full set of sites.

A Web site has been established that describes GRTS and many example applications (www.epa.gov/nheerl/arm). Algorithms both for creating GRTS designs and analyzing resulting data are available through the Web site. The algorithms are written in the statistical language "R," a freely available statistical software package (R Development Team 2006); the Web site contains information on how to download and install R.

The role of statistical surveys in combining data from different surveys

Among the many advantages of well-designed statistical surveys is that they can allow the integration or rolling-up of data across several surveys that might have been conducted for disparate purposes but monitored the same indicators. This integration is facilitated if several design rules are followed. One is that the surveys use a common representation (or frame) of the target resource from which the set of sites is selected or representations that can be matched. For stream surveys, a common representation of the stream network is the U.S. Geological Survey digital stream network at the 1:100,000 scale. A second design rule is that randomization is used in the selection of sites with the result that the probability that any particular site is selected is known. A third design rule is that the part of the frame that is to be monitored is clearly described. For example, one stream survey might only be interested in headwater or first-order streams. Another might be interested in all first- through third-order streams. Data from the two surveys could be combined if the above rules were followed by recalculating the site inclusion probabilities for the overlapping parts of the surveys. The investigator interested in headwater streams could build in the first-order stream data from the second survey. The investigator interested in the first- through third-order streams could add the first-order stream data from the first survey to the data set. In both cases, the effect is to increase sample sizes—and therefore improve precision of the resulting estimates.

Status and trends and the use of rotating panel designs

The term "status" implies a snapshot of condition during a specified time interval (e.g., how many adult spawners were present in Oregon's coastal streams during the 2006 spawning season, or what was the habitat condition during the years from 2004 to 2006). Precision of status estimates depends in part on sample size: the more sites visited, the greater the precision of the estimate.

Evaluating trends implies study of change over some time interval. If spatial patterns in the response tend to persist through time, then revisiting the same sites during the time interval of interest is optimal for trend detection. Estimating trend from the revisited sites eliminates some site-specific components of variation, so the resulting trend estimates are more precise. Balancing the need for more distinct sites for status estimation and revisiting individual sites for trend detection creates a design challenge that is resolved by the implementation of panel designs (Kish 1987; Skalski 1990; McDonald 2003). A panel consists of a set of sites visited on a specified pattern over time. For example, one kind of panel design might consist of a set of sites visited every year (annual panel), a set of sites visited every 3 years beginning with year 1 (year-1 panel), a set of sites visited every 3 years, beginning with year 2 (year-2 panel), and a set of sites visited every 3 years, beginning in year 3 (year-3 panel). The sites can be selected from a GRTS sample list (e.g., the first 25 sites on the list could be allocated to the annual panel, the second 25 to the year-1 panel, and so forth). Selected this way, each panel is a spatially balanced sample, with the result that each year has a spatially balanced sample consisting of 50 sites.

Although panel designs sacrifice some trend detection capability by devoting more effort to status estimation, their trend detection sensitivity tends to catch up to a strictly annual visit design after three cycles (Urquhart et al. 1993, 1998; Urquhart and Kincaid 1999). In the previous example, after 9 years, the trend detection capability of the panel design would be approximately the same as if 50 sites were visited every year. However, with the panel design, a total of 100 sites are visited, improving the estimates of status. Clearly, in designing surveys investigators must establish priorities for objectives. One design is not optimum for all purposes.

Master sample concept

When multiple organizations agree to use the same field sample protocols (i.e., the same response design) to conduct studies within the same geographic region (e.g., the Pacific Northwest), the next natural question is whether the individual designs can be integrated so that they can take advantage of sampling others have done. The concept of a master sample (Yates 1953) is an answer to this question. The basic idea is to select an equal probability sample over the entire geographic region with a sufficiently large sample size so that the sample size requirements for almost all studies conducted within the region will be satisfied. Such a dense, spatially balanced, ordered list of sites can be selected using the GRTS site selection algorithm.

For example, a set of sites covering the coastal stream networks in Oregon, spaced on average 1 km apart, could be selected with GRTS. This master sample can be classified in a variety of ways such that spatial balance is maintained if the original order of sites is maintained within the class. A target sample size can be selected by proceeding down the ordered list within the selected class. One investigator might be interested in a broad regional survey consisting of 100 sites across the full coastal stream network. Selecting the first 100 sites from the master sample covering that domain provides that investigator with a spatially balanced sample. A second investigator might be interested in sampling only part of that domain—the Nestucca watershed, for example—with a sample size of 50. Selecting the first 50 sites from the part of master list of sites in the Nestucca provides the second investigator with a spatially balanced sample. If both investigators measure the same stream attributes, their data can be easily combined in a statistically sound way. The technical details involve recalculating each site's likelihood of inclusion when the studies are combined.

A master sample for a broad region can serve multiple purposes. As outlined above, it can be used as the basis for designing surveys whose results can be rolled up in a statistically sound way to create a whole picture of status or trend from the sum of its parts, extending the utility of data collected for individual purposes beyond the setting in which they was collected. A master sample is also useful as a tool by which investigators can easily explore allocation of fixed sampling effort (e.g., can only afford to sample 200 sites) in different ways before settling on an optimal allocation. Are a sufficient number of samples allocated to relevant strata? Is the target domain reasonably covered? Numerous alternative designs can easily be created to explore and evaluate alternative designs yet maintain the basic survey design principles of spatial balance and randomization in site selection.

Perhaps a more far-reaching use of a master sample is to facilitate the integration of numerous monitoring programs across a broad region like the Columbia River basin. Currently there are tens of agencies, including federal, state, tribal, and private agencies, conducting monitoring programs within the Columbia River basin. Although these agencies have their own specific goals and objectives, they often are interested in similar riverine and riparian attributes, including estimating various fish population sizes and their physical and chemical habitat. Even though their spatial scales might differ (for example, across the entire Columbia River basin or statewide to the much finer scale of a small watershed), the responsible agencies might benefit by combining data from the other monitoring efforts that are conducted in their domains or at least by knowing where other agencies might be conducting fieldwork. Central housing of a master sample could serve as a design kiosk by which an agency requests a set of sites to meet its specific design requirements. In return, the agency could receive a list of sites along with an indication of which sites are already included in other agencies' monitoring plans. Duplication of effort could be avoided, integration of data derived by common protocols could be facilitated, site selection would follow design principles allowing valid statistical combination of data, and communication among agencies could increase. The next step could be institutional adoption of the concept of a master sample and the formation of a center that provides survey design assistance and coordination in the delivery of survey sites.

Conclusion

Determining the number of fish in streams and rivers or estuaries, their spatial patterns, and how those numbers and patterns change over time poses numerous challenges—not the least of which is determining where to collect fish. Although counting all the fish might be a desirable goal, it is only occasionally feasible;

therefore, statistically sound methods are necessary to estimate fish numbers and patterns. These methods fall into the broad category of statistical sample surveys.

Survey sampling has a long history in a great variety of areas and is built on a strong theoretical foundation. In this chapter, we argue for the incorporation of survey sampling techniques to monitor salmonid populations if a census cannot be conducted. We briefly review the variety of design approaches available, including simple and stratified random designs and systematic designs, and their shortcomings for environmental sampling. We also describe a newly developed, flexible approach that overcomes these shortcomings. We describe the concept of a master sample and suggest its role as a central organizing principle to facilitate integrated monitoring across the numerous agencies now conducting salmonid population monitoring.

Acknowledgments

Dana Foley motivated this essay and contributed regular editorial and formatting suggestions. David H. Johnson provided additional comments. Jennifer O'Neal and Jeff Rodgers reviewed the manuscript. We are grateful to all four for their rapid and timely assistance. This document was prepared at the U.S. EPA National Health and Environmental Effects Research Laboratory, Western Ecology Division, in Corvallis, Oregon, through Cooperative Agreement CR831682-01 to Oregon State University. It has been subjected to the agency's peer and administrative review and approved for publication. Mention of trade names or commercial products does not constitute endorsement for use.

Literature Cited

Cochran, W. G. 1987. Sampling techniques. 3rd edition. John Wiley & Sons, New York.

Converse, J. M. 1987. Survey research in the United States: roots and emergence 1890–1960. University of California Press, Berkeley.

Gilbert, R. O. 1987. Statistical methods for environmental pollution monitoring. Van Nostrand Reinhold, New York.

Jacobs, S. E., and T. E. Nickelson. 1998. Use of stratified random sampling to estimate the abundance of Oregon coastal coho salmon. Oregon Department of Fish and Wildlife: Final Report Fish Research Project # F-145-R-09, Portland.

Kish, L. 1987. Statistical design for research. John Wiley & Sons, New York.

Lohr, S. L. 1999. Sampling: design and analysis. Duxbury Press, Pacific Grove, California.

McDonald, T. 2003. Review of environmental monitoring methods: survey designs. Environmental Monitoring and Assessment 85(3):277–292.

Olsen, A. R., J. Sedransk, E. Edwards, C. A. Gotway, and W. Liggett. 1999. Statistical issues for monitoring ecological and natural resources in the United States. Environmental Monitoring and Assessment 54:1–45.

R Development Core Team. 2006. R: A language and environment for statistical computing. R Foundation for Statistical Computing, Vienna, Austria.

Särndal, C. 1978. Design-based and model-based inference for survey sampling. Scandinavian Journal of Statistics 5:27–52.

Skalski, J. R. 1990. A design for long-term status and trends monitoring. Journal of Environmental Management 30:139–144.

Stehman, S. V., and W. S. Overton. 1994. Environmental sampling and monitoring. Pages 263–306 in G. P. Patil and C. R. Rao, editors. Handbook of statistics, volume 12. Elsevier Science, Amsterdam, The Netherlands.

Stevens, D. L., Jr., and A. R. Olsen. 2004. Spatially balanced sampling of natural resources. Journal of the American Statistical Association 99:262–278.

Stevens, D. L., Jr. 2003 and A. R. Olsen. Variance estimation for spatially balanced samples of environmental resources. Environmetrics 14:593–610.

Stevens, D. L., Jr., and N. S. Urquhart. 2000. Response designs and support regions in sampling continuous domains. Environmetrics 11:13–41.

Thompson, S. K. 1990. Adaptive cluster sampling. Journal of American Statistical Association 85:1050–1059.

Thompson, S. K. 1991. Stratified adaptive cluster sampling. Biometrika 78:389–397.

Thompson, S. K. 1992. Sampling. John Wiley & Sons, New York.

Thompson, W. L. 2004. Sampling rare or elusive species. Island Press, Washington, D.C.

Thompson, W. L., G. C. White, and C. Gowan. 1998. Monitoring vertebrate populations. Academic Press, Inc., San Diego, California.

Urquhart, N. S., and T. M. Kincaid. 1999. Designs for detecting trend from repeated surveys of ecological resources. Journal of Agricultural, Biological, and Environmental Statistics 4(4):404–414.

Urquhart, N. S., W. S. Overton, and D. S. Birkes 1993. Comparing sampling designs for monitoring ecological status and trends: Impact of temporal patterns. Pages 71–85 in V. Barnet and K. F. Turkman, editors. Statistics for the environment. John Wiley and Sons, London.

Urquhart, N. S., S. G. Paulsen, and D. P. Larsen. 1998. Monitoring for policy-relevant regional trends over time. Ecological Applications 8:246–257.

Yates, F. 1953. Sampling methods for censuses and surveys. Hafner Publishing Company, New York.

Data Management: From Field Collection to Regional Sharing

This chapter has been written in two parts. Part I covers field data collection and emphasizes working with local-scale, observation-based data; it frames key data concepts and structures that allow local data to be connected subsequently with larger, distributed regional data systems. Part II deals with regional data sharing and focuses on the features of an effective network of regional data centers; the function of these centers is the generation, repository, and dissemination of data as a resource to people and organizations interested in regionwide, national, international, and global natural resource issues.

Introduction

The purpose of collecting any data is to help answer a question! Thus, developing and using protocols may result in a data-rich endeavor; however, this is not a guarantee that the necessary information will be available to answer a local or regional question(s). With the Information Age upon us, technology plays a large and important role in gathering, compiling, and synthesizing data. Today's issues and their complexities may potentially overwhelm resource managers in a sea of data; yet when resource agencies are presented with a concern or issue, managers may find themselves confronting a lack of information. The need to analyze data over time and space today requires an increased use of technologies, including their integration into research and monitoring studies as well as evaluation strategies. Resource managers must understand that data standards and protocols help refine the quality of data being collected, enhance its usability, and clarify its purpose. Without standards and protocols, resource managers will have only disparate data sets that contain various kinds of information to answer increasingly more complex questions at various scales (e.g., site, watershed, subbasin, and basin levels).

To this end, fishery managers face an urgent need to standardize information; postponing it only exacerbates the problems. Managers need high-quality, real-time data that can be shared by others. As our data management capabilities expand in unexpected ways from year to year, practitioners face the tremendous challenge of keeping up with what's available; broadening our horizons to consider new ways to manage data can be daunting, but we have to rise to the challenge. More than ever, emerging technologies are outgrowing old templates. Certainly we have abundant material on data collection at the local level, but our technology permits us to go beyond the local to the regional and global scales. This essay confronts the necessity of collecting field data along with creating designs for regional data structures and explicit management questions; it may also enlighten us as to how we can create international legislation to foster global data management systems.

The protocols described in this book allow us to open the door to advanced data management systems that can have regional and global applications; however, to start, we need to begin with field data. The development of this chapter began in 2004 with a 2-day meeting in Welches, Oregon, convened with the sole purpose of providing guidance for the design of observation-based data (OBD) system, especially for collecting fish-related data. Data management systems must be designed from end to end (i.e., data collection through data entry to database management, report production, and data sharing). To clarify these mechanics, we have framed this chapter in two parts.

Part I: Field Data Collection

Stewart Toshach, Richard J. O'Connor, Thomas A. O'Neil, Bruce A. Patten, Cedric Cooney, Bill Kinney, Paul Huffman, and Frank Young

In developing this guidance for field data collection, we recognize the following assumptions:

- Local data systems that are developed will feature free and open access.
- Data collection programs will have a principal investigator (PI) who understands the business and scientific logic of the effort and that data management support is needed from a data manager (DM).
- The DM for the project knows that the PI is responsible for identifying the program goals, the data collection deliverables, and the data products. The relationship between the PI and the DM is a critical one, in which success depends on excellent communication across different organizational disciplines, as well as through the planning, execution, and documentation stages of the project.
- Data will be shared, which means that the role of data owner and data steward must be identified and that the data itself must be sharable. With respect to these roles the data owner (who may be the PI) gives permission to the data steward (who may be the DM) to provide access to other users of the data based on agreements.
- Mutual benefit to sharing data is realized by encouraging participants to both use resources and contribute data, information, and knowledge.
- Data contributors should have full right to attribution for any uses of their data, information, or knowledge and the right to ensure that the original integrity of their contribution is preserved. Users of the data are expected to comply, in good faith, with terms of uses specified by contributors or data stewards.

Collecting data and developing small, localized databases do not necessarily pose problems, because they exist to meet the user's immediate and near-term needs; however, most of these data collectors and developers have little familiarity with data concepts and structures that would allow their contribution to a larger, distributed application. To deal most effectively with data systems designed to support the retrieval and integration of primary data, let us begin by establishing a local data set that lends itself to a common management schema.

Establishing a common management schema

To maximize the usefulness of data collected in the field, database architecture should incorporate the ability to apply data at various scales, such as connecting local data sets into larger regional systems. But before going regionally or globally, we need to establish a local data structure that can be moved into a data system. To do this, we need to have an overview of the process:

(1) Outline core questions that we would like to ask of a data structure.
(2) Define terms of data being collected.
(3) Create a consistent process for developing an OBD project (specifically, preparing a needs assessment and writing a data collection and management plan).
(4) Determine office and field procedures needed for entering data, along with Quality Assurance/Quality Control (QA/QC) protocols, analysis, reporting, and maintaining data.
(5) Design data forms.
(6) Create data fields (elements) and identify those that are required or recommended.
(7) Categorize fields or elements as either data or mapping.

1. Outline core questions

Developing core questions begins with determining the data fields you will need to help answer a local problem or concern. Remember that local data sets drive the types of questions that can be asked at a regional level. When developing your local data fields, be aware of regional needs and how your data set might fit in to help answer another question. An example of some local data questions that might occur within a watershed are

- What are the fish species?
- Where can I find them?
- What fish are listed as threatened, endangered, or of concern, and where do they occur?
- Which fish are native and which are exotic?
- How are the fish populations or selective stocks of interest faring (e.g., how many are there relative to historic populations sizes)?
- Why is a specific fish population declining or increasing (i.e., what are the limiting factors or main causes affecting the fish population(s) and their habitat(s))?
- How much fish habitat exists?
- Where does the fish habitat occur?
- What is the condition of the fish habitat (e.g., what is the assessment of watershed and marine ecosystem health)?
- Who is conducting the research, monitoring, or a management project, and where is this work occurring? Is there a need for public outreach and education of this work?

At the regional level you would follow a process that incorporates review of and consensus on the local data sets, which by necessity should be general in design to maximize the use of the information with other data sets by capturing key data elements. This does not mean that specific questions should not be listed; rather, it only means that they should be a subset under a general one. In other words, do not start out too narrowly focused, because you may miss important aspects that could help answer the specific question(s).

There are two approaches for determining questions and implementing data capture: (1) list the questions along with the data elements needed to answer them; and (2) look at the data sets within a region and determine what data elements are common, and then develop the questions that can be answered. Realistically, a combination of both approaches is probably most efficient. Both approaches call for agreement on common data elements and definitions so that data elements collected in one area have the same meaning within another part of the region.

To maximize the use of a regional data network based upon local data, the general questions need to be carefully articulated to allow for the eventual collection of as much detailed baseline information as possible. Therefore, the best tactic is to ask general questions that can be refined. As an example of this regional, top-down approach, the National Water Quality Monitoring Council (2004) developed a list of data elements considered important for reporting water quality. These elements were then sent to individual state agencies and organizations, which were encouraged to use the information. Along with the data elements, several suggested approaches for incorporating these data elements into their specific data programs were made. They included:

- at the local level, consider using all the data elements or as many as possible in your next water quality monitoring project or in developing your next water quality database;
- focus initially on categories of water quality data elements (WQDE) that you most want to improve in the near term and progressively expand the data elements included over time; and
- plan to include these WQDE in database modernization or updating; and in combination with other approaches, program field electronic devices for on-site entry of field data to download directly to your database.

2. Define basic terms

Defining terms for the data being collected helps ensure a common language in the creation of OBD systems. Here are some basic terms and definitions regarding OBD systems:

- Observation-based data: information generated during an activity performed by participants in which observations are collected about a subject following a methodology at a location during a period and for a particular purpose
- Activity: what you are doing (e.g., fish counting, habitat survey)
- Participants: the individuals or organizations performing roles associated with the activities (e.g., observer, pilot, data recorder)
- Methodology: how you are performing the activity (e.g., beach seining)
- Location: where you are performing the activity (e.g., port name, universal transverse mercator coordinate, latitude/longitude coordinate)
- Period: when you are performing the activity (e.g., 2004-03-11, 1320–1456, 2004 March 11/12:30 p.m., week 34)
- Purpose: why you are performing the activity (e.g., assessment of stock B, environmental impact study, escapement estimate) (This is usually defined within the project description and may be reflected as a title on a data form.)

- Observations: the details about a subject composed of characteristics that have values and may have a unit of measure
- Subject: what you are observing (e.g., a fish, mammal, boat, stream reach, rock)
- Characteristics: type of data element/detail (e.g., length, weight, age, color)
- Values: the category or measure of the detail (e.g., 3, blue, 4.859, XZ2)
- Unit of measure: unit used to report the value(s), where appropriate (e.g., centimeters, pieces, grams)

3. Create a consistent process

The more transparent and consistent the process is, the more reliable the data. To create this reliability for an OBD project, we recommend that the DM and PI prepare a needs assessment, write a data collection and management plan, review the field and office setup, and decide on the methods for data recording, data entry and quality review, data analysis, and data reporting. the steps to estalibsh these are

Planning

Creating a needs assessment: identify and document
- the people involved;
- data management budget;
- expected dates of data management deliverables;
- data outputs needed;
- activities to be done;
- what data will be collected;
- what data will be recorded (see *Data recording* below for guidelines);
- whether existing protocols and legislation are applicable (e.g., are there any data collection or reporting standards that must be met?);
- needed quality assurance/quality control (QA/QC) measures and responsibilities;
- who has user rights to create, read, update, and delete data elements (handle/manipulate data) and when these rights are granted;
- data security needs;
- when and how data and/or analysis results will be made accessible to requesters and for general access;
- data work flow;
- whether existing software and hardware are adequate;
- equipment needed to support data recording and storage (e.g., Personal Digital Assistant (PDA) or other equipment) and supplies;
- responsibility for completing the needs assessment document.

Writing a data collection and management plan

The PI will now be able to write a plan that can deliver the identified and documented needs. The data collection and management plan should identify the people responsible for the different tasks and what they will do. Once completed, submit this plan along with your documented needs for review and approval by the appropriate resource program and information systems staff in your organization.

4. Determine office and field procedures for data entry and handling

Field considerations

- Develop data collection form(s) and test with user group, if necessary.
- Acquire the data collection and management equipment and supplies identified in planning.
- Identify maps, global positioning systems (GPS), or other resources that will assist in accurately recording field location information.
- Train field staff in use of field forms, definitions, equipment, and field techniques.

Office considerations

- Install/develop and test any needed or existing data management hardware and software.
- Train staff in use of hardware and software.

Field and office considerations

- Test data collection system from end to end with representative users.

Data recording

Assumes that the researcher/observer has completed data collection following appropriate methods. (For example, the researcher has used a fyke net to capture salmon fry and is ready to record the observations such as the number of fish captured). Following the guidance document, the researcher should then

- complete data recording tasks, following data recording instructions;
- complete any QA/QC tasks (e.g., read the completed data recording to check for completeness, legibility or obvious errors);
- complete in-field data backup;
- deliver data to the appropriate person in appropriate form.

Data entry and quality review

- Enter data as soon as possible.
- Maintain strict control of file and version names.
- Archive this version of the data by storing a copy in a secured and preferably separate location.
- Perform all specified QA/QC tasks (e.g., double entry of the data can be used to perform data verification while value range checking can be used to perform data validation).

- Report recording discrepancies or errors to data collectors for resolution.
- Fix the records in the working version.
- Archive a copy of the working version.
- Repeat the specified QA/QC tasks (testing, fixing, and archiving) until errors are eliminated.

Data analysis
- Complete data analysis tasks using the current version of the data (e.g., develop the necessary statistical reports, charts, maps, and summary data sets). All derived data sets should be maintained and archived and subject to the same version control guidelines applied to primary data sets.
- If errors in current data set are found, perform all fixes and QA/QC tasks described in the Data Entry and Quality Review section (above).

Data reporting
- Complete a metadata document containing the necessary metadata information, including any limitations of the data. Completing a good metadata record is essential!
- Deliver data and/or analysis results according to formats, protocols, and instructions in the needs assessment document.
- If data or analytical errors are discovered, go back and correct a new copy of the appropriate archived version and label it as the current version to be used, using the current date.
- Audit data collection and management effort (for cyclical data collection projects) for inclusion in future data management updates.

5. Design data collection forms

The PI must ensure that the forms used to record or enter data, whether paper-based or electronic, are readily legible in the environment in which they will be used.

Consideration for data forms
- Appropriate font sizes, colors, or graphics should be used. Where possible, allow the data to be recorded or entered in an intuitive fashion (from top to bottom and from left to right). Electronic forms should use tab-key advance to navigate from one field to the next.
- Some systems may have to be created using more than one software or computer language.
- Clear instructions and definitions should be developed to support the system and be readily available to users. Table 1 gives an example of elements to consider when developing a field data form. If the needed data element will be coded, then it will be necessary to create a table of the codes and the corresponding data elements and definitions. If an electronic system will be used, these tables should be an integral part of that system and should be visible using that system's Help functions.

A growing approach is entering data on Web-enabled forms via PCs and/or wireless devices, due to the ability to provide immediate data validation and storage in a central location for records entered from a distant (field) location. While there are advantages of electronic reporting, there are also many lessons to be learned by developing the early field reporting prototypes using paper-based systems. This allows for the testing of data collection and the logical and mechanical reporting concepts before committing to the more expensive tasks of coding electronic reporting devices and developing databases. While changes to reporting systems are common during development, they can be minimized with careful design and iterative testing.

6. Create the "universe" of relevant data elements

- There is a minimum set of data elements that must be included in OBD collections to ensure that adequate value is derived from the collection effort.
- There are additional data elements that may be included, depending on the nature of the study and other factors.
- All the data elements that will be collected must be identified and described in a formal table.

Table 1 shows a list of the *minimum* set of data elements that must be recorded during OBD collection activities. A minimum set of data elements must be included to ensure adequate value is derived from the collection effort. Additional data elements may be included, depending on the nature of the study and other factors. All the data elements to be collected must be identified and described in a formal table. For each element, provide the data element name, definition, and format. These three columns should be part of a larger data description table (see Table 2), which identifies, categorizes, and maps data elements.

Some data managers may also want to create columns about the needed data element—whether it is required (yes/no) or conditional, based on the provision of other data, and for defined units of measure. The format column can also be used to report the length of characters that will be accepted by the database together with any decimal points. (See Table 2.1 for an example of how salmon spawning stock survey data elements found on an actual field sampling form were listed.)

7. Categorize and map data elements

Each element must be categorized (to ensure that no vital aspects of the study have been overlooked) and mapped (to determine where to record that element) with the help of an appropriately organized data description table.

Categorization
Each data element is assigned to one of the following OBD categories:

- Participants (who?)
- Methodology (what or how?)
- Location (where?)
- Temporal (when?)
- Purpose (why?)
- Subject (of an observation)
- Characteristics (of a subject)

Careful review of these assignments is done to ensure that every element has been assigned. If there are any categories not represented by at least one element, then critical information relevant to your study is not being recorded. Add elements as needed to represent all OBD categories. (See Table 2.2 for an example of salmon spawning stock survey data elements assigned to their appropriate OBD categories. Note that in this example, methodology (how) and purpose (why) categories were not represented by any data elements present on the field form.)

Mapping

All information related to an OBD collection activity must accompany your observations. Each data element needs to be recorded somewhere in your system. Data elements are commonly recorded in one of three locations:

- Project Documentation
- Field data form—header section
- Field data form—detail section

Use the following criteria to determine where to record a data element:

- Project documentation: data element value is fixed throughout the entire study
- Field form—header section: data element value is fixed for a single activity
- Field form—detail section: data element value changes from observation to observation

Aspects of certain data elements can be recorded in more than one place. For example, you may want to enter the time that all sampling started in the form header and enter times of individual observations in the form detail. (See Table 2.3 for an example of salmon spawning stock survey data elements mapped to their appropriate places.)

TABLE 1.—Minimum data for observation-based data collection.

Data Element Name	Data Element Definition	Format	Examples
Activity Name	Brief description of activity	Text	Chinook redd survey project
Name of data collector(s)	Name/s of individuals who are responsible for data collection	Text	Bruce C. Patten; RJO
Date of data collection activity	Calendar date expressed as YYYY/MM/DD	Alphanumeric	2003/05/21*
Time of data collection activity	Time of day expressed in hours, minutes, seconds local time (24-hour clock)	Alphanumeric	06:30:23; 15:15
Location of data collection activity—detailed	Location expressed as a point, line, or polygon with latitude/longitude expressed in decimal degrees (4 decimal places) or equivalent	Alphanumeric	−123.4890, 45.8734
Name of data collection activity location	Name of body of water (lake, stream), place name, or land unit	text	Ward Creek; Lake Washington; Bonneville Dam

Data Element Name	Data Element Definition	Format	Examples
Geographic code for data collection activity location	Standard code for a body of water (LLID) or land unit (fourth field HUC)	Alphanumeric	LLID: 1242059430208 (Ward Creek, Coos Basin, Oregon) HUC: 17030003 (Yakima River—lower and tribs)
The method(s) used during the data collection activity	Identification of the specific method(s) used to collect the data.	Alphanumeric	Method 27; Purse seine; Redd survey protocol (2006)

* Welches Working Group initially recommended MM/DD/YYYY; subsequently the Northwest Environmental Data-Network recommended using the International Organization for Standardization (ISO) format YYYY/MM/DD.

Below are examples of recommended data elements, and if they are needed they should be part of a larger Data Description.

Location:
    Sample/survey length/area:

Subject:
    Species:
    Run (for fish species):
    Sub-run (if applicable):
    Life stage:

Participant(s):
    Agency:

Methodology:
    Sampling method (gear):
    Target (if any):
    Photo available (Y/N):

Characteristics:
    Habitat type:
    Air temperature:
    Water chemistry/quality (temperature, clarity, pH, DO):
    Weather conditions:
    Waterbody physical attributes:

Miscellaneous:
    Page sequence (page _ of _) (especially important for paper records):

TABLE 2.—Example data description table with its three component sections: 1) identifying data elements, 2) categorizing data elements, and 3) mapping data elements.

TABLE 2.1.—Data description table: Identifying data elements (example entries are from the Daily Salmon Spawning Stock Survey Field Form [Duffy 2003]).

Section 1: Identifying Data Elements		
Data element name	Data element definition	Units of measure
Example		
Stream		
TRS		
Lat./long.		
Quad.		
Drainage		
County		
Starting location		
Ending location		
Feet/miles surveyed		Feet/miles
Date of survey		Daily
Water clarity		Feet
Water temp.		
Weather		
Air temp.		
Time		Hours/minutes
Crew		
Live fish observed		
Chinook adults		pieces
Chinook grilse		pieces
Coho		pieces
Steelhead		pieces
Unknown		pieces
Carcasses examined		
Chinook M Fk L		
Chinook F Fk L		
Coho M Fk L		
Coho F Fk L		
Tag numbers		
Other clips observed		
Skeletons observed		
Chinook		pieces
Coho		pieces
Steelhead		pieces
Unknown		pieces
Redds		pieces
Comments		pieces

Salmon spawning Survey	
Not on form	
Visual surveys	
Purpose	

TABLE 2.2 — Data description table: Categorizing data elements (example entries are from the Daily Salmon Spawning Stock Survey Field Form (Duffy 2003).

	Section 2: Categorizing data elements							
	OBD categories							
Data element name	Participants	Methodology (what)	Methodology (how)	Location	Period	Purpose	Subject	Characteristics
Example								
Stream				x				
TRS				x				
Lat./long.				x				
Quad.				x				
Drainage				x				
County				x				
Starting location				x				
Ending location				x				
Feet/miles surveyed								x
Date of survey					x			
Water clarity								x
Water temp.								x
Weather								x
Air temp.								x
Time					x			
Crew	x							
Live fish observed								
Chinook adults								x
Chinook grilse								x
Coho								x
Steelhead								x
Unknown								x
Carcasses examined								
Chinook M Fk L								x
Chinook F Fk L								x

Coho M Fk L			x
Coho F Fk L			x
Tag numbers			x
Other clips observed			x
Skeletons observed			
Chinook			x
Coho			x
Steelhead			x
Unknown			x
Redds observed			x
Comments			x
Salmon spawning survey	X		X
Not on form			
Visual surveys		X	
Purpose			X

TABLE 2.3 — Data description table: Mapping data elements (example entries are from the Daily Salmon Spawning Stock Survey Field Form (Duffy 2003).

	Data Mapping		
Data element name	Project documentation	Activity (field form—header)	Observation (field form—detail)
Example			
Stream		X	
TRS		X	
Lat./long.		X	
Quad.		X	
Drainage		X	
County		X	
Starting location		X	
Ending location		X	
Feet/miles surveyed		X	
Date of survey		X	
Water clarity		X	
Water temp.		X	
Weather		X	
Air temp.		X	
Time		X	
Crew		X	

Live fish observed

Chinook adults		X
Chinook grilse		X
Coho		X
Steelhead		X
Unknown		X
Carcasses examined		
Chinook M Fk L		X
Chinook F Fk L		X
Coho M Fk L		X
Coho F Fk L		X
Tag numbers		X
Other clips observed		X
Skeletons observed		
Chinook		X
Coho		X
Steelhead		X
Unknown		X
Redds		X
Comments		X
Salmon spawning survey	X	
Not on form		
Visual surveys	X	
Purpose	X	

Part II: Regional Data Sharing

Thomas A. O'Neil, Stewart Toshach, and Wayne Luscombe

As we embark on our own monitoring efforts, we need to keep regional data systems in mind; yet we acknowledge that our monitoring concerns will begin at a more local level. The creation of a data management system comes from a question-and-answer process. Regional protocols or standards help refine the quality of data being collected and enhance its usability, as well as clarify its purpose; however, having local or regional protocols or standards is not a substitute for the establishment of an effective regional data network that can answer broader questions about our natural resources. Leadership and an administrative framework are also needed to guide the development of protocols, standards, and guidelines for collection, compilation, and reporting of component data. It is with the establishment and coordination a network of regional data centers (directed towards the generation, compilation, and dissemination of accurate and complete data) that the major benefit occurs. These regional data centers are responsible for the generation and repository of special databases and for the dissemination of these databases as a resource to people and organizations interested in components of the regionwide natural resource data.

To assist resource managers with this task, this section explores aspects to consider when establishing a regional or global data network. The process of regional data sharing involves establishing an administrative framework, regional data centers, and a data management system, but first and foremost is the need to agree to work together and establish standards, which begins by identifying the principal or core questions that the regional or global data network is expected to answer. These components are considered essential:

- develop consistent data standards and protocols within and across types of monitoring;
- establish close working relationship for data consistency across data sources;
- identify and document specific data needs of the region;
- develop and recommend data collection standards and information to be shared across programs;
- share requirements and results with regional data networking entities;
- test the collection protocols, sampling methods, and data sharing mechanisms;
- implement coordinated solutions;
- embed common analysis capabilities and reporting capacity;
- provide public access sections or linked Web sites.

Establishing an administrative framework

A survey completed in 1995 for African nations identified nearly 100 different environmental information system (EIS) related activities. This number was estimated to be only 10%–20% of the total EIS activities in Africa at that time (Prévost and Gilruth 1997). Prévost and Gilruth's post-UNCED (United Nations Conference on Environment and Development) report estimated that over the prior three decades, support for EIS activities in Africa "from bilateral aid agencies, international organizations and NGOs is probably in the order of US $500 million." This is an example of why standards or protocols by themselves will not advance a regional data network without first establishing leadership and an administrative framework. Because each group may have its own standards and protocols, the need for information is compounded by the potential multiple of uses; hence, some coordination among groups is required.

To ensure its success and sustainability, a number of factors should be considered when designing and developing the framework for a regional data network. These include but are not limited to (1) establishing clear objectives, (2) coordinating initiatives and avoiding duplication of efforts, (3) developing a data strategy, (4) developing a realistic cost estimate, (5) building institutional support for the initiative, and (6) establishing management and technical steering committees. Stewart Toshach from the National Oceanic and Atmospheric Administration (NOAA), and Peter Paquet from the Northwest Power and Conservation Council, who co-lead the Northwest Environmental Data Network, wrote, "The overall goal is to materially and demonstrably improve the quality, quantity, and availability of data and related information in the Columbia Basin."

They then list the initial steps needed to establish an administrative framework:
- Distribute and discuss the draft memorandum of agreement (MOA) with regional stakeholders to gather input with a goal of expanding participation and creating agreement on a common regional MOA.
- Distribute and discuss the draft administrative framework with regional stakeholders with a goal of reaching agreement on an accountable regional administrative mechanism for a regional data network.
- Arrange for the existing project team and coordinating committee to be consolidated into one project team.
- Complete further coordination with other programs serving regional information management needs.
- Make information about the proposal for a regional data network publicly available and continue to solicit public input.
- Proceed to develop a detailed work plan and costs for phase II (to adopt/develop data network protocols and standards).

Developing an administrative framework is a necessary precursor to collecting, analyzing, and disseminating information from a regional data network. This framework is critical for the development of strategies by federal, state, and provincial organizations to evaluate, conserve, and protect our natural resources. One of the foremost steps that needs to occur in any administrative framework is standardizing the use of data elements or "establishing the use of common data elements" among organizations. It reflects agreement on representations, formats, and definitions of common data and metadata and their definitions. Determining standards begins by collaboratively identifying the principal or core questions that the regional or global data network is expected to answer.

Establishing regional data centers

All information is local, and its care and maintenance should remain in local hands. As a regional or global data network wants to be responsive to fish and wildlife needs within a specified area, it must ultimately participate or cooperate with the entities that collect and use natural resource information at the local level. To do this efficiently, identify regional data centers that can be a focal gathering point of pertinent local information needed to answer core questions. FishBase (<www.fishbase.org>) and the Interactive Biodiversity Information System (IBIS, <www.nwhi.org/index/ibis>) are excellent examples of multiple partnerships forged to address regional information needs. Each has more than 40 partners supporting their development. But in addition to establishing collaborative and cooperative working relationships within the region, the principal roles that a regional data center should address are:

Accessing data of local entities and experts

To sustain geographical and environmental information initiatives, building capacity and infrastructure is necessary to take full advantage of the available data. The old adage that the data is only as good as the people collecting it may hold some truth here. The experience of local people, their knowledge of an area, familiarity with the plants and animals, and understanding of customs and accessibility to sites all play an important role in how and what type of data is collected. Local experts may be more willing to bring forward their information when they understand how and for what purpose it would be applied. Thus, if questions arise regarding how and what was collected and why, contacting local sources or experts will help achieve a more reliable data set. Regional data center personnel will need to work collaboratively with local entities and experts because a primary purpose of any regional data center is to collect and disseminate the best information available.

Ensuring data quality

Data quality is always an important consideration because the computer doesn't "care" about the accuracy, reliability, or source of the digital data; it is all treated the same way. For example, many of the early efforts to convert standard topographic series and line maps into digital formats basically only created graphic layers of information with no internal topology, thus rendering them of limited use in geographical analysis. People who collect the data have a better understanding of how to use it and what the limitations are; therefore, a primary charge of any regional data center is to assure the quality of the information that is disseminated by continuing to connect with local experts and, as necessary, have them review the information.

Institutionalizing arrangements

Frequently, initiatives have not been developed with close organizational links to senior-level decision makers. Without these close associations, the initiatives have often been poorly understood by senior managers and therefore have not been given the financial resources or the political support needed for them to succeed. Uninvolved or poorly informed managers are less able to provide direction and guidance concerning the initiative's objectives and functions. The regional data centers need to champion their cause with local stakeholders and organizations to support the program.

Ensuring adequate funding

A successful initiative depends on adequate funding for installation, implementation, data development, and long term maintenance and operation. Many initiatives have been less than successful because adequate funding was not ensured for either the initial implementation or the long-term operation and maintenance. When preparing initial budgets, funding was often identified only for hardware and software components of a system. A general rule often used as a guideline for estimating relative costs is what has become known as the 20–80 rule, which suggests that 20% percent of the total cost is for system hardware and software and 80% is for data development activities, institutional costs, and other operating expenses. Financial support for the needs of regional data centers can be addressed in one of two ways: (1) the administrative framework, and/or (2) developing a demand-driven approach. For the latter method to be successful and sustainable, however, the center has to be able to respond more directly to actual demands and to be more closely and directly integrated into the decision-making processes. A demand-driven approach helps ensure that adequate funding is available because the decision support provided by the system becomes crucial to the decision-making process. The support becomes particularly indispensable if it can prove its economic value by avoiding costly errors in decisions and by helping decision makers arrive at the most cost-effective solutions.

Training

For geographic and environmental information systems to have a significant impact in a region requires more than a sophisticated technical capacity. Human resources must be developed through education and training programs to enable in-region agencies and organizations to take advantage of the information support tools available. In preparing education and training guidelines for sub-Saharan African countries, Van Genderen (1991) suggests that nine different groups should be targeted: decision makers and planners, opinion leaders, managers, resource surveying personnel, technical support staff, research workers, teachers, students, and the general public.

Establishing common data elements

A common data element is a set of information with the same attribute and an agreed-to definition. A set of standard data elements is then a group of common data elements that all have common definitions and are used to record, monitor, or describe situations or conditions associated with a specific activity (NWQMC 2004). Agreeing to use standard or common data elements allows sharing of information, thus enhancing the potential for increased use of recorded data (both spatially and temporally) within and among organizations. (Appendix A offers examples of common data elements from the Darwin Core Program, which is operated by Global Biodiversity Information Facility (GBIF) and was developed to make the world's biodiversity data freely and universally available on the Internet.)

A data strategy should be established with input from the various agencies that are involved in data development and management. It is only through a multiagency effort that a consistent approach can be developed. It is unwise for any one agency to assume independent responsibility for setting data standards associated with information from sectoral agencies such as forestry, agriculture, and transportation. When single agencies attempt to avoid that responsibility, it usually builds resentment and reluctance for other agencies to cooperate in a data standards effort.

Geographic and environmental information initiatives have greatly improved chances of success if the underlying data sets are both technically and thematically compatible. Harding and Wilkinson (1996) suggest that "interoperability of data and software is ... a particularly important issue, underpinning many of the other [issues related to successful information development]. Interoperability [applies] at many levels, not only across different database systems but also across hardware platforms, sites and disciplines, and involving interaction between processes and data of many types." Ensuring that databases are interoperable at the outset by establishing protocols for data collection will help avoid additional costs, inconveniences, discrepancies, and duplicate data sets created by different agencies. Data sets that are integration-ready promote and facilitate analysis and help reduce the overall costs of data development.

Well-documented data have standards and thus become more valuable with time, while undocumented data quickly erode. As mentioned, establishing standards allows for data sharing that enables data sets to increase in size, provide greater statistical power, and ultimately have a higher degree of confidence associated with them. Also, because of the multiple needs for information, data sharing increases the likelihood of more accurate and/or comprehensive assessments, because the meaning of each data set and how they fit into a given context are better understood. Last, the individual data set increases in value through use of common data elements because they increase the potential of using the data for purposes other than what was originally intended (NWQMC 2004). This potential can be quickly assessed in the metadata developed for each data set.

Most data collected for fish and wildlife have common elements like point of contact/collector name, date and time, species, and location. Table 1 lists the minimum elements that should be included on OBD collection forms. Of these elements, location is often of keen interest. When dealing with spatial data elements, there are several recent technical factors that help with standardizing data elements. They are the common hardware and software used to capture, record, and display the data. To capture and record data points, lines, or polygons, GPS units (especially handheld ones) have come into favor. It is important to understand GPS and how it can be used, because it is currently the most common way of acquiring a detailed location. GPS units help land, sea, and airborne users locate where they are on earth 24 hours a day by triangulating earth-orbiting satellites; typically, three satellites are needed to obtain a triangulation, but four or more render more accurate results. The GPS unit is actually a receiver that measures distance using the travel time of radio signals. There are several self-help guides (Letham 1998; Anderson 2002) if one desires a more in depth understanding.

To acquire, display, compile, and interpret data that has been recorded in the field, a geographic information system (GIS) is commonly used. GIS technology allows for multiple projections with some ease in converting from one to another, and records captured by GPS can quickly be georeferenced. Multiple spatial data layers such as administrative boundaries for state(s) and counties can reside within a GIS; if deemed important, they can be used as checks in the accuracy and data quality process. Spatial technologies provide tools to incorporate and analyze large data sets in a meaningful manner with the production of useful information. Data can be converted or displayed by locations or across a landscape and displayed as charts, drawings, or maps. These technologies provide a means to handle complexities such as incorporating scale and hierarchy concepts into ecosystem-based management approaches (O'Neill et al. 1996). These technologies also allow others to see how decisions are made, thus leaving a footprint(s) in the decision making process to follow.

Spatial technologies and mapping are as important to the manager as calculators and vehicles. Appendices 2–6 give examples of the spatial components for location and time data elements associated with differing levels of project complexity. Thus, spatial technologies can provide timely information in usable formats for decision makers. Spatial technologies like GIS are frequently described in terms of hardware (computers and workstations) and software (computer programs). Typically more computing power (speed and memory) and large data storage (disk space) are preferred. Workstations do most of the heavy lifting in handling large and/or complex data sets; to write and transfer data effectively in and out of these systems requires peripherals like tape storage and retrieval systems, CD-ROM, and DVD-RAM writers. To become familiar with developing applications using these technologies, see O'Neil et al. (2005).

Emerging Global Data Systems

In the following paragraphs, we provide an overview of some of the key global data systems relevant to salmonid researchers and managers.

The Ocean Biogeographic Information System (OBIS) is a Web-based provider of globally georeferenced information, a project of the Census on Marine Life, which plans to explain global biological diversity over 10 years with the collaborative efforts of 1,000 researchers in 73 nations. GBIF plans to be the "most-used gateway to biodiversity and other biological data on the Internet" by 2011. OBIS requires only that each data set contain the following four common fields: latitude, longitude, taxonomic name, and date/time of last modification; GBIF's requirements are similar.

State of the Salmon, a joint project of Wild Salmon Center and Ecotrust, is undertaking a salmon monitoring data inventory throughout the North Pacific; it focuses on large polygons such as Salmon Ecoregions and Hydro 1K (<www.stateofthesalmon.org/pattern/page.asp?pID=62>); because information is digitized, smaller data sets can easily be rolled into these larger regions. For example, State of the Salmon's database of salmon distribution at the Pacific scale incorporated data from the Raincoast Conservation Society, which works along the central coast of British Columbia at a very fine scale. Raincoast surveyed streams at specific field stations to record presence/absence of several species; since Raincoasts's data were collected with latitude/longitude coordinates, State of the Salmon easily overlaid the point coverage on top of its distribution coverage to make corrections to its database.

Through its volunteer membership of 7,000 species conservation experts, the Species Survival Commission (SSC) of the World Conservation Union (IUCN) holds what is probably the world's most complete body of information on the status and distribution of species threatened with extinction. Although abundant, the data and information contained within the SSC network is widely dispersed and sometimes difficult to access. More than half of SSC's members reside in the developing world and experience constraints in ability to store, share, and analyze their data. The Species Information Service (SIS) is being developed as SSC's data management initiative to address this problem. The SIS aims to become a worldwide species information resource (interlinked databases of species-related information managed by SSC's network of specialist groups). It will be easily accessible to the conservation and development communities, including scientists, natural resource managers, educators, decision makers, and donors, and will contribute to integrated biodiversity conservation products.

The core business of the World Conservation Union (IUCN) is generating, integrating, managing, and disseminating knowledge for conservation. This knowledge is used to empower people and institutions to plan, manage, conserve, and use nature and natural resources in a sustainable and equitable manner. IUCN is undertaking a new communications initiative: the Green Web. This will be IUCN's portal to all of its information resources, with SIS as the backbone. The SIS enables and empowers IUCN expert networks to bridge the scientific digital divide. In turn these networks empower educational institutions, nongovernmental orgainzations, and local communities to make use of their information. In addition, through the Green Web, IUCN's regional and country offices and specialist groups will provide facilities for under-resourced groups and communities to access the Internet and IUCN's knowledge base.

Despite the significant advances in local, regional, and global data systems, disparities exist among countries and among groups within countries regarding access to and use of information and communications technology. This results in a digital divide among scientists and other experts working to conserve the world's biodiversity. The establishment, coordination, and long-term support for a network of regional data centers will significantly advance our collective conservation efforts.

Literature Cited

Anderson, B. 2002. GPS afloat--gps navigation made simple. Fernhurst Books, West Sussex, UK.

Duffy, W. 2003. California Coastal Salmonid Restoration Monitoring and Evaluation Program—interim restoration effectiveness and validation monitoring protocols. B. W. Collins, editor. California Department of Fish and Game, Fortuna, California.

Harding, S. M., and G. G. Wilkinson. 1996. A strategic view of GIS research and technology development for Europe. Space Applications Institute--Environmental Mapping and Modeling Unit, report of the expert panel convened at the joint research centre, Ispra, Italy.

Letham, L. 1998. GPS made easy. Second edition. The Mountaineers, Seattle.

NWQMC (National Water Quality Monitoring Council). 2004. Revised guide for data elements for reporting water quality monitoring results for chemical, biological, toxicological, and microbiological analytes. Methods and Data Comparability Board of the National Water Quality Monitoring Council, NWQMC Technical Report.

O'Neill, R. V. 1996. Recent developments in ecological theory: hierarchy and scale. Gap analysis – a landscape approach to biodiversity planning. American Society for Photogrammetry and Remote Sensing, Bethesda, Maryland.

O'Neil T. A., P. Bettinger, B. G. Marcot, B. W. Luscombe, G. T. Koeln, H. J. Bruner, C. Barrett, J. A. Pollock, and S. Bernatus. 2005. Application of spatial technologies in wildlife biology. Pages 418–447 in C. Braun, editor. Wildlife techniques manual, 6th edition, Bethesda, Maryland.

Paulus, J., and S. Toshach. 2006. Best practices for recording location and time related data. White Paper–Northwest Environmental Data-Network. Northwest Power and Conservation Council, Portland, Oregon.

Prévost, Y. A., and P. Gilruth. 1997. Environmental information systems in sub-Saharan Africa. Towards environmentally sustainable development in sub-Saharan Africa—building blocks for Africa 2025. The World Bank, UNDP Post-UNCED Series Paper 12, Washington, D.C..

van Genderen, J. L. 1991. Guidelines for education and training in environmental information systems in sub-Saharan Africa: some key issues. The World Bank International Advisory Committee Guidelines Series 2, Washington, D.C.

Appendix A: Global data elements within the Darwin Core project

Element	Description	Can be NULL	Type	Min. Value	Max. Value
Record-level elements					
Global unique identifier	A universal resource name for the global unique identifier for the specimen or observation record	No	String		
Date last modified	Last time the data for the record was modified (e.g., June 5, 1994, 8:15 a.m., U.S. Eastern Standard Time)	No	Date-Time		
Institution code	Code or acronym identifying the institution administering the collection	No	String		
Collection code	Code or acronym identifying the collection within an institution in which the record is cataloged	No	String		
Catalog number	The alphanumeric value identifying an individual organism record within the collection	No	String		
Taxonomic Elements					
Scientific name	Full name to lowest taxon that the organism can be identified	No	String		
Kingdom	Name of kingdom where organism is classified	Yes	String		
Phylum	Name of the phylum (or division) of organism.	Yes	String		
Class	Name of the class of organism	Yes	String		
Order	Name of the order of organism	Yes	String		
Family	Name of family of organism	Yes	String		
Genus	Name of genus of organism	Yes	String		
Locality Elements					
Continent	Full name of the continent	Yes	String		
Water body	Full name of the body of water	Yes	String		
Island	Full name of the island	Yes	String		
Country	Full name of the country	Yes	String		
State/province	Full name of the state, province, or region	Yes	String		
Locality	Description of the locality where collection occurred	Yes	String		
Minimum elevation in meters	Minimum altitude above (positive) or below (negative) sea level	Yes	Double		
Maximum elevation in meters	Maximum altitude above (positive) or below (negative) sea level	Yes	Double		
Minimum depth in meters	Minimum depth below the surface of the water	Yes	Double		

Element	Description	Can be NULL	Type	Min. Value	Max. Value
Maximum depth in meters	Maximum depth below the surface of the water	Yes	Double		
Geospatial Elements					
Decimal latitude	Latitude of the collection location shown in decimal degrees.	No or could use other locational information	Double	-90	90
Decimal longitude	Longitude of the collection location shown in decimal degrees.	No or could use other locational information	Double	-180	180
Geodetic datum	Geodetic datum the latitude and longitude refer	Yes	String		
Collecting Event Elements					
Year collected	4-digit year in the Common Era calendar	Yes	Gregorian Year		
Month collected	2-digit month of year in the Common Era calendar	Yes	Integer	1	12
Day collected	2-digit day of the month in the Common Era calendar	Yes	Integer	1	31
Time collected	Time of day of collection or observation	Yes	Double	0	< 24
Julian day	Ordinal day of the year	Yes	Integer	1	366
Collector	Name(s) of collector(s)	Yes	String		
Biological Elements					
Sex	The sex of a biological individual	Yes	String		
Life stage	The age class, reproductive stage, or life stage of the organism	Yes	String		
References Elements					
Image URL	Digital images associated with the specimen or observation	Yes	String		
Related information	References to information	Yes	String		

Appendix B: Spatial Components for General Location and Time Data Elements Associated with Project—Level 1: General Project Information

Logical name	Name definition	Element name (for example only)	Element code, code range, or description
Project	A project is a unit of work defined by an organization or entity. A project may include one or more sites or one or more types and number of activities. Unique system identifier	PRJ_ID	Examples • Skagit River Habitat Restoration Project • Okanogan Water Quality Sampling Project • Oregon North Coast Nearshore Monitoring Project • Deschutes River Flow Monitoring Project
Project location description	Term that best describes the field location in relation to the surrounding environment	PRJ_LOC_DESC	Text field Examples: • Okanogan watershed • ESA Region • SW 1/4 of Section 36 of Township 29 Range 01
Project location latitude coordinate	Distance north or south of the equator. Decimal equivalent to the degrees-minutes-seconds latitude value	PRJ_LOC_LAT_COORD	Float, 2 places, 6 decimals; (4 decimals minimum) E.g. Range for WA: 45.000000 – 49.999999
Project location longitude coordinate	Distance east or west of the Central Meridian (Greenwich, England). Decimal equivalent to the degrees-minutes-seconds longitude value	PRJ_LOC_LONG_COORD	Float, 3 places, 6 Decimals, will accommodate signed values (4 decimals minimum); E.g. Range for WA: -116.000000 – -125.999999
Project horizontal datum	Model used to match the horizontal position of features on the ground to coordinates and locations on a map. (Note: When taking GPS measurements, it is very important to record your datum!)	PRJ_HORZ_DAT	01 - N. American Datum 1927 (NAD27- used on many USGS quad maps or NOAA charts); 02 - N. American Datum 1983 (NAD83 or 91 Adj.—based on Earth and satellite observations, similar to WGS84 but specific to North America.); 03 - High Accuracy Reference Network (HARN—similar to NAD83, but more accurate per GPS observations); 04 - World Geodetic System of 1984 (WGS84—world datum, based on Earth and satellite observations); 99 - unknown.
Project location collection method	Technique used to collect the horizontal coordinates of a Location	PRJ_LOC_COLL_MTH	1 - Address Matching - Block Face; 2 - Address Matching - House Number; 3 - Address Matching - Street Centerline; 4 - Address Matching - Unknown; 5 - Aerial Photography - Rectified; 6 - Aerial Photography - Unknown; 7 - Aerial Photography - Unrectified; 8 - Cadastral Survey (conventional land survey); 9 - Census Block 1990 Centroid; 10 - Census Block Group 1990 Centroid; 11 - Conversion from STR; 12 - Digital or manual raw photo extraction; 13 - Digitized off CTR screen/digitial data; 14 - Digitized - paper map; 15 - GPS carrier phase (employs the satellite Code's carrier signal to improve accuracy); 16 - GPS code phase (measurements based on pseudo random code broadcast by satellite); 17 - GPS kinematic (tracking location while moving using carrier phase); 18 - GPS (Unknown); 19 - Hand-measured - paper map (interpolation); 20 - LORAN-C; 21 - Orthophotography - digital; 22 - Orthophotography - paper; 23 - Satellite Imagery - Landsat MSS (Multi-Spectral Scanning); 24 - Satellite Imagery - Landsat TM (Thematic Mapper); 25 - Satellite Imagery - Other; 26 - Satellite Imagery - SPOT Panchromatic; 27 - Satellite Imagery - SPOT Multi Spectral; 28 - Zip Code Centroid; 29 - GPS (Code/Differential); 30 - Estimated Value 99 - unknown

Logical name	Name definition	Element name (for example only)	Element code, code range, or description
Project start date	The date that the project activity commenced		Date, YYYY/MM/DD format. (Only if applicable) E.g. 2003/03/12. Use a date of 1800/01/01 to indicate that the Start Date is not specified or is unknown.
Project end date	The date that the project activity ended		Date, YYYY/MM/DD format. (Only if applicable) E.g. 2004/03/12. Use a date of 1800/01/01 to indicate that the End Date is not specified or is unknown.

(This detail may not be necessary for all reporting purposes)

Note the information presented here is from Paulus and Toshach (2006).

Appendix C: Spatial components for location and time data elements associated with project—Level 2: project tracking at specific or numerous sites over time

Logical name	Name definition	Element name (for example only)	Element code, code range, or description
Project	The place where site activities associated with a project occur or the area where the work is done. Each site will pertain to just one project, but there can be more than one site for any given project. Location of the site or activities where work is conducted (on-the-ground activities) Unique system identifier	PRJ_SITE_ID	Note to readers: These elements need to be defined based on the type of project site work that is being done • Skagit River Habitat Restoration Site—2 stream reaches • Okanogan Water Quality Sampling Site—4 monitoring sites in study • Oregon North Coast Nearshore Monitoring Site—3 coastal reaches in project • Deschutes Flow Monitoring Site—2 gauging stations in project
Project location description	Term that best describes the site location in relation to the surrounding environment. Information that describes the place a Location exists	PRJ_SITE_LOC_DESC	Text field Example: • 200 yards north of the cattle crossing on Laumann Road, north of the intersection with Heidi Road
Project location latitude coordinate	Distance north or south of the equator. Decimal equivalent to the degrees-minutes-seconds latitude value	PRJ_SITE_LOC_LAT_COORD	Float, 2 places, 6 decimals; (4 decimals minimum) Eg., Range for WA: 45.000000-49.999999
Project location longitude loordinate	Distance east or west of the Central Meridian (Greenwich, England). Decimal equivalent to the degrees-minutes-seconds longitude value	PRJ_SITE_LOC_LONG_COORD	Float, 3 places, 6 decimals, (4 Decimals minimum); will accommodate signed values; Eg., Range for WA: -116.000000 – -125.999999
Project horizontal datum	Model used to match the horizontal position of features on the ground to coordinates and locations on a map (Note: When taking GPS measurements, it is very important to record your datum!)	PRJ_SITE_HORZ_DAT	01 - N. American Datum 1927 (NAD27- used on many USGS quad maps or NOAA charts) 02 - N. American Datum 1983 (NAD83 or 91 Adj.—based on Earth and satellite observations, similar to WGS84 but specific to North America) 03 - High Accuracy Reference Network (HARN—similar to NAD83, but more accurate per GPS observations) 04 - World Geodetic System of 1984 (WGS84—world datum, based on Earth and satellite observations) 99 - unknown
Project location collection method	Technique used to collect the horizontal coordinates of a site location	LOC_COLL_MTH	See Appendix 2 for potential list
Project start date	The date that the project activity commenced		Date, YYYY/MM/DD format. (Only if applicable) E.g. 2003/03/12. Use a date of 1/1/1800 to indicate that the Start Date is not specified or is unknown.
Project end date	The date that the project activity ended		Date, YYYY/MM/DD format. (Only if applicable) E.g. 2004/03/12. Use a date of 1800/1/1 to indicate that the End Date is not specified or is unknown.

(This detail may not be necessary for all reporting purposes)

Note the information presented here is from Paulus and Toshach (2006).

Appendix D: Spatial components for location and time data elements associated with project—Level 3: complex projects that track specifically measured features at a site

Logical name	Name definition	Element name (for example only)	Element Code, Code Range, or Description
Site Feature	The structure, form, or appearance of what is being tracked, measured or observed at any given project site. Within any give project site there may be various features represented as single points, linear features or aerial extents. Unique system identifier	SITE_FEA_ID	Note to readers: This needs to be defined based on the type of scientific/field information that is being collected *Example Code Tables:* • Transect measurement point • Fence • Wells • Fish hatchery raceway • Reach segments *Examples of Site Features:* Water sampling well locations • Individual gauging station location Location of addition to spawning gravel
Site feature location description	Term that best describes the feature location in relation to the surrounding environment. Information that describes the place a Location exists	SITE_FEA_LOC_DESC	Text field Example: 200 yards north of the cattle crossing on Laumann Road, north of the intersection with Heidi Road
Site feature location latitude coordinate	Distance north or south of the equator. Decimal equivalent to the degrees-minutes-seconds latitude value	SITE_FEA_LOC_LAT_COORD	Float, 2 places, 6 decimals; (4 decimals minimum) Eg. Range for WA: 45.000000 – 49.999999
Site feature location longtitude coordinate	Distance east or west of the Central Meridian (Greenwich, England). Decimal equivalent to the degrees-minutes-seconds longitude value	SITE_FEA_LOC_LONG_COORD	Float, 3 places, 6 decimals, (4 decimals minimum); will accommodate signed values; e.g., Range for WA: -116.000000 – -125.999999
Logical site feature horizontal datum	Model used to match the horizontal position of features on the ground to coordinates on a map	SITE_FEA_HORZ_DAT	01 - N. American Datum 1927 (NAD27- used on many USGS quad maps or NOAA charts) 02 - N. American Datum 1983 (NAD83 or 91 Adj.—based on Earth and satellite observations, similar to WGS84 but specific to North America.) 03 - High Accuracy Reference Network (HARN—similar to NAD83, but more accurate per GPS observations) 04 - World Geodetic System of 1984 (WGS84—world datum, based on Earth and satellite observations) 99 - unknown
Site feature start date	Technique used to collect the horizontal coordinates of a feature location.	SITE_FEA_STR_DT	Date, YYYY/MM/DD format. (Only if applicable) E.g. 2003/03/12. Use a date of 1800/1/1 to indicate that the Start Date is not specified or is unknown
Site feature end date	The date that the feature activity (sample collection, field measurement, field observation) ended. If a feature activity is essentially instantaneous, a Feature End Date is often not specified.	SITE_FEA_END_DT	Date, YYYY/MM/DD format. (Only if applicable) E.g. 2004/03/12. Use a date of 1800/1/1 to indicate that the End Date is not specified or is unknown
Site feature start time	The time that the feature activity began, for example the time of sampling	SITE_FEA_ST_TM	Feature start time (time the collection, measurement, observation started -using a 24hr clock at local time) (hhmmss) e.g., 164322 (only if applicable)
Site feature end time	The time that the feature activity ended, for example the end of sampling	SITE_FEA_END_TM	Feature end time (time the collection, measurement, observation ended -using a 24hr clock at local time) (hhmmss) e.g., 175231 (only if applicable)

(This detail may not be necessary for all reporting purposes)

Note the information presented here is from Paulus and Toshach (2006).

Appendix E: Optional elevation data associated with projects, sites, or a feature

Logical name	Name definition	Element name (for example only)	Element code, code range, or description
Elevation	The measure of the elevation of the project site above a reference datum	PRJ_SITE_VERT	Float, will accommodate signed values
Elevation units	The unit of measurement used to describe the elevation value	PRJ_SITE_VERT_UNIT	Text field; example Meters Feet
Elevation datum	The code for the reference datum used to determine the vertical measure	PRJ_SITE_VERT_DAT	Navd88 Ngvd29 Mean Sea-Level Local tidal datum Other
Elevation collection method	The technique used to establish the elevation or depth of the sampling site	PRJ_SITE_VERT_COLL_MTH	GPS carrier phase static relative position GPS carrier phase kinematic relative position GPS code (pseudo range) differential GPS code (pseudo range) precise position GPS code (pseudo range) standard position (Sa off) GPS code (pseudo range) standard position (Sa on) Other Altimetry Precise leveling-bench mark Leveling-non bench mark control points Trigonometric leveling Photogrammetric Topographic map interpolation

Note the information presented here is from Paulus and Toshach (2006).

Appendix F: Examples of location and time reporting for different types of features

Feature Name	Examples of location/time reporting detail from independent data collectors	Examples of location/time reporting detail from corporate data collectors
Install fish screen	• Location of screen (lat./long. dec. degree) • Date of install: YYYY/MM/DD	• Location of screen (lat./long. dec. degree) • Date of install: YYYY/MM/DD
Stream bank stabilization	• Start and end point of stabilization (lat./long. dec. degree) • Date of stabilization: YYYY/MM/DD	• Polygon of stabilization area (lat./long. dec. degree) • Date of stabilization: YYYY/MM/DD
Riparian area treated	• Start and end point (lat./long. dec. degree) • Date of treatment: YYYY/MM/DD	• Polygon of area treated (lat./long. dec. degree) • Date of treatment: YYYY/MM/DD
Road obliteration project	• Start and end point (lat./long. dec. degree) • Length of treatment • Date of obliteration: YYYY/MM/DD	• Line detail of road treatment (lat./long. dec. degree) • Date of obliteration: YYYY/MM/DD
Sediment control basin	• Centroid of basin (lat./long. dec. degree) • Date of sediment control: YYYY/MM/DD	• Polygon of basin (lat./long. dec. degree) • Date of sediment control: YYYY/MM/DD
Wetland creation project	• Centroid (lat./long. dec. degree) • Date of wetland creation: YYYY/MM/DD	• Polygon (lat./long. dec. degree) • Date of wetland creation: YYYY/MM/DD
Invasive species treatment	• Centroid of treatment area • Date of treatment: YYYY/MM/DD	• Polygon of treatment area (lat./long. dec. degree) • Date of treatment: YYYY/MM/DD
Hatchery fry/smolt release	• Location of point of release (lat./long. dec. degree) • Date and Time of release YYYY/MM/DD, hhhh/mm/ss	• Location of point of release (lat./long. dec. degree) • Date and Time of release YYYY/MM/DD, hhhh/mm/ss
Sampling site	• Location (lat./long. dec. degree) • Date and time of sample: YYYY/MM/DD, hhhh/mm/ss	• Location (lat./long. dec. degree) • Date and time of sample: YYYY/MM/DD, hhhh/mm/ss
Livestock exclusion fencing	• Start and end point (lat./long. dec. degree)	• Line detail (lat./long. dec. degree)

Note the information presented here is from Paulus and Toshach (2006).

Methods

This handbook was inspired by a vision to provide standard methods for the capture and counting of salmonids and to serve as the foundation for consistent regional and global data sets describing salmon populations.

The handbook aims to establish standard methods for four sampling objectives: (1) abundance, (2) distribution, (3) population trends, and (4) fish/habitat relationships. The initial step for this resource entailed a literature review of published and unpublished protocols for all commonly used sampling methods in freshwater habitats. The literature search focused on state and federal agencies, universities, and tribes within the Pacific Northwest and elsewhere in North America. Experts were identified and asked to provide available protocols for fish capture and counting. At universities, we approached faculty and research staff directly and conducted searches of university library resources.

We searched existing databases with protocols such as those maintained in British Columbia and the Klamath Resource Information System. Reference sections from available protocols were also used to track down original documents. After assembling more than 375 documents, we used a specific set of criteria to conduct a coarse screening on the content and value of each document. The screening process, based on work by Oakley et al. (2003), used the checklist below to determine whether protocols met a minimum standard for further consideration.

Essential Elements of Protocols

Background and objectives

Background—history, resources being addressed

Rationale—justification of selecting a given resource to inventory or monitor

Objective—list of measurable tasks

Sampling design

Site selection—defining boundaries or "populations" sampled; selecting sampling locations; stratification; and spatial design

Sampling frequency and replication—recommended number and location of sampling sites; frequency and timing of sampling; level of change that can be detected for the amount/type of sampling

Field/office methods

Setup—field season preparations and equipment setup (including permitting/compliance procedures)

Events sequence—sequence of events during field season or during preparation of a monitoring plan

Measurement detail—details of taking measurements, with examples of field forms

Sample processing—postcollection processing of samples (e.g., lab analysis, preparing voucher specimens)

Data handling, analysis, and reporting

Metadata procedure—database fields and sizes; sample collection information; site description; quality assurance

Database design—overview of database design and structure illustrating relationships between tables

Data entry—data entry procedures; verification and editing of data

Data summaries—data summaries and procedures for conducting statistical analyses

Report format—recommended report format with examples of summary tables and figures

Trend analysis—recommended methods for trend analysis

Archival procedures—data archival procedures

Personnel requirements and training

METHODS

Background and objectives
Responsibilities—crew and project roles and responsibilities
Qualifications—prior experience for paid and volunteer staff
Training—availability, locations, timing, and procedures
Operational requirements
Workload and schedule—factors to facilitate chronological planning
Equipment needs—list of equipment, materials, and facilities needed
Budget considerations—calculation guidelines
Literature cited

The coarse review occurred over several months, with several fish biologists reviewing documents. Each protocol described in a document was evaluated against the above criteria and was rated as "fully covered," "partially covered," or "not covered at all." The criteria and ratings for each protocol were stored in digital data sheets and organized in a searchable database along with author and citation information. The ratings revealed desirable components of each protocol and determined which were recommended for standardization.

From the coarse review, more than 375 documents describing protocols were reduced to 74 of the strongest evaluated. These "strongest 74" reflected the highest ratings for essential elements of protocols and were distributed among the different methods identified as those commonly used for capture and counting of salmonids and other freshwater fish. Through this process, the reviewers affirmed that for most methods no single document contained information meeting all criteria. Thus, to get protocols with the full range of elements, the strongest parts of existing documents would need to be highlighted and combined. If none of the documents reviewed were found to have essential elements, the authors decided to develop them from scratch.

Following literature review and protocol screening, a number of international experts in fish capture and population monitoring were invited to attend a workshop in Welches, Oregon, in March 2004. Participants were selected based on their professional and geographic areas of expertise from around the Pacific Rim and grouped into three work groups—field practitioners, biometricians, and data managers. The groups received a set of 375 protocols identified from the literature review and were asked to identify the most promising and pertinent sections. Each group drafted protocols relevant to its area of specialization. For example, the field practitioners drafted protocols for several data collection methods, the biometricians focused on data analysis methods, and data managers prepared a core set of variables and units for those variables critical to any fish population monitoring effort. The workshop contributed immensely to validating and refining best practices for an array of data collection, analysis, and management methods. The panelists also highlighted methods (e.g., counting towers, cast netting) not reflected in the initial literature search that were subsequently added to the final set of protocols and reviewed.

Workshop results were turned over to the core project team for further editing, peer review, and writing. A substantial task then confronted the project team: many of the protocols needed significant development and refinement. Teams of authors, consisting of individuals recognized as experts in their respective methods, including some workshop participants, were enlisted to further develop and refine the protocols. This process produced a set of draft 13 primary protocols, 5 supplemental techniques, and several essays, all of which were sent out for formal peer review.

Literature Cited

Oakley, K. L., L. P. Thomas, and S. G. Fancy. 2003. Guidelines for long-term monitoring protocols. Wildlife Society Bulletin 31:1000–1003.

METHODS

Carcass Counts
Bruce Crawford, Thaddeus R. Mosey, and David H. Johnson

Background and Objectives

Background
Most adult anadromous salmon die shortly after returning from the sea to spawn in their natal streams. Their bodies contribute significantly to the nutrients needed by the next generation of salmonids (Wipfli et al. 2003), a wide array of wildlife species (Cederholm et al. 2001), and the larger watershed ecosystems in which they play an integral part (Wipfli et al. 1999; Bilby et al. 2001; Stockner 2003) (see Figure 1). This protocol deals with the counting and biological sampling of recently spawned anadromous salmonid carcasses. We have drawn heavily from Heindl (1989) in the development of this protocol.

Salmon carcasses provide important information to scientists, including scales, tissue samples, length measurements, and population sex composition data. Scales and otoliths are used to determine age and offer insights into the population's age characteristics. Fish lengths (i.e., length from postorbital to hypural plate—POHL) provide length–frequency information. Although snout to fork (fork length) is one measurement option, POHL is more useful because the tails of spawners, especially females, are frequently so worn that the fork is difficult or impossible to locate. It is also important to use POHL to avoid length distortion caused by jaw development (kype). Some scientists have taken postorbital to fork length, which we consider less useful because of the need for subsequent conversion to either POHL or fork length (S. Young, Washington Department of Fish and Wildlife, personal communication). Marks (fin clips and/or tags) provide key information on the fish's origins. A clipped adipose fin indicates a hatchery-origin fish, which many also have coded wire tags (CWT) implanted in their snouts. In the field, a coded wire wand or detector can be used to determine the presence of a CWT in the snout of the fish; without field detection, it is recommended to cut off the snout to assess the CWT information at a later point. Other types of markers and tags as well as radio telemetry units may be collected from carcasses where specific studies are involved. In addition to scales, tissue samples from hard body parts (e.g., fin rays, vertebrae, otoliths) may be collected for age determination. Soft tissue samples (e.g., liver, heart, eyes) may be collected for genetic analysis. Field staff use both foot and boat surveys to locate carcasses.

Combined with redd counts and live fish counts, carcass counts can be used to assess escapement as well as carcass-derived nutrient contributions to the ecosystem. Although data from this protocol can be used to support other assessments, this document does not focus on counts of salmonids that have undergone mass die-offs due to other causes (e.g., water temperature or oxygen level issues, strandings, poisons).

CARCASS COUNTS

FIGURE 1.—Chinook salmon *Oncorhynchus tshawytscha* carcass. The marine-derived nutrients accumulated in salmon play an integral part in a healthy ecosystem. (Photo by Todd N. Pearsons, Washington Department of Fish and Wildlife.)

Rationale

Salmon carcasses are a source of biological samples (e.g., scales, other tissues) and crucial demographic measurements (e.g., body size, sex ratios, age). With marked fish, carcass recoveries offer a mechanism to support mark–recapture population assessments. Carcasses are also a means to assess the relative number of wild versus hatchery-based fish in the system. By estimating the numbers of spawning fish in a total redd count, one may determine the number of carcasses that need to be recovered (sampled) to evaluate key demographic and genetic aspects of hatchery and supplementation programs.

The ocean-derived nutrients salmon bring to spawning grounds are significant and play a substantial role in the functioning of a healthy ecosystem. Carcasses of adult hatchery fish are often placed in the upper reaches of the watershed for the nutrients they provide (Stockton 2003; Sanderson et al. 2004). Specific guidance for distributing salmonid carcasses has been developed in British Columbia (<www.bccf.com/steelhead/pdf/Carcass 2002 Final.pdf>) and Washington (<http://wdfw.wa.gov/hab/ahg/shrg_t11.pdf>). Carcass counts are often done in conjunction with foot-based visual counts of spawning fish and with redd counts.

Objectives

- Determine the length, sex, age, phenotypes, and genotypes of spawning fish by collecting a representative sample of the target species' carcasses.
- Determine ratios of hatchery origin and wild fish contributing to spawning populations.
- Recover carcasses in support of mark–recapture and other demographic studies.
- Determine contribution of nutrient amounts to ecosystem.

Sampling Design

Site Selection

Site selection for carcass surveys is based on the timing and spawning locations of adult anadromous salmonids. Streams should be divided into survey reaches prior to sampling; typically these are the same survey reaches that are used for redd counts. This can be done either through a randomized sampling protocol or through previous surveys that identify where the target salmon are known to spawn.

Sampling Frequency and Replication

Replication of carcass counts is necessary to estimate the total or relative number of spawners over the spawning season. Carcass counts are often done in conjunction with redd counts and/or counts of live spawning fish. As such, the expected spawning season should be identified for each survey. Surveys should be scheduled to begin before the first spawning takes place and to continue until after the last spawner completes the spawning process. Such counts are generally conducted every 7–10 days throughout the spawning season for the focal species.

Sample size for a given objective will be determined based on precision goal. For CWT analysis, the generally accepted sample size is 20% of the population of interest. Determination of estimates for proportion of gender, length, and age composition of a given population is dependent upon the precision requirements for those estimates. Estimates of the age composition of each gender of natural spawners and the overall proportion of each gender should be made so that all the estimated fractions are within ±5 percentage points of their true values 95% of the time. To estimate age composition using scales, the number of scale samples collected needs to be sufficient to estimate the mean length of major age classes (comprising >5% of the runs) so that all estimates are within ±10 mm of their true values 95% of the time.

We offer the following example, using chinook salmon. For age and gender, proportions from carcass sampling data, minimum sample sizes are as follows: the estimated escapement was approximately 4,000 females, 6,000 males ≥ age 3, and 1,500 males of age 2. Estimates based on redd count methodologies were lower than the estimates but would lead to smaller recommended sample sizes due to finite population correction. According to multinomial sampling guidelines in Thompson (1987), with an infinite population a sample size of 510 or greater will be adequate to ensure that proportions are within ±5 percentage points of the true values 95% of the time. We then applied the correction for finite population of $m=n/[1+(n-1)/N]$ where n is the initial sample size, m is the adjusted sample size, and N is the approximate population size (Zar 1984). Because not all collected chinook scales are readable (due to regeneration, etc.), we need to further adjust the sample size. Assuming >80% readable we then divide the sample size by 0.80 to get the final recommendation. Table 1 shows the target sample size of scale samples to be collected.

TABLE 1.—Sample sizes for achieving stated precision objectives for carcass and live fish recovery (either ±5 or ±10 percentage points at 95%). The starting points are 510 and 128, respectively, and are adjusted for finite population size and then for readable scales; ±10% is the stated goal.

Chinook	Pop. size	Corrected for			
		finite population		readable = 80%	
		±5%	±10%	±5%	±10%
Females	4,110	454	124	568	156
Males (over 50 cm, ≥age 3)	6,127	471	126	589	157

TABLE 2.—Numbers of fish to examine (C) to derive Petersen mark–recapture estimates with 95% confidence intervals of 10%, 25%, and 50% (100*p) of N across a range of population abundance (Robson and Regier 1964). A p value of 0.25 will yield 95% confidence intervals such that the point estimate is within ±25% of the true values 95% of the time.

Estimated number in population (N)	Total number marked (M)	Percent marked	Number of fish to examine (C)		
			$p=0.50$	$p=0.25$	$p=0.10$
5,000	100	2	965	2,017	3,967
5,000	200	4	524	1,244	3,265
5,000	300	6	355	889	2,756
5,000	400	8	266	685	2,371
5,000	500	10	210	552	2,069
5,000	600	12	173	460	1,825
10,000	100	1	1,946	4,059	7,951
10,000	200	2	1,068	2,527	6,577
10,000	300	3	731	1,824	5,590
10,000	400	4	553	1,421	4,848
10,000	500	5	443	1,159	4,269
10,000	600	6	368	976	3,805
10,000	800	8	273	735	3,107
10,000	1,000	10	215	585	2,608
10,000	1,200	12	176	482	2,233
20,000	200	1	2,155	5,092	13,198
20,000	400	2	1,128	2,892	9,798
20,000	600	3	759	2,007	7,758
20,000	800	4	569	1,529	6,398
20,000	1,000	5	453	1,230	5,427
20,000	1,200	6	375	1,026	4,699
20,000	1,600	8	277	763	3,679
20,000	2,000	10	217	602	2,999
20,000	2,400	12	177	494	2,514

Source: Chinook Funding Proposal. "Assessment of Chinook Salmon Spawning Escapement in the Green River via Mark–recapture and Redd Surveys, 2002." ("New Project": One Year Study, July 1, 2002 through June 30, 2003; Mark–recapture Studies Funded in 2000, 2001; Redd Assessments Funded in 1999, 2000, 2001.) Submitted May 9, 2002, by Tom Cropp, Steve Foley, Pat Hanratty, Peter Hahn, Biometrician, Washington Department of Fish and Wildlife. Submitted to U.S. Section Office, Northwest Region, National Marine Fisheries Service. Table 1, page 14.

It is important to be aware that there are differential recovery rates of carcasses that may bias sampling (e.g., females tend to end up closer to the redds, and lighter males may wash farther outstream [Zhou 2002]). Additionally, small fish may be underrepresented using carcass sampling because they are more easily scavenged and harder to see (Cropp et al. 2002).

Field/Office Methods

Setup

Prior to Arrival
Before a survey is undertaken, the surveyor must be well aware of what reach to survey, when the spawning season begins and ends, the time interval for the survey, what data are needed, and what equipment is required. The basis for the sampling design, including the source of information on spawning location, should be clearly and thoroughly documented in the project description metadata.

Equipment
The foot surveyor should be equipped with hip boots or waders, dark clothing or raingear to minimize disturbance to spawners, polarized sunglasses, appropriate carcass survey forms, maps, biological sampling forms, snout ID tabs and bags, and a handheld global positioning system (GPS) device. In some situations it is valuable for the surveyor to carry a CWT detection wand. Additional equipment for boat or canoe access includes life vests, dry bags, survival kit (containing matches, food, whistle, emergency blanket, etc.), helmets, and a throw line. See the equipment list on p. 68 for a full inventory.

Events/Sequence

Survey Description
Carcass surveys are usually undertaken at the same time that live fish or redd surveys are conducted. Be aware that carcasses can display evidence of numerous factors that affect the population being surveyed. Carcasses may be discovered at virtually any location in the stream and on the streambank. While they are most often located in slack water areas below spawning riffles, depending on the stream or river system, they may be distributed widely across bars, islands, and floodplain areas. When sampling a carcass, first examine it for marks, fin clips, operculum punches, tags, or telemetry units, and record each one noted. Next, measure the POHL (Figure 2) and record the result. The hypural plate forms the last and largest vertebra in the spinal column and is located in the caudal peduncle. The obvious flex point of the tail at the posterior edge of the hypural plate is the point to which measurements are made. Finally, open the abdominal cavity to determine the fish's sex and to ascertain the extent to which it spawned. If the carcass is a female, presence of only a few (<50 or so) loose eggs in the body cavity indicates a completely spawned fish. Larger numbers of retained eggs probably indicate incomplete spawning; in such cases, it will be necessary to estimate the percent of the eggs remaining and the percent of eggs deposited in the stream.

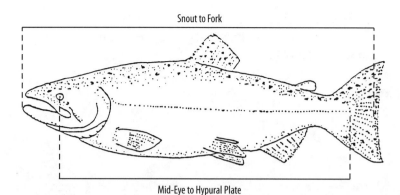

FIGURE 2.—Fish length measurements (from Heindl 1989). Note: The hypural plate forms the last and largest vertebra in the spinal column and is located in the caudal peduncle. The obvious flexpoint of the tail at the posterior edge of the hypural plate is the point to which measurements are made.

Other desired samples (e.g., scales, tissue) can then be removed and their collection noted on the applicable data forms. Scale sampling techniques are described in the following section. If the carcass carries a CWT (as detected by a CWT reader), remove its snout by cutting across the head, straight down behind the eyes, until reaching a point even with the line of the mouth. Next, place the knife in the fish's mouth and cut back toward the first incision until the snout is cut free. Finally, place the snout in a plastic bag with a numbered tag that identifies its source and enter the tag number on the data form. When sampling has been completed, cut the tail from the carcass to identify it as having been previously counted and sampled, and discard it.

In summary, the following steps are undertaken when conducting carcass counts:

1. Determine the species and whether it is a target species of the survey.
2. Assign the carcass a sample number and record the number on the data card.
3. Examine the surface of the fish for visible marks or tags. A missing fin will usually indicate a mark. The most common missing fin will be the adipose fin because it is used to indicate a hatchery-origin fish. A fish with a clipped adipose fin may also be tagged with a CWT in the snout. Record whether a mark has been observed on the appropriate form. A tag indicates that the fish is involved in a mark–recapture project to estimate overall population size.
4. Measure the carcass standard length to the nearest millimeter (postorbital to hypural plate).
5. Cut open the carcass and observe the gonads to determine the sex and record the information on the form.
6. If it is a female, evaluate the amount of eggs retained in the body cavity and estimate the egg retention percentage. This will require some previous training and familiarity with the fecundity of the target species. Record the percentage on the form.
7. Pass the CWT detection wand over the fish to determine if there is a CWT embedded in the snout. If a tag is detected, cut the forward end of the head (snout) off and place in a plastic bag. Number the snout sample with the same number as the carcass number.

8. Upon completing the process, cut the caudal (tail) fin off of the carcass to indicate that the carcass has been processed.
9. Repeat steps 1–8 for each carcass encountered or according to the established subsampling protocol.

Scale Sampling
Scale samples are collected to gain insight into the age and origin of the fish involved. Hatchery fish may be thermal-marked, a technique involving changes in water temperature at the hatchery that are reflected in the differential growth ring patterns shown in the otoliths. The scales most useful for analytical purposes (i.e., those containing complete growth records) are located along the sides of the fish, within the preferred area (Figure 3) (Clutter and Whitesel 1956; INPFC 1963). (The preferred area lies on either side of the fish, two to three scale rows above the lateral line and within six scale columns on either side of a diagonal line running from the posterior base of the dorsal fin to the anterior base of the anal fin.) Using a forceps or a small hemostat, remove six scales—three per side—from each fish sampled. If scales are missing from one side, take all samples from the other (and note this in the comments section of the data form).

If no acceptable scales can be found, do not collect any. Note that a good scale appears oval and well formed and, when held up to the light, shows a distinct central focal point with obvious concentric markings. Because they adhere tightly to their scale pockets, scales are frequently difficult to remove from carcasses. As a result, care must be exercised to collect only the scale and not other tissue parts. Examine each extracted scale for damage or regeneration (a consequence of previous scale damage).

Before mounting any scales, identify the scale card by labeling it with the sampling location, date, and card number. Put the fish's left-side scales *above* the cell number and its right-side scales below (see Figure 3). When mounting scales, it is important to orient each one similarly to facilitate subsequent reading and measuring in the lab. Also, place scales on the card in the manner they grew on the fish (with their exterior surfaces facing up). If in doubt whether the scale has been properly mounted, test it with a pencil point: a scale's rough outer surface will accept a pencil mark.

* About the size of this block

CARCASS COUNTS

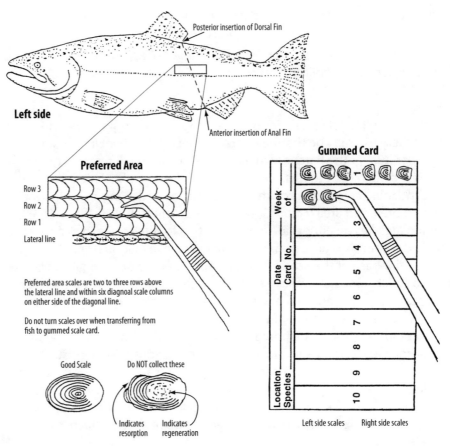

FIGURE 3.—Scale sampling procedure (from Heindl 1989).

Tissue Samples for DNA Analysis

From the U.S. Geological Survey lab, Anchorage, Alaska (<www.absc.usgs.gov/research/genetics/sampling.doc>).

- Ensure that your collection kit contains numbered microtubes (i.e., 1–100), a sample collection form, and a permanent marker.
- Record data associated with the respective sample (e.g., species, sample ID number, collection location, collection date, sex and age if known, name of the collector). Additional data may be required, depending upon study.
- Label the outside of the box (i.e., species, collection location, collection dates, collector name, and contact telephone number).

Fin Tissues
Tissue/1 vial/ethanol/room temperature

At least 50 fin samples for population genetic analysis and 3 or 4 reference samples (whole fish) for phylogenetic analysis need to be collected from each location. There are two methods for sample collection: dry or in vials with 100% ethanol (EtOH). The preferred method of collection is to store the tissue in a vial with 100% EtOH. If this is not possible, the following dry method can be used.

Tissue/1 container/dry/room temperature
Use clean scissors or a clean scalpel blade to cut a small piece of tissue from one of the fins of the live fish. Tissue size should be approximately 5 mm^2 *; a wedge from the upper or lower lobe of the tail fin works well. Although some DNA can be extracted from adipose fins, they are not easy targets because the tissue contains a lot of complex lipids. Eroded fins from dead salmon carcasses are highly degraded, and DNA is usually not readily extracted from such tissue. A well-dried 5-cm^2 piece of skin tissue works best under these conditions.

The date of collection, fish species and stock, type of collection method, and fish length, sex, and age (young of the year, juvenile, adult will suffice) should be collected with each fin where possible.

If samples are to be sent through the mail, ethanol should be drained from the samples immediately prior to mailing; the samples will be rehydrated upon receipt at the lab.

Dry Sample Collection (Alternative)
For the dry method, Whirl-Pack bags, Cryo-Tubes, or scale envelopes lined with high-quality filter paper work well. Either in the field after collection or in the office immediately upon return from the field, samples should be air-dried on filter paper or paper towels until all mucus and moisture in the fin have evaporated and the fin feels dry to the touch. Sun drying in the field works best and can be done quickly. Drying fins inside usually takes 18–24 hours at room temperature. Fungus and bacteria immediately invade the fins upon collection; these factors break down the cell walls of the tissue, and the DNA exudes into the surrounding medium, making DNA extraction in the lab difficult if not impossible. If drying is to be delayed for more than 4 hours, samples do best when packed on ice. DNA from moist-stored fins can often be all right for 6–8 hours, depending on the original condition and size of the fin clip.

Dried fin clips should be repackaged separately (make sure the baggy or envelope is also dry) and attached to field notes for shipment. Dry samples can be sent via surface mail without special packaging.

Hard tissue (bone or teeth/1 envelope/dry/room temperature)
Hard tissue samples such as bone or teeth should be kept as dry as possible. Because DNA is extracted from tooth pulp, the whole tooth is preferred. These can be stored in containers or envelopes.

Muscle (tissue/1 vial/tissue preservation buffer/room temperature)
Muscle tissue samples are the preferred samples for work that includes mtDNA analyses along with nuclear DNA analyses, particularly for birds. Among muscle tissue samples, heart is the most preferred, since the mtDNA yield is very high relative to nuclear yield. DNA can also be extracted from skin, teeth, and bone. Soft tissue samples can be stored at room temperature in the field in the tissue preservation buffer. Any muscle or skin tissue will work and can be stored in this buffer solution. It is important to ensure that the storage buffer completely covers the tissue sample. Also, make sure to clean instruments between sampling (to prevent cross-contamination) using a 10% bleach solution followed by a water rinse. A sample about the size of a pencil eraser is all that is needed, but make sure the sample is entirely submersed in the buffer.

Tags and Markers

Mark–recapture studies often employ various markers and tags (Guy et al. 1996). Floy ("spaghetti") tags are thin, colored pieces of plastic implanted into the body of the fish just beneath the dorsal fin. All tagged fish typically receive two tags of like color at the time of their trapping. Typical tag colors are blue, yellow, orange, grey, and brown. Jaw tags are another example of colored plastic tags; they are attached through the mouth region of the fish.

One technique of marking live fish for mark–recapture studies involves capturing fish and punching a small hole through their operculum (ODFW 2005), the bony flap covering the gills on both sides of the fish's head. Punches are typically made with a wide-gap paper punch and may be made in the left operculum, the right operculum, or both. For each side, there are three possible punch locations: above center, below center, and center; there are five possible hole shapes: square, round, rectangle, crescent, and triangle. Marks are stratified by Julian week and identified by the side of fish and area of operculum punched. The second capture event consists of recovering fish on the spawning grounds. All fish will be sampled for length, scales, and sex and checked for marks. Careful attention should be directed towards identifying opercula punches. Carcasses may fungus up, or the punch may skin over, making it difficult to see the mark. The gill cover may erode into the punch, changing the round hole to a crescent-shaped mark. Any remnant of an opercula punch will be recorded as a marked fish. Each fish will be sampled once and the tail removed.

Analytical Methods for Mark–recapture Data

The Chapman modification of the Petersen estimator can be used to estimate the population size of adult fish in each stream or river as long as all assumptions are met. From Seber (1982) the population size is estimated by

$$\hat{N} = \frac{(\hat{n}_1 = 1)(n_2 + 1)}{(m_2 + 1)} - 1 \qquad \text{(eq 1)}$$

$\hat{n}_1$ = estimated number of marked fish on the spawning grounds,
n_2 = the number of fish inspected for marks on the spawning grounds,
m_2 = the number of marked fish recaptured on the spawning grounds,

where

$$\hat{n}_1 = n_1 - mort - h - f \qquad \text{(eq 2)}$$

n_1 = the total number of fish marked at the seining or trapping site
$mort$ = the number of tagged fish that were found dead within a few days of tagging
h = the number of tagged fish recovered
f = the number of tagged fish found (or extrapolated from radio tagging) downstream of the tagging sites or out of basin.

The assumptions for use of the Petersen estimator (Seber 1982, restated) are

1. all fish have an equal probability of being caught and marked at the first-capture site; *or*
2. all fish have an equal probability of later being inspected for marks (i.e., the second sample is a simple random sample); *or*
3. marked fish mix completely with unmarked fish in the population between events;

and

4. there is no recruitment to the population between capture events (population is closed);
5. there is not trap-induced behavior;
6. fish do not lose their marks and all marks are recognizable (and reported on recovery in the second sample);
7. survival is equal for marked and unmarked fish.

Data Handling, Analysis, and Reporting

Data from carcass counts should be entered into a database for data management. Separate counts for redds, live spawners, or carcasses should be maintained separately in the database. Table 3 provides metadata for variables that should be collected during the carcass surveys. Examples of field data forms for carcass count data are shown in ODFW 2005 and for the collection of genetic tissue samples in Appendix 1.

TABLE 3.—Sample parameters of adult anadromous salmonid carcass count data.

Description	Metric	Format	Comments
Species	Text	Text	Note the species sampled
River	Text	Text	Record the stream name and section being surveyed
Reach length	Meters	XXXX.X	Record the total distance of the survey reach
Dead	#	XXXX.	Number of carcasses observed
Days sampled	#	XXXX.	Record the total number of days sampled
Sample number	Text	XXXX.	Assign a number to the carcass sample. This number will correlate with the scale number and the snout number so that age and CWT information can be tied to the correct carcass
Carcass length	Cm	XXX.X	Record carcass length (POHL)
Sex	M/F	X	Record carcass sex
Carcass tag/mark	None, AD, CWT, RP, LP, D, RV, LV, AN, JT, ST	1–4	Record whether carcass had a mark or contained a coded wire tag. CWT = coded wire tag. AD = adipose clip, RP = right pectoral clip, LP = left pectoral clip, D = dorsal fin clip, RV = right ventral clip, LV = left ventral clip, AN = anal fin clip, JT = jaw tag, ST = spaghetti tag, None = no tag or mark
Scale sample	#	XXXX.	Record the scale envelope/card number
Egg retention	%	XX.X	Record the percentage of eggs retained in the carcass postspawning
CWT snout collection	Bag #	XXXX.	Record the bag number
Reach	Text	Text	Record the reach description and reach code
Scale position	Text	Text	Record the location where the scales were taken.
Date	Date		Record the date of the survey
Samplers	Text		Record the last name of the samplers
Start latitude	D, M, S	XX,XX,XX	Record the latitude (or universal transverse mercator [UTM] and zone) of the beginning survey point
Start longitude	D, M, S	XX,XX,XX	Record the longitude (or UTM and zone) of the beginning survey point

CARCASS COUNTS

Description	Metric	Format	Comments
End latitude	D, M, S	XX,XX,XX	Record the latitude (or UTM and zone) of the end survey point
End longitude	D, M, S	XX,XX,XX	Record the longitude (or UTM and zone) of the end survey point
Temp	Degrees C	XX.X	Record the stream temperature at the time of the survey
Time	Time	Time	Record the time the survey began
Conditions	Text	Text	Record the water conditions and weather at the time of the survey
Other	Text	Text	Describe other samples taken and remarks

Data analysis procedures will vary with the objectives of the survey. Generally, however, data are evaluated with respect to the abundance or relative abundance of spawners throughout the spawning season. Other methods, such as area-under-the-curve (English et al. 1992), may be used to extrapolate the total number of spawners/redds/carcasses from the survey data.

Analytical Methods for Age and Gender Composition

We offer the following example, using chinook salmon:

The proportion of the spawning population composed of a given age within the single size group ≥ 50 cm in length is estimated as a binomial variable from fish sampled from each of the stock components, (a) hatchery fish and (b) natural spawner groups (equation 3). (The natural spawner group may be split into upper and lower river and into males and females.) Those less than 50 cm in length are usually 100% age 2 "jack" males (often very low sample size), whereas those ≥ 50 cm in length are age 3 and older.

$$\hat{p}_j = \frac{n_j}{n} \qquad (eq\ 3)$$

where $\hat{p}_j$ is the estimated proportion of the population of age j; n_j is the number of chinook salmon of age j; and n is the number of chinook salmon ≥50 cm in length taken on the spawning grounds.

Sample variance is calculated as

$$v(\hat{p}_j) = \frac{\hat{p}_j(1-\hat{p}_j)}{n-1} \qquad (eq\ 4)$$

Numbers of spawning fish by age are estimated as the summation of products of estimated age composition and estimated abundance within size categories i

$$\hat{N}_j = \hat{p}_j \hat{N} \qquad (eq\ 5)$$

with a sample variance calculated according to procedures in Goodman (1960)

$$v(\hat{N}_j) = v(\hat{p}_j)\hat{N}^2 + v(\hat{N})\hat{p}_j^2 - v(\hat{p}_j)v(\hat{N}) \qquad (eq\ 6)$$

Sex composition and age-sex composition for the entire spawning population and its associated variances are also estimated with the equations above by first redefining the binomial variables in samples to produce estimated proportions by sex $\hat{p}_k$, where k denotes gender (male or female) such that $\sum_k \hat{p}_j = 1$, and by age-sex $\hat{p}_{jk}$, such that $\sum_{jk} \hat{p}_{jk} = 1$.

Personnel Requirements and Training

Responsibilities
A crew of two carcass count surveyors should be used. They can split up part of the survey area if necessary, but two persons should be employed for safety reasons. The project leader supervises the crew and must ensure that each surveyor or team carries all necessary maps and equipment.

Qualifications
The staff conducting a spawning survey should have been properly trained by an experienced field biologist and should have 1 year of experience in sampling fish. Volunteers can be used when carefully trained and evaluated.

Training
Crews should be trained in the classroom first with illustrations of sampling techniques, equipment needed, process for location of survey reaches, and so forth. This should be followed up with at least one survey with an experienced instructor who can assist the student in the field sampling techniques of carcass counts and scale/tissue sampling.

Operational Requirements

Field Schedule
The field schedule for carcass counts is determined by the spawning season of the target species.

Equipment List

Item	Comments	Cost
Data forms and backing	e.g., clipboard, digital data device	
Pencils		
GPS handheld device	Used to record survey latitude and longitude or UTM	
Cell phone and/or Motorola FM 2-way TalkAbout radios	Useful for emergencies and for maintaining contact with other crews	
Polarized glasses		
Wading gear with belt	Not recommended if rafting	
Metric measuring tape		
Forceps or hemostat	For taking scale samples	
Scale cards/envelopes	For collecting scales	
Rubber bands	For holding cards/envelopes together	
Knife and sharpener		
Machete		
Water thermometer		
Scalpel	For removing soft tissue	
Small saw or wire cutters	For cutting off snouts	
Collectors permits	State and/or federal as needed	

Item	Comments	Cost
Tissue sample containers	Glass or plastic vials with caps	
Labels	To identify tissue samples	
Plastic bags	For carrying snouts, etc.	
Small ice chest and dry ice	If gonadosomatic index samples are collected	
First-aid kit		
Colored flagging or spray paint	For marking multiple survey reaches	
Indelible markers	For marking flagging	
Day pack	For carrying supplies, lunch, etc.	
Extra clothing	Jacket, raingear, hat	
Rubber raft	If floating on a river	
Air pump		
Spare oar or paddle		
Spare oarlock		
Tool kit	Pliers, crescent wrench, wire, heavy tape, bolts, nuts, washers, etc.	
Rope		
Waterproof bag		
Flotation vest		
Wet suit		

Budget

The following guidelines can be used to calculate budget.

Activity/item	Cost
Equipment	variable
Staff time for two biologists to walk the stream reach.	3 hours
Travel time	variable
Preparation time	1 hour
Training	1 hour
Lab workup	2 hours
Data analysis	8 hours

Literature Cited

Bilby, R. E., B. R. Fransen, J. K. Walter, C. J. Cederholm, and W. J. Scarlett. 2001. Preliminary evaluation of the use of stable isotope ratios to establish escapement levels for Pacific salmon. Fisheries 26:6–14.

Cederholm, C. J., D. H. Johnson, R. Bilby, L. Dominguez, A. Garrett, W. Graeber, E. L. Greda, M. Kunze, J. Palmisano, R. Plotnikoff, B. Pearcy, C. Simenstad, and P. Trotter. 2001. Pacific salmon and wildlife -ecological contexts, relationships, and implications for management. In D. H. Johnson and T. A. O'Neil, managing directors. Wildlife-habitat relationships in Oregon and Washington. Oregon State University Press, Corvallis.

Clutter, R., and L. Whitesel. 1956. Collection and interpretation of sockeye salmon scales. International Pacific Salmon Fisheries Commission Bulletin 9.

Cropp T., S. Foley, P. Hanratty, and P. Hahn. Chinook Funding Proposal. "Assessment of Chinook Salmon Spawning Escapement in the Green River via Mark–recapture and Redd Surveys, 2002." ("New Project": One Year Study, July 1, 2002 through June 30, 2003; Mark–recapture Studies Funded in 2000, 2001; Redd Assessments Funded in 1999, 2000, 2001.) Submitted May 9, 2002, by Tom Cropp, Steve Foley, Pat Hanratty, Peter Hahn, Biometrician, Washington Department of Fish and Wildlife. Submitted to U.S. Section Office, Northwest Region, National Marine Fisheries Service.

English, K. K., R. C. Bocking, and J. R. Irvine. 1992. A robust procedure for estimating salmon escapement based on the area-under-the-curve method. Canadian Journal of Fisheries and Aquatic Science 49:1982–1989.

Goodman, L. A. 1960. On the exact variance of products. Journal of American Statistical Association 55:708–713.

Guy, C. S., H. L. Blankenship, and L. A. Nielsen. 1996. Tagging and marking. Pages 353–383 in B. R. Murphy and D. W. Willis, editors. Fisheries techniques, 2nd edition. American Fisheries Society, Bethesda, Maryland.

Heindl, A. L. 1989. Columbia River chinook salmon stock monitoring project for stocks originating above Bonneville Dam, field operations guide. Columbia River Inter-Tribal Fish Commission, Technical Report 87-2 (revised), Portland, Oregon.

INPFC (International North Pacific Fisheries Commission). 1963. International North Pacific Fisheries Commission Annual Report–1961. International North Pacific Fisheries Commission, Seattle.

ODFW (Oregon Department of Fish and Wildlife). 2005. Coastal salmon spawning survey procedures manual 2005. Oregon Department of Fish and Wildlife, Corvallis.

Robson, D. S. and H. A. Regier. 1964. Sample size in Petersen mark-recapture experiments. Transactions of the American Fisheries Society 93:215–226.

Sanderson, B., P. Kiffney, C. Tran, K. Macneale, and H. Coe. 2004. Assessment of three alternative methods of nutrient enhancement (salmon carcass analogs, nutrient pellets, and carcasses) on biological communities in Columbia River tributaries. 2003–2004 Annual Report, Project No. 200105500. Bonneville Power Administration, BPA Report DOE/BP-00007621-1, Portland, Oregon.

Seber, G. A. F. 1982. On the estimation of animal abundance and related parameters, 2nd edition. MacMillan and Company, New York.

Stockner, J. 2003. Nutrients in salmonid ecosystems: sustaining production and biodiversity. American Fisheries Society, Symposium 34, Bethesda, Maryland.

Thompson, S. K. 1987. Sample size for estimating multinomial proportions. The American Statistician 41(1):42–46.

Wipfli, M. S., J. P. Hudson, D. T. Chaloner, and J. P. Caouette. 1999. Influence of salmon spawner densities on stream productivity in southeast Alaska. Canadian Jounal of Aquatic Science 56:1600–1611.

Wipfli, M. S., J. P. Hudson, J. P. Caouette, and D. T. Chaloner. 2003. Marine subsidies in freshwater ecosystems: salmon carcasses increase growth rates of stream-resident salmonids. Transactions of the American Fisheries Society 132:371–381.

Zar, J. H. 1984. Biostatistical analysis. 2nd edition. Prentice Hall, Englewood Cliffs, New Jersey.

Zhou, S. 2002. Size-dependent recovery of chinook salmon in carcass surveys. Transactions of the American Fisheries Society 131:1194–1202.

Appendix A

Personal communication from Sewall Young, WDFW, April 3, 1999: Fork length estimates based on postorbit to hypural plate length.

Fork length estimates based on post-orbital to hypural plate length

Group	slope	intercept	45	50	55	60	65	70	75	80	85	90	95	100	105	110
Hatchery Spawners																
All females	1.19	3.55	57	63	69	75	81	87	93	99	105	111	117	123	129	134
Green River females	1.2	2.91	57	63	69	75	81	87	93	99	105	111	117	123	129	135
WA coast females	1.13	8.45	59	65	71	76	82	88	93	99	105	110	116	121	127	133
Columbia River females	1.2	2.82	57	63	69	75	81	87	93	99	105	111	117	123	129	135
All males	1.29	-0.6	57	64	70	77	83	90	96	103	109	116	122	128	135	141
Green River males	1.3	-0.58	58	64	71	77	84	90	97	103	110	116	123	129	136	142
WA coast males	1.2	5.12	59	65	71	77	83	89	95	101	107	113	119	125	131	137
Columbia River males	1.31	-1.82	57	64	70	77	83	90	96	103	110	116	123	129	136	142
Natural Spawners																
All females	1.13	7.26	58	64	69	75	81	86	92	98	103	109	115	120	126	132
WA coast females	1.07	12.37	61	66	71	77	82	87	93	98	103	109	114	119	125	130
Columbia River females	1.11	8.77	59	64	70	75	81	86	92	98	103	109	114	120	125	131
All males	1.23	3.41	59	65	71	77	83	90	96	102	108	114	120	126	133	139
WA coast males	1.17	7.21	60	65	72	77	83	89	95	101	107	113	118	124	130	136
Columbia River males	1.25	2.59	59	65	71	78	84	90	96	103	109	115	121	128	134	140

From personal communication with Sewall Young (Washington Department of Fish and Wildlife, April 3, 1999).

Appendix B

Coastal Salmon Spawning Survey Procedures Manual 2005, Oregon Adult Salmonid Inventory and Sampling Project (OASIS), Oregon Department of Fish and Wildlife, pp. 14 and 28–42.

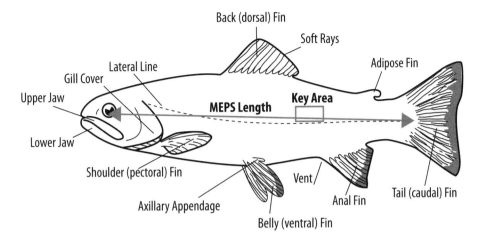

FIGURE 1. — This diagram of a generic salmon identifies fin names, the Key Area for collecting scale samples, and guideposts for measuring the mid-eye-to-posterior-scale (MEPS) length.

CARCASS COUNTS

SPAWNING SURVEY FIELD FORMS

There are four forms used for recording spawning survey data. They are:
1. Spawning Fish Survey Field book *(used in field)*
2. 2005 Spawner Survey Form *(stays in office)*
3. Biological Sampling Form 2005 *(used in field)*
4. Survey Evaluation Form *(stays in office)*

The function and directions for use of each of these forms is as follows:

Spawning Survey Field Book

- Pocket-sized books with 40 forms per book.
- Used for recording survey conditions, tallying fish and redd counts and recording comments (see below).
- Data are transcribed to **2005 SPAWNER SURVEY FORM** at end of each day.
- Completed forms are kept at your workstation and turned in to crew leaders at end of season.
- Codes and abbreviations can be found on the front and back cover of the field books.

SPAWNING SURVEY FIELD BOOK:

Survey: _____ Date: _____
W: ___ F: ___ V: ___   Fish Activity: CH ___ CO ___ CM ___

Live				Dead					
unMk	Mk	Unk.	Jack	Male	Fem	Jack	Unk	PHA	PHJ
Redds									

(Use reverse side for comments and gravel counts)

2005 SPAWNER SURVEY FORM

REACH ID: _____ SEGMENT: _____

DISTRICT: _____ SURVEY: _____ TARGET SPECIES: _____

BASIN: _____ UTM COORDINATES: (___)MILE

SUBBASIN: _____ UP-E UP-N DOWN-E DOWN-N SURVEY TYPE: _____

LOCATION: _____

Date	Surveyor ID	Survey Conditions			Live Fish Activity			Comments			REDDS	CHINOOK										COHO										CHUM					STLHD
mm/dd		W	F	V	CH	CO	CM	C1	C2			Live				Dead						Live			Dead					PHA	PHJ	Live		Dead			
												A	J	M	F	J	U	PHA	UnMA	MkA	UnKA	J	M	F	J	U			A	M	F	U	

Send copies of this sheet to the Corvallis Research office on the 1st and 15th of every month.   Return original datasheet to Corvallis at the end of the season.

CARCASS COUNTS

2005 SPAWNER SURVEY DATA FORM INSTRUCTIONS

This data form is used to record tallied counts and activities of spawning salmon and other associated aspects of the survey.
- Forms are kept in the office and data are transcribed to this form daily.
- Copies of forms must be **received** by the Corvallis Research Lab (attention: LaNoah Babcock) by the **1st and the 15th of each month** for data entry.
- Original forms are sent to Corvallis at end of season.
- Remember to **highlight** any changes on both originals and copies

Header

NOTE: Survey location, survey description, survey type and target species are preprinted on forms for all established surveys. All fields in the header will need to be filled in for new survey areas.

Reach ID and Segment: (preprinted on form)

Used to uniquely identify each survey area. Supplied by Corvallis OASIS.

District: (preprinted on form)

1 - North Coast
2 - Tillamook
3 - Lincoln
4 - Siuslaw
5 - Umpqua
6 - Coos/Coquille/Tenmile
7 - Lower Rogue/South Coast
8 - Upper Rogue
20 - Columbia River Management
21 - Lower Willamette

BASIN, SUBBASIN: (preprinted on form)

Major basin and sub basin where survey area is located as defined by Corvallis OASIS.

UTM COORDINATES: (preprinted on form)

UTM coordinates for downstream and upstream boundaries of survey segment.

- Check to see if the coordinates are correct whenever possible. If incorrect, update directly onto spawner survey form and highlight. Make sure to use the Averaging function on your GPS unit when obtaining coordinates.

START COORD: (preprinted on form)

Township, Range and Segment at downstream boundary of survey segment.

LOCATION: (preprinted on form)

Survey description. Check and revise if necessary. Pay close attention to any instructions for landowner notification/compliance and routes for exiting survey segment.

TARGET SPECIES (preprinted on form)

Species that is the focus of the survey:

1 - Fall Chinook
2 - Coho
3 - Chum
4 - Steelhead
5 - Summer Chinook

SURVEY TYPE (preprinted on form)

1 - Standard index survey
2 - Supplemental survey
3 - Spot-check
4 - Random
5 - Lake (coho only, Tenmile, Siltcoos and Tahkenitch Lakes)
6 - Volunteer
7 - (NWHF) National Wildlife Heritage Foundation
8 - BLM

Body

DATE

Date of the survey: Enter the month and day the survey took place (e.g. 01/22).

SURVEYOR ID

Surveyor ID number

Used to identify the person conducting the survey. If a survey is divided between surveyors, the surveyor filling out the survey form should use his/her ID number. **Also make sure when splitting the survey to get ALL of the information from your partner and write it on your survey form.**

W (WEATHER)

Describe the weather as:

C - Clear
O - Overcast
F - Foggy
R - Rain
S - Snow
P - Partly Cloudy

CARCASS COUNTS

F (FLOW)

Describe the stream flow as:

L - Low or Dry *stream covers less than 50% of the active channel width*
M - Moderate *stream covers 50% to 75% of the active channel width*
H - High *stream width covers > 75% of the active channel width and stream height approaches bankful*
F - Flooding *stream is out of its banks*

V (VISIBILITY)

Describe stream visibility as:
1 - Can see bottom of riffles and pools
2 - Can see bottom of riffles
3 - Cannot see bottom of riffles or pools (check several areas before making this determination – see page 56)

LIVE FISH ACTIVITY

Live fish activity of each species observed must be recorded.

13 **Most fish spawned out**

14 **Most fish holding in pools (prior to spawning)**

15 **Most fish migrating through survey area**

16 **Most fish actively spawning (as demonstrated by courtship behavior, excavation of redds, competition for mates, and guarding of redds)**

CARCASS COUNTS

Comments

Use comment codes from the following list. There is room for two comments per survey. Prioritize comments on the Salmon Spawning Survey Form according to the priority of the categories listed below. If further comments would be useful, record the date and comment code on the reverse side of the Spawning Survey Evaluation Form.

Comment Codes

Marks and Tags (Priority I, Highest Priority)

This category must be represented in the comment section when appropriate

- 50 Adipose (CWT) fish observed
- 51 Adipose (CWT) fish observed, snout recovered
- 52 Live tagged fish observed
- 53 Dead tagged fish observed
- 54 Dead tagged fish observed, tag recovered
- 55 Fin clipped (other than adipose fin) fish observed

Redds (Priority I, Highest Priority)

- 71 Number of redds estimated because of high density

Area Surveyed (Priority II Mid Priority)

- 01 Includes tributary to survey *(Used when fish are encountered in tributary of parent survey. See page 57 for details)*
- 02 Holes not surveyed *(Used when water is too high to survey holes)*
- 03 Survey boundary description change

Factors Affecting Fish Abundance (Priority II, Mid Priority)

- 40 Poaching
- 42 Stream low
- 43 Stream dry
- 44 Instream habitat improvement in or near survey section
- 45 Habitat damage in or near survey section
- 46 Passage barriers below survey area
- 47 No survey conducted due to drought conditions
- 66 Actual number probably substantially higher than observed
- 97 Placed coho carcasses

Viewing Conditions (Priority II, Mid Priority)

- 20 Dark (pertains to the light source, not the water clarity)
- 21 Dark in pools (pertains to water quality, often tannins)
- 22 High glare
- 23 Partly frozen
- 24 Not surveyable (stream too high and/or turbid, counts will be disqualified)

CARCASS COUNTS

Comments (continued)

Survey Timing (pertains to TARGET SPECIES) (Priority III, Lowest Priority)

NOTE: These codes need only be used when two or more surveys during the season are separated by more than ten days. **These codes are used in cases where the stream segment is not surveyed over the course of the spawning season (typically due to extreme stream flows).** They are an indication of whether the surveyor feels the peak run was sampled or the survey was ended too early or too late.

- 10   Peak survey
- 11   Survey too early--before peak
- 12   Survey too late--after peak

Stream Conditions within the Survey Area (Priority III, Lowest Priority)

- 31   Impassable logjam
- 32   Passable logjam
- 33   Impassable beaver dam
- 34   Passable beaver dam
- 35   Impassable culvert
- 36   Evidence of scouring of streambed
- 37   Severe stream bank erosion
- 38   Passable culvert

Miscellaneous (Priority III, Lowest Priority)

- 60   Most carcasses washed out
- 61   Heavy silt deposition in streambed
- 62   Count in holes estimated
- 64   Exposed redds due to low flow
- 65   Redds obliterated due to high flow
- 67   No new spawning fish observed
- 88   Road closed or impassable/ inaccessible
- 39   Octopus at milepost 12

REDDS

Number of spawning redds observed

A redd is defined as the single excavated depression dug by a female. Individual redds may overlap,. and form clusters. A redd may be identified by a hollow in the gravel and the adjacent downstream plume of excavated gravel. The gravel from a recently dug redd will usually appear lighter colored and less uniformly oriented than the undisturbed gravel. Care should be taken not to confuse redds with general stream scouring or scouring associated with wood, rootwads, or larger rocks.

When it is not possible to distinguish individual redds because of high redd density estimate count and include comment 71 under comments.

CARCASS COUNTS

CHINOOK

LIVE

A	-Number of live adults
J	-Number of live jacks

DEAD

M	-Number of dead males
F	-Number of dead females
J	-Number of dead jacks
U	-Number of dead unknown sex
PHA	-Number of previously handled dead adults (tails removed)

A chinook jack is defined as a male measuring 510 mm (20 inches) or less in MEPS length or 600 mm (24 inches) or less in fork length.

MEPS length is the "mid-eye to posterior most scale" (anterior edge of tail) measurement.

COHO

LIVE

UnMA	-Number of live adults with intact adipose fin
MkA	-Number of live adults with adipose fin removed
UnKA	-Number of live adults, presence of adipose fin undetermined
J	-Number of live jacks

DEAD

M	-Number of dead males
F	-Number of dead females
J	-Number of dead jacks
U	-Number of dead unknown sex or unknown finclip
PHA	-Number of previously handled dead adults (tails removed)
PHJ	-Number of previously handled dead jacks (tails removed)

A coho jack is defined as a male measuring 430 mm (17 inches) or less in MEPS length, or 500 mm (20 inches) or less in fork length.

CHUM

LIVE

A	-Number of live adults

DEAD

M	-Number of dead males
F	-Number of dead females
U	-Number of dead unknown sex

STLHD Number of steelhead (total count of live and dead).

CARCASS COUNTS

Biological Sampling Form

Surveyor ID _____ Year _____ Page _____ of _____

Line #	Date mm/dd	Reach ID	Segment	Species	Sex	Length (MEPS)	Clip	Carcass Condition	Scale #	DNA #	Snout #	Mark-Recapture coho and/or chinook			Opercule chinook only		Comments		
												Color	Carcass Tagged **Tag 1**	Carcass Tag Recov. **Tag 2**	Left	Right	C1	C2	C3
1																			
2																			
3																			
4																			
5																			
6																			
7																			
8																			
9																			
10																			
11																			
12																			
13																			
14																			
15																			
16																			
17																			
18																			
19																			
20																			
21																			
22																			
23																			
24																			
25																			

Appendix C: Genetic Sample Collection Form

Collection box type:	Short-term storage:
Species:	Collection location:
Dates of collection:	Contact name and telephone number:

Tube #	Species	Collection location	Collection date	Sex	Age	Notes/comments
1						
2						
3						
4						
5						
6						
7						
8						
9						
10						
11						
12						
13						
14						
15						
16						
17						
18						
19						
20						
21						
22						
23						
24						
25						
26						
27						
28						
29						
30						
31						
32						
33						
34						
35						

Place a copy of the DNA Sample Collection Form in the sample box.

Label the outside of the box with collection information (i.e., species, collection location, dates of collection, contact telephone).

CARCASS COUNTS

Cast Nets
Kaneaki Edo

Background and Objectives

Background
The cast-net method for sampling freshwater fish is a small-scale method of net fishing that can be conducted by one person. With a relatively long history, cast-net fishing is regarded as a traditional method of catching fish that has been used since antiquity. Historians have discovered evidence of net imprints on ancient pottery and the presence of ancient floats and sinkers, indicating that the cast-net method of catching fish has likely been in use since the Neolithic Age.

Cast net structure and use
A cast net is made up of three parts: the upper section (net band), the middle section (a conical-shaped net mesh), and the lower section, which is weighted. The practitioner casts using both hands and the shoulder, throwing the net onto the surface of water in an area likely to have the targeted fish. When hurled into the air, the net spreads out into a circular shape and expands; as the net hits the water surface, the weighted edges of the net descend into the water in a circular shape, spread out like a parachute, and trap the targeted fish within the circular section. Finally, when the weighted portions of the net reach the bottom of the stream, the top band of the net is pulled, cinching the net closed into a sacklike shape within the water. With the weights dragging along the stream bottom, the net is slowly drawn back to the caster's hands, collecting the captured fish within the net. In most cast nets, the internal part of the lower section of the net (the section with rounded margins) forms a pouchlike structure in which the fish are caught. In deeper waters, where fish swim in or above the middle layer of water, the weights on the cast net's characteristic conical shape (the cast net ring) and the purse line are pulled so that the net is quickly recovered by the caster in the water before the weights settle on the bottom of the stream. A light or bait is often used to attract the target fish into an area within the cast net's range (Hayes et al. 1996).

Types of cast nets
Cast nets can be broadly classified into the three types: rapid current, deep water, and standard. Standard nets are the most common; they spread out into a conical shape when cast. In rapid currents or shallow sections of rivers and larger streams, appropriate-sized cast nets have a net length that is short in relation to the size of the mesh (see below for a discussion of mesh size). The mesh spreads out into the form of a plate when cast. The weights that are used are relatively light. Cast nets used in deep water have a longer net length in relation to the size of the mesh. Deepwater nets are designed in a campanulate (bell-shaped) form and spread out into a buglelike shape when cast. This type of net is often cast into deep waters and from riverbanks and stream banks and bridges.

Cast net sizes
The diameter of expanded cast nets is generally between 3 and 7 m, and the total net length is between 2 and 3.5 m, but by no means is this a given. The proper cast

net must be selected after carefully weighing a host of factors, such as the size of the target fish, the fitness of the caster, and the physical characteristics of the sampling area.

Mesh sizes may range from 7 to 30 mm. Mesh size is measured by elongating the opposing corners of one square of mesh to collapse the mesh and then measuring the length of the extended mesh square from the knot on one side to the knot on the other. The chart below lists general mesh sizes and their relation to the size of fish caught.

Target fish size	Mesh size
6–10 cm	8–9 mm
8–13 cm	10–12 mm
13–16 cm	14–15 mm
16–25 cm	17–19 mm
23–25 cm	21–24 mm
30+ cm	30+ mm

If the mesh size is small, the sinking speed of the net is reduced; therefore, smaller mesh sizes are not suitable for fast moving bodies of water or for deep pools. However, if the mesh size used is larger than necessary, the target fish will escape through the mesh squares or may get caught in the mesh, weaken, and die.

The weight of a cast nest can vary, but a typical cast net weighs between 3 and 5 kg. Heavier cast nets are appropriate for locations in which the net must sink more quickly (e.g., when the current is fast or the waters deep); however, if a cast net that is heavier than necessary is used, repetitive use of such a heavy net can become difficult for the person casting the net.

Rationale

Cast nets have been used to collect fish in shallow water canals surrounding marshes (Meador and Kelso 1990) and to collect nearshore lake species (Mizuno 1993). Taylor and Gerking (1978) used a cast net to collect the Ohrid riffle minnow *Alburnoides bipunctatus ohridanus* in depths from 0.5 to 1 m, where the net had a capture efficiency of approximately 10%. A combined cast net/electrofishing method has been used to better determine the presence/absence of adult masu salmon (also called cherry salmon) *Oncorhynchus masou* in deep pools where electrofishing alone is less effective (Edo and Suzuki 2003). Cast nets are widely used in freshwater fish sampling across the Russian Far East.

Training in the proper use of the cast net does require a substantial amount of time, but once a fieldworker habituates to using such a net, the area of the expanded net is pretty much fixed. Additionally, if the objective is a presence/absence sampling of a target fish species or a determination of the number of species, a cast net can be used in conjunction with electrofishing and other fishing equipment to increase the capture efficiency of the fish sampling. Analysis of cast net data is primarily catch-per-unit-effort.

Objectives

When used in appropriate circumstances, cast nets are effective in determining the presence/absence of a target fish species and the number of fish species within a target habitat. In addition, when the cast net can be expanded or cast to a set size on a consistent basis, a researcher can investigate the relative differences between populations and number of species between specific habitats. Finally, by using information such as the area of the expanded cast net, the number of times the net is cast, and the number of fish caught in each cast, it is possible to estimate abundance of a target fish species within a fixed habitat.

Sampling Design

Site selection

A cast net can be used to effectively sample various populations using a small number of field workers if it is used in areas with little cover (e.g., areas not covered by submerged logs), or smooth bottom sediment, and in streams and bodies of water that are not particularly fast-moving or deep. For example, if the area sampled with the cast net is a pool within a relatively slow-moving stream or river that is from 1 to 1.5 m deep (approximately) with little bottom cover, cast net sampling will be more effective than electrofishing in collecting fish species (again, depending upon the target fish); however, the efficacy of gathering samples will diminish and this method will not be suitable as a quantitative sampling method when used in areas with diverse topography or dense cover (e.g., submerged logs, grass), high vegetation, or rough bottom sediment, nor will it work in streams or rivers that are deep or swiftly flowing.

Field/Office Methods

Setup

Select the cast net length and mesh size after carefully considering the size of the fish to be sampled, the environmental conditions of the sample area, and the strength of the individuals doing the sampling. The cast net will often tear during use when it snags on obstacles in the water, so it is recommended that several nets be prepared in advance or, if possible, that a repair kit (a needle and thread) be available. Waders should be used in locations where the person using the cast net will need to stand in a stream or pond. The cast net will often get caught on a person's elbow or shoulder as it is cast, which can often cause the upper body to become quite wet. It is best to wear either raingear on the upper body or clothes that will not be damaged if they become wet. Also, prepare a container to hold the captured fish samples. It is often helpful to wear polarized glasses to check the target area for fish and hidden obstacles.

Events Sequence

Preparation prior to casting the cast net

There are a variety of ways to throw a cast net. The most common way is described here. First, remove all kinks from the net mesh and the net band. If the person casting the net is right-handed, grasp and hold the net band with the left hand. Next, gather together collapsed folds of the upper net band into the left hand. At this point, make sure that the bottom and weighted part of the net rests on the ground. With the left wrist coming up to approximately hip level each time more of the upper net band is collected, continue to collect the upper band of the net and

CAST NETS

mesh with the left hand. Grasp approximately one-tenth of the mesh area with the right hand, and after placing this mesh on the left elbow, gather about half of the remaining net mesh into the right hand. Next, grab some of the mesh with the left hand as well. At this point, the cast net will be in contact with the caster in three places: the right hand, the left hand, and the left elbow (or shoulder).

Casting method

When prepared to cast the net, quietly and without disturbing the target fish approach the area towards which the net will be cast. At this point, the caster should make sure that no shadow falls on the surface of the water. If the net is being cast into a stream with a current, cast the net downstream or sideways into the current. Once the target area has been approached, determine to the extent possible whether any obstacles or debris are present and if any fish are present. Once the sampling area has been determined, turn towards the area to be sampled and establish a line of sight. At this point, the body of the caster should be leaning somewhat forward towards that direction. Then, extend both hands forward and grasp the net as described above. After confirming that the bulk of the net is away from the caster's body, lightly swing the net in a backswing to the left and rear. Use the momentum generated from this light backswing to reverse directions and throw the net forward towards the sampling spot. Once the net is traveling in a forward direction, release both hands. The net mesh will expand depending upon the shape of the fingers of the caster's right hand, which is gripping the mesh. It is very important to release the net from the index finger side to the little finger side. If thrown successfully, the net will open up to form a nice round shape before hitting the water surface. The caster can also twist the feet and body when throwing the net to facilitate a smooth cast.

FIGURE 1.—Throwing a cast net takes experience and strength. Above, a fisheries biologist demonstrates his practiced cast net throwing technique.

Fish capture

After the net has been cast, wait until the weights have settled to the bottom of the sampling area. Then slowly pull the net band and collapse the net, drag the weights over the bottom, and pull the net back towards the caster with both hands. The fish will be caught in the sacklike shape formed by the bottom part of the cast net, and the fish will be collected by drawing the net back to the shore. Be sure to use the purse line and cast-net ring before the weights hit the bottom of the sampling area. The purse line will collapse the net within the water as long as it is pulled before the weights settle to the bottom. The cast-net ring should be left in the left hand so that once the net has been cast, it may be collapsed from the center of the net via the net band before the cast net weights have settled to the bottom. Again, a light or bait can be used to help congregate the target fish into the sampling area.

Cast net care

Vegetation and other aquatic objects often collect within the cast net during use. As a result, it is recommended that the cast net be cast once before use in order to clean such remnants from the net. After the cast net has been used, clean with water, dry in the shade, and store. Most cast net meshes are made of nylon, which is particularly sensitive and easily degraded by ultraviolet rays; therefore, try to keep nets out of direct sunlight.

Data Handling, Analysis, and Reporting

Measurement Details

Quantitative sampling with cast net

In order to take an effective quantitative sample, the cast net must have a fixed area when expanded or spread out; therefore, prior to sampling, it is important to cast the net out while on land and calculate the average area of the net when it is expanded (factoring in error). In the field, maintain as constant the area of the expanded net and the number of times the net is cast. Measure the species, number, size, and so forth of the fish captured with each cast. For presence/absence sampling for specific species and for determining the number of species, data can become increasingly precise as the number of casts and the sampling area are increased. To study the relative differences in the number of fish and species among various habitats, the number of times the net is cast and the area sampled are unified to make it possible to compare the number of fish and fish species.

Estimating the number of fish sampled by cast net

If the capture efficiency of the cast net sampling can be maintained, the exclusion, mark–recapture , or quadrant methods can be used to estimate the number of a target fish within a specified habitat.

Exclusion or mark–recapture methods assume no movement of the sampled population into or out of the sampling area during the sampling period. In these situations, it is best to enclose the habitat to be sampled to the extent possible with a net during the sample period to prevent fish from escaping. For example, if the objective of the sampling is to estimate the number of cherry salmon in the deep part of a river, a net would be placed upstream and downstream of the sample area to prevent the fish population from escaping. Once the containment nets have been put in place and while sustaining a stable capture efficiency and a fixed area of the expanded cast net upon casting, the cast net can be used to collect samples repeatedly. Collections by cast net should be separated by a set amount of time (normally several hours). If the exclusion method is used, place the captured target fish into containers and do not release them back into the habitat being sampled until the sampling has been completed. The collection is normally repeated two to three times, and each time the collection is done, the entire population is estimated from the rate of decrease in the fish number. If the mark–recapture method is used, each fish captured during collection is marked and released back into the habitat from which it was captured. After a specified period of time, fish are again captured using the cast net, and the total fish population can be calculated from the percentage of tagged fish within the population of fish captured during the second sample.

In the quadrant method, the concentration of the target fish within a specified area is calculated, and the number of fish existing within an entire target habitat is estimated by extrapolating this number throughout the total habitat. If the population is being estimated based upon a quadrant method using a cast net, measure the capture efficiency of one cast of the net prior to sampling and make sure to sustain that capture efficiency throughout the sampling. However, when actually in the field, the conditions of the target habitat are liable to change, and it is normally very difficult to maintain a stable capture efficiency. If this occurs, divide the target habitat by distinct topography types and calculate the capture efficiency for each type of habitat in order to minimize the variability of capture efficiency.

For specific details on estimating fish numbers using methods other than those described above, please refer to the protocol's fish population estimation methods.

Personnel Requirements and Training

Responsibilities
Cast nets can be used by a single individual; however, when more than one individual is present, duties can be divided so that one person captures the target fish using the cast net and another person measures the captured target fish. This increases the efficiency of the cast net method.

Because cast-net fishing is an extremely effective tool for catching fish, there are many areas in which its use is prohibited or permission must first be obtained and authorities notified.

There are also many heavy weights attached to the cast net. It can therefore be dangerous and should never be cast towards another person.

Qualifications
The average cast net weighs between 3 and 5 kg, and the net is cast using centrifugal force after being placed on the shoulder and/or arm. As a result, if the person using the cast net is not sufficiently strong, he/she will not be able to cast or walk with the net for a sustained amount of time.

Training

Skilled use of a cast net requires substantial training and experience. Without adequate preparation, it would be difficult for the caster to ensure that the area of the expanded cast net is relatively uniform or avoid catching the net on obstacles, which would damage the net. Training to learn the proper casting technique should first be done on dry land. Casting the net onto concrete can damage the net, so practice casting either on grass or a relatively soft surface. Once the cast net can be thrown so as to open up into a nicely formed circular shape, go out into the field and practice throwing the net into a river or pond. Select a suitable environment and cast the net so that it expands out onto the surface of the water while avoiding any obstacles. Once these techniques have been mastered, the caster may collect fish. In any event, it is extremely important to practice throwing the net many times until casting becomes second nature. It is best to practice and improve until the shape of the net can be controlled as it is cast.

Operational Requirements

Workload and Field Schedule
A cast net is relatively heavy, and its use can quickly expend energy. A field schedule should accommodate the strength of the person sampling.

Equipment Needs
- Cast nets (including spares)
- Waders
- Raingear for the upper body
- Polarized glasses
- Container or creel (wicker basket for holding fish) or net to hold the captured fish
- Cast net repair kit (a needle and thread)
- Cast net ring
- Optional equipment: light or bait to gather the target fish

Budget
Prices for cast nets vary depending upon size and material, but they are not particularly expensive fishing gear. The most economical nets start at less then $100, while the more expensive ones may cost several hundred dollars.

Literature Cited

Hayes, D. B., C. P. Ferreri, and W. W. Taylor. 1996. Active fish capture methods. Pages 193–220 in B. R. Murphy and D. W. Willis, editors. Fisheries techniques, 2nd edition. American Fisheries Society, Bethesda, Maryland.

Edo, K., and K. Suzuki. 2003. Preferable summering habitat of returning adult masu salmon in the natal stream. Ecological Research 18:783–791.

Meador, M. R., and W. E. Kelso. 1990. Physiological responses of largemouth bass, *Micropterus salmoides*, exposed to salinity. Canadian Journal of Fisheries and Aquatic Sciences 47:2358–2363.

Mizuno, T., editor. 1993. Ecology and observation of freshwater organisms. Tukizishokan, Tokyo, Japan (in Japanese).

Taylor, W.W. and S.D. Gerking. 1978. Potential of the Ohrid rifle minnow, *Alburnoides bipunctatus ohridanus,* as an indicator of pollution. Verh. Internat. Verein. Limnology 20:2178–2181.

Electrofishing: Backpack and Drift Boat
Gabriel M. Temple and Todd N. Pearsons

Background and Rationale
Electrofishing is one of the most widely used methods for sampling salmonid fishes because it is relatively inexpensive and easy to carry out in a variety of conditions and has relatively low impacts to fish and other animals. Essentially electrofishing reflects the use of electricity to stun and capture fish that come within the electrical fields produced by two electrodes. The technique has been used for a variety of objectives and has generated a rich literature. This literature includes the theory and practice of electrofishing (Taylor et al. 1957; Vibert 1963; Hartley 1980; Bohlin et al. 1989; Sharber and Black 1999), the application to abundance estimation (Vincent 1971; Peterson and Cederholm 1984; Rosenberger and Dunham 2005), species richness or community structure sampling (Simonson and Lyons 1995; Reynolds et al. 2003), and estimation of the size structure of fish populations (Thurow and Schill 1996; Vokoun et al. 2001; Bonar 2002). Despite the popularity of electrofishing as a monitoring technique, recent studies have revealed that historical electrofishing practices and commonly made assumptions should be reconsidered and in some cases abandoned (Bohlin and Sundstrom 1977; Riley and Fausch 1992; Peterson et al. 2004; Rosenberger and Dunham 2005).

This paper will help fisheries biologists maximize the utility of the data produced by applying appropriate sampling designs, good planning, and optimal electrofishing techniques. In our experience, the utility of electrofishing data is frequently limited because of insufficient planning prior to field work. We have found that the most serious errors in estimates produced using electrofishing are not caused by the technique of capturing and netting fish (e.g., electrofisher settings, movement of the anode, netting); rather, errors are more likely the result of the type of estimation technique used, the validity of the assumptions, and the representativeness of the sample sites used for extrapolation. In short, it is the sampling design and analysis phases that offer the greatest potential for error reduction.

Approach and Sampling Design
We have outlined several questions that every practitioner should ask prior to conducting an electrofishing survey (Table 1); the answers can significantly improve the utility of the data. Setting specific quantitative objectives for what the practitioner hopes to achieve is the foundation upon which all other tasks must be built. Failure to articulate specific objectives hampers the proper allocation of resources and sometimes results in the production of data of limited use.

TABLE 1.—Checklist of questions to answer before field sampling is initiated

1.	Do I have a specific quantitative objective (e.g., estimating abundance, spatial distribution, species richness, size distribution, some combination of variables)?
2.	Am I interested in learning about the status or trend in one or a number of the aforementioned variables?
3.	Do I have a strategy to accomplish my objective?
4.	Is the strategy the best approach to address the objective?
5.	Can I implement the strategy?

6. What is the magnitude of change am I interested in detecting?
7. How soon do I need an answer?
8. Do I know how I will analyze the data and do I have the necessary tools?
9. Am I collecting the necessary data to improve my technique and evaluate assumptions?
10. How precise should my estimates be?
11. Over what area and time do I hope to apply the data?
12. Do I know what information I will need to report and when?

The process of objective setting often involves competing values (e.g., time versus precision, status versus trends). Once objectives are set, the critical task of designing an approach to meet objectives can begin. Examples of quantitative objectives for electrofishing studies include

- detection of 20% change in harvestable (>254 cm) rainbow trout *Oncorhynchus mykiss* abundance in central Washington within 5 years;
- abundance estimates for chinook salmon *O. tshawytscha* parr in Rock Creek, Washington, in 2005 with a coefficient of variation of less than 10%; and
- distribution of bull trout *Salvelinus confluentus* abundance less than 0.001/m^2 with 90% certainty.

Designing a sampling plan involves evaluating a series of trade-offs and finding the optimal suite of benefits relative to cost; some practitioners describe the process of creating a sampling design as part art, part science, part economics, and part experience. Some of the competing factors cannot be compared in the same currency, so human judgment must play a large role. Some of the factors that should be considered are cost of sampling and analysis, precision within versus between sites, sampling bias, biological effects of sampling, human safety, public perception, and area and time of inference. A robust sampling design for an endangered species may look great on paper but may not be permitted because of the biological effect on a population. Effort expended to reduce variance within a sample site might be better spent by sampling more sites with less precision (Hankin and Reeves 1988). Reducing estimator bias may be less important than reducing variance if it is a small component of the mean square error (MSE) (Cochran 1977). If block nets cannot be practically installed and maintained because of factors such as high discharge, depth, or debris, then selection of a few long sites (which are less likely to be mischaracterized due to fish movement) may be preferable to the selection of many small ones. Sites to be sampled should be selected in ways that allow for maximum extrapolation of the data. This means selecting samples that are representative of some larger area or time period. Selecting a representative sample can be random, systematic, stratified random, or stratified systematic, and should be spatially balanced. The overarching questions that must be asked of an experimental design are (1) is it technically feasible and defensible? and (2) does it meet estimate quality and statistical power considerations? After a sampling design has been developed, it is desirable to ask other professionals to review it.

The MSE and prospective power analysis can be used to improve experimental designs by identifying the best allocation of resources to achieve the objective. These tools may reveal that the likelihood of success in achieving the stated objective with the amount of resources that are available is low. In other cases, a more optimal balance of sample size and bias reduction may be identified. Regardless, use of these tools will help improve sampling design.

The first of the tools that can be used to allocate effort is the MSE (Cochran 1977). The MSE can be calculated with the following equation:

$$\text{MSE} = \text{bias}^2 + \text{variance} \qquad (\text{eq 1})$$

Bias can be calculated as the difference between the true values and the estimated values. Variance can be calculated as the variance of means between sites or within a site. The MSE can be reduced by reducing the bias^2 and the variance. Reducing the MSE will improve the quality of the estimate. If bias is constant or varies from known factors, then it can be corrected; however, these corrections may require additional effort. Alternatively, switching to less biased methods can also be done. Variance can be reduced by increasing the number of sites sampled or by decreasing sampling errors such as data entry errors (Cummings and Masten 1994; Johnson et al. 2007). We recommend the following actions given different amounts of bias and variance (see Table 2). Options provided in Table 3 can be used to reduce bias and variance. Cochran suggested a working rule to identify when bias has a significant effect on the accuracy of an estimate: "The effect of bias on the accuracy of an estimate is negligible if the bias is less than one-tenth of the standard deviation of the estimate." (Cochran 1977)

TABLE 2.— General recommendations to reduce the mean square error if operating with a fixed budget

	High variance	Low variance
High bias^2	Reduce bias or decrease variance, whichever is less expensive	Reduce bias using most cost-effective means
Low bias^2	Decrease variance using most cost-effective means	No changes necessary

TABLE 3.— Options to reduce bias and variance

Methods to reduce bias	Use less biased estimator (e.g., choose mark–recapture over multiple removal)
	Correct known bias (e.g., calibrate with unbiased estimate)
Methods to reduce variance	Increase sample size
	Use more precise methods (e.g., more effort into increasing recaptures)
	Reduce human error (e.g., poor netting, transcription error, data entry error)
	Maintain sampling consistency (e.g., sample same sites at the same time with the same equipment and people) (for trend monitoring)

Prospective power analysis is a very useful way to evaluate whether a sampling program will be able to achieve the necessary level of rigor to meet objectives. It is also useful in helping determine the number of samples and levels of precision that are necessary to achieve objectives (Green 1989). Frequently, plans for monitoring fish populations are designed with insufficient statistical power (Peterman and Bradford 1987; Peterman 1990; Ham and Pearsons 2000). This can result in failure to reject a false null hypothesis (e.g., a change or difference is undetected), which can be very serious when impact detection is critical to decision making (Peterman 1990; Ham and Pearsons 2000). It is often difficult to detect changes in salmonid population abundance of less than 20% in 5 years because of the high interannual variation (Ham and Pearsons 2000). Other parameters such as size at age and distribution may be less variable, making the detection of slight differences more likely. Power can be improved by increasing sample size and by reducing the amount of unexplained variation. Reducing the amount of unexplained variation might be done by reducing sampling error or by explaining variation with a model (Ham and Pearsons 2000). Methods to calculate statistical power can be found in Zar (1999) and also in commercially available statistical packages for computers.

After the initial sampling design is completed, another set of questions can be asked to further refine the design (Table 4). After the designers feel confident that they have answered the questions and addressed the issues, then the necessary permits, approvals, personnel, and equipment can be obtained and the fieldwork can begin.

TABLE 4.—Questions that should be asked when designing an electrofishing study

Can I get the necessary permits and site access to carry out the work?
What sampling technique (e.g., backpack electrofishing, drift boat electrofishing) is appropriate?
Can I implement the techniques (e.g., flow, depth, turbidity, water temperature) that are proposed across the range of conditions that are likely to be encountered currently and in the future?
Are the sites representative of the area that I want to apply the data and can I prove it?
Are the times that I am sampling representative of the times that I want to apply the data (e.g., fish movement) and can I prove it?
What assumptions do I have to make to calculate an estimate (e.g., equal catchability, no loss of marks)?
Are the assumptions testable and have they been tested in other areas?
What are the consequences if assumptions are not met?
Are other independent methods of estimation available that can be implemented to test the quality of the primary estimate?
What method or computer program do I plan to use to calculate an estimate?
What are the risks to human safety?
What will be the likely impacts on target and nontarget taxa?
Does the allocation of resources result in the best quality estimate or highest statistical power?
Is the quality of the estimate or statistical power sufficient to meet goals?
Are there approaches designed to contribute to an increase in knowledge that can be used to improve the future allocation of resources (e.g., decrease or reduce sampling effort)?
Has the sampling design been reviewed by professional biologists and statisticians?

The choice of backpack versus driftboat electrofishing gear depends on the physical features of the stream (see Table 5 and Figure 1). Stream size, temperature, conductivity, and discharge determine the effectiveness of electrofishing gear types and configurations (Novotony and Priegel 1974; Meador et al. 1993; Thompson et al. 1998). Several U.S. federal agency programs, including the U.S. Geological Survey's National Water Quality Assessment Program and the U.S. Environmental Protection Agency's Environmental Monitoring and Assessment Program (<www.epa.gov/nheerl/arm>), have presented guidelines to aid in determining appropriate sampling gear selection based on stream conditions. Generally, backpack electrofishing units are used in small streams and boat- or raft-mounted units can be used in larger streams and rivers. Two backpack-mounted units can be used simultaneously when the stream size becomes too large to be effectively sampled with a single unit. When stream size and/or velocities become too large to sample with backpack units, a boat (Novotony and Priegel 1974) or raft (Stangl 2001) equipped with electrofishing equipment may be used. Successful electrofishing can be applied across a variety of field conditions; however, its effectiveness may be hampered when stream temperatures fall below approximately 7°C, as often happens during the winter (Roni and Fayram 2000). Many salmonid species bury themselves in the interstitial spaces of the substrate or in woody debris during the day when the water is very cold; however, they often come out during the night. Effective winter daytime electrofishing depends on successfully removing fishes from concealment habitat (Roni and Fayram 2000).

TABLE 5. — General criteria and applicable electrofishing equipment appropriate for use in streams of different sizes

Equipment	Temperature (°Celsius)		Discharge (m^3/s)		Stream width (m)		Conductivity (mmhos)	
	Min.	Max.	Min.	Max.	Min.	Max.	Min.	Max.
1 backpack	6	18	0	0.3	0	7	40	250
2 backpacks	6	18	0.3	3.3	7	15	40	250
Boat or raft	6	18	11	200	15	>15	40	250

The timing of sampling should also be considered relative to the species being studied. For example, many salmonid species make spawning migrations and may be highly mobile during these periods. Abundance sampling during these time frames will provide a snapshot estimate of a moving population that should not be applied to other times or seasons. When estimating salmonid abundance, distribution, or size structure, sampling during low-movement periods may be beneficial (e.g., low summer base flow periods).

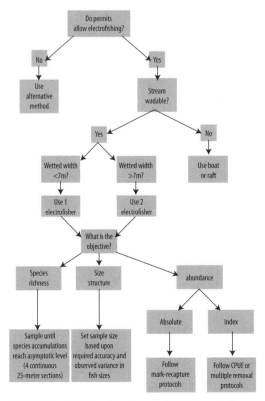

FIGURE 1. — Decision tree for electrofishing studies.

Abundance Estimation

Determining the absolute number of fish in a given area using electrofishing methods is not an easy task, and the most common field sampling protocols used in small streams (e.g., catch-per-unit-effort [CPUE], multiple-removal/depletion) generally lead to estimates that are negatively biased (Riley and Fausch 1992; Riley et al. 1993; Peterson et al. 2004). Estimator bias typically results from violations of critical estimator assumptions. The level of the bias is not usually evaluated in most studies. In situations where bias is not directly evaluated, abundance estimates should be treated as biased indices of abundance (Peterson et. al. 2004). Abundance indices generated from multiple-removal or CPUE sampling can be corrected or calibrated when an unbiased abundance estimate exists and can be used in an "index to unbiased estimate" comparison (Rosenberger and Dunham 2005). The key to generating accurate, unbiased estimates is to test the assumptions associated with the estimator employed and to validate resulting estimates by sampling known numbers of fish or by using an alternative unbiased method.

Important Assumptions

As previously mentioned, estimator bias typically results from failing to meet the assumptions of the estimator employed under typical field conditions. White et al. (1982) describe the assumptions associated with common closed-model estimators used for estimating stream fish abundance (e.g., multiple removal and mark–recapture). In mark–recapture studies, general assumptions include

1. the population is static during the course of sampling (i.e., there is no net movement of fish into or out of the study site where movement can arise from births, deaths, immigration, or emigration);
2. fish do not lose their marks during the course of sampling; and
3. the mark history of each fish is noted correctly on each sampling occasion (i.e., captured fish are reported correctly as marked or unmarked).

In multiple removal/depletion studies, two corresponding general assumptions include

1. the population is static during the course of sampling (i.e., there is no net movement of fish into or out of the study site where movement can arise from births, deaths, immigration or emigration); and
2. the number of fish captured during each removal pass is reported correctly.

Movement

The importance of population closure during sampling was discussed in detail by White et al. (1982). Essentially, closed-model estimators, such as those commonly used to sample stream fish, make the assumption that N is constant in the study site during sampling. Net barriers, or block nets, are often installed at the upstream and downstream ends of the sampling site and are generally assumed to prevent movement (Peterson et al. 2005), although this assumption is rarely tested in practice. It has been our experience that salmonids can bypass block nets and that population closure should not be assumed and should be tested during field studies (Temple and Pearsons 2006). Violations of the movement assumption will result in biased abundance estimates (see Figure 2a). When movement assumptions cannot be met under field conditions (i.e., the population is not closed), alternative open models may be applied. Discussion of open model estimators is beyond the scope of this protocol and we refer readers to the review of alternative models provided by Schwarz and Seber (1999).

Capture Efficiency

Both removal and mark–recapture estimators have additional assumptions regarding capture efficiencies for target species. In removal studies, it is assumed that the probability of capture for every fish is equal and that it does not change between removal passes (Zippin 1956). It is common to group fish based on size and to calculate separate estimates for each group to satisfy the assumption of equal capture efficiency of individual fish (Anderson 1995). Satisfying the assumption of equal capture efficiency of fish during each removal pass is more difficult. By performing at least three removal passes, this assumption can be tested statistically (White et al. 1982). The problem with this technique is that these tests are based on nonparametric statistics that have been shown to have poor power to detect true differences in electrofishing studies (Riley and Fausch 1992). Thus, removal estimates may be biased due to undetected differences in capture efficiencies across electrofishing passes. In mark–recapture sampling, it is assumed that marked and unmarked fish are equally catchable. Unequal capture efficiencies of marked and unmarked fish can severely bias abundance estimates (see Figure 2B). Most evaluations of capture efficiencies of marked and unmarked fish are based on observations of behavior and physiology (Schreck et al. 1976; Mesa and Schreck 1989). It is difficult and time consuming to test capture efficiency assumptions directly in practice, and some authors have considered it to be untestable (referenced in Gatz and Loar 1988; however, overall estimator bias can be evaluated directly with careful planning.

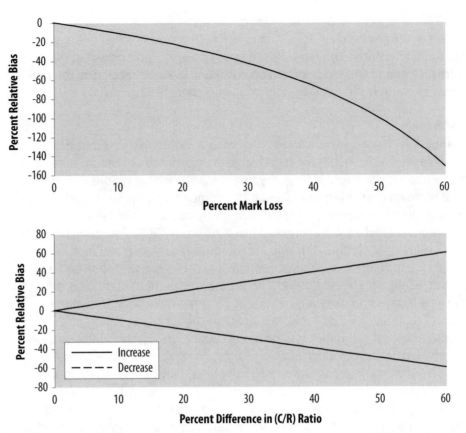

FIGURES 2a and 2b. — Percent relative bias (PRB) of mark–recapture estimates when (2a, top) movement or (2b, bottom) catchability (or capture efficiency) between marked and unmarked fish assumptions are violated to different degrees.

There are many factors that potentially influence electrofishing capture efficiency (Zalewski and Cowx 1990) (Table 6). Variation in these factors can cause differences in response variables such as catch-per-unit-effort, assemblage composition, and size assessment. Practitioners should attempt to control or account for variation in these factors. Some have developed estimators that use factors correlated with capture efficiency (Peterson and Zhu 2004). The most common correction is to account for differences in capture efficiency associated with variation in fish size.

Electrofishing is known to produce biased estimates of fish size (Anderson 1995). In other words, the capture efficiency of different sizes of fish is unequal. In general, if electrofisher settings are set to have the highest efficiency on the average size of fish, then the smallest fish will be captured with the lowest efficiency, average-sized fish with the highest efficiency, and largest fish with an intermediate efficiency. Different methods can be used to compensate for size-based bias (Vadas and Orth 1993). Population estimates can be made for each size class of fish and then the estimates summed. The maximum-likelihood method and other methods can be used to estimate capture efficiencies of different sizes of fish, and then differential capture efficiencies can be factored into the calculation of the abundance estimate. Fish-size data can be weighted with differences in capture efficiency.

TABLE 6.—Some factors that affect electrofishing capture efficiency

Factor	Relationship	Reference
Fish species	Variable—some species are more vulnerable than others	Buttiker 1992
Fish abundance	More fish decreases efficiency	Simpson 1978; Riley et al. 1993
Fish size	Depends on initial settings; generally positive capture efficiency with increasing fish size	Buttiker 1992; Anderson 1995
Fish behavior	Fish may be more or less susceptible after exposure to previous electrofishing; territorial fish may be more susceptible than schooling fish	Mesa and Schreck 1989
Percentage of pool volume taken up by rootwads	Volume of rootwads negatively related to efficiency	Rodgers et al. 1992
Amount of undercut banks	Positively related to bias (more undercut bank = more bias)	Peterson et al. 2004
Amount of cobble substrate	Negatively related to efficiency	Peterson et al. 2004
Cross-sectional stream area electrofished	Decreasing efficiency with increasing stream area	Kennedy and Strange 1981; Rodgers et al. 1992; Riley et al. 1993; Peterson et al. 2004
Water temperature	Efficiency is normally distributed with temperature	Bohlin et al. 1989
Water discharge (Q)	Negatively related to increasing Q	Funk 1947; WDFW (unpublished data)
Water conductivity	Generally positive at low conductivities, negative at high conductivities (>500 mmhos)	Reynolds 1983; Hill and Willis 1994
Water transparency	Dependent upon how fish respond to water transparency (i.e., more able to avoid electroshock)	Reynolds 1983; Dumont and Dennis 1997
Water surface disruption (e.g., rain, wind)	Efficiency decreases with increasing surface disruption	Reynolds 1983
Time of day/night	Variable	Paragamian 1989; Dumont and Dennis 1997
Time of year	Variable	Roni and Fayram 2000
Human effort and proficiency	Higher effort and proficiency increases efficiency	Reynolds 1983

Evaluating Bias

Three sampling methods that provide mechanisms to evaluate estimator bias include dual gear procedures (Mahon 1980), stocking fish (Rodgers et al. 1992), or sampling known numbers of marked fish (Peterson et al. 2004). Dual gear procedures assume that one gear produces accurate and precise estimates. Employing a second gear type that is 100% efficient would be ideal for this comparison. In reality, however, very few sampling gears can achieve 100% efficiency. Examples of common applications have included the use of toxicants as the second gear type to generate accurate and precise estimates for comparison with electrofishing estimates (Mahon 1980). Sampling with toxicants is lethal to the sample fish and may not be acceptable for sampling rare or listed species, not appropriate for long-term monitoring, and unlikely to be 100% efficient (Bayley and Austen 1990). Stocking known numbers of fish into selected areas and subsequently sampling them has also been proposed as a means to evaluate gear bias (Rodgers et al. 1992); however, it may be counterproductive to artificially stock fish into areas because it is not known if parameter estimates generated from artificial stocking will be representative of the wild population, and stocked fish may not exhibit normal behaviors. Capture–recapture approaches may be the most appropriate methods for sampling wild fish in wadable, coldwater streams. With the considerations of bias, we will focus our discussion on mark–recapture sampling to generate estimates from known numbers of marked fish. We do this because the commonly used alternative, multiple-removal/depletion sampling, has been documented to produce biased estimates under typical field conditions (see Table 7) (Cross and Stott 1975; Riley and Fausch 1992; Peterson et al. 2004); however, recall from discussions of the MSE, that when estimator bias is small relative to sampling variance, multiple-removal sampling may still produce useful information because the error associated with estimator bias is only a small portion of the total MSE. Thus, using estimators or expansions based on multiple-removal/depletion based efficiencies, such as single-pass (Jones and Stockwell 1995) or CPUE (Simonson and Lyons 1995) sampling, might be adequate to index abundance in many situations. For studies that require accurate estimates or that do not have prior knowledge of sampling variance and estimator bias under typical field conditions, we recommend following mark–recapture protocols to generate abundance estimates.

TABLE 7. — Species (Spp) abbreviations include coho salmon *Oncorhynchus kisutch* (COH), brook trout *Salvelinus fontinalis* (EBT), brown trout *Salmo trutta* (BT), rainbow trout *O. mykiss* (RBT), bull trout (BULL), and cutthroat trout *O. clarkii* (CUT).

Reference	Location	Spp	Bias
Cross and Stott 1975	Rectangular ponds	Gudgeon roach	34%
Petersen and Cederholm 1984	Small Washington streams	COH	9%
Riley and Fausch 1992	Small Colorado streams	EBT BT RBT	9%
Rodgers et al. 1992	Small Oregon streams	COH	33%
Riley et al. 1993	Newfoundland streams	Atlantic Salmon parr	23%
Peterson et al. 2004	Small Idaho and Montana streams	BULL CUT	88%
Temple and Pearsons 2005	Small Washington streams	RBT	18%

Catch-per-unit-effort (CPUE)

Generally, increasing and/or validating the accuracy and precision of abundance estimates requires increased time, effort, and money (Bohlin et al. 1989). In some instances, a simple index of abundance is acceptable when study objectives do not require accurate and precise estimates. In such cases, simple CPUE indices may be appropriate. CPUE indices assume that the rate of catch in a sample is proportional to stock size (Thompson et al. 1998). Examples of backpack electrofishing CPUE indices have included single-electrofishing pass sampling to index abundance (Jones and Stockwell 1995; Mitro and Zale 2000). In some instances CPUE sampling may provide abundance estimates that are as reliable as those from traditional removal sampling (Jones and Stockwell 1995; Kruse et al. 1998), particularly when calibrated with unbiased estimates; however, caution should be used when interpreting CPUE indices because the level of bias in absolute terms usually will not be known (Bohlin et al. 1989).

CPUE sampling is perhaps the least labor-intensive electrofishing method available for indexing abundance. Sampling generally consists of selecting the study site(s), establishing the metric of effort, and sampling fish. Effort is often recorded as the amount of time (seconds or minutes) the electrofishing unit supplies electricity to the water and abundance indices are presented as fish/time (e.g., fish/minute). Other measurements of effort include establishing and sampling a specific length of stream or establishing a particular number of fish to sample. CPUE sampling can be performed following the mark–recapture electrofishing protocols presented in Table 9 in the following sequence: Perform steps 1–3, 6–10, 12–13, 15–16, and 23. Calculate the abundance index as the number of fish captured per unit effort.

One problem associated with CPUE sampling is that capture efficiency is assumed to be independent of field conditions (e.g., turbidity, discharge, temperature, depth, habitat complexity). This may be an appropriate assumption if conditions are very similar across comparisons. If it is not, corrections may be necessary. Correcting CPUE indices can be performed by calibrating them with known estimates (Fritts and Pearsons 2004).

Multiple Removal/Depletion

A common method used for enumerating salmonids in small streams is based on the multiple removal/depletion technique proposed by Moran (1951) and by Zippin (1958). Under the removal model, the declining catch of fish between multiple electrofishing passes is used to calculate capture efficiencies and abundance estimates. A common protocol requires a section of a stream to be isolated with block nets and a minimum of two removal electrofishing passes to be performed, although three or more removal passes are generally recommended so that catchability assumptions can be statistically tested (White et al. 1982). Practitioners should be aware that the tests used to evaluate equal catchability assumptions generally have low power to detect true differences (Riley and Fausch 1992). The Zippin removal estimator appears sensitive to violations of catchability assumptions (Bohlin and Sundstrom 1977). To circumvent this, Otis et al. (1978) proposed the generalized removal estimator that allows for unequal catchability of fish between removal passes. This estimator requires conducting four or more removal passes; a protocol that could be costly and time consuming. In addition, completion of four removal passes is no guarantee that the resulting removal estimate will be unbiased (Riley and Fausch 1992). The most efficient removal electrofishing protocol will minimize the number of removal passes that must be performed and will maximize the utility of the data. Conducting only two removal passes may satisfy effort requirements but likely produces biased estimates with wide confidence intervals (Riley and Fausch 1992). Additionally, two pass estimates fail when the number of fish captured on the second pass is greater than the number captured on the first electrofishing removal pass. To correct for this, some studies have recommended pooling catch data from multiple sites (Heimbuch et al. 1997). To satisfy effort and utility requirements, Connolly (1996) developed charts for field use based on stringent removal guidelines that produced the most reliable estimates with a minimum number of removal passes. In situations where accurate estimates are not necessary, removal sampling can be a useful method to index abundance. When accurate estimates are necessary, removal estimates should be interpreted with caution unless calibrated with an unbiased estimate, or critical assumptions are tested and validated under field conditions.

Multiple removal sampling protocols are similar to the mark–recapture protocols presented in Table 9, with the exception that fish do not need to be marked and fish captured after each electrofishing pass are enumerated and held in live wells (preferably outside the sampling section) until at least two but preferably three or more electrofishing passes are performed. Tables such as those presented by Connolly (1996) should be used as an aid to determine the number of removal passes to perform based upon the removal patterns observed in the field and the precision required by the study. Multiple removal sampling can be performed following the mark–recapture electrofishing protocols presented in Table 9 in the following sequence: Perform steps 1–13, then repeat steps 6–13, holding fish captured in the first electrofishing pass live wells. Consult removal tables presented by Connolly (1996) to determine if another electrofishing pass is required. If so, repeat steps 6–13. If not, consider step 20 and perform steps 21–23. Calculate the abundance index following the equations presented in Seber and LeCren (1967) or Zippin (1956) or consult Table 13 for online sources for computer programs to aid in computations.

Mark–Recapture

Mark–recapture (capture–recapture) electrofishing protocols have been shown to be useful means to measure the sampling efficiency of known numbers of marked fish (Rodgers et al. 1992; Peterson et al. 2004) and to generate population estimates (White et al. 1982); however, one drawback of this technique is logistical constraints that are commonly assumed to be associated with the recovery period—the time delay between the marking and recapture sampling. Recovery periods lasting longer than a single workday limit the utility of the mark–recapture method for short sites (i.e., 100 m) in tributaries because fish movement is difficult to control during longer periods, thereby violating the closed population assumption of the estimator. In some streams with high flows or debris loads, installing and maintaining block nets for long periods is extremely difficult. Thus, small stream electrofishing protocols extending beyond one working day are impractical for fisheries practitioners in many locations.

Most authors have recommended a minimum 24-hour recovery period between mark and recapture sampling. Peterson et al. (2004) suggested that sampling protocols utilized for electrofishing in coldwater areas should allow a recovery time of between 24 and 48 hours. This recommendation was assumed to provide a balance between the violations of movement and equal catchability assumptions. Similarly, Mesa and Schreck (1989) observed that wild cutthroat trout resumed normal behaviors only after a 24-hour recovery period between marking and recovery. Schreck et al. (1976) questioned the validity of mark–recapture estimates when the recovery period was shorter than a working day. These studies based their recovery periods on behavioral observations (Mesa and Schreck 1989) or physiological response to electrofishing (Schreck et al. 1976). Regardless of the behavioral or physiological differences between marked and unmarked fish, the appropriate recovery period should satisfy the critical assumption that marked and unmarked fish are equally catchable.

The appropriate recovery period to provide between mark and recapture sampling can be evaluated by comparing the catchability of marked fish that have recovered for at least 24 hours versus those that have recovered for shorter periods. The length of the shorter recovery period should allow both mark and recapture sampling to be completed in a single day. In a case study in the Yakima River, Washington, there was no significant difference in the catchability of rainbow trout that had recovered for 24 h versus a 3-h period. Thus, a 3-h recovery between marking and recapture proved to be a sufficient recovery period to allow marked rainbow trout to recover from handling (Temple and Pearsons 2006). This finding should be tested in other areas under different conditions before it is widely applied.

Similar to multiple removal sampling, mark–recapture sampling can produce biased estimates when the assumptions of the estimator are not met under field conditions (Table 8). The severity of the bias is generally lower for mark–recapture estimates of stream fish than reported for removal sampling (Peterson and Cederholm 1984; Rodgers et al. 1992). In contrast to removal estimates, mark–recapture estimates have the benefit of being fairly robust to potential bias introduced from poor crew experience and variable environmental conditions (e.g., channel type, habitat complexity, stream size). Nevertheless, the assumptions associated with mark–recapture estimates should be tested under field conditions.

TABLE 8.—Examples of published literature documenting negative bias in mark–recapture estimates of roach and coho (COH) salmon abundance.

Reference	Location	Spp	Bias
Cross and Stott 1975	Rectangular ponds	Roach	12%
Petersen and Cederholm 1984	Small Washington streams	COH	5%–6%
Rodgers et al. 1992	Small Oregon streams	COH	15%

Mark–Recapture Backpack Electrofishing Protocol

The step-by-step protocol is presented in Table 9 for quick reference. We will provide more information in the text for each step. Sequential numbers in Table 9 correspond to the following paragraphs for easy reference.

1. Select sampling site. Sampling sites should be representative with respect to habitat parameters and fish abundance if the sample will be extrapolated over any spatial scale greater than the site.

2. Identify appropriate length of stream to sample. See Establishing Linear Site Length—Backpack Electrofishing section, p. 112.

3. Measure site length along the contours of the stream channel. Minimize walking in the stream channel to prevent spooking fish.

4. Install blocking nets at the upstream and downstream boundaries of the site perpendicular to the flow of the water. Ensure that the size of the net mesh is small enough to prevent movement of the smallest fish that will be estimated. When determining net placement, nets may be moved upstream or downstream if the site boundary is unsuitable for net placement. For example, deep pools, swift water, or large debris piles may be difficult areas to secure nets. Stretch net across the stream channel. Small rocks may be used to secure the bottom of the net to the streambed. Once the net is secured along the stream channel, the ends may be secured to standing timber using ropes or string or may be anchored to the stream bank using large boulders. Branches or sticks may be used to prop the net up across the stream channel (see Figure 3).

5. We recommend installing an additional block net in the middle of the site to provide a means to evaluate movement rates and potential violations of movement assumptions. If evaluating movement rates, follow procedures 6–16 and then consider step 17.

6. Place live well buckets along the stream margin to hold captured fish. Conventional 5-gal buckets make convenient live wells that are cheap, durable, and easy to carry. We suggest using buckets colored similarly to the stream substrate to prevent substantial pigment changes in captured fish. The number of buckets needed may depend upon site length and fish density. We recommend separating sites into 10–25 m intervals. This allows replacement of fish close to their point of capture. Avoid overcrowding fish in live wells (see Figure 5).

7. Measure and record stream temperature and water conductivity prior to electrofishing. Conductivity, or the water's potential to carry electricity, is temperature-dependent and can vary throughout a single day. Electrofisher settings should be adjusted to the manufacturer's recommended guidelines based on water conductivity and temperature. In our experience, straight, unpulsed DC produces minimal injury to salmonids (McMichael et al. 1998) and efficient capture of salmonids. Generally, straight DC outputs ranging between 200 and 400 Volts are effective for capturing salmonids when water conductivities range between 150 mmhos and 40 mmhos.

8. Conduct safety check prior to electrofishing. Ensure electrofishing equipment is functioning properly. All crew members who will participate in electrofishing should be outfitted with waterproof waders and rubber or neoprene gloves to insulate against electric shock. Polarized field glasses should be worn by all field personnel to minimize glare and to protect the eyes from bright conditions. This also facilitates fish capture.

9. Clear or reset seconds timer on electrofishing unit.

10. Begin electrofishing. When sampling salmonids, we recommend electrofishing in an upstream direction (see Figure 6). When stream size requires two electrofishing units, both should operate in tandem (side by side), moving upstream at the same pace. Begin electrofishing at the bottom block net, thoroughly checking for fish that may be impinged on the net. Systematically progress upstream, taking care to electrofish all habitat in the stream channel. Complex habitat such as debris jams and deepwater areas will require more effort than homogenous habitats. One or two crew members should be outfitted with dip nets to facilitate the capture of stunned fish. It is often effective for netters to remain downstream from the operator and to keep dip nets within 1 m of the anode. All fish observed should be netted and swiftly removed from the water to prevent injury. Captured fish should be immediately placed into the appropriate live well held at the stream margin. It may be convenient for one crew member to carry a live well bucket during sampling to facilitate transfer of netted fish to the bucket.

    a. In homogenous habitats it is often effective for the electrofisher operator to move the anode in a W-shaped pattern across the stream channel while wading upstream. Netters should be prepared at all times to net stunned fish.

b. In complex habitats such as debris jams and undercut banks, it is often effective for the operator to insert the uncharged anode into the debris, depress the electrofisher switch, and slowly move the anode into open water areas. Fish will often be "pulled" from the debris into the open water where netters can capture them. Complex areas such as these often conceal several fish and should be thoroughly sampled until no additional fish are captured.

c. In deepwater areas such as pools or deep runs, it may be difficult to capture fish. One effective technique may be to "chase" fish into shallow water areas where they can be easily captured. The operator can keep the electrofisher charged while moving it back and forth across the channel and up and down in the water column. Netters should attempt to capture fish that become stunned but should remain conscious of the water depth to avoid submersion of hands or arms in the water. Systematically electrofish the entire pool area and slowly move upstream. Fish will often flee the deep water moving upstream.

d. In fast water areas, it is often effective for the operator to insert the anode into the water an arm's reach upstream, depress the electrofisher switch, and move the anode downstream at approximately the same velocity the water is traveling. Netters should have their nets pinned against the substrate in the fast water areas. Constriction points in the streamflow, such as between two large rocks, make good areas for dip-net placement. The operator can move the anode downstream into the dip net and then release the electrofisher switch to discontinue shocking. The net should then be immediately removed from the water and inspected for fish. Often, shocked fish will be pushed into the net by the stream flow or drawn into the net by the anode. Netters should check their nets frequently in these areas because fish will often become impinged in them without the crew's knowledge.

11. Continue electrofishing upstream until the upstream block net is reached. Thoroughly sample the substrate along the bottom of the block net for fish that may have moved upstream during sampling. Check downstream net for fish that may be impinged.

12. Record the amount of time electricity was supplied to the water (from seconds timer on unit).

13. Live wells can now be retrieved and fish can be anesthetized, identified, enumerated, and measured. Fish length is reported in a variety of ways and commonly measured as the maximum standard length, maximum total length, or fork length (Anderson and Gutreuter 1983). We recommend fork length for its ease of use. There are numerous conversions presented in the literature, to convert between length measurements (Carlander 1969; Ramseyer 1995).

14. For short-term sampling experiments, we recommend clipping or notching a small portion of one of the fins to identify marked fish. There

are several fish marking techniques that can be employed to mark fish (Parker et al. 1990), but for simple batch marking, fin clipping is a simple, cost-effective mark that is easy to apply and to identify in field settings.

15. Allow fish to recover from anesthesia and handling. If anesthetics are used to reduce handling stress, follow the manufacturer's guidelines for dosage information and appropriate recovery time. Fish will generally regain their equilibrium and begin to swim upright when recovering from anesthesia. Avoid releasing fish that have not fully recovered from the anesthetic. Do not release marked fish that are injured until sampling is complete because if they do not swim or behave normally, they may have a different catchability than their unmarked counterparts. Injured fish that are not marked and released should be accounted for in the final estimate by simple addition.

16. Release marked fish. Note the release time. We recommend releasing marked fish close to their point of capture. Subdividing the site into 10–25 m sections facilitates returning fish close to their capture point.

17. Allow fish to recover for a predetermined length of time. Avoid disturbing the stream section during the recovery period. Some authors recommend at least 24 hours for fish to recover from electrofishing, handling, and marking (Mesa and Schreck 1989); however, there are few investigations that evaluate appropriate recovery periods in terms of catchability. In the Yakima River Basin, we found 3 hours to be a sufficient amount of time for fish to recover and exhibit catchabilities similar to their unmarked counterparts. Appropriate recovery periods should be tested for individual studies.

18. Electrofish the sampling site again and capture all observed fish following steps 6–13. It is best to attempt to apply an effort similar to what was employed during the first electrofishing pass, but this is not critical in a Petersen-type mark–recapture protocol.

19. Captured fish can now be anesthetized, identified, enumerated, and measured. Pay particular attention to identify and enumerate marked fish and unmarked fish on the field data sheet. Fish can be released near their point of capture after fully recovering from handling. An example data sheet is presented in Appendix A.

20. If evaluating movement rates, repeat steps 6–16 in the adjacent stream section established by installing a block net in the middle of the site (step 5). This adjacent section can be sampled during the recovery period established for the section sampled first. Apply a different fin clip than that used in the adjacent section. Differentially marking fish between sections will allow the identification of fish to their upstream or downstream origin after taking the recapture sample. Release fish as in step 16. Perform steps 18–19 in the first section sampled after the minimum recovery period established in step 17 has been satisfied. Finally, perform steps 18–19 in the adjacent stream section after the minimum recovery period established in step 17 has been satisfied.

21. Release all fish near their point of capture at the end of sampling.

22. Remove block nets.

23. Sampling is complete. Return to office.
24. See Data Analysis section.

FIGURE 3. — Block nets used to isolate stream section.

TABLE 9. — Step-by-step backpack electrofishing mark–recapture protocol

1. Select site location.
2. Identify appropriate section length to sample.
3. Measure site along stream contours.
4. Install block nets at upstream and downstream boundaries.
5. If evaluating movement, install additional block net in the middle of the site.
6. Distribute live well buckets at 10–25 m intervals along the stream margin.
7. Record stream conductivity and temperature.
8. Perform safety check.
9. Reset seconds timer on electrofishing unit.
10. Electrofish in an upstream direction and net all fish observed within the site.
11. Thoroughly check all nets for fish.
12. Record electrofishing seconds.
13. Retrieve live wells, anesthetize, identify, enumerate, and measure fish. Record data on waterproof data sheets.
14. Mark fish (e.g., fin clip). Record marking data on waterproof data sheets.
15. Allow fish to recover from handling until they swim and behave normally (generally 5–10 min.).
16. Distribute and release marked fish back into site.
17. Allow appropriate recovery period (3–48 hours; avoid disturbing site during this period).
18. Perform recapture electrofishing pass (following steps 6–12).
19. Retrieve live wells, anesthetize, identify, enumerate, measure, and identify all fish. Record mark history of all fish (marked or unmarked). Record data on waterproof datasheets.
20. If evaluating movement, perform steps 6–19 in adjacent netted stream section.
21. Redistribute and release all fish at the end of sampling.
22. Remove block nets.
23. Sampling is complete. Return to office.
24. Perform analysis to generate abundance estimate.

Species Richness/Community Structure/Size Structure Sampling

Electrofishing techniques have commonly been used to capture fish for describing species richness/community structure (Lyons 1992; Angermeier and Smogor 1995; Patton et al. 2000; Reynolds et al. 2003) and the size structure of fish populations (Paragamian 1989; Thurow and Schill 1996). Species richness studies often utilize single-pass electrofishing, or CPUE sampling, to estimate species presence or absence and relative abundance. When using presence/absence sampling to identify species richness, rare fish distributions, or simple presence or absence of a species at a particular geographical locale, considerations of sampling efficiency should be taken into account (Bonar et al. 1997; Bayley and Peterson 2001). Failure to identify an individual species at a location does not demonstrate that it does not exist there and may be the result of poor sampling efficiency. Failure to identify an individual species will demonstrate that it is not present only when the probability of detecting it when it is present is 100%—a highly improbable scenario (Bayley and Peterson 2001). Thus, there is some level of uncertainty when concluding that a species is not present because it was not captured, and that is related to sampling efficiency.

A common assumption that is made when estimating relative abundance is that capture efficiencies are the same for different species and/or for different age-classes of fish. This assumption is unlikely to be true, particularly for species of different sizes that use different habitats. Some have tried to rectify this by generating capture efficiencies for each species observed (Bayley et al. 1989; Angermeier and Smogor 1995; Bayley and Austen 2002).

Establishing Linear Site Length—Backpack Electrofishing

One important consideration should be the amount of stream to sample when using backpack electrofishing to capture stream fish. Sampling too long of a stream section may become labor intensive and could be expensive. Sampling too short of a stream section may not provide a true representation of the population parameter under study and could lead to poor data quality, particularly if the sample is intended to be extrapolated to estimate some population level parameter. We acknowledge that there are many different ways in which electrofishing may be used to study fish and their communities, and sampling methods may be very different based on study objectives. Commonly, stream sections of equal length are selected (based on statistically valid techniques) and sampled to estimate abundance or density. Hankin (1984) advised against selecting and sampling stream sections of equal length and proposed sampling in stream sections of varying length based upon breaks in natural habitat units. This technique can increase the precision of estimates, particularly when the objective is to estimate abundance (Hankin 1984); however, in very small streams, selecting stream sections based on habitat units that are extremely small may not be appropriate for estimating abundance because common abundance estimators (e.g., removal or mark–recapture) are based on a large sample theory that may not apply if there are very few fish in each small habitat unit sampled. Thus, the minimum amount of stream to sample will be dependent on the type of stream, habitat characteristics, and the population density in the units (Bohlin et al. 1989).

In a case study in the Yakima basin, Washington, estimates of rainbow trout density and of the size structure of the population were independent of stream width, suggesting that sites based on set linear stream distances were appropriate for monitoring these parameters. Our observations of cumulative abundance and size structure estimates of rainbow trout and the temporal variability associated with 25 m incremental increases in sampling distances suggested that longer sampling distances generally increased the accuracy of the estimates; however, abundance estimates were more variable than size estimates with increasing stream area. We found that sampling 200 m sites provided an acceptable balance between effort requirements and estimate precision for long-term monitoring of these variables. Of course, practitioners will need to balance their required level of accuracy with the costs associated with the sampling required to achieve that level of accuracy.

We found that species richness estimates were not independent from stream size. Similar to Cao et al. (2001), our data indicates that species accumulations will reach an asymptotic level at shorter linear stream distances in smaller streams than in large streams. Thus, large streams will require longer sampling sections than small streams when the primary objective is to assess community structure. Sampling effort requirements have been typically reported as the number of channel widths that must be sampled to collect a large proportion of the species present in a given stream reach at some predetermined level of accuracy (Lyons 1992; Angermeier and Smogor 1995; Patton et al. 2000; Reynolds et al. 2003). We found that 27–31 channel widths was the minimum sampling distance required to detect 90% of the species present in our streams. For generating accurate species richness estimates, Lyons (1992) recommended sampling stream lengths comprised of a minimum of 35 channel widths. Patton et al. (2000) found that stream lengths of 12–50 times the mean wetted stream width should be sampled to capture 90% of species when electrofishing small streams in the Great Plains region of the United States. In western Oregon streams, Reynolds et al. (2003) found that electrofishing linear stream distances of 40 times the mean channel width captured 90% of the species present. It may be more efficient and accurate to construct species accumulation curves to help determine the amount of stream to sample to collect a given percentage of species present (see Figure 4). For instance, study sections can be sampled in 25-m increments and cumulative species totals determined for each 25 m sampled. Hypothetically, we will assume research goals require 90% of species in the site be detected. Using Figure 4, we sample contiguous 25 m increments until no new species are captured for four contiguous 25 m sections. In this case, a 250-m stream section will be the minimum acceptable linear sampling distance required to capture 90% of the species present in the stream section (judged from the mean line in Figure 4). In some streams, 90% of species present will be captured at short linear stream distances and in others, long sections must be sampled. Initially, a long stream section must be thoroughly sampled to determine the true numbers of species present and to serve as the benchmark for calculating the percentage of species detected in each 25 m increment.

Since abundance and size variables can be effectively monitored using standard linear distances and species richness is more effectively monitored based upon variable site lengths, monitoring programs may need to adopt different sampling effort strategies for monitoring different response variables. When trout abundance or size variables are of interest, sampling effort requirements may be based on predetermined sampling distances. When species richness is the primary variable of interest, linear sampling distances should be based upon multiples of the mean wetted width. When all variables are of interest, a hybrid approach may be used in which set linear distances are sampled for abundance information and additional stream area sampled for species richness estimates. The appropriate amount of effort will be a balance between the program objectives and a fixed budget. When stream size was less than 40 channel widths, Reynolds et al. (2003) suggested that 150 m was necessary to ensure that sufficient numbers of individual fish were captured for estimating species richness in Oregon streams. We attempted to maximize our sampling to collect abundance, size, species richness, and rare fish distribution and found 200 m to be an appropriate site length for our needs.

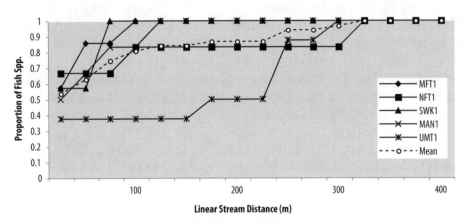

FIGURE 4. — Species accumulation curves for five sampling sites and the mean of the five sites.

Permitting and Equipment

Prior to initiating electrofishing, appropriate research and/or sampling permits should be obtained from the appropriate authority. For waters containing sensitive or listed species in the United States, research permits may be obtained from NOAA Fisheries (formerly National Marine Fisheries Service) for anadromous salmonids or from the U.S. Fish and Wildlife Service for resident salmonids; additional sampling permits may be required from state governments. Practitioners will need to facilitate good relations with private landowners to maintain access to sites located on private lands. Often landowners are willing to grant access to sites via their property when field personnel take the time to communicate the purpose of the study and maintain a professional and open dialogue with them. With the appropriate permits in hand and after all sampling design issues have been considered, practitioners will be ready to begin sampling fish. We recommend that equipment checklists be developed to aid in preparing field gear for sampling activities. An example equipment list for a typical mark–recapture backpack electrofishing sample is presented in Table 10. Examples of mark–recapture data sheets are provided in Appendices A and B and should be printed on waterproof paper. When all field gear has been assembled for a stream survey, crews can travel to the sampling sites and begin sampling.

TABLE 10. — Sample equipment checklist for typical backpack electrofishing mark–recapture survey

Sampling permits
Backpack electrofisher
Spare electrofisher batteries (or fuel for gas-powered units)
Pole mounted anode (and spare)
Dip net(s)
Block nets
Fish sampling gear (balance, measuring board, anesthetic)
Sharp scissors
Buckets or live wells
Thermometer and conductivity meter
Data sheets and pencils (examples provided in appendices A and B)
Waterproof gloves and waders
Polarized field glasses
Other gear:
Global positioning system (GPS) unit
Rain gear

Boat Electrofishing

When environmental conditions such as stream size or discharge (see Table 5) prevent effective electrofishing with backpack-mounted electrofishers, larger boat- or raft-mounted electrofishing units may be more effective for capturing fish for estimating abundance. Most of our experience includes using a drift boat-mounted electrofishing unit; we will discuss mark–recapture protocols in the context of drift electrofishing. However, much of this discussion will apply to other boat electrofishers in streams as well. There are several anode and cathode arrangements utilized on electrofishing boats.

There are other electrofishing boat designs, including motor-powered boats, that may require different specific protocols; however, the underlying principles in mark–recapture estimation remain the same—with the primary objective being to capture, mark, release, and recapture fish. The Petersen-type estimator is commonly used in large rivers because multiple removal techniques are impractical in them. The assumptions of the Petersen-type mark–recapture estimate are the same for both boat and backpack electrofishing techniques, although there may be additional considerations for practitioners to consider when sampling large streams and rivers with boat- or raft-mounted units.

FIGURE 5.—Overcrowded fish in livewell.

FIGURE 6.—Electrofishing a wadable stream.

Movement

Sampling in large streams and rivers becomes more problematic due to their large size. It is often not practical or feasible to isolate the study site with block nets to prevent movement of fish, and thus movement may be difficult to control during sampling; however, it is practical to sample longer linear stream sections, and as the sampling reach length is increased and the number of fish sampled is sufficiently large, the proportion of fish that potentially move in or out of the site at the margins becomes a small proportion of the population and will have a minimal effect upon the resulting estimate. Although mark–recapture estimates assume there is no movement during sampling, note that equal immigration or emigration of unmarked fish at the reach margins will not have any negative effect upon a mark–recapture estimate. Violations of movement assumptions will not negatively affect mark–recapture estimates as long as the effective ratio of marked/unmarked fish in the sample is not altered during the course of sampling. Thus the basic Petersen-type mark–recapture estimate is sensitive only to emigration of marked fish or unequal immigration or emigration rates of unmarked fish. Violations of movement assumptions that may bias the estimate can be tested directly using traps or weirs, or by some other direct estimate (Gatz and Loar 1988). If movement assumptions cannot be met, practitioners may consider using a mark–recapture model that does not assume population closure such as those discussed by Schwarz and Seber (1999).

Capture Efficiency

It is not generally recommended for both mark and recapture sampling to be completed in a single day when drift electrofishing. Captured fish are put in live wells kept on board the boat and transported downstream as the boat drifts downstream and are therefore relocated. Fish that are relocated will need time to redistribute and randomly mix with the unmarked fish. Releasing marked fish at several points within the site will facilitate redistribution. We recommend subdividing the sampling site into smaller reaches, and captured fish should be identified, measured, marked, and released at those points within the site. Vincent (1971) recommends that sampling sites should be at least 1,000 feet (305 meters) in length—or long enough to ensure a minimum sample size of 150 fish. It is best to release fish in shallow-water areas that are not in close proximity to deep pools, side channels, or river mouths to ensure that marked fish redistribute themselves and stay in the sampled river channel. If fish are released in deepwater areas such as deep pools, they have the potential to move to the bottom of the pool and effectively become uncatchable during the recapture sample. Releasing fish in several shallow-water areas and allowing at least 2 days between mark and recapture sampling (Vincent 1971) may help satisfy the assumption of equal catchability of marked and unmarked fish. In some instances, night electrofishing can increase sampling efficiency (Paragamian 1989), and boats or rafts should be outfitted with halogen lamps to aid navigation if sampling at night (see Figure 8).

Drift Boat Electrofishing Protocol

While actively electrofishing for bank-oriented species, close proximity to bank structures should be maintained while keeping the boat or raft at an angle of about 45° to the direction of the flow. In this position, the cathode and anode are in an effective fishing position while giving the rower the greatest ability to respond to changes in bank structures. It may not be necessary to avoid overhanging trees if the netter can duck down, brace him/herself, or otherwise be prepared and the boat can pass underneath without making contact. Avoid close proximity to sweepers or low-hanging structures in fast water and alert the net person of low-hanging branches or potential collisions. Keep the anode ring and droppers just far enough off the shore to minimize its contact with the bank or stream bottom. In areas with gradual sloping and/or uniform bottoms, try to keep the anode suspended about 0.7 m from the stream bottom.

Finally the rower should maintain a speed equal to that of the water. If drifting too fast, fish will be missed as the boat passes over them before the netter can net them. If moving too slowly, fish will be pulled downstream ahead of the boat before the netter can catch them. The rower can often adjust the boat speed and position ahead of time to present the gear where the fish are most likely to be. The step-by-step protocol is presented in Table 11 for quick reference. We will provide more information in the following text for each step. Sequential numbers in Table 11 correspond to the following paragraphs for easy reference.

1. Determine site location, boundaries, and site length.
2. Launch drift electrofishing boat at the upstream access point.
3. Start generator and turn on electrofishing equipment while floating downstream with the current.
4. Check that voltage and amperage output settings do not exceed 7 A and 450 V (McMichael et al. 1998).
5. Rower maintains boat at a 45° angle towards the bank.
6. Netter nets all visible target fish and places in on board live well.
7. Drift to predetermined workup station, disengage electrofishing equipment, and anchor or beach the boat.
8. Identify, measure, weigh, and mark target fish. For short-term sampling experiments, we recommend clipping or notching a small portion of one of the fins (caudal, ventral, or anal) to identify marked fish. There are several fish marking techniques that can be employed to mark fish (Parker et al. 1990), but for simple batch marking, fin clipping is a simple, cost-effective mark that is easy to apply and to identify in field settings. We recommend using fish anesthetic to facilitate fish handling. An example data sheet for recording fish data is provided in Appendix A.
9. Allow a minimum of 15 min for fish recovery from handling. If anesthetic is used, allow the appropriate recovery period per manufacturer's recommendation.
10. Release marked fish in shallow water 15 m upstream from boat to facilitate redistribution and prevent reshocking marked fish. Do not release injured fish or fish that have not recovered from anesthesia. Account for fish that are not released during analysis.
11. Quickly return to boat, pull anchor, and resume electrofishing.
12. Repeat electrofishing procedures 3–11 until the end of sampling site is reached. Drift to boat launch.
13. Electrofishing run is complete. In large rivers, it may be necessary to perform two marking runs to successfully capture fish on both riverbanks.
14. Allow fish to recover and redistribute from capture and marking before resampling the same stretch of stream/river.
15. Return to site. Repeat steps 2–13. Substitute fish marking with examining for marks and recording marked and unmarked fish in the sample. In large rivers, it may be necessary to perform two recapture runs to capture fish successfully on both riverbanks.
16. Data can be analyzed considering the information presented under the Data Analysis section.

TABLE 11. — Drift boat electrofishing mark–recapture protocol

1.	Select site.
2.	Launch drift boat.
3.	Turn on electrofishing equipment.
4.	Check that voltage and amperage output do not exceed 7 A and 450 V.
5.	Rower maintains boat at a 45° angle towards bank.

6. Netter nets all target fish and place in on board live well.
7. Navigate to station workup, disengage electrofishing equipment, and anchor boat.
8. Identify, measure, weigh, and mark target fish.
9. Allow fish minimum of 15 min to recover from handling.
10. Release marked fish 15 m upstream from anchored boat.
11. Quickly return to boat, pull anchor, turn on electrofisher, and resume electrofishing.
12. Repeat procedures 3–11 until end of sampling site is reached and float to boat launch.
13. Remove boat from river. Electrofishing run is complete.
14. Allow fish to recover and redistribute between sampling runs.
15. Repeat steps 2–13 for recapture run. Substitute fish marking with enumerating marked and unmarked fish in the sample.
16. Analyze data.

FIGURE 7. — Navigation through challenging obstacles should be carefully planned.

FIGURE 8. — Night drift boat electrofishing crew.

Personnel Requirements and Training

The required crew size for effective backpack electrofishing depends on the size of the stream sampled. Effective capture of salmonids can be obtained in most instances with a three-person crew when sampling with a single backpack-mounted unit. This allows one person to be dedicated to operating the electrofisher and at least one person dedicated to netting fish, while a second person carries a portable live well and provides additional back up netting. When two backpack units are used in tandem, an additional operator and net person should be used (i.e., a total of five persons). In many instances, budgetary restraints may limit crew sizes; in these situations, volunteer help may be a valuable resource (Leslie et al. 2004). Under no circumstance should electrofishing be performed alone.

Drift boat and raft electrofishing requires a minimum of two crew members; on larger rivers, three or four are appropriate. Under most conditions, two are sufficient. One person is dedicated to maneuvering and navigating the boat or raft while the other nets fish from the bow of the craft. Crew members may alternate rowing and netting to avoid fatigue. Both crew members should be familiar with the stream section that will be sampled to avoid encountering any unforeseen obstacles (see Figure 7). Drift electrofishing should not be performed by personnel inexperienced in navigating drift boats or rafts in the riverine environment. In most cases, the river will be large enough to be dangerous and crews should carefully read the section on safety before venturing out.

All personnel who work in the general vicinity of an electrofishing operation must be thoroughly trained about electrofishing theory, application, and safety concerns before electrofishing begins. Every electrofishing crew should have at least one principal operator or crew leader who is experienced or certified in electrofishing in charge during the electrofishing operation (Reynolds 1983). The U.S. Fish and Wildlife Service's National Conservation Training Center provides formal electrofishing training and certification courses; course information and scheduling is available online at <http://training.fws.gov/BART/courses.html>. In addition, some equipment manufacturers also provide training to familiarize practitioners with their equipment.

Safety

Backpack Electrofishing Safety

The safety of the crew is always the top priority. All crew members should reserve the right to refuse electrofishing without fear of reprimand if conditions appear unsafe. Perform a safety check prior to electrofishing to ensure that the equipment is functioning properly following the equipment manufacturer's guidelines. Do not electrofish with equipment that is not functioning properly. All crew members should be outfitted with waterproof waders (e.g., neoprene) to insulate against electric shock. Waterproof rubber lineman-style gloves should be worn by all crew members anytime electrofishing is underway. Electrofishing should not proceed when environmental conditions are unsafe (e.g., deep water, high velocities, excessive rainfall).

Drift Electrofishing Safety

As previously stated, crew safety is paramount. Unfortunately, due to the variability between river characteristics, experience of the crew rather than specific environmental guidelines is ultimately the deciding factor on whether or not a survey should be undertaken. For example, high flow has the potential of pushing fish towards the riverbanks, thereby increasing capture efficiency; however, choosing to perform a drift boat or raft electrofishing survey under high-flow conditions and in an area where stream width is not sufficient to allow maneuverability, or where there may not be adequate time to maneuver around obstacles, would exhibit very poor judgment.

The application of electricity into water to induce the capture of fish is inherently dangerous work. The voltage and amperage produced by most boat or raft mounted units is significant enough to cause injury to personnel, and safety precautions should be employed to minimize the risk of electrocution. Crew members should wear waterproof waders and dry, rubber, lineman-style gloves to insulate against electric shock. The dip net used should be constructed with a nonconductive fiberglass handle. The use of a personal flotation device (PFD) is mandatory for all crew members anytime electrofishing is underway. All crew members should familiarize themselves with the location of the power shutoff switches associated with the electrofishing unit. The netter stationed on the bow should always utilize a dead-man switch and should attach it to his/her PFD. This switch is designed to shut off the electrical current should the netter fall out of the boat or raft. The electricity should be shut off anytime the boat or raft is beached or anchored and whenever a crew member exits the craft.

Crews should expect to hit submerged objects while drift electrofishing. The rower should use good judgment while navigating and always maintain control. If it appears that a collision is imminent, the best tactic is to point the bow at the object and try to slow down as much as possible to minimize the force of impact. The netter can also use the net handle to help push off the obstacle. Constant communication among the crew members is critical. The netter should alert the person rowing of hidden dangers and the person rowing should alert the netter of dangers as well. Always steer clear of potentially hazardous areas where the safety of the crew and boat may be jeopardized. Safe operation of the boat or raft and crew should be the top priority.

There may be occasions when the crew is faced with working in inclement weather or under adverse conditions. Mild drizzle, light rain, or snow usually poses little threat to the crew. Electrofishing can proceed as long as the vital electrical components are protected from the rain; however, should heavy rains form a continuous sheet of water on the boat or raft, crew, and electrical components, electrofishing should be discontinued until more favorable conditions develop. Electrofishing trips should be aborted anytime a thunderstorm produces lightning in the general vicinity of the crew.

Safe towing of the boat or raft to and from work locations should also be a concern of the crew. Make sure the trailer is in good operating condition and that it is properly hitched to the tow vehicle with the ball, hitch, light wiring, and safety chains. The trailer wheel bearings should be checked and greased frequently. The tread wear and tire pressure on the trailer tires should be carefully monitored while traveling. The operation of the trailer lights should be periodically checked. Finally, remember that a trailer extends the length of the tow vehicle and requires a greater stopping distance, wider cornering radius, and generally more cautious driving.

Fish Handling

Fish handling procedures are a critical element of sampling, especially in systems with threatened or endangered species. As researchers and managers, we are obliged to handle the fish we are sampling responsibly (Nickum et al. 2004). This is an important topic that should be considered by all practitioners sampling fish. Detailed fish handing guidelines are available from Nickum et al. 2004, the Canadian Council on Animal Care (CCAC) 2005, and Stickney (1983).

When mark–recapture sampling, it is critical that handled and marked fish return to normal activities before the recapture sample begins to help satisfy the equal catchability assumption associated with mark–recapture estimators. There are several examples of things we can do to reduce the handling stress of sampled fish (see Table 12). For instance, the zone 0.5 m from an electrofishing anode has been termed the potential zone of injury; thus fish in close proximity to the anode should be quickly removed from it while electrofishing (NMFS 2000). After removal from the water, captured fish should be held for the shortest time possible in live wells with recirculated or aerated water that has approximately the same temperature as the ambient water temperature they were removed from. An innovative live well design used for holding captured stream fish was proposed by Isaak and Hubert (1997) for stream sampling and by Sharber and Carothers (1987) for boat or raft units. Perhaps the most important consideration is to minimize handling of fish during any research or sampling activity.

TABLE 12.—Examples of activities that help minimize stress in sampled fish

Use approved anesthetic when handling.
Use aerated live wells or frequently change water.
Use live well colors that mimic the natural environment and include cover in them (e.g. rocks, leaves).
Minimize handling time.
Minimize sampling at water temperatures above 18°C.
Avoid overcrowding fish in live wells.
Avoid drastic differences in water temperature between live wells and ambient stream water.
Segregate large predatory fish from small fish to prevent predation in live wells.

Electrofishing Injury

Concerns over the potential for electrofishing to harm fish populations, particularly in areas with endangered species, have been presented in the literature (Snyder 1995; Nielsen 1998). It is clear that electrofishing can harm individual fish, particularly large individuals or susceptible life stages or species (McMichael et al. 1998). Injuries can range from bruising to broken backs; however, in most cases, the proportion of fish within populations that are actually sampled is small, and even large injury rates at the sample scale may be acceptable for monitoring and conservation-oriented studies (McMichael et al. 1998). Furthermore, electrofishing-induced injury rates evaluated from the population level may be negligible when considering natural mortality rates of stream fish (Schill and Beland 1995). Repeated electrofishing from long-term monitoring did not appear to have detrimental population level effects on salmonids in small Colorado streams (Kocovsky et al. 1997). As with any sampling technique, the value gained by sampling must be weighed against the ecological impact to the population.

Data Analysis

There has been much literature published on the theoretical and mathematical properties of both open- and closed-model mark–recapture estimators, including the classic works of Ricker (1975), Otis et al. (1978), and Seber (1986, 1992). Schwarz and Seber (1999), building on the work of Seber (1982, 1986, 1992), provide a comprehensive review of the literature related to estimation of animal population parameters. There have been many statistical advances in evaluations of mark–recapture data through the years. Many recent advances include variations on the basic mark–recapture abundance estimator to allow for various violations of underlying estimator assumptions. Discussing the statistical developments associated with evaluating mark–recapture data is beyond the scope of this protocol. We refer readers to the detailed review and references in Schwarz and Seber (1999).

We will provide the formula for the basic Petersen abundance estimator so that field practitioners without access to the cited references or computer software can calculate an abundance estimate based on simple mark–recapture field data. We direct readers to Ricker (1975) for more detailed discussion of the Petersen estimator. We would also like to make readers aware that the Petersen estimator is a simple case (approximation) of what are known as maximum likelihood estimates (White et al. 1982). When more than a single marking and a single recapture sampling occasion are performed, abundance estimates cannot be computed with the simple form of the Petersen estimator presented and iterative calculus techniques and likelihood functions should be used to calculate abundance. We refer readers to Otis et al. (1978) for detailed discussions regarding likelihood functions and maximum likelihood estimation. Researchers from Montana Fish, Wildlife and Parks have generated easy-to-use software that will calculate likelihood estimates quickly and efficiently (see Table 13).

The basic premise of mark–recapture estimates is that the ratio of marked/unmarked fish collected in a sample where M fish are marked is the same as the ratio of marked fish in the total population (M/N). An estimate of abundance from mark–recapture data can therefore be calculated from equation 2,

$$N = MC/R \qquad \text{(eq 2)}$$

where N = the population estimate, M = number of fish marked during the mark run(s), C is the total number of fish in the recapture sample(s), and R is the number of marked fish captured in the recapture sample.

Bailey (1951) and Chapman (1951) presented mathematical corrections to the Petersen estimate when it was recognized that it may be biased when sample sizes are low. Chapman's modification of the Petersen estimate is provided in equation 3:

$$N = \frac{(M+1)(C+1)}{(R+1)} - 1 \qquad \text{(eq 3)}$$

Robson and Reiger (1964) suggest that an unbiased estimate of N can be generated from the Chapman modification of the Petersen estimate when one or both of the following conditions are met:

1. The number of marked fish M plus the number of fish captured during the recapture sample C must be greater than or equal to the estimated population N.
2. The number of marked fish M multiplied by the number of fish taken during the recapture sample C must be greater than four times the estimated population N.

Calculations providing approximate 95% confidence intervals in the population estimate (N) are summarized by Vincent (1971) and can be calculated using equations 4 and 5:

$$\text{Estimate} \pm 2\sqrt{\text{Variance}} \qquad (eq\ 4)$$

where

$$\text{Variance} = \frac{(\text{Population Estimate})^2(C - R)}{(C + 1)(R + 2)} \qquad (eq\ 5)$$

Equations 2–5 allow for hand calculation of population abundance estimates and confidence limits when all assumptions are met during sampling. There are many more complex estimators that can be used to estimate population abundance when assumptions cannot be tested in field settings due to budget or time constraints. Many of these estimators are available through Internet resources and are often free or inexpensive (see Table 13). These resources contain information as well as Internet links to computer software programs for estimating various population and community parameters beyond simple abundance calculations. Many computer programs contain complex procedures that will select appropriate population estimators based on the observed field data. Other programs use iterative calculus techniques to produce maximum likelihood estimates based on observed mark–recapture data.

TABLE 13.—Population analysis online information and software links available on the Internet

Web address	Affiliation
www.mbr-pwrc.usgs.gov/software	U.S. Geological Survey
www.warnercnr.colostate.edu/~gwhite/software.html	Colorado State University
www.fisheries.org/units/cus	American Fisheries Society computer user section
www.cs.umanitoba.ca/~popan/	University of Manitoba
http://fwp.state.mt.us/default.html	Montana Fish, Wildlife and Parks

Closing Remarks

Electrofishing will likely continue to be one of the dominant methods of sampling freshwater fishes until improved sampling techniques are developed. Although the practice of electrofishing is relatively easy, designing the right approach to achieve objectives is quite complicated. Many trade-offs and assumptions must be evaluated to ensure that the best design can be implemented. In addition, flexibility to make improvements in long-term monitoring should be anticipated as new information, techniques, and equipment become available. This may involve implementing new methods and conventional methods at the same time so that historical data sets can be calibrated to new ones.

Literature Cited

Anderson, R. O., and S. J Gutreuter. 1983. Length, weight, and associated structural indices. Pages 283–300 in L. A. Neilsen and D. L. Johnson, editors. Fisheries techniques. American Fisheries Society, Bethesda, Maryland.

Anderson, C. S. 1995. Measuring and correcting for size selection in electrofishing mark–recapture experiments. Transactions of the American Fisheries Society 124:663–676.

Angermeier, P. L., and R. A. Smogor. 1995. Estimating number of species and relative abundances in stream-fish communities: effects of sampling effort and discontinuous spatial distributions. Canadian Journal of Fisheries and Aquatic Sciences 52:936–949.

Bailey, N. T. J. 1951. On estimating the size of mobile populations from recapture data. Biometrika 38:293–306.

Bayley, P. B., R. W. Larimore, and D. C. Dowling. 1989. Electric seine as a fish-sampling gear in streams. Transactions of the American Fisheries Society 118:447–453.

Bayley, P. B., and D. J. Austen. 1990. Modeling the capture efficiency of rotenone in impoundments and ponds. North American Journal of Fisheries Management 10:202–208.

Bayley, P. B., and D. J. Austen. 2002. Capture efficiency of a boat electrofisher. Transactions of the American Fisheries Society 131:435–451.

Bayley, P. B., and J. T. Peterson. 2001. An approach to estimate probability of presence and richness of fish species. Transactions of the American Fisheries Society 130:620–633.

Bohlin, T., S. Hamrin, T. G. Heggeberget, G. Rasmussen, and S. J. Saltveit. 1989. Electrofishing–theory and practice with special emphasis on salmonids. Hydrobiologia 173:9–43.

Bohlin, T., and B. Sundstrom. 1977. Influence of unequal catchability on population estimates using the Lincoln index and the removal method applied to electro-fishing. Oikos 28:123–129.

Bonar, S. A., M. Divens, and B. Bolding. 1997. Methods for sampling the distribution and abundance of bull trout/Dolly Varden. Washington Department of Fish and Wildlife, Research Report RAD97-05, Olympia.

Bonar, S. A. 2002. Relative length frequency: a simple, visual technique to evaluate size structure in fish populations. North American Journal of Fisheries Management 22:1086–1094.

Buttiker, B. 1992. Electrofishing results corrected by selectivity functions in stock size estimates of brown trout (*Salmo trutta L.*) in brooks. Journal of Fish Biology 41:673–684.

Cao, Y., D. P. Larsen, and R. M. Hughes. 2001. Evaluating sampling sufficiency in fish assemblage surveys: a similarity based approach. Canadian Journal of Fisheries and Aquatic Sciences 58:1782–1793.

Carlander, K. D. 1969. Handbook of freshwater fishery biology, volume 1. Iowa State University Press, Ames.

CCAC (Canadian Council on Animal Care). 2005. Guidelines on: the care and use of fish in research, teaching and testing. Canadian Council on Animal Care, Ottawa. Available: www.ccac.ca/en/CCAC_Programs/Guidelines_Policies/PDFs/Fish Guidelines English.pdf (February 2006).

Chapman, D. G. 1951. Some properties of hypergeometric distribution with applications to zoological sample censuses. University of California Publications in Statistics 1:131–159, Berkely.

Cochran, W. G. 1977. Sampling techniques, 3rd edition. John Wiley & Sons, New York.

Connolly, P. J. 1996. Resident cutthroat trout in the central coast range of Oregon: logging effects, habitat associations, and sampling protocols. Doctoral dissertation. Oregon State University, Corvallis.

Cross, D. G., and B. Stott. 1975. The effect of electric fishing on the subsequent capture of fish. Journal of Fish Biology 7:349–357.

Cummings, J., and J. Masten. 1994. Customized dual data entry for computerized data analysis. Quality Assurance: Good Practice, Regulation, and Law 3(3):300–303.

Dumont, S. C., and J. A. Dennis. 1997. Comparison of day and night electrofishing in Texas reservoirs. North American Journal of Fisheries Management 17:939–946.

Fritts, A. L., and T. N. Pearsons. 2004. Smallmouth bass predation on hatchery and wild salmonids in the Yakima River, Washington. Transactions of the American Fisheries Society 133:880–895.

Funk, J. L. 1947. Wider application of the electrical method of collecting fish. Transactions of the American Fisheries Society 77:49–60.

Gatz, A. J., Jr., and J. M. Loar. 1988. Petersen and removal population size estimates: combining methods to adjust and interpret results when assumptions are violated. Environmental Biology of Fishes 21:293–307.

Green, R. H. 1989. Power analysis and practical strategies for environmental monitoring. Environmental Research 50:195–205.

Ham, K. D., and T. N. Pearsons. 2000. Can reduced salmonid population abundance be detected in time to limit management impacts? Canadian Journal of Fisheries and Aquatic Sciences 57:17–24.

Hankin, D. G. 1984. Multistage sampling designs in fisheries research: applications in small streams. Canadian Journal of Fisheries and Aquatic Sciences 41:1575–1591.

Hankin, D. G., and G. H. Reeves. 1988. Estimating total fish abundance and total habitat area in small streams based on visual estimation methods. Canadian Journal of Fisheries and Aquatic Sciences 45:834–844.

Hartley, W. G. 1980. The use of electric fishing for estimating stocks of freshwater fish. EIFAC Technical Paper 33:91–95.

Heimbuch, D. G., H. T. Wilson, S. B. Weisberg, J. H. Volstad, and P. F. Kazyak. 1997. Estimating fish abundance in stream surveys by using double-pass removal sampling. Transactions of the American Fisheries Society 126:795–803.

Hill, T. D., and D. W. Willis. 1994. Influence of water conductivity on pulsed AC and pulsed DC electrofishing catch rates for largemouth bass. North American Journal of Fisheries Management 14:202–207.

Isaak, D. J., and W. A. Hubert. 1997. A live-bucket for use in surveys of small streams. North American Journal of Fisheries Management 17:1025–1026.

Johnson, C. L., G. M. Temple, T. N. Pearsons, and T. D. Webster. 2006. Comparison of error rates between four methods of data entry: implications to precision of fish population metrics. In T. N. Pearsons, G. M. Temple, A. L. Fritts, C. L. Johnson, and T. D. Webster. Yakima River species interactions studies: Yakima/Klickitat fisheries project monitoring and evaluation. Bonneville Power Administration Annual Report 2005, Portland, Oregon.

Jones, M. L., and J. D. Stockwell. 1995. A rapid assessment procedure for the enumeration of salmonine populations in streams. North American Journal of Fisheries Management 15:551–562.

Kennedy, G. J. A., and C. D. Strange. 1981. Efficiency of electric fishing for salmonids in relation to river width. Fisheries Management 12:55–60.

Kocovsky, P. M., C. Gowan, K. D. Fausch, and S. C. Riley. 1997. Spinal injury rates in three wild trout populations in Colorado after eight years of backpack electrofishing. North American Journal of Fisheries Management 17:308–313.

Kruse, C. G., W. A. Hubert, and F. J. Rahel. 1998. Single-pass electrofishing predicts trout abundance in mountain streams with sparse habitat. North American Journal of Fisheries Management 18:940–946.

Leslie, L. L., C. E. Velez, and S. A. Bonar. 2004. Utilizing volunteers on fisheries projects: benefits, challenges, and management techniques. Fisheries 29(10):10–14.

Lyons, J. 1992. The length of stream to sample with a towed electrofishing unit when species richness is estimated. North American Journal of Fisheries Management 12:198–203.

Meador, M. R., T. F. Cuffney, and M. E. Gurtz. 1993. Methods for sampling fish communities as part of the national water-quality assessment program. U.S. Geological Survey open file report 93–104, Raleigh, North Carolina.

Mahon, R. 1980. Accuracy of catch-effort methods for estimating fish density and biomass in streams. Environmental Biology of Fishes 5:343–360.

McMichael, G. A., A. L. Fritts, and T. N. Pearsons. 1998. Electrofishing injury to stream salmonids; injury assessment at the sample, reach, and stream scales. North American Journal of Fisheries Management 18:894–904.

Mesa, M. G., and C. B. Schreck. 1989. Electrofishing mark–recapture and depletion methodologies evoke behavioral and physiological changes in cutthroat trout. Transactions of the American Fisheries Society 118:644–658.

Mitro, M. G., and A. A. Zale. 2000. Predicting fish abundance using single-pass removal sampling. Canadian Journal of Fisheries and Aquatic Sciences 57:951–961.

Moran, P. A. P. 1951. A mathematical theory of animal trapping. Biometrika 38:307–311.

NMFS (National Marine Fisheries Service). 2000. Guidelines for electrofishing waters containing salmonids listed under the Endangered Species Act. Available: www.nwr.noaa.gov/ESA-Salmon-Regulations-Permits/4d-Rules/upload/electro2000.pdf (October 2005).

Nickum, J. G., H. L. Bart, Jr., P. R. Bowser, I. E. Greer, C. Hubbs, J. A. Jenkins, J. R. MacMillan, F. W. Rachlin, R. D. Rose, P. W. Sorensen, and J. R. Tomasso. 2004. Guidelines for the use of fishes in research. American Fisheries Society, American Society of Ichthyologists and Herpetologists, and the American Institute of Fishery Research Biologists. Available: www.fisheries.org/afs/publicpolicy/guidelines2004.pdf (February 2007).

Nielsen, J. L. 1998. Electrofishing California's endangered fish populations. Fisheries 23(12):6–12.

Novotony, D. W., and G. R. Priegel. 1974. Electrofishing boats: improved designs and operational guidelines to increase the effectiveness of boom shockers. Wisconsin Department of Natural Resources, Technical Bulletin 73, Madison.

Otis, D. L., K. P. Burnham, G. C. White, and D. R. Anderson. 1978. Statistical inference from capture data on closed animal populations. Wildlife Monographs 62:1–135.

Paragamian, V. L. 1989. A comparison of day and night electrofishing: size structure and catch-per-unit-effort for smallmouth bass. North American Journal of Fisheries Management 9:500–503.

Parker, N. C., A. E. Giorgi, R. C. Heidinger, D. B. Jester, Jr., E. D. Prince, and G. A. Williams, editors. 1990. Fish-marking techniques. American Fisheries Society, Symposium 7, Bethesda, Maryland.

Patton, T. M., W. A. Hubert, F. J. Rahel, and K. G. Gerow. 2000. Effort needed to estimate species richness in small streams on the Great Plains in Wyoming. North American Journal of Fisheries Management 20:394–398.

Peterman, R. M., and M. J. Bradford. 1987. Statistical power of trends in fish abundance. Canadian Journal of Fisheries and Aquatic Sciences 44:1879–1889.

Peterman, R. M. 1990. Statistical power analysis can improve fisheries research and management. Canadian Journal of Fisheries Management 47:2–15.

Peterson, N. P., and C. J. Cederholm. 1984. A comparison of the removal and mark–recapture methods of population estimation for juvenile coho salmon in a small stream. North American Journal of Fisheries Management 4:99–102.

Peterson, J. T., and J. Zhu. 2004. Cap-Post: capture and posterior probability of presence estimation, users guide, version 1. U.S. Geological Survey, Athens, Georgia.

Peterson, J. T., R. F. Thurow, and J. W. Guzevich. 2004. An evaluation of multipass electrofishing for estimating the abundance of stream-dwelling salmonids. Transactions of the American Fisheries Society 133:462–475.

Peterson, J. T., N. P. Banish, and R. F. Thurow. 2005. Are block nets necessary?: movement of stream-dwelling salmonids in response to three common survey methods. North American Journal of Fisheries Management 25:732–743.

Ramseyer, L. J. 1995. Total length to fork length relationships of juvenile hatchery-reared coho and chinook salmon. The Progressive Fish Culturist 57:250–251.

Reynolds, J. B. 1983. Electrofishing. Pages 147–143 in L. A. Neilsen and D. L. Johnson, editors. Fisheries techniques. American Fisheries Society, Bethesda, Maryland.

Reynolds, L., A. T. Herlihy, P. H. Kaufmann, S. V. Gregory, and R. M. Hughes. 2003. Electrofishing effort requirements for assessing species richness and biotic integrity in western Oregon streams. North American Journal of Fisheries Management 23:450–461.

Ricker, W. E. 1975. Computation and interpretation of biological statistics from fish populations. Fisheries Research Board of Canada, Bulletin 191, Ottawa.

Riley, S. C., and K .D. Fausch. 1992. Underestimation of trout population size by maximum-likelihood removal estimates in small streams. North American Journal of Fisheries Management 12:768–776.

Riley, S. C., R. L. Haedrich, and R. J. Gibson. 1993. Negative bias in removal estimates of Atlantic salmon parr relative to stream size. Journal of Freshwater Ecology 8:97–101.

Robson, D. S., and H. A. Regier. 1964. Sample size in Petersen mark–recapture experiments. Transactions of the American Fisheries Society 93:215–226.

Rodgers, J. D., M. F. Solazzi, S. L. Johnson, and M. A. Buckman. 1992. Comparison of three techniques to estimate juvenile coho populations in small streams. North American Journal of Fisheries Management 12:79–86.

Roni, P., and A. Fayram. 2000. Estimating winter salmonid abundance in small western Washington streams: a comparison of three techniques. North American Journal of Fisheries Management 20:683–692.

Rosenberger, A. E., and J. B. Dunham. 2005. Validation of abundance estimates from mark–recapture and removal techniques for rainbow trout captured by electrofishing in small streams. North American Journal of Fisheries Management 25:1395–1410.

Schill, D. J., and K. F. Beland. 1995. Electrofishing injury studies. Fisheries 20(6):28–29.

Schreck, C. B., R. A. Whaley, M. L. Bass, O. E. Maughan, and M. Solazzi. 1976. Physiological responses of rainbow trout (Salmo gairdneri) to electroshock. Journal of the Fisheries Research Board of Canada 33:76–84.

Schwarz, C. J., and G. A. F. Seber. 1999. A review of estimating animal abundance III. Statistical Science 14:427–456.

Seber, G. A. F. 1982. The estimation of animal abundance and related parameters, 2nd edition. Edward Arnold, London.

Seber, G. A. F. 1986. A review of estimating animal abundance. Biometrics 42:267–292.

Seber, G. A. F. 1992. A review of estimating animal abundance II. International Statistical Review 60:129–166.

Seber, G. A. F., and E. D. LeCren. 1967. Estimating population parameters from catches large relative to the population. Journal of Animal Ecology 36:631–643.

Sharber, N. G., and S. W. Carothers. 1987. Submerged, electrically shielded live tank for electrofishing boats. North American Journal of Fisheries Management 7:450–453.

Sharber, N. G., and J. S. Black. 1999. Epilepsy as a unifying principal in electrofishing theory: a proposal. Transactions of the American Fisheries Society 128:666–671.

Simonson, T. D., and J. Lyons. 1995. Comparison of catch per effort and removal procedures for sampling stream fish assemblages. North American Journal of Fisheries Management 15:419–427.

Simpson, D. E. 1978. Evaluation of electrofishing efficiency for largemouth bass and bluegill populations. Masters thesis. University of Missouri, Columbia.

Snyder, D. E. 1995. Impacts of electrofishing on fish. Fisheries 20(1):26–27.

Stangl, M. J. 2001. An electrofishing raft for sampling intermediate-size waters with restricted boat access. North American Journal of Fisheries Management 21:679–682.

Stickney, R. R. 1983. Care and handling of live fish. Pages 85–94 in L. A. Neilsen and D. L. Johnson, editors. Fisheries techniques. American Fisheries Society, Bethesda, Maryland.

Taylor, G. N., L. S. Cole, and W. F. Sigler. 1957. Galvanotaxic response of fish to pulsating direct current. Journal of Wildlife Management 21:201–213.

Temple, G. M., and T. N. Pearsons. 2005. Estimating the populations size of rainbow trout in small coldwater streams: the influence of estimator selection and recovery time. Pages 71–87 in T. N. Pearsons, A. L. Fritts, G. M. Temple, C. L. Johnson, T. D. Webster, and N. H. Pitts. Yakima River species interactions studies: Yakima/Klickitat fisheries project monitoring and evaluation. Bonneville Power Administration Annual Report 2004, Portland, Oregon.

Temple, G. M., and T. N. Pearsons. 2006. Evaluation of the recovery period in mark–recapture population estimates of rainbow trout in small streams. North American Journal of Fisheries Management.

Thompson, W. L., G. C. White, and C. Gowan. 1998. Monitoring vertebrate populations. Academic Press, Inc., San Diego, California.

Thurow, R. F., and D. J. Schill. 1996. Comparison of day snorkeling, night snorkeling, and electrofishing to estimate bull trout abundance and size structure in a second-order Idaho stream. North American Journal of Fisheries Management 16:314–323.

Vadas, R. L. Jr., and D. J. Orth. 1993. "A new technique for estimating the abundance and habitat use of stream fishes." *Journal of Freshwater Ecology* 7:149–164.

Vibert, R. 1963. Neurophysiology of electric fishing. Transactions of the American Fisheries Society 92:265–275.

Vincent, R. 1971. River electrofishing and fish population estimates. The Progressive Fish Culturist 33:163–169.

Vokoun, J. C., C. F. Rabeni, and J. S. Stanovick. 2001. Sample-size requirements for evaluating population size structure. North American Journal of Fisheries Management 21:660–665.

White, G. C., D. R. Anderson, K. P. Burnham, and D. L. Otis. 1982. Capture–recapture and removal methods for sampling closed populations. Los Alamos National Laboratory, LA–8787–NERP, Los Alamos, New Mexico.

Zalewski, M., and I. G. Cowx. 1990. Factors affecting the efficiency of electric fishing. I.G. Cowx and P. Lamarque, editors. Fishing with electricity. Fishing News Books, Oxford, UK.

Zar, J. H. 1999. Biostatistical analysis, 4th edition. Prentice-Hall, New Jersey.

Zippin, C. 1956. An evaluation of the removal method of estimating animal populations. Biometrics 12:163–189.

Zippin, C. 1958. The removal method of population estimation. Journal of Wildlife Management 22:82–90.

Appendix A: Field data sheet for backpack electrofishing

Columns: pass | time | segment | spp | age: YJA | # | length | weight | mark | M B O B G R S C C H E I H S S E I E E EP
                                                                              R S S   U L K F E O S B O C C L P G
                                                                                                      L R F

Minnows: CH RSS
Sculpins: MSC TSC
Suckers: BLS RBT
Salmonids: EBT MW

LND NPM PSC LSS CUT SPC

SPD SSC MNS BUL COH

M — mortality
BR — bruise
OS — orange slash
BS — black spot disease
G — green
R — ripe
S — spent
ECU — eroded caudal / upper
ECL — eroded caudal / lower
IHK — injury / hooking

IEF — injury / electrofishing
IO — injury / other
HS — hooking scar
SCB — scar / bird
SCO — scar / other
EIL — eye injury / left
EIR — eye injury / right
EPF — external parasites / fins
EPG — external parasites / gills

Notes:

H2O TEMP: _____ C@ __:__ (end) C@ __:__ (start) C@ __:__ (noon) C@ __:__ _____ mmhosmhos

INITIALS:

Page ___ of ___

ELECTROFISHING: BACKPACK AND DRIFT BOAT

Appendix B: Electrofishing mark–recapture datasheet

Mark - Recapture Electrofishing Form
WDFW- Ecological Interactions Team (YSIS, rev. 9/99)

STREAM: _____ SECTION: _____

DATE: ____/____/____ FLOW: L / M / H INITIALS: _____ PAGE: _____

TRIP TYPE: MARK / RECAP , LEFT / RIGHT WATER TEMP: _____ @ ____:____ _____ mmhos

#	SPP	FL mm	WT g	MARK CODE*	MARK TYPE**	SCALES	TAG NUMBER	REMARK CODES	COMMENT
1									
2									
3									
4									
5									
6									
7									
8									
9									
10									
11									
12									
13									
14									
15									
16									
17									
18									
19									
20									
21									
22									
23									
24									
25									

REMARK CODES >>

* MARK CODE: 0 = UNMARKED
                     1 = MARKED

** MARK TYPE:
- LC = LOWER CAUDAL
- UC = UPPER CAUDAL
- AN = ANAL
- LV = LEFT VENTRAL
- RV = RIGHT VENTRAL

// AT WORK-UP STATION,
// RECORD THE STATION
// NUMBER AND E.F. TIME
// IN COMMENTS AS:
// "sta # (####)"

- M = mortality
- BR = bruise
- OS = orange slash
- I_ = injury + EF(=electrofishing) / O(ther)
- SC_ = scar + B(ird) / O(ther)
- EI_ = eye injury + L(eft) / R(ight)
- EP_ = external parasites + F(ins) / G(ills)
- EC_ = eroded caudal + U(pper) / L(ower)
- HS_ = hooking scar + L(eft) / R(ight)
- G_ = green + M(ale) / F(emale)
- R_ = ripe + M(ale) / F(emale)
- S_ = spent + M(ale) / F(emale)

Hydroacoustics: Rivers
Suzanne L. Maxwell

Background and Objectives

Hydroacoustic methods are typically used to assess abundance in migrating fish populations when other methods are not feasible (i.e., the river is too wide for weirs or too turbid for observation towers). In many instances, hydroacoustic systems may be preferable to more intrusive devices such as nets or traps. This protocol addresses the use of hydroacoustic systems in rivers from fixed, nearshore positions, although down-looking, mobile methods have also been used to assess migrating fish populations (Xie et al. 2002).

Several types of sonars have been used to assess fish populations in rivers. The simplest was the single-beam sonar. In the early 1960s a single beam, echo counting system—a Bendix counter—was developed to enumerate adult sockeye salmon *Oncorhynchus nerka*. The Bendix counter has been an important tool for in-season management of predominantly sockeye salmon and chum salmon *O. keta* in many commercial fisheries in Alaska (Dunbar 2001, 2003; McKinley 2002; Westerman and Willette 2003; Dunbar and Pfisterer 2004). Dual-beam systems first provided target strength information for individual targets, allowing quantitative estimates of fish abundances using either echo counting or echo integration procedures in the late 1960s and early 1970s (Dragesund and Olsen 1965; Craig and Forbes 1969; Forbes and Nakken 1972). Dual-beam systems have largely been used for lake surveys to assess both adult and juvenile fish (Schael et al. 1995; Vondracek and Degan 1995) but have also been used at fixed, nearshore positions to assess migrating adult salmon in rivers (Gaudet 1990; Enzenhofer et al. 1998; Pfisterer and Maxwell 2000). Later, split-beam systems (see Figure 1) were shown to provide more accurate information on fish position—and thus a more accurate measure of target strength. Split-beam systems are currently used to enumerate migrating adult salmonids in several rivers (Daum and Osborne 1998; Miller and Burwen 2002; Xie et al. 2002). Most split-beam systems are used at sites where fish are relatively spread out and fish passage estimates are relatively low (i.e., less than 2,000 fish/h). Sockeye salmon runs tend to be large and concentrated in time, and split-beam sonars were found to do a poor job of assessing sockeye salmon at passage rates higher than 2,000 fish/h (Biosonics Inc. 1999a, 1999b). Echo-counting, dual-beam, and split-beam sonar methods, along with basic sonar information, are described in Brandt (1996) and MacLennan and Simmonds (1992).

A new sonar—a dual-frequency identification sonar (DIDSON) (see Figure 2)—was recently tested in Alaska and found to work well for enumerating fish at high passage rates (i.e., more than 2,000 fish/h) (Maxwell and Gove 2004). The DIDSON is a high-frequency, multibeam sonar with a unique acoustic lens system designed to focus the beam to create high-resolution images (Belcher et al. 2001, 2002). The evaluation of the DIDSON included comparisons of sockeye salmon counts from the DIDSON, Bendix sonar, and split-beam sonars against visual observations in a clear river; range tests using an artificial target acoustically similar in size to sockeye salmon in a highly turbid river and deployment of the DIDSON on rocky river bottoms and artificial substrates to observe fish behavior at these sites (Maxwell and Gove 2004).

Note: In this protocol, we recommend fixed-location (riverside) DIDSON hydroacoustic equipment, as it has proven to be a significant advance over other fixed-location hydroacoustic equipment used in rivers. Readers planning to utilize mobile (e.g., boat-mounted) hydroacoustic equipment are directed to the Hydroacoustic Protocol for Lakes and Reservoirs (pp. 153–172).

With any sonar type, there are many conditions that need to be met before a sonar system can be used. Practitioners attempting to employ any of the hydroacoustic methods should consider consulting an experienced acoustician to evaluate their particular application. For sonar to be successful in enumerating migrating fish the following conditions need to be met:

(1) Fish need to be actively migrating. If fish are regularly traveling back and forth across the beam, they will be counted multiple times.

(2) Fish must be traveling within the detection range of the sonar system, which needs to be tested at each site.

(3) The river bottom profile must be mostly linear with laminar current flow. With the DIDSON, it is possible to ensonify a region where the slope changes from steeper near shore to a flatter slope offshore. The DIDSON contains a background subtraction algorithm that will subtract out the static background and leave the moving targets visible. If the slope starts out flatter and then grows steeper, the sonar beam may reach the fish but will not be reflected back to the transducer.

(4) Either the species of interest is the only species present or an alternative technique is used to apportion the sonar numbers to species.

Common alternative techniques include drift gill netting (McKinley 2002; Pfisterer 2002) or fish wheels (Westerman and Willette 2003). Although the DIDSON is capable of providing length measurements of fish, because of beam spreading, the subcentimeter accuracy is only possible out to approximately 12 m using the DIDSON's high-frequency mode. To use length measures to determine fish species, the length frequency curves cannot be overlapping. An experiment with broadband acoustics was used successfully to determine sizes of free-swimming rainbow trout *O. mykiss* and Arctic charr *Salvelinus alpinus* (McKeever 1998).

FIGURE 1. — Split-beam system equipment.

FIGURE 2.—A dual-frequency identification sonar (DIDSON) system.

Rationale

Using hydroacoustic systems to enumerate fish passage is preferable in many instances to more intrusive devices. There is no handling or mortality associated with acoustic sampling. Unlike earlier fixed-location, single- and dual-beam sonar systems (Gaudet 1990; Mesiar et al. 1990), the split-beam technique provides three-dimensional positioning for each returning echo, along with electronic data on the direction of travel and specific location for each passing target (Ehrenberg and Torkelson 1996; Steig and Johnston 1996; Daum and Osborne 1998); however, when mixed species occur at sampling sites, hydroacoustic estimates will need to be apportioned to the array of species using gillnetting, beach seining, fish wheels, or other capture methods. These capture techniques are not used for abundance estimates at these sites because they are confounded by variable water levels and fish densities.

Objective

Use fixed-station hydroacoustic techniques to enumerate migrating fish in rivers to obtain abundance estimates or indices of abundance.

Sampling Design

Site Selection

Sampling sites are selected to optimize data collection and minimize problems with operations of the sonar. The primary criteria for selecting a riverine sonar site include single channel; sand, mud, or small gravel substrate; uniform, nonturbulent flow; linear bottom profile; downriver from known spawning areas; active fish migration past the site (no milling behavior); and upriver from tidal influences (Daum and Osborne 1998). For long-range sampling (i.e., greater than 50 m), the slope needs to be steep enough to fit an acoustic beam from near shore to the maximum range of fish passage.

Equipment Selection

The DIDSON technology is easy to use and provides higher resolution images compared to other systems. The DIDSON is available in two versions: a standard version with the dual frequencies 1.8 and 1.1 MHz and a long-range version with frequencies of 1.2 and 0.7 MHz. The standard version has a maximum range of 30 m and can be used only in applications where fish migrate within this range. Adult salmon have been detected on the long-range DIDSON out to 50 m. Split-beam sonars are capable of detecting fish at a range beyond 50 m from the transducer and are used to assess fish out to 250 m (Pfisterer 2002). The DIDSON's high frequency beam is divided into 96 0.3°–12° beams with range settings up to 12 m. The 1.1 MHz beam is divided into 48 0.6°–12° beams with range settings up to 40 m. If the DIDSON's multiple beams are positioned horizontally, the field of view is 29° for both versions, and the vertical beam is 12°. Although the large vertical beam may not fit the narrow river, the excess beam can be pushed into the river bottom. Split-beam systems are available in multiple elliptical and circular beam sizes. The river column must be measured to determine the beam that will fit best. Water fluctuations throughout the sampling period also need to be addressed prior to making a beam size selection. Depending on where in the river fish migrate, multiple transducers may be required to adequately ensonify the river profile. For split-beam systems, the beam cannot interact with the river bottom unless the substrate is nonreflective. Split-beam systems are also available in many frequencies. For fish enumerating, traditionally either 420 kHz or 200 kHz (Daum and Osborne 1998; Xie et al. 2002; Pfisterer 2002) have been used for fish assessment. The lower frequencies (200 kHz or lower) will be the most effective at detecting fish at longer ranges (beyond 50 m). Range limitations of sonar systems are dependent both on frequency and power of the unit.

From a statistical perspective, the sampled population is considered "open" and consists of fish passing a site over a fixed time period. Data are usually gathered throughout the entire fish migration to estimate the migrating population. Depending on fish density and human constraints, a daily subsampling schedule may be needed to provide in-season passage numbers in a timely manner. A stratified sample design is often used when each riverbank is sampled separately (Daum and Osborne 1998).

Designs for mobile down-looking equipment

More complex designs are often used in deep rivers where the targets of interest migrate off the bottom. In these cases additional sampling strata are added to account for the additional transducer aims (Xie et al. 2002). Although it is likely moving to a DIDSON-type system, the Pacific Salmon Commission has used a downward-looking single-beam echo sounder (Biosonics Model 105) to estimate the passage of migratory sockeye salmon and pink salmon *O. gorbuscha* in the Fraser River at Mission, British Columbia, since 1977 (Woodey 1987). This program consists of two components: transect and stationary soundings. These two operational modes are designed to provide data on two aspects of migratory salmon abundance: the transecting vessel collects information on fish density across the river; the stationary sounding acquires statistics on migration speed. The data from this program is processed using a duration-in-beam method (Thorne 1988), which leads to daily estimates of salmon movement past the survey site over a 24-h period. The statistical handling of the data has been refined and improved recently (Banneheka et al. 1995); however, the accuracy of the estimates depends on the validity of a number of assumptions. The following assumptions are the most important:

(1) All fish swim upstream along trajectories parallel to the riverbanks;

(2) The majority of the fish are distributed in the midwater portion of the water column with few migrating near the river surface, where they would not be ensonified, or near the river bottom, where target returns would not be discernible from the bottom echo;

(3) Few fish migrate in the shallow water, where the survey vessel is incapable of sampling;

(4) Fish swimming speed is negligible relative to the transecting speed of the survey vessel; and

(5) Fish behavior is not altered by the presence of the survey vessel.

The first four assumptions define fish behavior scenarios that are ideal for a downward-looking acoustic sampling of fish targets; the last is related to a fish–vessel interaction. Violation of these assumptions could result in bias in the daily estimates of abundance, as described in Banneheka et al. (1995); therefore, it is essential that these assumptions be assessed in the river using an independent measuring system (Xie et al. 2002).

Sampling Frequency and Replication

Many salmon species migrations have clumped distributions, which vary with species, site, and density. In these situations, whether using the DIDSON or split-beam sonar, subsampling may be required to assess accuracy and precision in counts due to the high-data loads produced with these methods. A viable option is initially (first year) to conduct random sample 24-h periods throughout the season and then subsample from these 24-h periods to determine the time period required to generate the accuracy and precision stated for the project. Estimates based on 10 min/h samples have proved useful for enumerating sockeye salmon from towers (Seibel 1967). Reynolds et al. (In press) also reference a 10 min/h systematic sampling design for salmon counts. This same subsampling framework could be applied to sonar methods.

Field/Office Methods

Setup

Preseason tasks

1. Map the river bottom in the selected area. This can be done with down-looking sonar coupled to a global positioning system (GPS). Or if the river is shallow, directly measure the depth from shore to shore at predetermined intervals (i.e., 1 m or less). A cross-river profile can then be plotted.

2. Determine species mixture at the site. Frequently, a netting program or alternative method is chosen to determine the extent of the migratory corridor and the types of species present. If mixed species are a concern,

an apportionment program also needs to be employed. Current sonar projects use drift gillnetting, beach seining, purse seining, or fish wheels to apportion the sonar estimates to species.

3. Select equipment (see page 146).

Events Sequence
1. Select overall site.
2. Create a bathymetry map of site.
3. Select equipment—type of sonar system, beam size, and so forth.
4. Choose best river-bottom profile.
5. Build mounts for the sonar systems.
6. Deploy and aim equipment.
7. Field calibrate the sonar.
8. Set up an organized structure for collecting and storing data.
9. Begin sampling.
10. Count fish from either the DIDSON images or split-beam echograms.
11. Expand counts for time not sampled.

Measurement Details
In this section, we offer basic measurement details for down-looking hydroacoustic efforts; the remainder of this section will detail the cross-channel aspects of the DIDSON and split-beam equipment.

Down-Looking Equipment
In the Mekong River (Lao PDR), hydroacoustic data were collected from hired local boats at a cruising speed varying between 3 and 5 knots, using a SIMRAD EY 500 scientific echo sounder (Kolding 2002). The split-beam transducer, model ES70-11, operating at 70 kHz, was mounted on a fixed structure on the fore-port side of the boats at a depth of 0.1–0.3 m below the water surface, and echo recording started 1 m from the transducer (i.e., 1.3 m below the surface).

In the Fraser River (British Columbia, Canada), the electronic components of the hydroacoustic system operated by Xie et al. (1997) and Xie et al. (2002) were set up inside a cabin of a boat that was tied to a log-boom piling. Acoustic transducers were mounted on a tripod deployed on the river bottom and linked to the echosounder via underwater cables. The transducer's aiming was controlled manually through a rotator controller and monitored with the underwater positioning sensor. They used the following hardware: Hydroacoustic Technology, Inc. (HTI; Seattle, Washington) Model 240 split-beam digital echo sounder; two acoustic transducers; HTI Model 340 split-beam digital echo processor; ASL 10-kHz Sounder Digitizer; HTI Model 660 remote rotator controller; PT 25 Dual Axis rotator; underwater positioning sensor and tripod; and a Communication Systems International Model GBX-8A differential GPS receiver.

DIDSON and split-beam systems
Methods to enumerate migrating fish are described below for both DIDSON and split-beam systems. In addition to collecting the passage estimates, several

environmental measures should be collected at the acoustic site. These measures include

1. daily water level—measured from a permanent marker so inter-year data can be compared,
2. daily turbidity (i.e., if turbidity affects signal propagation),
3. daily water temperature—input into sonar software to calculate range, and
4. reverberation level—set sonar to lowest possible threshold and record signal when no fish are present. This represents the background reverberation (see Figure 3).

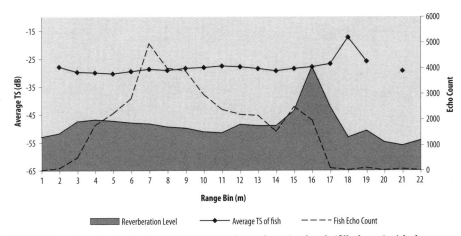

FIGURE 3.—Sample plot illustrating background reverberation levels (filled area) with the average target strength of fish plotted within 1 m range bins (solid line). The dashed line shows the number of echoes used to determine target strength at each range. The background reverberation level was collected with split-beam sonar at very low threshold levels (–150 decibels). The peak at 16 m is a combination of volume reverberation and reverberation from the river bottom.

DIDSON

Methods described below on how to use the DIDSON to enumerate migrating adult salmon are drawn predominately from Maxwell and Gove (2004).

Deployment and Aiming

For the DIDSON, an attitude sensor that provides absolute pitch angle should be attached to the DIDSON transducer prior to deployment. The attitude sensor should be aligned with the transducer and leveling the transducer onshore prior to deployment using a bubble level. Any change from level can be used to adjust the pitch data obtained from the sensor. The height of the transducer should be determined by plotting the beam on a plot of the river-bottom profile and surface and determining the best height and pitch that will reduce shadowing effects and cover the needed range requirements (see Figure 4). Maxwell and Gove (2004) mounted the lower edge of the transducer 36 cm from the river bottom and pitched –8.0° from level, matching the pitch of the bottom slope. This position facilitated coverage of the area directly in front of the DIDSON transducer and reduced shadowing from fish traveling close to the transducer. The DIDSON beam can be pushed into the river bottom if necessary, but the beam should not interact with the river's surface. The bottom subtraction algorithm built into the software works well to remove static bottom reflections from the image.

A weir (e.g., 5 × 10 cm mesh) is erected just downstream of the site to prevent fish from swimming behind and in the nearfield of the transducer. The weir should extend a minimum of 1 m beyond the face of the transducer and ideally be placed at a slight upriver angle to direct fish offshore more gradually. With the weir angled, fish are less likely to make a sharp turn around the weir in an attempt to return nearshore. Fish moving upstream and close to the shore would encounter the weir, be forced to move offshore, and then pass through the sonar beam. For best detection, passing fish should be perpendicular to the beam when ensonified.

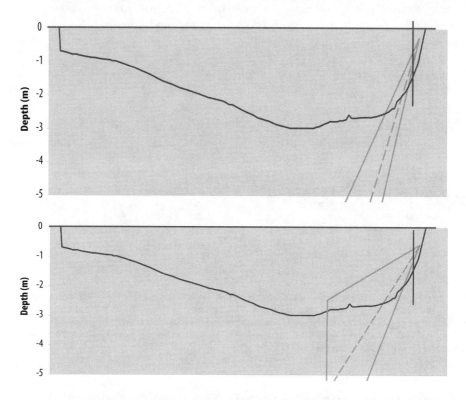

FIGURE 4.—Two extreme riverside transducer aims illustrate how to reduce shadowing effects (top) and increase range (bottom). The optimal beam geometry will be between these two positions.

Settings

The only settings on the DIDSON that affect data collection are the start range, end range, frame rate, and frequency. The frequency is automatically set based on the range settings but can be overridden as long as the ping requirements are still met. Using the minimum range setting to meet the sampling needs will result in the choice of frequency. The 1.8 MHz ranges out to 12 m, the 1.0 MHz ranges to approximately 40 m, and the 0.7 MHz ranges to approximately 90 m (with fish perpendicular to the beam). Higher frequencies (dependent on range) and higher frame rates (dependent on range and computer RAM and speed) will produce higher resolution images. Settings controllable on playback that do not affect data collection can be adjusted to maximize fish detection. Primary settings include threshold, intensity, and background subtraction.

Data Collection and Processing

Once the transducer is adequately aimed and settings are chosen, the current DIDSON software allows for subsampling. Subsampling can be set up for any portion of each hour. Sound Metrics (DIDSON Vendor) has added the option of subsampling hours or minutes within the day. (It has also created added the option of including an attitude sensor to the DIDSON, which can be retrofitted to an existing unit.) The sizes of DIDSON files depend on the frequency. The middle frequency produces files approximately 26 megabytes/min, with the higher frequency being twice that and the lower frequency roughly half that. These files can be saved directly to external drives to avoid one data transfer step.

Once saved, the files can be replayed using the same DIDSON data acquisition software. Technicians manually count upstream traveling fish on one hand using a counter and downstream fish on the other. Files can be sped up or slowed depending on passage rates. Cross-river range distributions can be obtained by mouse clicking on fish (with the fish count function checked) as the fish cross the image. The clicked positions are saved to a file that includes the fish number, time, range, length, and thickness of the image (length and thickness are only produced if the software is able to recognize the fish image). To date, the automated fish counter included in the DIDSON software package is not effective, largely because of shadowing effects. Work is in progress to solve the auto-tracking issues. Counts of fish images can be input onto a Microsoft Excel worksheet.

Split-beam

Recommended split-beam methods for enumerating migrating salmon are drawn from Daum and Osborne (1998) and Xie et al. (2002).

Deployment and aiming

To enumerate migrating fish at a riverine site, a transducer is placed on either bank of the river and positioned near the shore. A weir is erected downstream of the site to prevent fish from swimming behind and in the nearfield of the transducer. Ideally, the weir should extend beyond the range of the transducers nearfield (dependent on both frequency and beam size). The transducer should be deployed as low to the river bottom as possible (a boot width from the bottom to the lower edge of the transducer). The beam is then aimed along the river bottom matching the slope. Prior to aiming, a profile should be created or extracted from bathymetry data. The river bottom profile and surface should be plotted and sample beams drawn to determine the optimal aim and beam size. Once this has been completed, the aim is fine-tuned using the techniques on page 146. If the bottom substrate is fairly reflective, the beam will need to be raised until minimal or no bottom reflections are received at the sampling threshold level (see Figure 5). If the bottom is absorptive, as is common, the beam is aimed into the river bottom, positioning the axis of the beam just above bottom. According to Daum and Osborne (1998), on the Chandalar River, where the dominant species is chum salmon, precise aiming is critical because most of the fish travel close to the bottom. Once an aim has been selected, real-time electronic echograms can be monitored to ensure that the aim remains constant. Any changes in the nonfish reverberation or vertical position of passing fish should alert the operator to a change in the aim. The transducer should be reaimed if this occurs. Also, monitoring the real-time target strength of passing fish can detect a change in aim (i.e., the same size fish become acoustically smaller due to orientation changes in the beam).

To aim the split-beam transducer
1. Measure
   a. Distance from the river bottom to the bottom of the transducer.
   b. Distance from river bottom to water's surface at the transducer.
   c. Distance from transducer to shore.
   d. Distance from transducer to the end of the weir.
2. Start aiming process with this aim obtained from the beam profile plots.
3. Wrap a salmon-size target (10-cm-diameter sphere partially filled with copper BBs) in a mesh bag using a 23 kg or heavier monofilament line. Tie a loop on the end of the line, far enough up so that the knot will be above water level when the target is near the river bottom. (Note: this target has been shown to be the approximate acoustic size of a sockeye or chum salmon. If other species are being assessed, find a new acoustic target that returns an echo that is similar in strength to the fish.)
4. Attach the salmon-size target to an extension pole and extend it in front of the transducer beyond the nearfield (1 m for a 6° × 10° 201 kHz split-beam sonar). (Note: a loop can be tied on the end of the line to the extension pole; then, the target's loop can be drawn through the pole's loop, making it easier to remove and add targets.)
5. Position the target so that a line drawn from the transducer mount to the target would perpendicularly bisect a line parallel to the river's current; then lower the target to approximately 10 cm off the river bottom.
6. Adjust the aim of the split-beam transducer until the target appears in the center of the beam horizontally and in the central portion of the lower half of the vertical beam. If the river bottom consists of a hard substrate, the transducer beam may have to be raised so that the target rests closer to the lower edge of the beam. If the river bottom is soft, the transducer may be lowered slightly, moving the target closer to the central axis of the beam. Use the Alt + Print Screen command to copy a picture showing the position of the target in the two-dimensional graphs of HTI's Department of Environmental Protection program, and then paste it either to a drawing program or PowerPoint presentation to document the aim. (Note: if fish targets are present, it may be necessary to raise and lower the target until the operator is assured that the echoes are coming from the target.)
7. Use the target to check the aim of the beam at several ranges. If the target is not visible at multiple ranges, it may be necessary to use a multiple transducer system.

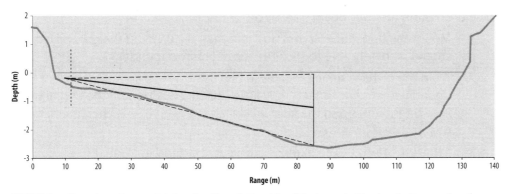

FIGURE 5.—Transducer beam plot showing the relationship of the beam to the river bottom and surface. In this situation, the beam is lightly touching the river bottom.

If the current is strong, migrating fish will most likely be found near the river bottom. There are situations in which fish rise up off the bottom and pass over the beam. In these cases, a more complex sampling structure is needed. This can be achieved by either creating multiple aims or adding additional transducers to ensonify more of the water column. Xie et al. (2002) used multiple aims to adequately ensonify the regions of the water column utilized by migrating fish.

Settings
In Daum and Osborne (1998), the HTI split-beam sonar settings used for sampling chum salmon passage were pulse width 0.10–0.38 ms; on-axis threshold –40 decibels (dB) (10 dB lower than the predicted target strength estimate [Love 1977] for the smallest chum salmon in the Chandalar River); horizontal off-axis criteria set to the half-power beam width; and vertical off-axis criteria increased beyond the half-power beam width so that echoes from fish traveling close to bottom would be accepted; maximum range was 11–16 m on one bank and 65–78 m on the opposite bank.

In Maxwell and Gove (2004), a Biosonics' 201 kHz, 6.4o circular, split-beam transducer and attitude sensor were deployed in water 36 cm deep and 12 cm above the river bottom (to the lower edge of the transducer), and pitched –4.4° from level to sample sockeye salmon. Other settings included 17.2 pings/s; 0.2 mS transmit pulse width; –50 dB data collection and editing threshold; 1.0–8.5 m range; and single target criteria, including a –50 dB target threshold, 0.02–0.6 pulse width acceptance measured 6 dB below the pulse peak, 10 dB maximum beam compensation, and 3 dB maximum standard deviation of the alongship and athwartship angles. A sound speed of 1,443 m/s (Del Grosso and Mader 1972) and absorption coefficient of 0.013068 dB/m (Francois and Garrison 1982) were calculated using a measured water temperature of 9°C.

Field Calibration
Daum and Osborne (1998) had the sonar system calibrated in a laboratory based on the comparison method (Urick 1983). Field calibrations were performed three times during the season with a calibration sphere (38.1 mm tungsten carbide sphere) to ensure that the systems' electronics were functioning properly. Other good calibration techniques are found in Robinson and Hood (1983), Foote (1990), and Simmonds (1990).

To field-calibrate the split-beam transducer

1. Mount the transducer so it is no more than 8–10 cm off the ground (you should be barely able to stick the toe of your boot under it).
2. Wrap the carbide sphere in a mesh bag using 10–12 kg monofilament line (monofilament line reflects very little signal). Tie a loop on the end of the line, far enough up so the knot will be above water level when the target is near the river bottom.
3. Attach the target to an extension pole and extend in front of the transducer just beyond the nearfield (1 m for a 6 × 10° 201 kHz split-beam sonar), lowering it to approximately midway between the river's surface and bottom to avoid reverberation interference from either surface. (Note: a loop can be tied on the end of the line to the extension pole, and then the target's loop can be drawn through the pole's loop, making it easier to remove and add targets.)
4. Position the transducer beam so the target is centered both vertically and horizontally.
5. Set the sonar parameters as you would for sampling, except the threshold should be set as low as possible. Collect 1,000 pings or more from the target. (Note: if fish targets are present, it may be necessary to raise and lower the target until the operator is assured that the echoes are coming from the target.)
6. Determine the average target strength of the target and compare to the laboratory calibration. Adjust the calibration parameters if necessary by changing the system gain. Document the target file name, the sonar parameters, and the average target strength in a logbook.

Sampling Processing/Data Analysis

Split-beam sonar data was displayed, auto-tracked, edited, and exported using SonarData's Echoview software with the integrated Blackman auto-tracking algorithm (Maxwell and Gove 2004). The split-beam sonar counts were obtained by visually counting the echogram traces (by manual count) and by auto-tracking using the Blackman algorithm. Other researchers use the HTI Trackman to mark and store tracked fish data from electronically produced echograms.

A second method of obtaining counts is to print charts (Pfisterer 2002) and mark and count individual fish traces. The counts can then be transferred to spreadsheets or a database.

There should be enough personnel to produce daily counts during the field season. Following the field season, the postprocessing responsibilities include assessing ancillary data such as cross-river distributions, fish swimming speed, and examinations of overall aspects of fish movements.

Personnel Requirements and Training

Responsibilities

Project leaders will be responsible for

1. purchasing and assembling needed sonar equipment.

2. deploying and aiming the sonar.
3. verifying the aim periodically through the season (split-beam) and conducting real-time monitoring of fish to detect changes in the vertical distribution of fish or the abrupt change in target strength of passing fish when species composition has not changed.
4. calculating thresholds (split-beam).
5. selecting optimal sampling configurations for the specific riverine environment (split-beam).
6. setting up the configuration files (split-beam).
7. developing a series of diagnostic tests to ensure that the sonar is functioning correctly (split-beam).
    a. performing threshold tests,
    b. monitoring signal from each channel separately, and
    c. testing the gain.
8. training technicians.
9. redeploying and reaiming if environment conditions require it (split-beam).

Technicians will be responsible for
1. camp logistics (if remote site),
2. setting up basic equipment,
3. hauling supplies to sampling site,
4. learning to aim, set up, and deploy the DIDSON,
5. redeploying and aiming if necessary (DIDSON only),
6. doing manual fish counts from either DIDSON images or split-beam echograms, and
7. recording counts on spreadsheets.

Qualifications

For DIDSON projects, project leaders should have a basic understanding of sonar principles. They can work with the vendor to obtain needed training to operate the DIDSON. Technicians should have experience in operating computers in a Windows environment and saving and transferring files.

For split-beam sonar projects, the project leader needs a more extensive background in sonar principles and a basic understanding in how the sampling parameters affect the data being collected. Technicians should have computer experience and some knowledge of sonar principles and the ability to quickly learn new software programs.

Use of both technologies requires a project leader with experience in budgeting, purchasing, project planning, data analysis, and report writing.

Training

Project leaders
We recommend classes or reading on basic sonar principles, training at multiple sites to understand how the sonar parameters affect the data collection, and extensive training on transducer aiming for split-beam operations. Several vendors have classes on sonar principles. Some classes may also be available through local colleges or universities.

Technicians
We recommend training, which is traditionally done on site. A few hours of training are usually necessary to become familiar with DIDSON operations. Once technicians are familiar with the setup, they may be allowed to move and redeploy and reaim if necessary. If they are to set up a split-beam site, extensive training is required. It is not recommended that they move or redeploy the sonar without oversight by trained personnel. Aiming is critical and if off, the estimates will become completely invalid. Training on how to select fish from nonfish items also needs to be covered. If printed charts are used, trained personnel should check their progress until they feel comfortable that the technician is able to distinguish fish from nonfish.

Operational Requirements

Workload and Field Schedule
Three to four technicians and one project leader are needed to operate a two-bank sonar operation for either system. The project leader will either need to be on-site or quickly available for split-beam operations. In addition, the site should be checked periodically to ensure that the system is operating correctly.

Equipment Needs
1. Sonar unit (either DIDSON or split-beam sonar)
2. Automated rotators if needed (automated tilt is very useful)
3. Attitude sensor (this is a must; without it aiming requires much more guesswork)
4. Controller computer for primary operations
5. Data processing computer (may be controller computer if budgets are tight but a backup is essential)
6. External drives or other means to store data
7. Permanent data storage devices (DVD writers plus disks)
8. Power; if site is remote, generator, solar panels, water turbines, etc., may be utilized singly or in conjunction with one another
9. Calibration equipment (tungsten carbide spheres)

Budget Considerations
A sample budget using 2003 costs in U.S. dollars is included below for the DIDSON system. The split-beam system would be the same with the exception of the

unit cost. Today a split-beam costs approximately $40,000. The costs listed are approximations and will change over time. The DIDSON, although more costly for the equipment, will be cheaper to operate as more experienced staff are only needed for the initial deployment.

Estimated costs of DIDSON and accessory equipment

Item	Estimated cost per unit
Sonar equipment	
DIDSON sonar w/cables	$80,000
Rotator and control	$15,000
Attitude sensor	$4,000
Transducer mount	$500
Laptop computer	$3,000
Misc. cables, USB, ethernet, etc.	$200
Electronic storage devices	$1,000
Power needs/per river bank	
Combination of small Honda generator, solar panels, batteries, and chargers	$5,000
Total	$108,700

Literature Cited

Banneheka, S. G., R. D. Routledge, I. C. Guthrie, and J. C. Woodey. 1995. Estimation of in-river fish passage using a combination of transect and stationary hydroacoustic sampling. Canadian Journal of Fisheries and Aquatic Sciences 52:335–343.

Belcher, E. O., B. Matsuyama, and G. M. Trimble. 2001. Object identification with acoustic lenses. Pages 6–11 in Conference proceedings MTS/IEEE oceans, volume 1, session 1. Honolulu, Hawaii.

Belcher, E. O., W. Hanot, and J. Burch. 2002. Dual-frequency identification sonar. Pages 187–192 in Proceedings of the 2002 International Symposium on underwater technology. Tokyo, Japan.

Biosonics Inc. 1999a. Alaska statewide sonar project results of 1998 field demonstrations, final report. Report to the Alaska Department of Fish and Game, Juneau.

Biosonics Inc. 1999b. Alaska statewide sonar project results of 1999 field demonstrations, final report. Report to the Alaska Department of Fish and Game, Juneau.

Brandt, S. B. 1996. Acoustic assessment of fish abundance and distribution. Pages 385–432 in B. R. Murphy and D. W. Willis, editors. Fisheries techniques, 2nd edition. American Fisheries Society, Bethesda, Maryland.

Craig, R. E., and S. T. Forbes. 1969. A sonar for fish counting. Fiskeridirektoratets Skrifter Serie Havundersokelser 15:210–219.

Daum, D. W., and B. M. Osborne. 1998. Use of fixed-location, split-beam sonar to describe temporal and spatial patterns of adult fall chum salmon migration in the Chandalar River, Alaska. North American Journal of Fisheries Management 18:477–486.

Del Grosso, V. A., and C. W. Mader. 1972. Speed of sound in pure water. Journal of the Acoustical Society of America 52:1442–1446.

Dragensund, O., and S. Olsen. 1965. On the possibility of estimating year-class strength by measuring echo-abundance of 0-group fish. Fiskeridirektoratets Skrifter Serie Havundersokelser 13:47–75.

Dunbar, R. 2001. Copper River hydroacoustic salmon enumeration studies, 2000 and 2001. Alaska Department of Fish and Game, Commercial Fisheries Division, Regional Information Report No. 2A01-3, Anchorage.

Dunbar, R. 2003. Anvik River sonar chum salmon escapement study, 2003. Alaska Department of Fish and Game, Division of Commercial Fisheries, Regional Information Report No. 3A03-14, Anchorage.

Dunbar, R., and C. T. Pfisterer. 2004. Sonar estimation of fall chum salmon abundance in the Sheenjek River, 2002. Alaska Department of Fish and Game, Division of Commercial Fisheries, Regional Information Report No. 3A04-10, Anchorage.

Enzenhofer, H. J., N. Olsen, and T. J. Mulligan. 1998. Fixed-location hydroacoustics as a method of enumerating migrating adult Pacific salmon: comparison of split-beam acoustics vs. visual counting. Aquatic Living Resource 11(2):61–74.

Ehrenberg, J. E., and T. C. Torkelson. 1996. Application of dual-beam and split-beam target tracking in fisheries acoustics. ICES Journal of Marine Science 53(2):329–334.

Faran, J. J., Jr. 1951. Sound scattering by solid cylinders and spheres. Journal of the Acoustical Society of America 23:405–418.

Foote, K. G. 1990. Spheres for calibrating an eleven-frequency acoustic measurement system. Journal du Conseil International pour l'Exploration de la mer 46:284–286.

Forbes, S. T., and O. Nakken, editors. 1972. Manual of methods for fisheries resource survey and appraisal, part 2: the use of acoustic instruments for fish detection and abundance estimation. FAO Manual in Fisheries Science 5, Rome.

Francois, R. E., and G. R. Garrison. 1982. Sound absorption based on measurements. Part II: Boric acid contribution and equation for total absorption. Journal of the Acoustical Society of America 72:879–890.

Gaudet, D. M. 1990. Enumeration of migrating salmon populations using fixed-location sonar counters. Rapports et Proces-Verbaux des Reunions, Conseil International pour l'Exploration de la Mer 189:197–209.

Kolding, J. 2002. The use of hydro acoustic surveys for the monitoring of fish abundance in the deep pools and fish conservation zones in the Mekong River, Siphandone area, Champassak Province, Lao PDR. Living Aquatic Resources Research Center, LAAReC Technical Report No.0009 (ISSN 1608-5612), Vintiane, Lao, PDR.

Love, R. H. 1977. Target strength of an individual fish at any aspect. Journal of the Acoustical Society of America 62:1397–1403.

MacLennan, D. N., and J. E. Simmonds. 1992. Fisheries acoustics. Chapman and Hall, London.

Maxwell, S. L., and N. E. Gove. 2004. The feasibility of estimating migrating salmon passage rates in turbid rivers using dual frequency identification sonar (DIDSON) 2002. Alaska Department of Fish and Game Regional Information Report No. 2A04-05, Anchorage.

McKeever, T. J. 1998. The estimation of individual fish size using broadband acoustics with free-swimming salmonids *(Oncorhynchus mykiss, Salvelinus alphinus)*. MS thesis. Memorial University of Newfoundland, St. John's, Newfoundland, Canada.

McKinley, W. L. 2002. Sonar enumeration of Pacific salmon into the Nushagak River, 2001. Alaska Department of Fish and Game, Commercial Fisheries Division, Regional Information Report No. 2A02-25, Anchorage.

Mesiar, D.C., Eggers, D.N., and Gaudet, D.M. 1990. Development of techniques for the application of hydroacoustics to counting migratory fish in large rivers. Rapports et Proces Verbaux des Reunions, Conseil International pour l'Exploration de la Mer 189:223-232.

Miller, J. D., and D. Burwen. 2002. Estimate of chinook salmon abundance in the Kenai River using split-beam sonar, 2000. Alaska Department of Fish and Game, Sport Fish Division, Fishery Data Series No. 02-09, Anchorage.

Pfisterer, C. T. 2002. Estimation of Yukon River salmon passage in 2001 using hydroacoustic methodologies. Alaska Department of Fish and Game, Division of Commercial Fisheries, Regional Information Report No. 3A02-24, Anchorage.

Pfisterer, C. T., and S. L. Maxwell. 2000. Yukon River sonar project report, 1999. Alaska Department of Fish and Game, Division of Commercial Fisheries, Regional Information Report No. 3A00-11, Anchorage.

Reynolds, J. H., C. A. Woody, N. E. Gove, and L. F. Fair. 2007. Efficiently estimating salmon escapement uncertainty using systematically sampled data. In C. A. Woody, editor. Sockeye salmon evolution, ecology, and management. American Fisheries Society, Symposium 54, Bethesda, Maryland.

Robinson, B. J., and C. Hood. 1983. A procedure for calibrating acoustic survey systems with estimates of obtainable precision and accuracy. FAO (Food and Agriculture Organization of the United Nations) Fisheries Report 300:59–62.

Schael, D., J. A. Rice, and D. J. Degan. 1995. Spatial and temporal distribution of threadfin shad in a southeastern reservoir. Transactions of the American Fisheries Society 124:804–812.

Seibel, M. C. 1967. The use of expanded ten-minute counts as estimates of hourly salmon migration past counting towers on Alaskan rivers. Alaska Department of Fish and Game, Commercial Fisheries Division, Informational Leaflet 101, Juneau.

Simmonds, E. J. 1990. Very accurate calibration of a vertical echo sounder: a five–year assessment of performance and accuracy. Rapports et Proces-Verbaux des Reunions Conseil International pour l'Exploration de la mer 189:183–191.

Steig, T. W., and S. V. Johnston. 1996. Monitoring fish movement patterns in a reservoir using horizontally scanning split-beam. ICES Journal of Marine Science 53:435–441.

Thorne, R. E. 1988. An empirical evaluation of the duration-in-beam technique for hydroacoustic estimation. Canadian Journal of Fisheries and Aquatic Sciences 45:1244–1248.

Urick, R. J. 1983. Principles of underwater sound. 3rd edition. Peninsula Publishing, Los Altos, California.

Vondracek, B., and D. J. Degan. 1995. Among and within transect variability in estimates of shad abundance made with hydroacoustics. North American Journal of Fisheries Management 15(4):933–939.

Westerman, D. L., and T. M. Willette. 2003. Upper Cook Inlet salmon escapement studies 2003. Alaska Department of Fish and Game, Commercial Fisheries Division, Regional Information Report No. 2A04-03, Anchorage.

Woodey, J. C. 1987. In-season management of Fraser River sockeye salmon *(Oncorhynchus nerka)*: meeting multiple objectives. Pages 367–374 in H. D. Smith, L. Margolis, and C. C. Wood, editors. Sockeye salmon *(Oncorhynchus nerka)* population biology and future management. Canadian Special Publication Fisheries and Aquatic Sciences 96.

Xie, Y., G. Cronkite, and T. J. Mulligan. 1997. A split beam echosounder perspective on migratory salmon in the Fraser River: a progress report on the split-beam experiment at Mission, B.C., in 1995. Pacific Salmon Commission Technical Report No. 8, Vancouver.

Xie, Y., T. J. Mulligan, G. M. W. Cronkite, and A. P. Gray. 2002. Assessment of potential bias in hydroacoustic estimating of Frasier River sockeye and pink salmon at Mission, B.C. Pacific Salmon Commission Technical Report No. 11, Vancouver.

Appendix A: Hydroacoustics: Rivers Data Sheet - Sonar Field Calibration

Project _____ Date _____
Location _____ Depth _____

Sphere _____   TS -39.5dB

Temperature _____   alpha _____

Salinity _____   Sound Velocity _____

15 Degree Circular Transducer
Calibration File: _____ G_140 _____ G_12

Acquisition File: _____

Observed TS: _____ Adjust Gains: Y / N G_140 _____ G_12

Acquisition File: _____

Notes: _____
_____
_____

4x10 Elliptical Transducer
Calibration File: _____ G_140 _____ G_12

Acquisition File: _____

Observed TS: _____ Adjust Gains Y / N G_140 _____ G_12

Acquisition File: _____

Notes: _____
_____
_____

Hydroacoustics: Lakes and Reservoirs
J. Christopher Taylor and Suzanne L. Maxwell

Background
Fisheries hydroacoustics uses transmitted sound to detect fish. Sound is transmitted as a pulse and travels quickly and efficiently through water. As the sound pulse travels through water it encounters objects that are of different density than the surrounding medium, such as fish, that reflect sound back toward the sound source. These echoes provide information on fish size, location, and abundance. The basic components of the acoustic hardware and software function to transmit the sound, receive, filter and amplify, record, and characterize the echoes. While there are many manufacturers of commercially available "fishfinders," quantitative hydroacoustic analyses require that measurements are made with scientific-quality echo sounders that have high signal-to-noise ratios and are easily calibrated.

Over the past three decades, vertical or down-looking hydroacoustics has become increasingly important to the assessment of anadromous and land-locked salmonids (Thorne 1971, 1979; Burczynski and Johnson 1986; Mulligan and Kieser 1986; Levy et al. 1991; Yule 1992; Parkinson et al. 1994; Beauchamp et al. 1997; Wanzenbock et al. 2003), and lake and reservoir fishes (Thorne 1983; Brandt et al. 1991; Degan and Wilson 1995; Schael et al. 1995; Vondracek and Degan 1995; Cyterski et al. 2003; Taylor et al. 2005). Hydroacoustics provide a repeatable, noninvasive method of collecting high-resolution (submeter scale), continuous data along transects in three dimensions (MacLennan and Simmonds 1992). MacLennan and Simmonds (1992) as well as Brandt (1996) give a thorough introduction in the use of hydroacoustics for measuring fish abundances and distributions.

The density and distribution of lake, reservoir and lowland river fishes varies by season and time of day and is influenced by a range of abiotic, biotic and behavioral factors such as temperature, oxygen concentration, and vertical distribution of predators and prey (Lucas et al. 2002). Schools of sockeye salmon *Oncorhynchus nerka* occurring in lakes and reservoirs disperse in midwater at night (Johnson and Burczynski 1985; Clark and Levy 1988; Parkinson et al. 1994; Beauchamp et al. 1997). Likewise, forage fishes occur in patches, typically aggregated during the day and more dispersed at night (Appenzeller and Leggett 1992; Schael et al. 1995). Under these dispersed or disaggregated distribution patterns, densities can be acoustically estimated using vertically oriented transducers as long as the fishes are a sufficient distance from the surface to permit detection.

In earlier studies, surveys were conducted with a single downward-oriented transducer. Unfortunately, acoustic estimates of surface-oriented fish gathered by down-looking transducers can be biased and lack precision because of limited sample volume near the apex of the cone (Burczynski and Johnson 1986). This limitation is problematic when assessing species known to be surface-oriented, such as rainbow trout *O. mykiss* (Wurtsbaugh et al. 1975; Stables and Thomas 1992; Warner and Quinn 1995) and cutthroat trout *O. clarkii* (Nilsson and Northcote 1981; Beauchamp et al. 1997; Knudsen and Saegrov 2002; Baldwin et al., in press). Several studies have demonstrated that this limitation can be overcome by sampling with a horizontally aimed (side-looking) transducer (Johnston 1981; Kubecka et al. 1992; Kubecka and Duncan 1994; Tarbox and Thorne 1996; Hughes 1998; Kubecka and Wittingerova 1998; Lyons 1998); however, other factors can limit the effectiveness of side-looking transducers.

Acoustic technology has become increasingly sophisticated, making synoptic down- and side-looking hydroacoustic assessments viable. The development of narrow-beam transducers with negligible side lobes allows depths between 1.5 and 5.0 m to be sampled with horizontal sonar (Kubecka 1996). The ability of split-beam transducers to measure angular locations of echoes in the ensonified volume has also improved measurements of in situ target strengths (Foote et al. 1986; Traynor and Ehrenberg 1990). Target tracking, or the assemblage of multiple echoes from a single scatter into an ensemble, has led to lower variance estimates of target strength and improved ability to resolve returns from single and multiple targets (Ehrenberg and Torkelson 1995). Finally, the advent of fast multiplexing, or alternating ping transmission between two or more transducers controlled by a single Echosounder, now allows near simultaneous data collection with multiple transducers (Thorne et al. 1992).

General equations relating target strength (measured in decibels [dB]) to total length have been developed for fish in dorsal aspect (Love 1971, 1977; McCartney and Stubbs 1971; MacLennan and Simmonds 1992; Brandt 1996). These equations are often used to convert mean target strengths to mean fish lengths, assuming that most fish are oriented dorsal-ventrally when sampled. Horizontal acoustic measurements of target strength in limnetic environments are less useful because there is no way to determine the orientation of the fish relative to the axis of the acoustic beam. The relationship between target strength and horizontal aspect has been studied under laboratory conditions, and equations relating fish lengths to target strengths in side aspect (Dahl and Mathisen 1983) and random orientation (Love 1977; Kubecka 1994; Kubecka and Duncan 1998) have been developed. Although these equations exist, few researchers have applied these algorithms to compare in situ measurements of fish length from horizontal beaming to measurements collected with an active sampling gear such as a purse seine.

For hydroacoustic assessments to gain wider acceptance from decision makers, it is important to show that sonar data can be corroborated with density, biomass or relative abundance data collected with an active sampling gear, such as purse seines (Yule 2000; Taylor et al. 2005) or a midwater trawl (Burczynski and Johnson 1986), electrofishing (e.g. Kubecka et al. 2000), or angler surveys (Frear 2002). Used in conjunction with hydroacoustics, these gears verify the species composition and sizes of fish in lake and reservoirs. Purse seining is effective at determining open-water species composition, developing length–frequency distributions, and measuring relative abundance of populations (Whitworth 1986). But seines and trawls only sample a small portion of the total surface area, and spatial heterogeneity in fish distributions can lead to high variation in catches.

Rationale

The ability to "see" and count what is under the surface of the water without disturbing the habitat or the fish is a key advantage of hydroacoustics. Hydroacoustics can sample the entire water column quickly, and detailed maps of fish densities and mean sizes can be obtained over large bodies of water. As more area is encompassed by a sample, many of the sampling problems created by the spatial patchiness of fish distribution are alleviated. Thus, there tends to be less variation in density estimates across acoustic transects compared to purse-seine hauls or other gear types. Also, the frequency band used in scientific sonars (typically 38 to 200 kHz) is not detectable by most fishes (except see Mann et al. 2001; Gregory and Clabburn 2003). There remain limitations in the type of data that can be collected using hydroacoustics. Currently, single frequency hydroacoustics cannot identify the target species, though broadband and multifrequency sonar systems are showing promise in discerning species in low-diversity systems (Fernandes et al. 2003). Side-looking mobile hydroacoustics cannot discern modes in length–frequency distributions unless large differences in length classes exist. When these limitations are recognized, hydroacoustic sampling efforts are cost effective, as estimates from creel surveys are expensive and labor intensive, and the estimates developed from catch-per-unit-effort (CPEU) measures are not necessarily directly proportional to fish density (Hubert 1996:158–159; Yule 2000). When used in concert with purse seining or other active sampling gears, hydroacoustics can provide a comprehensive survey method capable of providing valuable information on target size, population densities, and spatial distribution. Additional aspects of the strengths and limitations of acoustic surveys can be found in MacLennan and Simmonds (1992) and Brandt (1996).

Objectives

There are several levels of information that can be obtained from a mobile hydroacoustic survey in lakes, inland reservoirs, or lowland rivers. These levels range from simple species or object detection (presence/absence) to spatial (or temporal) distribution of individuals or groups (densities) to systemwide biomass estimates for the target species or guild. Care should be taken to clearly identify the objectives of the study to optimize a sample design in terms of timing of sample, staff hours of effort and data, and analytical methods that will be required to address the objectives. Below are examples of objectives for mobile hydroacoustic surveys, followed by examples of prior studies that have addressed similar objectives using mobile hydroacoustic surveys (see pages 159–60)

1. to determine spatial and temporal fish distribution in a water body;
2. to obtain density estimates for either adult or juvenile fish in lakes, reservoirs, or lowland rivers using down-looking or a combination of down- and side-looking hydroacoustic methods; and
3. to estimate systemwide fish biomass (e.g., forage fish) when hydroacoustics are combined with other sampling techniques

Events Sequence

Given the objectives of the study, a sequence of events is followed in order to optimize the sampling program:

1. Select a lake or river, for example, where fish estimates are needed;
2. Determine the level of information required for the study (e.g., presence/absence or biomass estimation);
3. Create or obtain a shoreline and bathymetric map of the lake or rivers;
4. Establish a spatial sampling design based on prior knowledge of target species distribution or statistical considerations;

5. Determine the best timing for the sampling based on diurnal or seasonal behavior of the species;
6. Determine the best down- and side-looking transducers deployment on either a boat mount or towed platform based on prior knowledge of vertical distribution of the species;
7. Determine the optimal acoustic parameters for sampling based on water conditions, target size, or other acoustical properties;
8. Perform an in situ calibration of the acoustic system using an object of known target strength and known location;
9. Perform the hydroacoustic survey;
10. Select software processing tools and analytical methods dictated by the objectives final output;
11. Perform quality checks on the data; and
12. Process data.

Sampling Design

Site Selection and Timing

Selection criteria for hydroacoustic sampling of lakes or reservoirs include sufficient water depth and known species composition. If the lake contains predominately one species, or if the target species can be distinguished from other species by depth or other spatial properties (e.g., littoral versus limnetic), a hydroacoustic survey can stand alone. If mixed species are present, an alternate method is needed to apportion the hydroacoustic estimates into individual density estimates for each species. Possible apportionment methods include purse seines, towed nets, and gill nets (Cyterski et al. 2003).

Transect sampling designs can include single paths following the main channel of a lake or reservoir, a single transect that zigzags from shore to shore, or several parallel transects that run perpendicular to the axis of the water body (see Figure 1; Yule 2000; Jolly and Hampton 1990). Using any of these transect designs results in hydroacoustic data that are typically auto-correlated (Schael et al. 1995; Vondracek and Degan 1995; Taylor et al. 2005). Abundance estimates are calculated from these data by extrapolating from blocked averages of depth-integrated (two-dimensional) data (Vondracek and Degan 1995) or are modeled using spatially explicit techniques such as geostatistics (e.g., Taylor et al. 2005). Typically, the transect design will dictate the analytical methods (or vice versa) that are used to assess the distribution pattern of fish populations.

HYDROACOUSTICS: LAKES AND RESERVOIRS

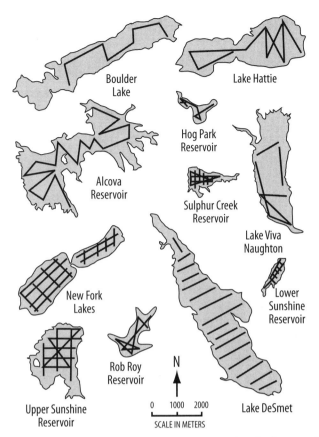

FIGURE 1.—Map of 11 study waters showing various hydroacoustic transect designs (from Yule 2000).

Objective 1: Characterizing spatial distribution
Schael et al. (1995) provide a description of evaluating patchiness in the distribution of shad in a reservoir. They used a patch recognition algorithm (Nero and Magnuson 1989) to analyze echo-integrated hydroacoustic data to define patches and patch characteristics (e.g., numbers, density, area, mean depth) for shad in Lake Norman, North Carolina. Their transects were 2.5 km long and 0.2 km apart, and extended across the lower main basins of the reservoir. During most surveys, they observed 12–16 patches/km with fish densities exceeding twice the average background density, and 1–2 patches/km with fish densities 50 times the average background density.

Objective 2: Obtaining density estimates of a fish population
Vondracek and Degan (1995) provide a thorough evaluation of among- and within-transect variability in estimates for shad populations in Lake Texoma, Texas–Oklahoma. They found that the within-transect variation was significantly higher during the day than at night. Coefficient of variation values decreased nonlinearly with increasing blocking intervals for day and night surveys; estimated values of 20% were achievable at interval lengths of about 150 m at night, whereas during the day, the minimum interval was greater than 210 m. They suggest the best approach in surveys of forage fishes in temperate reservoirs is a nighttime, stratified-random design of transects that incorporate large-scale gradients of fish density. Nighttime surveys were recommended since the shad species both

tended towards disaggregated distribution patterns during the evening and night (see also Schael et al. 1995). They also recommend block averages of transects of 150–200 m in length to minimize complications of spatial correlation and reduce within-transect variance. Gangl and Whaley (2004) used target tracking to locate individual targets. Transect subunits between 300 and 500 m in length produced densities that were statistically independent. Assuming statistical independence to employ arithmetic methods of density calculations will require analysis of the data to determine the spatial scale or distance of serial correlation. Subsampling distance or transect length will depend upon the spatial structure and distribution of the selected species. A method is presented below that does not require assumptions of statistical independence in subsamples and considers the spatial correlation implicitly in the estimates of density or abundance.

Objective 3: Estimating systemwide abundance and biomass

Taylor et al. (2005) compared both longitudinal and cross-channel sampling designs in Badin Lake (reservoir) in North Carolina in July 2000 and December 2001 and characterized both large- and small-scale spatial patterns in forage fish density. They found that sampling along longitudinal transects was a more efficient means to characterize spatial patterns of forage fish distribution and to estimate systemwide abundance and biomass, relative to data collected using both longitudinal and cross-channel sampling designs. They used geostatistics, and specifically kriging in their approach to estimate mean density and lakewide abundance. As previously mentioned, the sampling design must take into account the spatial distribution of the species as it may relate to other correlates such as water column depth or distance from shore.

Field/Office Methods

Setup and Measurement Details

To set up transects for the survey, bathymetry maps are helpful, but at minimum, an outline of the lake region is required. The sampling transects need to be contained within regions deep enough for the sonar and should include more intensive sampling in regions where fish are more concentrated (Jolly and Hampton 1990; Taylor et al. 2005). Adequate coverage of a water body is important to take advantage of the continuous nature of the data collection that occurs as part of hydroacoustic surveys. Several texts provide details on establishing optimal sampling programs to maximize system coverage with transects while not overextending staff (Cochran 1977; MacLennan and Simmonds 1992). The design of transects should take into account all these factors in addition to other logistical considerations such as navigability of the water bodies and workers' safety.

It is important to address both seasonal and diurnal movements and behavior of fish species prior to setting up the survey. Preliminary acoustic surveys or prior knowledge of a species' behavior and ecology can be utilized to obtain this information prior to setting up the survey to assess abundance (Lucas et al. 2002). Both time of day and light level have been found to alter fish behavior (MacLennan and Simmons 1992), and should be taken into consideration when planning a hydroacoustic survey. Yule (2000) sampled during the day and night and then chose to forego the daytime estimates because target species (rainbow and cutthroat trout) were either few in number or in schools, making density estimates difficult. Vondracek and Degan (1995) sampled both day and night, but the data was divided into two groups to account for the behavioral differences in their primary target species (threadfin shad *Dorosoma petense*). Shad displayed schooling behavior during the day and were mostly dispersed at night. Following advice from Vondracek and Degan (1995) and Schael et al. (1995), Taylor et al. (2005) sampled for forage fish at night, when local shad species were more disaggregated.

Appenzeller and Leggett (1992) reported on pelagic fish community abundance estimates obtained by acoustic methods for pelagic fish in Lake Memphremagog, Quebec. Reflecting diel light conditions, the fish were either in aggregated schools during the day or dispersed schools at night. Due to acoustic shadowing, densities were underestimated when fish were aggregated, with data suggesting that this bias could have been as large as 50%. When sampling juvenile sockeye salmon in lakes, surveys are traditionally done at night because the juveniles are more dispersed. In addition to considerations of time of day, seasonal patterns of distribution as well as other logistical considerations will likely influence the timing of a hydroacoustic survey. When sampling in temperate climates, surveys should be planned to avoid leaf-fall, wind, or rain that could affect surface interference, and boat traffic on navigable waterways can also affect the amount of noise that can disrupt the detection of target species in the water column.

Equipment Deployment

Recent developments in hydroacoustic technology have resulted in equipment that is generally portable and readily mobilized to even the most remote study site (see Figure 2). Echosounder placement on the survey vessel is usually determined by the user and likely includes such concerns as engine noise (both acoustic and electrical), comfort of operator, and location of power sources. Most commercially available scientific echo sounders are powered by 12-V power that is readily available on most boats. The power supply should be separate or otherwise isolated from that used by the vessel engine, as electrical interference can cause noise on the acoustic signal. Operators can use either deep-cycle 12-V batteries or gas-powered generators.

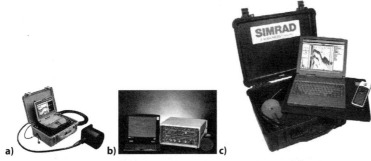

FIGURE 2.—Portable split-beam hydroacoustic systems from the major manufacturers: (a) Biosonics Inc., (b) Hydroacoustic Technology, Inc. (HTI), and (c) Simrad. Systems include an echo sounder contained in a rugged container or rack-mountable unit that has waterproof ports for attaching transducer cables, global positioning system cables, and network cables. Systems are controlled via wired or wireless LAN communication by a laptop computer.

Transducer deployment is specific to the survey and vessel design in addition to mounting requirements of the specific manufacturer. The transducer can be mounted under the hull of the boat, attached to the side of the boat (when a side-looking transducer is used), mounted in a towed body, or mounted on a rotator forward of vessel (see Figure 3). For horizontal work, it is important that the side-looking beam is on a stable platform and the beam direction and angle can be adjusted to account for interference caused by reflection of sound on the water–air interface. Sea-state or surface conditions usually dictate the best approach for deploying the horizontal-aimed transducer beam or whether a survey can be conducted (Gangl and Whaley 2004).

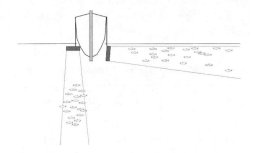

FIGURE 3.—Two transducers for simultaneous down-looking and side-looking deployments to sample fish targets throughout the water column.

Hardware settings and software controls

Most scientific grade echo sounders are controlled by a laptop computer connected via serial cable, wired, or wireless network connection. Echo sounder settings are selected through the user interface of data acquisition software provided by the manufacturer or third-party vendor. Setting the sonar parameters is site- and survey-specific and also depends on the manufacturer (see Table 1). General parameters would include speed of sound and sound loss or absorption, which is primarily determined by salinity/conductivity and water temperature. Thresholds are also set to accept returns from echoes that are above a given level or target strength. Thresholds need to be set low enough so that the echo returns from the target species can be observed on the edge of the nominal beam width. Target strengths of the surveyed species should be researched or calculated based on fish length (e.g., Love 1977). Ideally, the threshold should be set as low as possible; however, a signal to noise ratio of 12 dB or higher is desired and is usually the limiting factor when reducing the threshold for small species (e.g., 20 mm) (MacLennan and Simmonds 1992). Other environmental conditions also need to be considered for both setting threshold parameters. High conductivity can greatly attenuate the acoustic signal. Extremely high turbidity can scatter the signal, weakening the returning echoes. Under either condition, the power and gain settings may be increased effectively lowering the thresholds. Detection at the deeper levels can be greatly compromised in very deep lakes with high conductivity or turbidity due to signal spreading and attenuation.

Global positioning systems (GPS) are typically an integral part of mobile hydroacoustic surveys. Handheld to boat-mounted navigational systems can be integrated into the data acquisition system. The method of data transfer between the GPS and hydroacoustic system is dependent upon manufacturer specification, but usually involves latitude, longitude, speed, and directional information transfer in real time or as a separate time-stamped data file to be linked to hydroacoustic data.

TABLE 1. — Examples of recently published mobile hydroacoustic studies including specification of equipment settings and data processing procedures used

System	Family	Day/night	System	Freq (kHz)	Transducers	Mount	Vert./horiz.	Pulse rate (s−1)	Pulse width (ms)	Processing method
Reservoir[a]	Clupeid	Night	Biosonics, DB	200	1	Towed–side	Vertical	5	0.4	Integration
Lake[b]	Salmonid	Night	Biosonics, DB	420	1	Towed–bow	Vertical	2	0.4	Both
Lowland River[c]	Various	Day	Simrad, SB	120	1	Fixed–bow	Horizontal	Unknown	Unknown	Integration
Lakes and Reservoirs[d]	Salmonid	Night	HTI, SB	200	2	Fixed–side	Both	5	0.20	Tracking
Lake[e]	Salmonid	Day	Simrad, SB	200	1	Towed–side	Vertical	Unknown	0.4–1	Tracking
Lake[f]	Salmonid	Both	Simrad, SB	70	1	Unknown	Vertical	Unknown	Unknown	Tracking
Lake[g]	Coregonid	Night	Simrad, SB	120	1	Fixed–side	Vertical	2–10	0.1	Tracking
Lake[h]	Various	Day	HTI, SB	200	2	Fixed–side	Both	10	0.2	Tracking
Reservoir[i]	Clupeid	Night	HTI SB	200	2	Fixed–side	Both	10	0.18	Integration

[a]Vondracek and Degan 1995. [b]Beauchamp et al. 1997. [c]Kubecka et al. 2000. [d]Yule 2000. [e]Elliot and Fletcher 2001. [f]Mehner and Schulz 2002. [g]Encina and Rodriguez–Ruiz 2003. [h]Gangl and Whaley 2004. [i]Taylor et al. 2005.

Laboratory and field calibration

Setting threshold levels and determining target strength values of fish are dependent on a calibrated acoustic system. Without good calibration information, the results are invalid. Many project leaders send their sonar systems in for yearly laboratory calibrations. The advantage of yearly calibrations is that the vendor or specialist performing the calibration has the opportunity to verify that the electronics of the system and make sure all is working correctly. Finding out that something has gone wrong after the system is in the field can be very frustrating.

Regardless of whether a preseason laboratory calibration is performed, a field calibration is essential. Site-specific environmental conditions can determine the calibration technique. Two calibration methods are presented below. (Note: Many spherical objects can be used as targets for in situ system calibration; however, manufacturers suggest a standard calibration target of known target strength constructed of copper or tungsten carbide [Foote and MacLennan 1983].)

Yule 2000

The receiving sensitivity of the echo sounder was calibrated in the field periodically using a Dunlop long-life Ping-Pong ball (target strength of −39.5 dB). Results of field tests indicated agreement with laboratory calibration and consistent sensitivity between surveys. The pole mount was designed to adjust the vertical aiming angle of the six-degree transducer by worm gear. The initial metering of the worm gear was accomplished by sampling five Ping-Pong balls placed at known depths and set along a straight line. With knowledge of target depth, range, and angle of target passage through the beam, the orientation of the transducer axis was calibrated using trigonometry. Under slight chop, the vertical aiming angle was set to 7° below the surface, and this change was noted on field sheets.

Vondracek and Degan 1995

The hydroacoustic system was calibrated with U.S. Navy standards at the Biosonics laboratory in Seattle, Washington. Once in the field, the system was again calibrated before and after sampling with standard tungsten carbide reference calibration spheres (Foote and MacLennan 1983). If system calibrations were different than the expected target strength of the standard calibration sphere, the systems source level voltage was adjusted before analyses.

Data Handling, Analysis, and Reporting

Data handling and data management

Data collected using hydroacoustic equipment is stored in proprietary formats specific to the manufacturer on a computer hard drive, and for some systems as raw sound on digital audiotapes. File formats, file types, and the specific information included in each file are specific to the manufacturer. Users should consult manuals and be familiar with the structure of the data that is being collected. Regardless of the data format, accurate labeling of files and tapes is critical to reconstruct surveys during analysis and archive data for future analysis. These hydroacoustic systems and computer controllers are frequently exposed to less-than-ideal conditions, which can risk data loss due to damage. After completing a survey (or a leg of a survey), data should be archived in raw format to external or removable media such as a CD or DVD.

Data accumulation can be significant depending upon the aquatic systems being surveyed, temporal and spatial resolution of the data, and the specific configurations of the hydroacoustic system. It is not unusual to collect data on the order of hundreds of megabytes per survey day. In addition, it is very common for data to be collected in several files per survey, corresponding to a temporal duration based on sampling strategies (hourly, daily, etc.) or per transect, region, or system; therefore, consistent and logical file naming should be maintained throughout the survey. Many systems automatically assign a time stamp as part of the file names during each survey. This file name should be recorded on a field data sheet along with pertinent attributes such as more detailed descriptions of hydroacoustic system settings, regions being surveyed, time duration, personnel involved in data collection, and potential system errors, or other notes related to data quality during the survey. Copies of these field data sheets can then be stored with archived versions of the raw data files for future analysis. Where surveys require the use of many hydroacoustic systems, people, and complex sampling strategies, the nature of associating files with surveys may require additional data handling in the form of databases to maintain accurate records to link field notes with raw data and for data analysis and project report preparation. This can be accomplished in a simple spreadsheet or something as complex as a relational database or even a geographic information system (GIS).

Data processing

Data processing can incorporate any or all of the following three analysis components:

      a. Echo counting

      b. Target tracking or track counting

      c. Echo integration

Data analysis (as per study objectives) follows and is dependent on the question addressed with the mobile hydroacoustic surveys:

d. Characterizing spatial pattern
   e. Estimating and mapping densities
   f. Estimating systemwide biomass

Once the data files are collected, the next step is to process the files. The first step in file processing is to remove unwanted signals from bottom reverberation, boat engine noise, or other sources from the data. Most sonar systems come with editing software programs designed for this task.

Echo counting

Echo counting is the technique of counting individual echoes that pass through the acoustic beam during a mobile survey. This technique is used when fish densities are very low and echoes from individual fish do not overlap (due to close proximity of fish at the same distance from the transducer). Echo counting is based on user-determined criteria such as target strength threshold, which is equivalent to restricting to an expected size range of fish targets.

Echo/target tracking

This procedure accumulates individual echoes into tracks or traces corresponding to individual fish moving through the acoustic beam (see Figure 4). Additional characteristics such as average target strength and direction of movement (relative to the moving survey vessel) can be gleaned from this technique. This accumulation method can be done manually or automatically using manufacturer's software or third-party software described in detail on pages 164–165. For both echo counting and target tracking, corrections need to be made to account for the changes in detection of fish as the beam width increases and more water is sampled at great depth or distance from the transducer.

Echo Integration

When fish density is high and fish targets overlap, it can become difficult-to-impossible to isolate individual fishes or tracks (see Figure 4). In this case, it is no longer appropriate to use echo counting or target tracking. Instead, a process called echo integration is used. The procedure is based on the principle that the total acoustic energy returned from a sampled water volume is proportional to the number of fish in that volume. That is the total energy is equal to the sum of the acoustic energy from the individual targets. Knowledge of the acoustic size (target strength) of the individual targets is still needed as this information is used to scale the total acoustic energy returned to an estimate of fish density. Therefore, some form of target strength analysis or target tracking is typically performed on a subset of the data, where the targets are non-overlapping, or by using supplemental information from biological sampling to get an estimate of the average size of fish present.

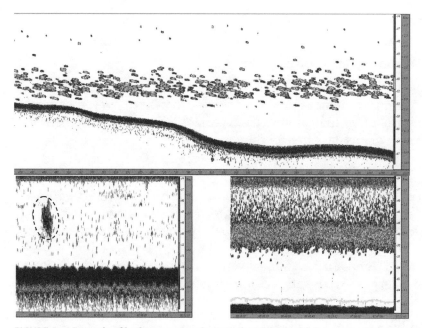

FIGURE 4. — Example of hydroacoustic echogram for individual fish targets when target tracking would be appropriate (top). Two examples of hydroacoustic echograms when target tracking would not be possible and when echo integration would be used: aggregated fish distributed in a school (bottom left) and layer (bottom right).

Available Software for Data Processing and Analysis

SonarData's Echoview (<www.SonarData.com>) now supports all the main scientific sonar system vendors. This software is expensive but very good for working with this type of data. The files are first imported into the editing software, calibration information is added, and then the echograms are ready for editing. All unwanted data is selected and labeled "bad data." The remaining data is then echo integrated using traditional integration methods (Ehrenberg and Kanemori 1978). The data is output both as a linear summed voltage and as 20-log of the summed voltage (dB). Additional progress is being made in hydroacoustic processing software. Packages such as Sonar 4 and Sonar 5 (<http://folk.uio.no/hbalk>) are showing great promise in handling data from numerous manufacturers and providing a wide range of analytical techniques.

Following the echo integration process, the single-target data is output based on user-set criteria. The output is in the form of average target strength values (average back-scattered cross section from individual fish) per cell. The target strength measures should be plotted both in range and time increments to determine how much variation exists. If fish differing in size are vertically stratified, then target strength values will vary according to range. If diel patterns exist in fish of different sizes, target strength will vary according to time. A possible averaged target strength matrix might be divided into 1 m depth bins and day/night temporal segments. The scaling of the integration data will be based on the matrix determined from the variability in the target strength values.

Taylor et al. (2005) processed their data for echo-integration and split-beam target tracking using Echoscape (v. 2.10, Hydroacoustic Technology, Inc., Seattle, Washington). Split–beam analysis was used to determine the acoustic size (target strength) of individual fish targets in decibels. Using equations for clupeiform species, target strengths from the down-looking (dorsal aspect) were converted to approximate fish size (Love 1977) and to wet weight. Volumetric densities were integrated throughout the water column to produce densities in two dimensions having units of fish/m². The database was incorporated into a GIS for data visualization and analyzed using S-PLUS (ver. 6.1, Insightful Corp.) for spatial structure and determination of abundance and biomass. Two statistical procedures were used to calculate densities and estimate systemwide abundance and biomass. First, an arithmetic mean and variance of densities assuming identical and independent data were determined for the entire survey and then for each survey design. These summary statistics were extrapolated across the surface area of the reservoir and summed to produce systemwide abundance and variance of this estimate. The second procedure involved empirically modeling the spatial structure of the data using geostatistics. This technique involved three steps: spatial detrending, variogram analysis, and kriging. This latter technique of using geostatistics resulted in similar average densities and improvements in the precision of abundance estimates based on approximated variance when compared to arithmetic averaging and extrapolation. Model-based density and abundance estimations are beneficial as they also provide information on the spatial structure and distribution of the species of interest in the aquatic system. These spatially explicit approaches have the added benefit of not requiring a prescribed randomized sampling plan as it implicitly models both large- and small-scale spatial variability (Rivoirard et al. 2000). Precision of abundance estimates still require adequate coverage of the water body as undersampling regions of high fish density can result in poor estimates and reduced precision (Taylor et al. 2005).

To obtain an abundance estimate for the lake or reservoir, the cell densities are expanded based on a ratio of the volume sampled to the volume of the water body. Further analyses can address issues such as among-transect variation (Vondracek and Degan 1995), diel patterns, depth distributions, and seasonal patterns. In the excerpt below, Yule (2000) describes the process used to obtain density estimates from target tracking.

Yule 2000

Side-looking fish density estimates by transect were calculated by dividing the numbers of detected fish by the volume of water sampled. Sample volume (m³) was calculated by multiplying travel distance (m) by the average side-looking range (m) by the average height of the cone (m). Sample volume was corrected for the inability to detect fish within 10 m of the transducer. Side-looking population estimates for each reservoir or lake were calculated by multiplying the mean density estimate (averaged across all transects) by the volume of water between the surface and a depth of 8 m.

With the down-looking transducer, sampling volume expands with increasing range. To standardize fish density estimates for increasing sample volume, detected fish were weighted back to a 1-m wide swath at the surface using the following formula:

$$F = 5\ 1/[2 \cdot R \cdot \tan (7.58)], w \qquad (eq\ 1)$$

Where F_w equals weighted fish, R equals range, and 7.5° equals one-half the nominal transducer beam width.

For example, at 3.8 m below the 15° transducer, the cone diameter $2 \cdot R$

[tan (7.5°)] is 1.0 m. It follows that a fish tracked at 3.8 m of range equaled one weighted fish at the surface (all fish were normalized to a 1-m transect width). At 20 m below the transducer, the cone diameter is 5.3 m, and a fish tracked at this range equaled 0.19 weighted fish. I derived estimates of fish densities (fish/m^3) by summing weighted fish by transect and dividing that by transect length. Fish detected by the down-looking transducer in the top 8 m of the water column were not processed to avoid overlap with side-looking density estimates (i.e., double counting). Down-looking population estimates for each reservoir and lake were calculated by multiplying the mean density (averaged across all transects) by the surface area.

Confidence intervals surrounding mean density estimates were calculated for both side-looking and down-looking acoustics. Each transect, regardless of length, was treated as a sample unit in the calculation of variability. Horizontal acoustic estimates of fish tracked during daylight surveys were partitioned to salmonids and nonsalmonids based on proportions captured by purse seining. Nighttime acoustic estimates of pelagic fish at Boulder Lake, New Fork Lakes, and Lake Viva Naughton were partitioned to salmonids and nonsalmonids based on overnight gill-net catches.

Exploratory data analysis should be conducted on preliminary surveys to determine the best approach for data analysis depending upon the chosen objective (as outlined above). If there is no indication of spatial autocorrelation in the transect survey data sets, a more simple arithmetic approach can be employed to calculated systemwide densities or total biomass or abundance. If data are spatially correlated, spatially explicit methods like geostatistics or spatial subsampling will be required. Further improvements in these spatially explicit, model-based estimation techniques are still the topic of much research and advances are continuously being made in statistical theory and software development.

Personnel Requirements and Training

Responsibilities

Project leader

1. Purchase and assemble needed sonar and ancillary equipment
2. Mount the sonar
3. Calculate thresholds and determe optimal sampling parameters
4. Set up the transect coordinates
5. Ensure that the boat operator is able to stay on the designated transects
6. Check weather conditions prior to setting up sampling dates
7. Perform all pre-season tasks needed for the project
8. Train technicians
9. Process final data; perform Quality Assurance and Quality Control of data; write report

Technicians

1. Acquire/develop detailed maps (including depth contours) of the lake or reservoir
2. Assist with data collection
3. Edit data
4. Export data for further processing
5. Operate the vessel

Qualifications

The project leader should have some background in basic acoustic principles and experience in operating the type of acoustic system selected for the study. In addition, the project leader should have experience in all aspects of operating a project, including budgeting, writing operational plans, coordinating the study, and operating boats. Technicians should be experienced in the operation of boats and have basic computer skills. The project leader and/or technicians should be familiar with the seasonal and diel behavior and ecology of the target fish species.

Training

Specialized training is required to use hydroacoustic techniques. Project leaders (at least) will need to be knowledgeable in how to use the equipment, understand the basic concepts, determine that applicability of this technology to their project, and be able to undertake the data survey design, analyses, and interpretation. Training on how to operate hydroacoustic systems is usually available from the vendor from which the system was purchased. The vendor should be contacted directly to obtain the location and timing of training schedules.

Operational Requirements

Workload and Field Schedule

The workload and field schedule are dependent on the study parameters. The size of the lake or reservoir and the number of transects required will determine the level of effort needed to complete the study.

Equipment Needs

1. Split beam echo sounder with one or two transducers (beam dimension should be considered based on sampling volume and expected water depth).
2. Mount or towable platform to attach transducers to boat.
3. Power to operate the echo sounder (battery or small generator).
4. Calibration equipment (calibration spheres/Ping-Pong balls).
5. Editing software programs.
6. Rotating device (optional).
7. Attitude sensor to record pitch of the side-looking transducer (optional).

8. Global Positioning System (GPS) linked to boat/echo sounder array.
9. Laptop computer and back-up hard-drives.
10. Deep-cycle 12-V batteries or gas-powered generator for power supply.

Budget Considerations

Purse-seine and horizontal acoustic assessments are rapid, and with good weather, a crew of six people can estimate salmonid numbers in a small impoundment (500–1,500 ha) in 1–2 d (Yule 2000). Similarly, a forage fish assessment in a 2,100-ha reservoir, using both horizontal and vertical acoustics, along with a purse seine, was accomplished during 2 nights (Taylor et al. 2005).

Fisheries hydroacoustic systems with one transducer cost approximately US$40,000 (as of 2005). Costs for all accessory equipment such as GPS, mounts, or tow-fish and laptops will need to be researched as prices for these technologies are becoming more cost effective.

Literature Cited

Appenzeller, A. R., and W. C. Leggett. 1992. Bias in hydroacoustic estimates of fish abundance due to acoustic shadowing: evidence from day-night surveys of vertically migrating fish. Canadian Journal of Fisheries and Aquatic Sciences 49:2179–2189.

Baldwin, C. M., D. A. Beauchanp, and J. van Tassell. 2000. Bioenergetic assessment of salmonid predator-prey supply, and demand in Strawberry Reservoir. Transactions of the American Fisheries Society 129:429–450.

Beauchamp, D. A., C. Luecke, W. A. Wurtsbaugh, H. G. Gross, P. E. Budy, S. Spaulding, R. Dillenger, and C. P. Gubala. 1997. Hydroacoustic assessment of abundance and diel distribution of sockeye salmon and kokanee in the Sawtooth Valley Lakes, Idaho. North American Journal of Fisheries Management 17:253–267.

Brandt, S. B., D. M. Mason, E. V. Patrick, R. L. Argyle, L. Wells, P. A. Unger, and D. J. Stewart. 1991. Acoustic measures of the abundance and size of pelagic planktivores in Lake Michigan. Canadian Journal of Fisheries and Aquatic Sciences 48:894–908.

Brandt, S. B. 1996. Acoustic assessment of fish abundance. Pages 385–419 in B. R. Murphy and D. W. Willis, editors. Fisheries techniques, 2nd edition. American Fisheries Society, Bethesda, Maryland.

Brown, M. J., and D. J. Austen. 1996. Data management and statistical techniques. Pages 17–61 in B. R. Murphy and D. W. Willis, editors. Fisheries techniques, 2nd edition. American Fisheries Society, Bethesda, Maryland.

Burczynski, J. J., and R. L. Johnson. 1986. Application of dual-beam acoustic survey techniques to limnetic populations of juvenile sockeye salmon *(Oncorhynchus nerka)*. Canadian Journal of Fisheries and Aquatic Sciences 43:1776–1788.

Clark, C. W., and D. A. Levy. 1988. Diel vertical migrations by juvenile sockeye salmon and the antipredation window. American Naturalist 131:271–290.

Cochran, W. G. 1977. Sampling techniques. 3rd edition. John Wiley & Sons, New York. 428 p.

Cyterski, M., J. Ney, and M. Duval. 2003. Estimation of surplus biomass of clupeids in Smith Mountain Lake, Virginia. Transactions of the American Fisheries Society 132:361–370.

Dahl, P. H., and O. A. Mathisen. 1983. Measurement of fish target strength and associated directivity at high frequencies. Journal of the Acoustical Society of America 73:1205–1211.

Degan, D. J., and W. Wilson. 1995. Comparison of four hydroacoustic frequencies for sampling pelagic fish populations in Lake Texoma. North American Journal of Fisheries Management 15:924–932.

Duncan, A., and J. Kubecka. 1993. Hydroacoustic methods of fish surveys. National Rivers Authority, R&D Note 196, Bristol, UK.

Ehrenberg, J. E., and T. C. Torkelson. 1995. The application of multibeam target tracking in fisheries acoustics. ICES Journal of Marine Science 53:329–334.

Ehrenburg, J. E., and R. Y. Kanemori. 1978. A microcomputer based echo integration system for fish population assessment. Proceedings of the I.E.E.E. (Institute of Electrical and Electronic Engineering) Conference on Engineering in the Ocean Environment 5:204–207.

Elliot, J. M., and J. M. Fletcher. 2001. A comparison of three methods for assessing the abundance of Arctic charr, Salvelinus alpinus, in Windermere (northwest England). Fisheries Research 53:39–46.

Encina, L., and A. Ridriguez-Ruiz. 2003. Abundance and distribution of a brown trout *(Salmo trutta)* population in a remote high mountain lake. Hydrobiologia 493:35–42.

Fernandes, P. G., P. Stevenson, A. S. Brierley, F. Armstrong, and E. J. Simmonds. 2003. Autonomous underwater vehicles: future platforms for fisheries acoustics. ICES Journal of Marine Science 60(3):684–691.

Foote, K. G., A. Aglen, and O. Nakken. 1986. Measurement of fish target strength with a split-beam echo sounder. Journal of the Acoustical Society of America 80:612–621.

Foote, K. G., and D. N. MacLennan. 1984. Comparison of copper and tungsten carbide spheres. Journal of the Acoustical Society of America, 75: 612–616.

Frear, P. A. 2002. Hydroacoustic target strength validation using angling creel census data. Fisheries Management and Ecology 9:343–350.

Gangl, R. S., and R. A. Whaley. 2004. Comparison of fish density estimates from repeated hydroacoustic surveys on two Wyoming waters. North American Journal of Fisheries Management, 24:1279–1287.

Gregory J., and P. Clabburn. 2003. Avoidance behaviour of Alosa fallax fallax to pulsed ultrasound and its potential as a technique for monitoring clupeid spawning migration in a shallow river. Aquatic Living Resources 16:313–316.

Hubert, W. A. 1996. Passive capture techniques. Pages 157–181 in B. R. Murphy and D. W. Willis, editors. Fisheries techniques, 2nd edition. American Fisheries Society, Bethesda, Maryland

Hughes, S. 1998. A mobile horizontal hydroacoustic fisheries survey of the River Thames, United Kingdom. Fisheries Research 35:91–98.

Johnson, R. L., and J. J. Burczynski. 1985. Application of dual-beam acoustic procedures to estimate limnetic juvenile sockeye salmon. International Pacific Salmon Fisheries Commission, Progress Report No. 41, New Westminster, B.C..

Johnston, J. 1981. Development and evaluation of hydroacoustic techniques for instantaneous fish population estimates in shallow lakes. Washington State Game Department, Fisheries Research Report 81-18, Olympia.

Jolly, G. M., and I. Hampton. 1990. Some problems in the statistical design and analysis of acoustic surveys to assess fish biomass. Rapports et Proces-Verbaux do Reunions Conseil International pour l'Exploration de la Mer 189:415–420.

Knudsen, F. R., and H. Seagrov. 2002. Benefits from horizontal beaming during acoustic survey: application to three Norwegian lakes. Fisheries Research 56(2):205–211.

Kubecka, J. 1996. Use of horizontal dual-beam sonar for fish surveys in shallow waters. Pages 165–178 in I. G. Cowx, editor. Stock assessment in inland fisheries. Fishing News Books, Oxford, UK.

Kubecka, J., and A. Duncan. 1994. Low fish predation pressure in the London reservoirs: I. Species composition, density, and biomass. International Review of Hydrobiology 79:143–155.

Kubecka, J., and A. Duncan. 1998. Acoustic size vs. Real size relationships for common species of riverine fish. Fisheries Research 35:115–125.

Kubecka, J., A. Duncan, and A. Butterworth. 1992. Echo counting or echo integration of r fish biomass assessment in shallow waters. Pages 129–132 in M. Weydert, editor. European conference on underwater hydroacoustics. Elsevier Applied Science, London.

Kubecka, J., A. Duncan, W. M. Duncan, D. Sinclair, and A. J. Butterworth. 1994. Brown trout populations of three Scottish lochs estimated by horizontal sonar and multimesh gill nets. Fisheries Research 20:29–48.

Kubecka, J., J. Frouzová, A. Vilcinskas, C. Wolter, and O. Slavík. 2000. Longitudinal hydroacoustic survey of fish in the Elbe River supplemented by direct capture. Pages 14–26 in I. G. Cowx, editor. Management and ecology of river fisheries. Blackwell, Oxford, UK.

Kubecka, J., and M. Wittingerova. 1998. Horizontal beaming as a crucial component of acoustic fish stock assessment in freshwater reservoirs. Fisheries Research 35:99–106.

Levy, D. A., B. Ransom, and J. Burczynski. 1991. Hydroacoustic estimation of sockeye salmon abundance and distribution in the Strait of Georgia, 1986. Pacific Salmon Commission, Technical Report 2, Vancouver.

Love, R. H. 1971. Dorsal-aspect target strength of an individual fish. Journal of the Acoustical Society of America 49:816–823.

Love, R. H. 1977. Target strength of an individual fish at any aspect. Journal of the Acoustical Society of America 62:1397–1403.

Lucas, M. C., L. Walker, T. Mercer, and J. Kubecka. 2002. A review of fish behaviours likely to influence acoustic fish stock assessment in shallow temperate rivers and lakes. R&D Technical Report W2-063/TR/1, Environment Agency, Bristol, UK.

Lyons, J. 1998. A hydroacoustic assessment of fish stocks in the River Trent, England. Fisheries Research, 35:83–90.

MacLennan, D. N., and E. J. Simmonds. 1992. Fisheries acoustics. Chapman and Hall, London.

Mann, D. A., D. M. Higgs, W. N. Tavolga, M. J. Souza, and A. N. Popper. 2001. Ultrasound detection by clupeiform fishes. Journal of the Acoustical Society of America 109(6):3048–3054.

McCartney, B. A., and A. R. Stubbs. 1971. Measurement of the acoustic target strength of fish in dorsal aspect including swim-bladder resonance. Journal of Sound and Vibration 15(3):397–420.

Mehner, T., and M. Schulz. 2002. Monthly variability of hydroacoustic fish stock estimates in a deep lake and its correlation to gillnet catches. Journal of Fish Biology 61:1109–1121.

Mulligan, T. J., and R. Kieser. 1986. Comparison of acoustic population estimates of salmon in a lake with a weir count. Canadian Journal of Fisheries and Aquatic Sciences 43:1373–1385.

Nero, R. W., and J. J. Magnuson. 1989. Characterization of patches along transects using high-resolution 70-kHz integrated acoustic data. Canadian Journal of Fisheries and Aquatic Sciences 46:2056–2064.

Nilsson, N. A., and T. G. Northcote. 1981. Rainbow trout *(Salmo gairdneri)* and cutthroat trout *(S. clarki)* interactions in coastal British Columbia lakes. Canadian Journal of Fisheries and Aquatic Sciences 38:1228–1246.

Parkinson, E. A., B. E. Rieman, and L. G. Rudstam. 1994. Comparison of acoustic and trawl methods of estimating density and age composition of kokanee. Transactions of the American Fisheries Society 123:841–854.

Rivoirard, J., J. Simmonds, K. Foote, P. Fernandes, and N. Bez. 2000. Geostatistics for estimating fish abundance. Blackwell Science, Oxford, UK.

Schael, D. M., J. A. Rice, and D. J. Degan. 1995. Spatial and temporal distribution of threadfin shad in a southeastern reservoir. Transactions of the American Fisheries Society 124:804–812.

Stables, T. B., and G. L. Thomas. 1992. Acoustic measurement of trout distributions in Spada Lake, Washington, using stationary transducers. Journal of Fish Biology 40:191–203.

Tarbox, K. E., and R. E. Thorne. 1996. Assessment of adult salmon in near-surface waters of Cook Inlet, Alaska. ICES Journal of Marine Science 53:397–401.

Taylor, J. C., J. S. Thompson, P. S. Rand, and M. Fuentes. 2005. Sampling and statistical considerations for hydroacoustic surveys used in estimating abundance of forage fishes in reservoirs. North American Journal of Fisheries Management 23:75–83.

Thorne, R. E. 1971. Investigations into the relations between integrated echo voltage and fish density. Journal of the Fisheries Research Board of Canada 28:1269–1273.

Thorne, R. E. 1979. Hydroacoustic estimates of adult sockeye salmon (*Oncorhynchus nerka*) in Lake Washington 1972–1975. Journal of the Fisheries Research Board of Canada 36:1145–1149.

Thorne, R. E. 1983. Application of hydroacoustic assessment techniques to three lakes with contrasting fish distributions. FAO (Food and Agriculture Organization of the United Nations) Fisheries Report 300:269–277.

Thorne, R., C. J. McClain, J. Hedgepeth, E. S. Kuchl, and J. Thorne. 1992. Hydroacoustic surveys of the distribution and abundance of fish in lower Granite Reservoir, 1989–1990. Contract Report of BioSonics, Inc. to Walla Walla District, U.S. Army Corp of Engineers, Seattle.

Traynor, J. J., and J. E. Ehrenberg. 1990. Fish and standard-sphere target-strength measurements obtained with a dual-beam and split-beam echo-sounding system. Rapports et Procés Verbaux des Réunions, Conseil International pour l'Exploration de la Mer 189:325–335.

Vondracek, B., and D. J. Degan. 1995. Among- and within-transect variability in estimates of shad abundance made with hydroacoustics. North American Journal of Fisheries Management 15:933–939.

Warner, E. J., and T. P. Quinn. 1995. Horizontal and vertical movements of telemetered rainbow trout *(Oncorhynchus mykiss)* in Lake Washington. Canadian Journal of Zoology 73:146–153.

Wanzenbock, J., T. Mehner, M. Shulz, H. Gassner, and I. Winfield. 2003. Quality assurance of hydroacoustic surveys: the repeatability of fish-abundance and biomass estimates in lakes within and between hydroacoustic systems. ICES Journal of Marine Science 60:486–492.

Whitworth, W. E. 1986. Factors influencing catch-per-unit-effort and abundance of trout in small Wyoming reservoirs. Doctoral dissertation. University of Wyoming, Laramie.

Wurtsbaugh, W. A., R. W. Brocksen, and C. R. Goldman. 1975. Food and distribution of underyearling brook and rainbow trout in Castle Lake, California. Transactions of the American Fisheries Society 104:88–95.

Yule, D. 1992. Investigations of forage fish and lake trout Salvelinus namaycush interactions in Flaming Gorge Reservoir, Wyoming-Utah. Master's thesis. Utah State University, Logan.

Yule, D. L. 2000. Comparison of horizontal acoustic and purse-seine estimates of salmonid densities and sizes in eleven Wyoming waters. North American Journal of Fisheries Management 20:759–775.

Fish Counting at Large Hydroelectric Projects
Paul G. Wagner

Background and Objectives

In the Pacific Northwest, salmon and steelhead *Oncorhynchus mykiss* are an integral part of tribal and nontribal cultures, supporting important subsistence, commercial, and recreational fisheries. During the past century, salmon stocks have declined on the U.S. Pacific Coast to the point where several populations are now listed as threatened or endangered under the Endangered Species Act. The reasons for the decline are many, including extensive development of hydroelectric projects. Because salmon and steelhead are such an important part of the Pacific Northwest economy, culture, and heritage, widespread salmon recovery efforts are currently underway. Accurate juvenile and adult passage estimates at large hydroelectric projects can be vital to monitoring the success or failure of these efforts and are also frequently used for a variety of other management applications.

Hydroelectric projects can impede anadromous salmonid passage at adult and juvenile life stages. Where possible, passage facilities such as adult fishways and juvenile (smolt) bypass systems are often included in hydroelectric project design or retrofitted to older existing structures. Juvenile bypass systems are intended to divert downstream migrating smolts away from turbine intakes through screening, louver, gulper, or surface collection systems. At large hydroelectric projects, these juvenile bypass systems often include collection facilities where fish may be held and counted, anesthetized and examined, used in research studies, or loaded into vehicles such as tanker trucks or barges to be transported downstream around other hydroelectric projects. Adult passage facilities are typically of pool and weir design and are intended to attract and provide safe passage for adult upstream migrants. Such facilities are usually equipped with count stations that allow enumeration of migrants by either direct observation or interrogation of time-lapse video. In addition, adult fishways often include a trap so that fish can be sampled, marked, or loaded into tanker trucks for transport upstream around other existing nonpassable hydroelectric projects to otherwise inaccessible spawning habitat.

The Columbia River basin has been called the most hydroelectrically developed river system in the world. Large hydroelectric project development began with the completion of Rock Island Dam in 1932, followed by much larger federal projects such as the Bonneville and Grand Coulee dams completed in 1938 and 1941, respectively (see the Center for Columbia River History Web site, <www.ccrh.org/river/history.htm#hydro>). Eleven large hydroelectric projects currently exist in the U.S. portion of the mainstem Columbia River; numerous other projects occupy the tributaries.

Because of this extensive hydropower development, the Columbia River basin also supports what could arguably be described as the most advanced fish passage technology and associated counting programs in the world. It should be noted that although similarities exist between sites, the passage facilities and associated counting protocols are often site specific. The objective of this chapter is to provide an overview of the protocols used to conduct adult and juvenile fish passage counts and to derive fish passage estimates in the Columbia River basin. While the data for this chapter is derived from hydroelectric projects located on the Snake and Columbia rivers of the U.S. portion of the Columbia River basin, the methods herein are intended to be applicable to all large hydroelectric projects, although some site-specific modification will likely be necessary.

Sampling Design

The counting procedures for both juvenile and adult salmonids described in this paper are to occur at large hydroelectric projects that are fixed or otherwise preselected sites and are therefore not subject to discretionary site selection. Populations sampled will consist of all anadromous fish populations that originate from upstream locations.

Juvenile Salmonid Passage Monitoring

Juvenile Passage Routes

Downstream migrating salmonids can pass hydroelectric projects by several routes (see Figure 1). At projects equipped with screening systems, such as those at most U.S. Army Corps of Engineers (USACE) projects in the Columbia River basin, submersible screens are placed in turbine intakes that guide a portion of the juveniles entering the intakes away from the turbines and through a bypass system. The proportion of fish that enter the intakes but are guided away from the turbines and through the bypass system is termed fish guidance efficiency (FGE). Unguided fish pass the project through the turbines. FGEs have been shown to be highly variable. Potentially, such variability can occur within and between project powerhouses by such factors as fish species, race, age, and size, by unit operation configuration, and as a result of other external environmental factors (e.g., water temperature, amount of sunlight). Because it is not feasible to measure or monitor FGE throughout each passage season at large hydroelectric projects, FGE is normally not factored into juvenile passage estimates at large hydroelectric projects and is essentially assumed to be 100%.

Bypass systems route fish either to the project tailrace or to a collection facility located downstream. Fish that arrive at the collection facility may be held and then transported via truck or barge around other downstream projects to a release location.

Substantial fish passage may also occur through the project spillways. Spillway passage efficiency is the proportion of fish per unit of water flow that pass over the spillway versus through the powerhouse. Similar to FGE, it is not feasible to monitor spillway passage efficiency throughout each passage season at large hydroelectric projects, and therefore, a 1:1 ratio is normally assumed (i.e., when an equal amount of flow is passed over the spillway and through the powerhouse, then an equal number of fish are assumed to pass the dam through each of the two routes).

Other potential routes of passage include navigation locks or adult fishways downstream; however, given the relatively small amount of project flow that is diverted through these later two routes, passage at these locations is usually considered insignificant and is not included in passage estimates. (The Dalles Dam on the mainstem Columbia River is unique among USACE projects in that the ice and trash sluiceway is used to bypass juvenile fish to the project tailrace.)

Juvenile Facilities Description

Guidance and Bypass Systems

The juvenile fish bypass systems in place at seven of the eight lower Columbia and Snake River dams operated by the USACE guide fish away from turbines by means of submerged screens positioned in front of the turbines in the intake structures. The juvenile fish are directed up into gatewells where they pass through orifices into a collection channel that runs the length of the dam (see Figure 2). The fish are then either routed or "bypassed" back out to the river below the dam, or, at the four dams with fish transport facilities, fish can be routed to a holding area for loading onto specially equipped barges or trucks for transport downriver (USACE <www.nwd.usace.army.mil/ps/colrvbsn.htm>).

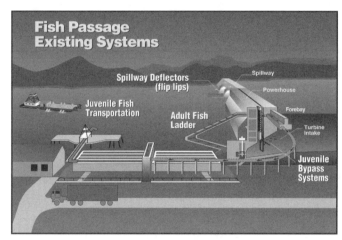

FIGURE 1. — Juvenile passage routes (Source: USACE <www.nwd.usace.army.mil/ps/colrvbsn.htm>).

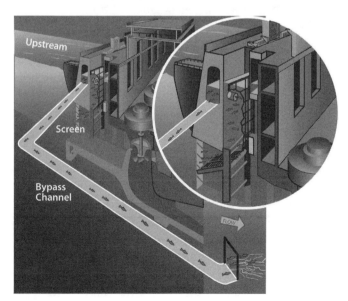

FIGURE 2. — Juvenile bypass system (Source: USACE <www.nwd.usace.army.mil/ps/colrvbsn.htm>).

Juvenile Fish Facilities

Fish-counting activities normally occur at existing juvenile collection facilities located within the bypass systems at these large dams (see Figure 3). Such facilities typically route fish from the powerhouse via a screening system through a collection gallery and water elimination system. Fish are carried through the collection gallery channel past a primary dewatering unit where all but a small portion of the water is removed. Fish and remaining water pass through a bar separator (see Figure 4) where debris may be removed and large fish are separated from smaller fish. Timed sample gates are located on the flumes exiting the separator, and a portion of the juvenile fish collection is diverted through an electronic fish counting system into sampling tanks (see Figure 5) at a predetermined sampling rate. The electronic fish counting system normally consists of holding or "counter" tanks that discharge through electronic fish-counter tunnels and into the sample tanks. Juvenile collection facilities at large dams are normally staffed 24-h a day to allow immediate response to potential equipment failures or debris blockages or to prevent overloading of holding tanks or raceways. The electronic fish counting system allows hourly monitoring of the approximate number of fish sampled, and through simple extrapolation based upon the sampling rate, an estimate of the total number of fish collected. Some level of count error is usually associated with these types of electronic fish counting systems (e.g., due to the presence of debris, air bubbles, or multiple fish passing simultaneously through the counting field), and therefore, the electronic counts are corrected by actual hand counts of fish diverted into the sampling tanks at the end of each 24-h period.

Sampling and Counting Overview
In the Columbia River basin, the juvenile collection facilities are operated and maintained by the USACE at federal projects or by public utility districts. Most of the juvenile salmonid counting projects conducted at large hydroelectric projects are through the Smolt Monitoring Program (SMP). This Bonneville Power Administration–funded program samples fish at 14 sites—6 traps and 8 large dams within the Columbia River basin anadromous corridor. Until recently, counting operations at these sites were coordinated through the Fish Passage Center, located in Portland, Oregon; fieldwork is carried out by various agencies. Data are gathered on juvenile migrant salmon and incidental species under a well-developed set of protocols that will be emphasized in this chapter. Counting activities are scheduled to encompass the primary smolt outmigration period for all species, which is primarily from March through the end of October. A wide range of information is collected in association with the monitoring program, including species-specific passage timing, migrant mark recovery, external injury, disease, descaling, and mortality. Passive integrated transponder (PIT) tagging is also carried out at some SMP sites to aid in gathering longer-term data such as travel time, passage timing, and survival for specific mark groups (e.g., Berggren et al. 2000). Juvenile collection facilities are typically equipped with PIT tag interrogation systems, which allow passive interrogation of all fish collected.

Sample tanks may include pre-anesthetic chambers, in which fish are initially sedated or pre-anesthetized prior to handling (see Figure 6). Small allotments of fish are routed directly from the pre-anesthetic chambers into a laboratory building and held in a recirculating anesthetic bath where they are examined and counted (see Figure 7). Here, project staff examine each fish to determine species, age, rearing disposition (hatchery or wild); check for injuries and descaling; and, in some cases, measure lengths and weights as necessary for calculating barge, truck, or raceway loading densities. The Fish Passage Center Data Entry Program Manual (Franzoni et al. 2004) provides detailed descriptions of data collected on individual fish and overall sample statistics such as mortality rate. Counts are usually tallied via a manual or electronic tabulation system located adjacent to the sampling trough (see Figure 8). After examination, anesthetized fish are routed to a recovery raceway (see Figure 9) where they are allowed to recover before being returned to the river or loaded onto transportation vehicles. To reduce the likelihood of predation losses, a 1-h minimum recovery time is recommended for juvenile fish that are to be returned to the river.

FIGURE 3. — McNary Dam juvenile fish collection facility.

FIGURE 4. — Juvenile fish facility bar separator at the McNary Juvenile Fish Facility.

FIGURE 5. — Juvenile fish facility sample tank at the McNary Juvenile Fish Facility.

FIGURE 6. — Sample tank pre-anesthetic chambers at the McNary Juvenile Fish Facility.

FISH COUNTING AT LARGE HYDROELECTRIC PROJECTS

FIGURE 7. — Fish counting station at the McNary Juvenile Fish Facility.

FIGURE 8. — Juvenile fish hand-tabulation system the McNary Juvenile Fish Facility.

FIGURE 9. — Recovery raceway at the McNary Juvenile Fish Facility.

Sampling Rates

The sample rate is based upon the projected collection for the sample period, which is in turn based upon the number of fish collected the previous day and the operator's best estimate of changes in passage. The goal is to sample between 500 and 800 juvenile salmonids per day. These numbers of fish reflect the need to maintain a coefficient of variation (cv = standard error of collection/estimate of collection) lower than 5%. Due to the nonuniform nature of emigration from the sample counting tanks, subsamples are recommended. The rates listed in Table 1 determine how often and what duration the sample gate will divert fish from the collection facility or bypass. As shown in Table 1, six subsamples per hour are preferred, with the exception of 100% sample rate or at very low sample rates. The Fish Passage Center recommends a minimum of two subsamples per hour with a minimum of 12 per sample to minimize edge effects during low rate sampling. Note that when more than 50,000 fish are expected, the Fish Passage Center further recommends a 1% sample rate—except when more than 75,000 fish are expected, in which case a lower sample rate may be necessary to avoid risk of capturing more fish than can be safely handled. Also, sampling to provide fish for research marking programs or for other purposes may require increases beyond the recommended sample rates. The presence of large numbers of nontarget species can also require that the sample rate be lowered to prevent exceeding holding density criteria. For example, at McNary Dam on the Columbia River, large numbers of juvenile American shad *Alosa sapidissima* in the sample during late summer operations often make it necessary either to reduce the sample rate or curtail sampling altogether, especially when transportation is ended early based on relatively small numbers of salmonids in the collection. Holding densities should not exceed 1.0 kg of fish per 16.5 L of water in volume, or 0.6 kg of fish per L/min in inflow (0.5 lbs of fish per gallon of water or 5 lbs of fish per gal/min in inflow) (USACE Northwestern Division 2005).

TABLE 1.—Fish Passage Center (Portland, Oregon) recommended sampling rates for Snake River dams

Estimated daily collection	Subsample rate %	Equivalent multiplier 1/sample rate	Sample s/h	Subsamples per hour	Subsample duration seconds	Estimated number of fish in sample
Emergency	0.50%	200	18	2	9	
>75,000	0.70%	143	25.2	2	12.6	>525
50,000–75,000	1.00%	100	36	3	12	500–750
35,000–50,000	1.50%	66.6	54	4	13.5	525–750
25,000–35,000	2.00%	50	72	6	12	500–750
16,500–25,000	3.00%	33.3	108	6	18	495–750
12,500–16,500	4.00%	25	144	6	24	500–660
10,000–12,500	5.00%	20	180	6	30	500–625
7,500–10,000	7.00%	14.3	252	6	42	525–700
5,000–7,500	10.00%	10	360	6	60	500–750
4,000–5,000	12.50%	8	450	6	75	500–625

Estimated daily collection	Subsample rate %	Equivalent multiplier 1/sample rate	Sample s/h	Subsamples per hour	Subsample duration seconds	Estimated number of fish in sample
3,000–4,000	15.00%	6.66	540	6	90	450–600
2,500–3,000	20.00%	5	720	6	120	500–600
1,500–2,500	25.00%	4	900	6	150	375–625
500–1,500	50.00%	2	1,800	6	300	250–750
<500	100.00%	1	3,600	1	3,600	<500

Field/Office Methods

Fish Anesthetization and Handling

Anesthetization and handling of fish are necessary components of a smolt monitoring program, and care should always be taken to minimize associated fish stress and mortality. Tricaine methanesulfonate (MS-222) is registered by the U.S. Food and Drug Administration for use as a fish anesthetic and is used at all SMP sites. A stock solution of MS-222 and river water is prepared prior to the daily handling of fish. Stock solution concentrations are typically 100 g/L. To minimize handling stress, small allotments of smolts are gently crowded into the sample tank pre-anesthetic chambers and initially introduced to anesthetic. The average concentration of MS-222 used at SMP sites varies between 40 and 85 mg/L and depends upon the number, species, and size of fish, water temperature, and water chemistry. Sites in the Snake River often have natural buffers in the water that make it necessary to use higher concentrations to be effective. Ideally, anesthetic induction time (time to which fish initially show visible signs of anesthetization, such as rolling over on their backs) should range from 2 to 3 min and should not be less than 1 min.

After most of the fish exhibit signs of anesthetization, they are then transferred from the pre-anesthetic chamber to a sorting trough where they can be examined and counted. At SMP sites, this "wet" transfer (i.e., fish remain in water at all times) is accomplished by opening an evacuation valve located at the bottom of the pre-anesthetic chamber, thereby delivering fish via gravity flow through a piping system to the sorting trough, which is located inside the laboratory building. Flushing water is turned on to ensure complete evacuation of the pre-anesthetic chamber and delivery of all fish to the sorting trough. The Fish Passage Center recommends that no more than 80 fish at one time be delivered to the sorting trough to minimize the amount of time that fish are held in anesthetic.

The sorting trough contains a recirculating MS-222 bath. The anesthetic concentration in the recirculation system should be held at 35 to 45 mg/L to allow fish to be examined and counted while they slowly recover from initial anesthetization. Because the recirculating anesthetic system is a closed system, it should include water-cooling capability. Water within the system should be maintained at a temperature appropriate for the species present and should be monitored frequently throughout the course of fish handling activities, especially during the summer months. System operators should also be aware that large numbers of fish can metabolize and reduce the anesthetic concentration within the recirculation system. Additional anesthetic may need to be added periodically when large numbers of fish are processed for programs such as research marking.

After fish are handled and counted, they are transferred via a gravity flow piping system to a recovery raceway or tank. In all cases involving the handling or counting of fish, wet transfer is recommended. The Fish Passage Center further recommends that the time between initial exposure to anesthetic in the pre-anesthetic chambers and delivery to the recovery raceway not exceed 15 min. All mortalities should be removed from the recovery raceway and counted daily so that daily mortality rates can be calculated to allow assessment of the effect of sampling and handling activities on the well-being of the sampled fish. Sampling activities should be adjusted to minimize mortality rates.

General strategies to minimize fish stress and mortality include the following

1. Minimize the time fish are held in a crowded condition to the extent possible.
2. Minimize the time fish are held in anesthetic to the extent possible.
3. Start with relatively low concentrations of anesthetic and then increase as necessary. Fish should be anesthetized no more than needed (i.e., just so they may be easily handled and the necessary information obtained). Note that there is a fine line that separates underanesthetized from overanesthetized, and both conditions are detrimental to fish. Overanesthetization will simply overdose and kill fish directly; however, underanesthetized fish can jump from examiners hands or require excessive squeezing during examination, which can elevate mortality rates as well. Anesthetists should communicate regularly with fish handlers during fish processing operations and procedures, or anesthetic concentrations should be adjusted as necessary to minimize detrimental effects.
4. Remove smaller or more sensitive species (or species listed under the Endangered Species Act) from anesthetic first, followed by larger, less sensitive species.
5. Leave fish in water to the extent possible during examination.
6. Wet transfer whenever possible.
7. Check water temperatures in closed systems regularly and adjust as necessary.
8. Flush closed piping fish delivery systems at least three times at the conclusion of daily activities to prevent fish from becoming stranded in the pipe once the flow is turned off for the day. Operators should be cognizant that underanesthetized fish or fish that have recovered early from anesthetization can hold in delivery pipes (such as from the pre-anesthetic chambers to the sorting trough or from the sorting trough to the recovery raceway). Increasing the flow to flush fish out of such systems is generally ineffective as juvenile salmonids tend to swim harder and more actively towards the flow source when the velocity is increased. The preferred and recommended method to remove fish from such systems is to shut the flow off, allow the pipe to drain momentarily, and then restore the flow long enough to allow fish the opportunity to turn downstream and exit the pipe. This procedure should be repeated at least three times or until operators are certain that all fish have been cleared from the pipe.

Data Handling, Analysis, and Reporting

Collection Estimates

Sample counts of juvenile fish by species and rearing disposition (when rearing type can be determined based on clips and marks) are tallied daily throughout the sampling season at large hydroelectric projects equipped with collection and sampling systems. Typically 6 samples per hour, over a 24-h period (often 7 a.m. to 7 a.m., not a calendar day), are diverted to the sample tank at smolt monitoring program sites. Collection estimates are then extrapolated from the daily juvenile fish samples at each site by dividing the sample count by the sample rate. This is typically reported by species and rearing type.

Passage Index

The passage index is a real-time statistic that can be used to measure in-season passage timing for management purposes. The passage index is the collection count expanded to account for the proportion of water passing through the powerhouse compared to other routes, such as spill. So for a given daily collection (C), average daily powerhouse discharge (PH), and average daily total discharge (Q), the daily passage index (PI) is calculated as shown in equation 1. Note that all values are daily averages based on the sample day, which typically runs from 7 a.m. to 7 a.m. at smolt monitoring program sites in the Columbia River basin.

$$PI = C*(Q/PH) \qquad (\text{eq 1})$$

As can be seen by equation 1 the passage index adjusts the collection upward to account for discharge through nonpowerhouse routes, especially spill; however, the passage index does not account for fish passing through turbines and assumes a 1:1 fish to water ratio for expansions. The passage index is not sensitive to changes in FGE or spill efficiency, the proportion of fish passing in spill per volume of water spilled. Based on analyses by the Fish Passage Center (FPC), the passage index has proven quite robust to changing operations at projects for any given year, and therefore, because of its minimal assumptions and ease of calculation, has been adopted for use in determining season passage timing for migrating juvenile salmonids. The passage index is typically smaller in magnitude than the overall out-migrating population because no attempt is made to estimate actual spill efficiency or turbine passage; therefore, the magnitude, or population size, of the run is often not fully accounted for by the passage index. Other methods must be incorporated to determine the total size of the run migrating past a given dam.

Population Index and Population Estimates

The simplest method to determine the overall size of the out-migration is to estimate seasonal collection efficiency for each species and then divide the total collection for that species by the collection efficiency as shown in equation 2.

$$N = \frac{C}{ce} \qquad (\text{eq 2})$$

where
N = estimated seasonal population size
C = estimated total seasonal collection
ce = estimated seasonal collection efficiency

If greater precision and statistical comparisons are required, the FPC recommends using methods similar to those described in Sandford and Smith (2000). Peven et al. (2005) provide important insights into specific methodologies for conducting juvenile salmonid survival studies.

Data Handling Validation (Quality Assurance and Quality Control)

There are three levels of quality assurance and quality control applied to the collection and dissemination of smolt monitoring program data in the Columbia River basin. Smolt data is accumulated 24 h per day and are summarized and transmitted to a central repository once per day. Each of these transmittals is considered a "batch" of data. Data are first collected on hand logs during the sampling day and then transcribed into an electronic form to complete the batch. The first level of quality control is applied to this transcription process. Each data element passes through an initial validation process as it is transcribed from paper to electronic form. These validation checks are built into the remote data entry systems and occur at the moment of data entry. User inputs are automatically examined to verify that they consist of realistic values and are internally consistent.

The second level of data quality control occurs at the other end of the data entry process, at the data repository. The smolt monitoring program conducts work at seven dams in the Columbia River basin. Monitoring at these sites occurs 7 d per week over periods varying from 21 weeks (Rock Island Dam) to 30 weeks (Lower Granite, Little Goose, Lower Monumental, John Day, and Bonneville dams) to 36 weeks (McNary Dam). Each smolt monitoring site sends one batch per day to a central repository, and once per week a second quality-assurance protocol is applied to data from each site to compare the hand logs (daily data sheets) to the electronic data summaries. The goal of this quality assurance and quality control (QA/QC) protocol is to cross-check enough daily batches with enough daily data sheets to assure that the potential discrepancy rate across the total batches for a given site is acceptably low. This QA/QC procedure requires cross-checking two of the seven daily batches per week at each of the monitoring sites. The FPC randomly selects the two batches to be examined each week. FPC personnel located in Portland, Oregon, is responsible for conducting the data cross-check and reporting the results back to the site personnel. The cross-check of a daily batch consists of manually verifying that the data entries in the SMP database match those from the hand logs. The data entries are arranged in several data tables: (1) catch summary, detailing the sample-related parameters and flow/spill entries; (2) catch detail, summarizing the fish counts per species, number of fish descaled, mortalities, and sample rates; (3) incidental catch detail, which enumerates the number of nontarget fish collected; (4) a mark detail table summarizing counts of fish with elastomer tags, photonic tags, spaghetti tags, and freeze brands; and (5) a transportation detail table showing the number of fish transported and bypassed at the collector dams.

If no discrepancies are found within the two batches examined for a given site, then the QA/QC procedure for that site is finished for that week and the process will begin again the following week. Under the condition that no discrepancies are found in any of the batches examined over the full season, the FPC will be 95% confident that the discrepancy rate across all batches for the season will not exceed approximately 5%. This estimation utilizes methods given in Cochran (1977). If for each site we let N = total number of batches, X = number of batches with discrepancies, n = number of batches checked, and x = number of checked batches with discrepancies, then we may use the hypergeometric distribution to determine the probability of finding x discrepant batches in the n batches examined when X discrepant batches actually exist in the total N batches. The number X that satisfies the probability statement $Pr(x = 0|X, N, n) = 0.05$ is the upper 95% confidence limit for the number of discrepant batches in the total N batches when no discrepant batches ($x = 0$) are found in the n sampled batches. In this case, there is a high probability that the seasonal discrepancy rate is less than X/N.

If a discrepant batch is found at a site during a given week, then the FPC randomly picks two additional batches from that week to be cross-checked. If neither of these new batches shows discrepancies, then the QA/QC procedure is finished for that week; however, if additional discrepancies are found, then selection of batches and cross-checking will continue until the site is back in compliance.

The third level of data quality assurance also occurs once a week. Each week, all the data that has been entered year-to-date at a site is sent back to the site as a large "validation" spreadsheet. These spreadsheets are then used by remote site personnel to compare to their own internal records of the data collected year-to-date, and any additional discrepancies between the data at the central repository and the data at the remote site are addressed at this time.

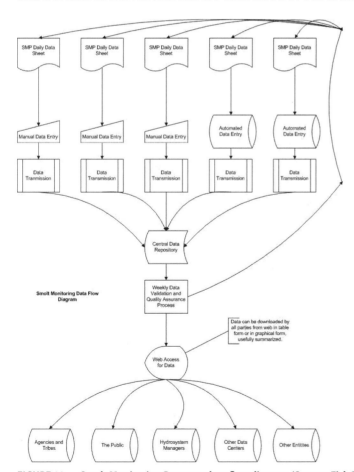

FIGURE 10. — Smolt Monitoring Program data flow diagram (Source: Fish Passage Center, Portland, Oregon).

Regional Integration

In the Columbia River basin, smolts originate from three states and cross multijurisdictional lines on their way to the Pacific Ocean. Because of this, smolt monitoring data collected at large mainstem hydroelectric projects must be regionally integrated to satisfy the needs of multiple managers.

Presently, the most useful data format for enabling regional data integration is the comma separated value (.CSV) format. The Federal Geospatial Data Committee has created a metadata format that is useful for regional integration, and the U.S. Geological Survey has further refined this metadata format for biological purposes and created a superset metadata format that is part of the National Biological Information Infrastructure. Another type of data that is useful for regional data integration is geographic reference data that can be used in geographic information systems.

When monitoring smolt passage across a large hydrosystem such as those in the Columbia River basin, one key concern for hydrosystem managers who review smolt data at this scale is whether the fish collected are actually active downstream migrants or true "smolts." Juvenile salmonids in the mainstem that have a fork length of 60 mm or less have been shown to be too small to be active downstream migrants, and these fish are therefore categorized as fry. While finer categorical distinctions (e.g. parr, fry, sac fry) may be useful to the life stage researcher studying juvenile salmonids in the tributaries, when juvenile passage data are summarized at the larger regional scale, the most useful distinction is simply between fry and smolts.

Personnel Requirements and Training

Responsibilities and Qualifications

The number of personnel required to conduct smolt monitoring activities at a dam site is ultimately dependent upon the number of fish to be handled on a daily basis. At a bare minimum, a three-person crew per sampling day is recommended: (1) a supervisory biologist, (2) an anesthetist, and (3) a fish handler/counter. Additional fish handlers/counters will be necessary to accommodate increased sampling needs such as to provide additional fish for research marking programs.

The supervisory biologist schedules, trains, and supervises staff, oversees and participates in handling and counting activities, communicates with the anesthetist and the fish handlers/counters to ensure proper coordination between anesthetization and counting activities, and monitors water temperatures and checks the recovery raceway periodically during sample processing for mortalities. The supervisory biologist is also responsible for coordinating with other activities, such as research marking programs, that may impact the operations of the smolt monitoring program. This person typically performs much of the electronic data entry, is ultimately responsible for ensuring that QA/QC protocols are followed at the site, and writes the annual report.

The anesthetist is responsible for preparing the stock anesthetic solution, watering up the recirculating anesthetic system at the start of the day and adding anesthetic, removing debris from the sample tank, crowding the sample fish into the pre-anesthetic chambers in the proper allotments and anesthetizing them at the proper dosage, flushing each allotment of sample fish to the sorting trough, dewatering the recirculation system at the end of each day, and periodically disinfecting the system. This person should also maintain a written log of daily anesthetic concentrations used for both the recirculation and pre-anesthetic systems.

The fish handlers/counters are responsible for identifying and counting the fish, hand-recording data for the supervisory biologist as necessary, and assisting the supervisory biologist with electronic data entry at the end of each sampling day.

Equipment

- Aquarium nets and dip nets
- Fish anesthetic (MS-222)
- Anesthetic dispenser
- Gram scale (for weighing fish and anesthetic)
- Buckets (20) (5 gal)
- Chlorine (for disinfecting the recirculation system)
- Rubber gloves and filtration mask (for mixing anesthetic or working with chlorine)
- Clipboards, mechanical pencils, and data forms
- Desktop computer(s), preferably with DSL or better data transfer capability
- Hard hats and steel-toed shoes (if required by dam operators to work at the dam site)
- Thermometer

Adult Passage Counts

Adult Passage Routes

Upstream passage of adult salmonids at large hydroelectric projects is largely restricted to the adult fishways. At projects so equipped, navigation locks may provide an additional route of passage, although this is normally considered negligible and not taken into consideration in adult passage estimates.

Fallback

Adult upstream migrants can also pass downstream, and these fallbacks can confound upstream passage estimates. Fallbacks are defined as prespawn upstream migrants that first pass upstream past a hydroelectric project and then fall back downstream. This is in contrast to kelts or members of *iteroparous* species such as steelhead that survive after spawning and pass back downstream as true downstream migrants. Fallbacks are upstream migrants that theoretically should be taken into consideration in upstream passage estimates; kelts are true downstream migrants and should not. Fallback may result in significant count error in terms of tracking the number of fish that remain above the dam after passing through the fishways. For example, 7%–10% of the adult steelhead that passed upstream through the adult fishways at McNary Dam were shown to fall back eventually through the juvenile bypass system prior to spawning (Wagner 1990; Wagner and Hillson 1991). Fallback routes include all those potentially available to juvenile downstream migrants as described earlier. Because it is usually not feasible to measure or monitor adult fallback comprehensively at large hydroelectric projects throughout each passage season, fallback, although mentioned here, is normally not taken into consideration in adult passage estimates.

Adult Facilities Description

Adult fishways at large hydroelectric projects typically are of a pool and weir design (see Figure 11) with attraction flows provided at the downstream entrances to help guide the adults to the fishways. This design creates a series of gradual steps that allow adult salmonids to pass from the project tailrace upstream to the forebay (see Figure 12).

Counting facilities are normally built into the fishway. The most common type of facilities feature picket leads, which effectively restrict fish passage to a narrow count slot. The spacing between the picket leads is 2.5 cm. The width of the count slot ranges from 46 to 56 cm; it is generally accepted that its width be no less than 46 cm. A large viewing window adjacent to the count slot allows personnel or cameras to view passing fish from an enclosed count room. The windows may be partitioned such that juvenile salmonids, jacks (precocious males), and adults can be readily distinguished and counted separately (see Figure 14). Placing a white backboard opposite the viewing window and controlling the source and direction of light within the count slot improves viewing conditions. Preferably, all sources of light originate within the count room and show through the viewing window and onto the white backboard. Lighting can affect fish passage and is best kept to a minimum.

A less common type of counting facility includes the same picket lead structure as previously mentioned, but the count slot includes a ramp that forces the passing fish very near to the surface. Personnel located above the count slot view the fish from above for identification. Again, controlling the source of light and painting all surfaces of the count slot white improves viewing conditions.

Last, there is a recently developed counting facility design for fishways that does not include overflow weirs. This design consists of a caisson installed into the ladder on the upstream side of a weir. The caisson includes a submerged viewing window that is perpendicular to the orifice in the weir. Fish are viewed as they pass through the weir.

In each type of facility, all fish ascending the fishway are counted either by direct observation by personnel stationed in the counting room or by interrogation of time-lapse video.

Trapping facilities may also be part of the fishway design; they allow operators to handle and examine fish directly for a variety of reasons, such as to collect tissue samples, tag fish, or remove fish for hatchery broodstock or transportation operations. PIT tag interrogation systems are becoming increasingly common in adult passage facilities. Such systems allow passive interrogation of all adult upstream migrants for PIT tags as the fish pass through the counting stations. PIT tag interrogation information can be automatically stored on local computer drives or transferred through the Internet to a central repository. Radio receivers may also be installed at adult fishways and allow counting of fish tagged for radio telemetry studies; however, these systems are usually installed to address a particular research need during a specific migration period and are usually not a permanent part of the adult counting system.

FIGURE 11. — Typical adult fishway (Source: USACE <www.nwd.usace.army.mil/ps/colrvbsn.htm>).

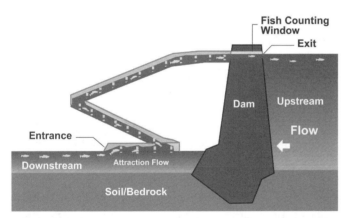

FIGURE 12. — Typical Columbia River fishway design (Source: USACE Fish Passage Plan 2005).

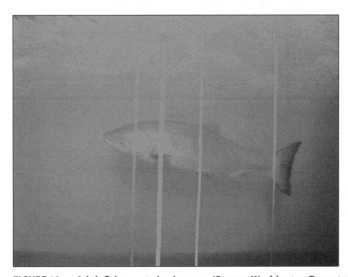

FIGURE 13. — Adult fish count viewing area (Source: Washington Department of Fish and Wildlife).

Sampling Rates

Virtually all upstream migrants—with the exception of those fish that are caught in the fishery, fall back, or die—must pass through the fishways and counting facilities at large hydroelectric projects and, thus, can be counted. Large river systems often support multiple species of upstream migrants, such as summer and winter steelhead; coho salmon *O. kisutch*; spring, summer, fall chinook salmon *O. tshawytscha*; sockeye salmon *O. nerka*, pink salmon *O. gorbuscha*, and chum salmon *O. keta*. These species—especially those that overwinter in river systems such as summer-run steelhead—can potentially pass hydroelectric projects 24-h per day each day throughout the year. Under such conditions, comprehensive accounting of all adult passage would require year-round, 24-h-per-day counting; however, a variety of constraints usually negate year-round counting, and the counting season is usually set to encompass the majority of each run. Fish that pass outside of the counting season are considered to be negligible in number.

Determination of an actual sampling rate to attain a specified sampling goal requires at least a site-specific understanding of both diurnal migration characteristics and seasonal migration periods for each species or race of concern. Counting periods should be adjusted to satisfy site-specific goals. In the Columbia River basin, the primary goal is to allow fish passage and minimize migration delay. Counting is therefore conducted during time frames that account for the majority of each run, without inhibiting fish passage.

The Washington Department of Fish and Wildlife is contracted to conduct adult passage counts at most of the USACE projects in the Columbia basin. The goal of the counting program is to account for at least 95% of the daily passage of Pacific salmon and steelhead (and at least 85% of the American shad) at each USACE project. It is based upon a 16-h-per-day, 7-d-per-week counting schedule, which is maintained throughout the primary migration periods for each species. An example of this schedule is provided in Table 2. The example schedule takes into consideration that adult passage occurs primarily during daylight hours, and with some-site specific adjustment, the migration season is generally April through December. It should also be noted, as further detailed in the Methods section, that direct ocular counts are conducted for 50 min each hour, and therefore, the true daily sampling rate is only 83.3% of the hourly counting periods identified below. Video interrogation counts occur continuously throughout each counting period.

TABLE 2.—Example counting schedule used at USACE Columbia River basin projects (Source: Washington Department of Fish and Wildlife)

Portland District	Walla Walla District
Bonneville Lock and Dam: Count both ladders.	McNary Lock and Dam: Count both ladders.
Jan 1–Mar 31: Video count 0400–2000 PST.	Mar 1–Mar 31: Video count 0400–2000 PST.
Apr 1–Oct 31: Visual count 0400–2000 PST.	Apr 1–Oct 31: Visual count 0400–2000 PST.
Nov 1–Dec 31: Video count 0400–2000 PST.	Nov 1–Dec 31: Video count 0400–2000 PST.
The Dalles Lock and Dam: Count both ladders.	Ice Harbor Lock and Dam: Count both ladders.
Feb 20–Mar 31: Video count 0400–2000 PST.	Mar 1–Mar 31: Video count 0600–1600 PST.
Apr 1–Oct 31: Visual count 0400–2000 PST.	Apr 1–Oct 31: Visual count 0400–2000 PST (Count south shore ladder by direct observation and north shore ladder by closed-circuit TV).

Portland District	Walla Walla District
Nov 1–Dec 7: Video count 0400–2000 PST.	Lower Monumental Lock and Dam: Count both ladders.
John Day Lock and Dam: Count both ladders.	Apr 1–Oct 31: Visual count 0400–2000 PST.
Feb 20–Mar 31: Video count 0400–2000 PST.	Little Goose Lock and Dam: Count one ladder (south).
Apr 1–Oct 31: Visual count 0400–2000 PST.	Apr 1–Oct 31: Visual count 0400–2000 PST.
Nov 1–Dec 7: Video count 0400–2000 PST.	Lower Granite Lock and Dam: Count one ladder (south).
	Mar 1–Mar 31: Video count 0600–1600 PST.
	Apr 1–Oct 31: Visual count 0400–2000 PST.
	Jun 15–Aug 31: Video count 24 hours per day (Interrogate videotape fish counts from 2000 to 0400 PST and only for the hourly 10-min breaks during visual counting 0400–2000 PST).
	Nov 1–Dec 15: Video count 0600–1600 PST.

Field/Office Methods

Counts by Direct Observation

Counts by direct observation are obtained instantaneously by fish counting staff during each daily count period. This counting technique simply requires that individual counting staff be present at each counting station to conduct direct counts of fish as they pass upstream through the viewing area. Counts are registered on hand tabulators appropriately labeled to represent each fish group of interest. An example of the fish groups counted at USACE projects in the Columbia River basin is provided in Table 3. Counts are registered for each fish that successfully passes upstream past the viewing area. Fish that pass back downstream through the viewing area are subtracted from the hourly count total.

TABLE 3. — Fish species counted at USACE projects in the Columbia River basin (Source: Washington Department of Fish and Wildlife)

1. Chinook salmon:
Adult—56 cm or longer in length.
Jack—30.5 to 55.9 cm in length.
2. Coho salmon:
Adult—46 cm or longer in length.
Jack—30.5 to 46 cm in length.
3. Sockeye salmon.
4. Pink salmon.
5. Chum salmon.
6. Steelhead.
7. Bull trout *Salvelinus confluentus*.
8. American shad *Alosa sapidissima*.
9. Pacific lamprey *Lampetra tridentata* (also known as *Entosphenus tridentatus*).

Counts by direct observation are normally conducted for 50 min each hour. This is done to allow fish counting staff the opportunity to transfer hourly counts from the tabulators to a daily summary form as well as to allow employee break periods. A 1.2 expansion factor is used to estimate fish passage during the 10-min period when counts are not taken. Daily forms are simply totaled, expanded by 1.2, and then delivered to the supervisory biologist.

Counts by Video Interrogation
The utilization of time-lapse video equipment to record the passage of fish has become increasingly popular in recent years as this counting technique does not require that personnel be present at the counting stations. Video cameras are installed at each counting station and continuously record fish passage through the viewing window. Required maintenance of this equipment is minimal. The typical video monitoring system consists of an analog video camera, digital video recorder (DVR), and monitor. The DVR is set to record 5 to 10 frames per second at the highest resolution possible for the duration of the count period. At a later time, a technician interrogates the video data for fish passage. Certain DVRs allow the technician to interrogate video data simultaneously while the DVR is recording current fish passage data. Other DVRs allow for the removal of a portable hard drive so the technician may interrogate the video data off site. In either case, manual review of videotapes is the most labor-intensive aspect of counting by video interrogation. Utilizing time-lapse video to perform video interrogation has several advantages over counting via direct observation:

1. Video interrogation is more accurate because it does not require count expansion to account for the 10-min-per-h break periods.
2. If desired, filming can occur 24 h per day with very little additional staff time required.
3. Video interrogation allows a greater level of quality control because the video data can be replayed and reanalyzed.
4. Video interrogation counts, including tape review, generally require less staff time than direct observation counts.

New digitally based video monitoring systems are now being developed and field-tested. Such systems consist of an area scan camera that is linked directly to a computer installed with a frame grabber card. The analog video signal is streamed to the PC from the camera. The generated images (~three frames per second) are filtered for the presence of fish using a motion-detection-based algorithm. Once the motion detector is triggered, the fish image(s) are captured and image file(s) are created and stored to the computer hard drive. Each fish image file is stored with an associated time-stamp image file. A photo-editing software program is used to sort through a series of fish images generated from fish passage events and the aforementioned data is recorded (Hubble 2000).

Digital imagery is a relatively new technology that is being applied to conduct adult fish passage counts at Rocky Reach Dam on the Columbia River and at the Prosser and Rosa counting facilities on the Yakima River (J. Hubble, Yakama Nation, personal communication). Digitally based systems provide sharp images, image enhancement capabilities, and PC-based image storage. Ethernet data transfer and downloading capability minimize the need for manual servicing at the data collection site. In addition, when the technology is further developed and refined, automated image editing capability will also become available. Because self-editing technology is still in development, manual editing of digital adult passage counts is currently required.

Data Handling, Analysis and Reporting

Data Handling
Manual count sheets are delivered to the supervisory biologist and all counts are expanded by 1.2 (to account for the 10 minutes per hour that counting is not conducted) and totaled for the day to provide a daily passage estimate for each species/group. In the case of video monitoring, videotapes are normally removed from the counting stations and reviewed at a central office, and fish passage is tallied for the day. In the Columbia River basin, daily adult fish passage counts are provided to such entities as the FPC, and this information is posted on the FPC Web site and made publicly accessible.

QA/QC
Quality control measures for direct observation count data are difficult to implement because these data are based upon live counts of fish passing through the viewing area. The supervisory biologist should regularly perform simultaneous fish counts with each technician to help identify problems. Technicians should be well trained in species identification and familiar with the equipment used in daily operations. Potential sources of error include species misidentification, missed observations due to the simultaneous passage of a large number of fish past the viewing area, and inadequate viewing conditions due to poor lighting or heavy accumulation of algae on the viewing window and count slot surfaces.

Counts via video interrogation are far easier to control for quality because these counts can be checked and rechecked as necessary to ensure accuracy. Although video counts do provide permanent records of fish passage, these records must still be manually reviewed and tabulated. As with direct observation counts, these manual tabulations should be checked for numeric accuracy at the conclusion of each count period by the supervisory biologist or by a second technician to minimize the likelihood of simple tabulation error.

Personnel Requirements and Training

Responsibilities and Qualifications
The number of personnel required each day to conduct adult counting activities at a given large hydroelectric project depends upon the number of adult fishways present and the duration of the daily count period. A supervisory biologist schedules, trains, and supervises staff, oversees the counting activities, performs QA/QC protocols, and writes the annual report. A total of four fish counters per day will be required to conduct 16-hour counts at two adult fishways. The fish counters are responsible for species identification and enumeration, preparation of daily count tally sheets, and review of video data at sites where video interrogations occur. They are also responsible for performing routine maintenance at the counting stations or informing responsible parties when such maintenance is needed (e.g., keeping the viewing window clean and algae free).

Equipment
- Hand tabulators
- Pencils, notepads
- Calculator
- Digital video recorder (e.g., Sanyo DSR 300)
- High-resolution 33-cm (13-in) monitor
- High-resolution 0.83-cm (1/3-in) 520 line video camera with V/F Lens

Acknowledgments

Dave Hurson (USACE) and Steve Richards (Washington Department of Fish and Wildlife) provided valuable information regarding adult passage facilities and counting programs at USACE projects in the Walla Walla District. Brad Eby (USACE) and Rosanna Tudor (Washington Department of Fish and Wildlife) provided helpful information regarding the juvenile bypass and sampling system configuration and smolt monitoring program procedures at McNary Dam. Jerry McCann (Fish Passage Center) provided a thorough description of the smolt monitoring program, much of which was adopted directly into this chapter.

Literature Cited

Berggren, T. J., L. R. Basham, M. DeHart, H. Schaller, O. Langness, E. Weber, and P. Wilson. 2000. Comparative survival rate study of hatchery PIT-tagged chinook. Status report for migration years 1996–1998. 2000 Annual Report. Report to Bonneville Power Administration, Contract No. 8712702, Portland, Oregon.

Cochran, W. G. 1977. Sampling techniques (3rd edition). John Wiley & Sons, New York.

Franzoni, H., D. Wood, and K. May, editors. 2004. FPC 32 Remote Site Data Entry Program. Manual available from Fish Passage Center Web Site: www.fpc.org.

Hubble, J. D. 2000. Adult fish video monitoring. Project Annual Report. Bonneville Power Administration, Project 88-120-5, Portland, Oregon.

Peven, C., A. Giorgi, J. Skalski, M. Langeslay, A. Grassell, S. Smith, T. Counihan, R. Perry, and S. Bickford. 2005. Guidelines and suggested protocols for conducting, analyzing, and reporting juvenile salmonid survival studies in the Columbia River basin. Available under its title at www.cbr.washington.edu/papers/index.html.

Sandford, B. P., and S. G. Smith. 2000. Estimation of smolt-to-adult return percentages for Snake River basin anadromous salmonids, 1990–1997. Journal of Agriculture, Biological and Environmental Statistics 7(2):243–263.

USACE (U.S. Army Corps of Engineers) Northwestern Division. 2005. Fish Passage Plan Corps of Engineers projects. U.S. Army Corps of Engineers, CENWD-PDW-R, Portland, Oregon.

Wagner, P. 1990. Evaluation of the use of the McNary Bypass System to divert adult fallbacks away from turbine intakes. State of Washington, Department of Fisheries, Habitat Management Division, Report to United States Army Corps of Engineers, Modification to Contract Number DACW-68-82-C-0077, Task Order Number 9, Walla Walla.

Wagner P., and T. Hillson. 1991. Evaluation of adult fallback through the McNary Dam Juvenile Bypass System. State of Washington, Department of Fisheriesm Habitat Management Divisionm Report to United States Army Corps of Engineersm Contract Number DACW-68-82-C-0077m Task Order Number 10, Walla Walla.

Redd Counts

Sean P. Gallagher, Peter K. J. Hahn, and David H. Johnson

Summary

The purpose of this protocol is to describe field methods for the consistent collection of salmonid redd abundance and subsequent estimation of adult salmonid breeding population size. We recommend surveys be conducted on predetermined, 3–5-km long stream reaches, using a spatially balanced rotating panel design. We suggest an annual draw of 10% of all reaches in the sampling universe as the target goal for monitoring; furthermore, to account for access problems and other barriers to sampling, we recommend that the initial sample draw should over-select reaches (sampling rate of 25%) to provide flexibility in the field. One field survey should occur prior to fish entering the spawning areas, with surveys thereafter conducted 7–14 d apart until new fish and redds are no longer observed. Surveyors will need to recognize that stream flows and/or weather conditions will have some bearing on the temporal aspects of surveys. All redds will be identified to species, measured, and georeferenced. Redd longevity and observer efficiency in redd detection will be estimated for each watershed by tracking the condition of individual redds measured during previous surveys. To document sex ratios, the sex of all live fish will be visually identified on behaviors at redds or other visual cues (dead fish will be identified, sexed, inspected for tags, and measured, per the carcass count protocol, page 59). In situations where multiple salmonid species overlap on a given spawning area, redd sizes will help differentiate the species involved.

Background and Objectives

Background

The family *Salmonidae* is characterized in part by most members being gravel nest spawners (Eddy and Underhill 1978). Female salmon and trout excavate a nest in gravel substrate, deposit eggs that are externally fertilized by one or more males, quickly cover these with gravel, and begin to dig another nest. A contiguous series of these nests is called a redd (Kuligowski et al. 2005). Nest or egg burial depth (DeVries 1997), redd size (Burner 1951; Orcutt et al. 1968), water depth, water velocity, and substrate preferences (Bjornn and Reiser 1991) vary among species. Burner (1951) divided redd building into three stages: prespawning, spawning, and postspawning. During prespawning, females excavate a nest pot and clear it of loose gravel and fine materials, leaving only larger cobbles with clean interstitial spaces for eggs to lodge. During spawning, the female alternately deposits eggs and digs to cover fertilized eggs with loose gravel as she moves upstream, digging more nests. The loose material dislodged and swept downstream is called the tail spill. During post-spawning, a female Pacific salmon has deposited all her eggs and, in what is termed "spent" condition, continues to dig gravels upstream of the nests until her death. Briggs (1953) wrote that both male and female steelhead *Oncorhynchys mykiss* drift downstream after spawning. Burner (1951) defines a mature redd as one in which all eggs have been deposited and some postspawning digging has occurred. Thus, a mature redd consists of a pot on the upstream end and a tail spill of excavated gravels covering the incubating eggs, with the downstream end of the tail spill consisting of excavated fine material not covering eggs (see figures 1–3). The shape of a completed redd may influence water movement through egg pockets (Thurow and King 1994). Newly formed redds appear lighter in color than the undisturbed channel, except in gravel of basaltic origin (Bjornn and Reiser 1991), and may remain discernable for a period of days to weeks, depending on stream flow and periphyton accumulation (Susac and Jacobs 1999; Gallagher and Gallagher 2005; Isaak and Thurow 2006).

As the product only of reproductive adults, counts of salmon redds provide an index of effective population size (Meffe 1986). Redd counts are widely utilized to provide indirect estimates or indices of spawning escapement on rivers that lack counting facilities (Beland 1996; Maxell 1999). For example, Isaak et al. (2003) used a 45-year chinook salmon *O. tshawytscha* redd count data series to examine metapopulation characteristics in Idaho. Redd counts have been used to monitor chinook salmon since 1947 and Chum salmon *O. keta* since 1998 on the Columbia River (Dauble and Watson 1997; Geist et al. 2002). Bull trout *Salvelinus confluentus* populations have been monitored in Idaho and Montana for more than 20 years using redd counts (Maxell 1999; Dunham et al. 2001). Redd counts have been used in California, Oregon, and Washington to monitor coho salmon *O. kisutch* (Lestelle and Weller 2002) and steelhead populations *O. mykiss* (Maahs and Gilleard 1993; Jacobs et al. 2001; Boydstun and McDonald 2005). Redd counts are the primary metric for monitoring salmonids in Washington and Oregon and are proposed for monitoring coastal salmonids in California (Boydstun and McDonald 2005). Atlantic salmon *Salmo salar* populations have been monitored in Maine and other parts of North America for many years (Beland 1996 and references therein). Redd counts are used in evaluating population trends (Rieman and Myers 1997; Maxell 1999).

Population growth rate (e.g., the number of recruits-per-spawner) (Isaak and Thurow 2006) is typically derived from data sets in which robust estimates of escapement and recruits are available (Beland 1996). Although mark–recapture experiments have been shown to be accurate and precise for salmonid population estimation (Minta and Mangel 1989), they require capture programs that are expensive to operate and maintain, are subject to mechanical failure, require that fish be handled and tagged and their movements impeded, and often require specific geomorphic and hydrological features for placement and operation (Gallagher and Gallagher 2005). With due respect given to the inherent observation error rates involved, redd counts offer a less intrusive and less expensive alternative to mark–recapture programs.

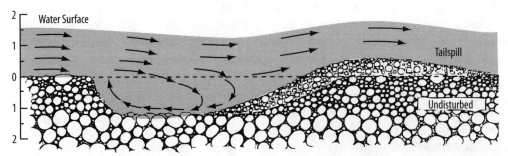

FIGURE 1.—Typical currents in a salmonid redd (Illustration: Andrew Fuller, from Burner 1951, 98)

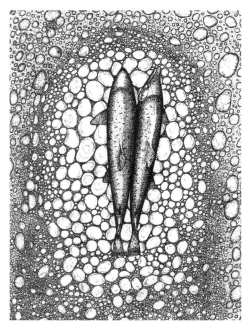

FIGURE 2. — A pair of spawning salmon on a redd (from Burner 1951, 99)

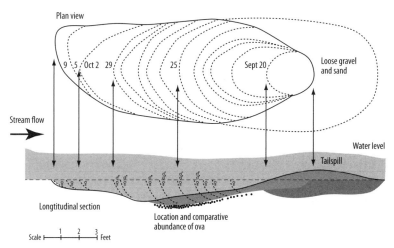

FIGURE 3. — Diagrammatic views of a fall chinook salmon redd measured daily (Illustration: Andrew Fuller from Burner 1951, 101)

Rationale

Fish populations are an integral component of any aquatic system in which they exist. Intermittent streams as well as perennial rivers and streams are important for salmon spawning and contribute to aquatic ecosystem health (Everest 1973; Erman and Hawthorne 1976). In many instances, salmon act as keystone species, and, by their presence, absence, or trends in abundance, they provide an indication of the overall health of a watershed (Cederholm et al. 2001). Salmon presence, absence, or trends in abundance may also indicate the condition of specific components of watershed health, such as water quality, quantity, or temperature.

Dunham et al. (2001) state that redd counts are less intrusive and expensive than tagging, trapping, underwater observation, weirs, and genetics for inventorying bull trout populations, and that with limited resources, more populations can be inventoried over a longer period. Counting redds can be done with relative ease, and the counts may serve as an index of adult spawning escapement (Beland 1996; Dauble and Watson 1997; Rieman and Myers 1997; Maxell 1999; Muhlfeld et al. 2006). If a fishery manager could estimate the level of adults returning to spawn, as measured by redd counts, and identify areas where egg deposition was not apparent in otherwise suitable habitat (i.e., under-seeded), the opportunities for improving management targets would be greatly enhanced (Beland 1996). Redd counts coupled with information on population age structure can be used to calculate population growth rate (e.g., number of recruits per spawner) and examine metapopulation dynamics (Isaak and Thurow 2006).

Despite the potential effects of counting errors and species (and life history type) differences in contributions to redd abundances, redd counts have been shown to be significantly related to independent estimates of escapement for a number of salmon species. Redd counts for Atlantic salmon have been shown to be positively correlated with the numbers of adult spawners (Hay 1984; Heggberget et al. 1986; Madden et al. 1993; Semple et al. 1994; Beland 1996). Similar relationships between adult abundance and redd counts have been shown for brown trout *Salmo trutta* and brook trout *Salvelinus fontinalis* in North America (Benson 1953; Beard and Carline 1991) and brown trout in New Zealand (Hobbs 1937). Gallagher and Gallagher (2005) found significant relationships between chinook and coho salmon *O. kisutch* and steelhead redd counts and independent escapement estimates. Steelhead redd counts and females released above a counting structure on Snow Creek, Oregon were significantly positively related (Susac and Jacobs 2002). On the Columbia River from 1964 to 1992, chinook salmon escapement over McNary Dam was significantly positively correlated with redd counts (Dauble and Watson 1997). Bull trout escapement was significantly related to redd counts in Idaho (Dunham et al. 2001), and expanded redd counts and mark–recapture estimates were found to be similar among basins in Oregon (Al-Chokhachy et al. 2005). These studies support the notion that redd counts serve as reasonable indices of salmon escapement. It should be noted that, while most of the above studies tried to ensure the best redd counts possible, few applied measures to reduce counting errors or address observer efficiency. Nonetheless, significant relationships between redd counts and escapement were observed in these studies. This suggests that, while reducing errors in redd counts should be part of any monitoring plan, it may not be fundamental to establishing basic relationships between redd counts and escapement.

Objectives

The objectives of employing redd count surveys for salmonids are to

1) index temporal abundance of spawners;
2) estimate total abundance of spawning females or, when other data is available (e.g., carcass surveys), the total spawning population;
3) determine spatial spawning distribution; and
4) determine temporal spawning distribution.

Tasks necessary to achieving these objectives under this protocol include

- accurately counting spawning salmon redds by species;
- determining redd life (i.e., longevity of redds);
- consistently measuring redd sizes and dimensions (1) for species identification, (2) to compare redds made by hatchery and natural-origin fish, and (3) to distinguish redd characteristics of resident and migratory forms that spawn at similar times and places (e.g., *O. mykiss*).

Accurate estimates of the number of adult fish escaping harvest (escapement) to spawn are essential for effective management and conservation of salmonids (Busby et al. 1996; McElhany et al. 2000; Chilcote 2001). Trends in the reproductive portion of a population are often the most important characteristic for species recovery and conservation (Al-Chokhachy et al. 2005).

The validity of redd counts for population monitoring depends on two critical assumptions: (1) redds are counted with minimal error (no double, over-, or undercounting errors), and (2) redd numbers reflect population status (Dunham et al. 2001).

Redd counts can be used to estimate the number of female spawners in a given year by assuming one redd per female or by multiplying redd counts by a constant such as 1.2 to account for multiple redds per female (Duffy 2005). Redd counts have also been used to estimate total escapement by multiplying redd counts by estimates of the number of fish per redd (Al-Chokhachy et al. 2005; Gallagher 2005a) or by using redd areas and estimates of the female-to-male ratio (Gallagher and Gallagher 2005).

Redd counts, assuming that counting errors are sufficiently reduced, are usually employed because they provide a cost-effective index of adult salmon abundance useful for population monitoring and trend detection (Schwartzberg and Roger 1986; Rieman and Myers 1997; Maxell 1999; Isaak et al. 2003). The accuracy and precision of any salmonid population monitoring technique should be critically evaluated (Al-Chokhachy et al. 2005) and, if employing redd surveys, should include a pilot study to evaluate bias in redd counts for population monitoring prior to initiating a large-scale, long-term program.

Sampling Design

Redd counts for population monitoring should be conducted following a sampling design appropriate for estimating annual abundance and population trends for the species of interest. The spatial scale of such a monitoring program might range from the reach level to the watershed or regional level. The geographic scale of the monitoring program will influence sampling scheme selection. Isaak and Thurow (2006) suggest that conducting redd counts with a spatially continuous, temporally replicated sampling design will reduce errors associated with simple random designs and provide more accurate ecosystem views. Rotating panel designs (Firman and Jacobs 1999) incorporate the need for high precision in annual estimates of abundance at a broad spatial scale with the need for a large number of repeat visits necessary for trend detection (Boydstun and McDonald 2005). In coastal Oregon, salmonids are monitored by spawning ground surveys where 10% of all habitats are surveyed annually. For redd counts in Oregon, individual stream reaches are selected using generalized random tessellation stratified sampling (GRTS), a type of spatially balanced rotating panel design (Stevens 2002). This approach has recently been proposed for monitoring California's coastal salmonids (Boydstun and McDonald 2005).

When conducting redd counts, counts in nonrandomly selected index areas should not be used. Index areas may not represent population dynamics of salmonids at a regional scale (Rieman and McIntire 1996) and may miss redds due to annual variation in spatial distributions or spawning activity (Maxell 1999; Dunham et al. 2001). Lack of randomization in site selection and shifts in fish distribution may bias results from index site monitoring and mask population trends (Isaak and Thurow 2006).

A complete census requires that all stream reaches known to support the fish species of interest be surveyed for redds. If financially and logistically feasible, a census gives the best information for total redds (and female spawners), including spatial and temporal parameters. Care must be taken to ensure that all possible spawning areas are surveyed. Although the variance of the total redd estimate is zero, appropriately, it does not mean that the true number of redds is known. Krebs (1989) wrote that total counts are often of dubious reliability. This can be due to factors that prevent surveyors from seeing redds, overlapped redds counted as single redds, counting natural scour features as redds, and not including all spawning areas within the survey. However, these same factors operate identically in any statistically designed sampling program and are not included in their variance estimates. A census design should include explicit subdividing of the stream network so that the quality of the data can be described (such as clustering of redds). If possible, mapping each redd or using a global positioning system (GPS) to record each redd position is most desirable. Annual redd counts from index site monitoring generally consist of only a single point estimate and often are simple counts of visible redds. Such counts have no estimate of the associated statistical uncertainty (Maxell 1999) because they have not been randomly selected.

In many or most situations financial and logistic constraints force a sample design to be the best method of estimating total redds. Gallagher and Gallagher (2005) used a stratified index approach to estimate escapement from redd counts for several coastal rivers and streams in California. Their stratified random approach (Irvine et al. 1992) provided redd counts with associated statistical uncertainty for individual streams. This variance is the uncertainty of a sample expanded to nonsampled areas, not the uncertainty inherent in counting redds themselves. Other targeted research is needed to quantify the biases that cause redd counts to deviate from the truth. Random systematic and adaptive sampling designs should also be considered for redd surveys because of redd clustering. These should be compatible with a GRTS design. A stratified index approach, as well as large-scale regional approaches (Firman and Jacobs 1999), requires that the entire extent of spawning habitat within the sampling universe is known or established. The stratified index estimates employed by Gallagher and Gallagher (2005) were a specialized form of block sampling where the stream segments were blocks and the entire length of spawning habitat in a stream was the census zone. The mean and variance around redd density was calculated from the blocks and multiplied by the length of the census zone to estimate escapement for each stream. This approach can be used for estimating redd abundance in cases where the area of interest is a stream reach or a tributary and is fundamentally the same approach used in rotating panels designs (Boydstun and McDonald 2005).

The choice of the physical method used for counting redds will influence the sample design for redd counts surveys, site selection, sampling frequency, field and laboratory methods, and equipment needs. For smaller streams with reasonable accessibility, walking upstream and marking and counting redds is common (Gallagher and Gallagher 2005; ODFW 2005). For larger or more remote streams, rafting or aerial surveys by fixed-wing aircraft or helicopters can be used (Dauble and Watson 1997; ODFW 2005). The accuracy of aerial surveys may differ from walking surveys (Jones et al. 1998) and boat surveys. In some cases, a combination of methods may be used (Gallagher and Gallagher 2005; Isaak and Thurow 2006). In streams where spawning occurs in deep water, SCUBA diving or underwater cameras may be used to count redds (M. Gard, U.S. Fish and Wildlife Service, Sacramento, California, personal communication).

Survey designs and evaluation of bias in redd counts will likely vary depending on the field counting method. In instances where aerial surveys are used, it may be possible to calibrate observers by comparing repeat counts from multiple individuals, by using computer simulation (Jones et al. 1998), or from aerial photography (Neilson and Geen 1981). Bias in redd counts should be examined, reduced, and reported regardless of the physical method of counting and especially if counts are made from more than one field method.

Site Selection

Prior to selecting sampling sites, the study objectives, the size or distribution of the target population and the indicators that will be measured must be defined. Generally, a project area is defined in terms of the geographic range of a fish population, stock, or, in the case of threatened and endangered species monitoring in the United States, by evolutionarily significant units (ESU). If the objective is to obtain a complete census of the spawning population, stock, or ESU, then it is important to determine the distribution of the target population. In the case of redd counts for population monitoring, this would entail establishing the upstream and downstream limits of spawning habitat within streams in the study area. Criteria used for defining the population distribution must be clearly articulated (e.g., upper extent is < 20% gradient maintained for 100 m, gravel composition < 20% fines, water temperature < 20°C). Modeling approaches using combinations of geology, rainfall, land use, and other variables may be useful in delineating the extent of spawning habitat within a sample universe (Steel et al. 2004; Agrawal et al. 2005). Consultation with persons knowledgeable about areas within the sampling area or review of documents in agency or academic files may also assist with defining the spawning habitat in a sampling universe. A Delphi-type assessment might be employed to develop a rough cut at the extent of spawning habitat in the study area and then refined by field reconnaissance or as part of a pilot study. To determine the entire length of spawning habitat in streams where the extent of spawning habitat is not known, it should be determined by surveying the entire area of suspected habitat during the first year with foot survey efforts continuing for about 1 km above the last redd observed or to assumed barriers. In Figure 4 we offer an example map showing a stream with multiple sample reaches. Included on the map are the stream name (i.e., Pudding Creek), segment/reach name, stream segment number (under the GRTS sampling scheme), and locations of reach breaks.

Typically, sampling sites within a study area are selected probabilistically to reduce bias (Firman and Jacobs 1999; Stevens 2002; Boydstun and McDonald 2005). It is important to select a method or combination of methods that increases the degree to which the selected sample represents the population. The Oregon Department of Fish and Wildlife uses the Environmental Protection Agency's Environmental Monitoring and Assessment Program to draw a spatially balanced sample from the universe of all possible spawning habitat in its study area (Firman and Jacobs 1999). Boydstun and McDonald (2005) suggest an annual draw of 10% of all 3–5 km reaches in the sampling universe as the target goal for monitoring California's coastal salmonids; to account for access problems and other barriers to sampling, they suggest that the initial sample draw should over select reaches (sampling rate of 25%) to provide flexibility in the field. Some physical characteristics of the study area, such as snow melt–driven versus rainfall–driven systems, dams, or water diversions, and the life histories of species of interest may influence the sampling scheme and sample draws.

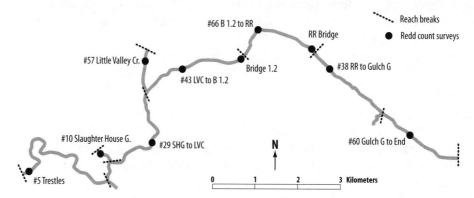

FIGURE 4. — Reach map for redd count surveys in Pudding Creek, California. Stream reach numbers reflect sample order under the generalized random tessellation stratified sampling (GRTS) scheme. (Note: Reach breaks are based on tributaries and other landscape features.)(Illustration: Andrew Fuller.)

Sampling Frequency

Redd counts for population monitoring should include a sampling design that is appropriate to the species of interest, should have sufficient replication for describing uncertainty, should have an established level of acceptable error, and should be conducted less than 14 d apart throughout the spawning run (ODFW 2004; Gallagher and Gallagher 2005). Surveys should begin prior to the onset of spawning of the species of interest and continue at least biweekly until spawning is complete. Redd count surveys should include marking newly made redds and recounting marked redds to estimate observer efficiency and reduce counting errors. If there is overlap in spawning among species, experiments and techniques should be established to differentiate between redd species. In some cases, and for some species, the duration of spawning may be short enough that one count at the end of spawning season can be used for an annual index, assuming that redds are not obscured by age, bedload movement, or scour, and therefore missed. In these instances, a pilot study should be conducted to determine if short duration of spawning activity exists and to examine potential bias in counts from one visit at the end of spawning. Muhlfeld et al. (2006) used statistical models of the error structure in redd counts to estimate observer efficiency for redd counts conducted once at the end of the spawning season. A redd count survey for such species, with timing of surveys dependent on some physical predictor of the onset of spawning, might employ the peak count method (Parken et al. 2003) to estimate redd abundance.

Error Rates and Reducing Observational Error
Counting errors may arise from redd species misidentification (Gallagher and Gallagher 2005), variation in habitat and redd characteristics (cover complexity, water depth and visibility, substrate composition, redd size, redd age, superimposition, or density), and if redds are obscured by periphyton or high discharge, spawning occurs outside the survey period, and spawning shifts from monitored to unmonitored reaches (Maxell 1999; Dunham et al. 2001). Redd counting errors may also occur due to variation in an individual observer's experience, training, energy, and enthusiasm (Dunham et al. 2001; Muhlfeld et al. 2006). Thus, counting errors may obscure important population trends (Beland 1996) and should be reduced to improve the power of redd counts for trend detection (Maxell 1999; Dunham et al. 2001).

The use of redd counts for population monitoring may be further complicated if female salmonids make more than one redd (Reingold 1965; Crisp and Carling 1989; Gallagher and Gallagher 2005; Kuligowski et al. 2005). The presence of "test" redds or false redds that are abandoned before eggs are deposited (Crisp and Carling 1989) may artificially inflate redd counts relative to escapement. The contributions of different life history forms of a species to redd abundance may influence redd count escapement relationships (Zimmerman and Reeves 2000, Al-Chokhachy et al. 2005). Overlap in spawning time and location among sympatric species (Fukushima and Smoker 1998; Geist et al. 2002; Gallagher and Gallagher 2005) may also confound redd count escapement relationships. An uneven sex ratio in the population of interest can complicate the use of redd counts for population monitoring.

A number of approaches for reducing redd counting errors due to species misidentification have been developed. Gallagher and Gallagher (2005) used logistic regression analysis to discriminate between chinook salmon, coho salmon, and steelhead redds in California and found that redd size (calculated from pot and tail spill measurements) and date were significant in predicting species. This approach reduced observer error in species identification from an average of 16% to 3.4%. In Alaska, sympatric sockeye salmon *O. nerka* and pink salmon *O. gorbusha* nests were classified by Fukushima and Smoker (1998) using discrimination analysis of depth, velocity, and stream gradient; however, their discrimination analysis misclassified more than 33% of the redds of these two species. Differences in chum and chinook salmon redd site selection within the same side channel of the Columbia River were attributed to differences in upwelling and downwelling of water in the hyporheic zone (Geist et al. 2002) and factors such as these might be important in differentiating redd species. Zimmerman and Reeves (2000) used stepwise discrimination to differentiate between redds of resident and anadromous steelhead, which correctly classified more than 64% of redds. Al-Chokhachy et al. (2005) attributed discrepancies between expanded redd counts and mark–recapture escapement estimates of bull trout to different contributions of migratory and resident forms, but they did not classify redds by life history type. Monitoring programs employing redd counts in systems in which sympatric species or different life history forms overlap should employ methods to reduce error in redd species identification. This should be examined in a pilot study as part of the development of a monitoring program.

Researchers have recently provided information relating to factors associated with errors in redd counts and improvements in survey design and data analysis methodologies that reduce counting errors in redd surveys. Dunham et al. (2001) did not find any relationship between habitat characteristics or observer experience level and redd count errors for bull trout redd counts in Idaho. They found that errors of both omission and misidentification were common among observers and that these errors tended to cancel each other out; furthermore, the researchers suggested that redd counts would be improved by better training of observers and use of more experienced personnel. Muhlfeld et al. (2006) also found that omission and false positive errors tended to cancel each other. They wrote that redd detection probabilities were high among observers, experienced observers made fewer mistakes, and redd counts can be used to accurately monitor bull trout populations. Muhlfeld et al. (2006) used models of observer error structure in redd counts to correct historic redd counts and thus improved their utility for population monitoring. Gallagher and Gallagher (2005) evaluated some of the bias in salmon and steelhead redd counts in California due to errors in redd species identification, detection of redds, and duration under variable survey conditions. They were able to reduce counting errors and produce reliable and precise redd counts by surveying biweekly, statistically differentiating redd species, uniquely marking redds and recounting all marked redds, calculating observer efficiency as the average of the percentage of known redds observed during each survey, having observers work in pairs, providing a "test" category for incomplete or questionable redds, and providing field and laboratory training. Gallagher and Gallagher (2005) presented models for predicting observer efficiency in redd counts from measurements of streamflow and water visibility. Monitoring programs using redd count surveys should investigate observer efficiency relative to the specific goals of the program and employ methodologies to reduce or eliminate errors in redd counts. This should include field and laboratory training of all observers, use of experienced observers whenever possible, pairing experienced and inexperienced observers, and providing specific written protocols to survey teams.

Field/Office Methods

Field season preparations
Depending on the species of interest, the geographic extent of the survey, and the desired precision in the data, the field season preparations for a redd count survey may require a few weeks or many months. In the case of a large-scale rotating panel survey design, defining the species of interest, the project goals and objectives, the sampling universe, and stream reach areas, creating a data matrix and conducting the sample draw, and identifying selected reaches on the ground might take several years. This process would likely need to include defining the extent of spawning habitat and access points, acquiring landowner permissions and necessary permits, and defining data flow and field survey logistics. Including issues such as the need to define the roles and responsibilities of the various entities involved, data needs, reporting requirements, and specialized permitting

and compliance issues with endangered or threatened species or state collecting permits could further prolong field season preparations. A good example of the complexities and field season preparation needs for redd count surveys is the Oregon Plan experience (<www.oregon-plan.org>). Preparation of the ODFW (2005) Coastal Salmon Spawning Survey Procedures Manual drew upon the decades of cumulative experience of many individuals and took a number of months to complete.

Prior to the start of sampling, the species of interest, geographic extent of the survey, and specifics of the data to be collected in relation to the desired precision of the results should be defined. In general, preseason preparation includes defining and selecting sampling reaches, identifying access points, acquiring landowner permission, developing survey maps and data forms, developing a survey schedule, preparing databases for data storage, identifying personnel and equipment needs, hiring and training survey crew members, and purchasing necessary equipment. Locating and setting up housing for remote area crews might also occur as part of preseason preparations. Holding coordination meetings, if necessary, with other entities involved in conducting redd count surveys within the study area should be done prior to the field season. Training of field crews prior to survey start should also occur as part of the preseason activities.

Events sequence during field season

We recommend weekly to biweekly surveys in a rotating panel or stratified random design beginning prior to the onset of spawning of the species of interest and continuing until spawning is complete. Further, we recommend that surveys be combined with marking newly made redds and recounting marked redds, estimation of observer efficiency, and reduction of counting errors. Crews should work in pairs for safety and have communication devices in case of emergency. Walking upstream, crews can generally cover about 3 to 5 km in a day, depending on redd density and stream complexity. Floating downstream in rafts or kayaks, crews can cover 8 to 10 km per day, depending on redd density, stream size, side channel complexity, and logistics of access points. Surveying by aircraft can increase the area covered in a day, but this type of survey is affected by stream topography and vegetation and limited to midday periods when sun reflection glare is low and visibility is good. The sequence of surveys should be such that all reaches are resurveyed less than 14 d apart. With multiple crews, large geographic areas, and stream flow or visibility limitations, the logistics of daily survey schedules can be quite complicated.

On each survey, surveyors will look for new and old redds and record information for flagged and newly created redds. Before crews depart for the day, they should check to see they have all the necessary field and safety equipment, maps, and data forms and know which reaches are being surveyed by which crew. The destination reaches, crew members, satellite phone number, vehicle(s) taken, and estimated completion time for each crew should be recorded on a checkout board each day.

Measurement details

Measurements taken during redd count surveys will depend on the species of interest, study design, required precision of the data, presence of more than one species or life history form, and the overall goal(s) of the monitoring program. Sufficient data should be collected to meet the primary criteria of establishing the validity of redd counts for population monitoring; these criteria are (1) redds are counted with minimal error (no double, over-, or undercounting errors), and (2) redd numbers reflect population status. This second criterion will require independent escapement estimation or will have to be accepted as true based on prior study or on other publications of redd population relationships. In the field, data should be entered on prepared forms (using write-in-the-rain paper) or into preprogrammed handheld computers.

The date, climatic conditions, stream name and reach identifier, surveyors, streamflow, and water visibility (quantified as the maximum depth the stream substrate is visible) should be recorded for each survey. All redds observed should be counted, measured (only if redd is completed), and uniquely marked with labeled flagging tied to the nearest solid object directly upstream of the pot to avoid double counting. All newly constructed redds observed should be identified to species, treated as unknown, or denoted as "test" (i.e., redds that appear incomplete to observers) or under construction; they then should be marked with flagging and counted during each visit. Test redds and redds under construction should be reexamined on consecutive surveys and reclassified appropriately based on their apparent completion.

During redd count surveys, individual redds should be counted, marked, and uniquely labeled on data forms and in the field to avoid double counting and to allow estimation of observer efficiency (Gallagher and Gallagher 2005). At a minimum, the date each redd was first observed, fish species, unique identifier number, and location should be recorded on the data form. Redds can be marked in the field by tying survey flagging securely to the nearest solid object near or above each redd. For each redd, the unique identification number, date first observed, location relative to the flag (distance and compass direction), and species should be recorded on the flag and on the data form. Redd locations should also be marked on topographic maps or otherwise georeferenced. It is important to keep track of redds on which fish are still active and redds that do not appear to be completed. In these cases, redds should be reexamined on the next survey. An indication that a redd should be reexamined should be added to the flag for that redd and noted on the data form so that its date of completion can be established. In some instances, redds identified in the field as not yet complete at the time of first observation are actually either "test" redds or stream features that are not actually redds. By keeping track of them in the data, it is possible to remove them from the final redd counts at the end of the season and thus reduce this source of overcounting error.

Knowledge about redd size (including shape, substrate, position in river, and size of river) is most useful to spawner survey staff while assigning redds to species. Spawning time for each species is also critical. For differentiating redd species, examining fish length and redd relationships (Crisp and Carling 1989), determining fish life residence times, and reporting known species redd sizes, it may be necessary to keep track of redds on which known species of fish were observed.

Filling Out Data Forms in the Field

We offer the following as guidance for filling out data forms in the field; readers may have additional redd data needs beyond those noted here.

Header of Data Form

Fill in redd and fish data forms header information for each survey even if nothing is observed (see Appendix A for blank and example data forms); if needed, use extra space for detailed notes on the back of the data form. Use the stream name, segment or reach name, reach number, and map number from the map of the segment you are surveying. If you are surveying a stream that has multiple reaches (reach numbers), use a new data form for each segment and write the section number in the section space in the data form header, even if no redds are observed. It is very important to keep the data for each reach separate and identify the reach each redd came from. For surveys in streams with multiple reaches, redds at the lower end of the section that are on the boundary line are not counted; those at the upper end of the section that are on the boundary line are counted. Record the date of the survey. If a survey section takes more than 1 day, note the date of the second day in the notes. The week number is the survey week; the first survey of the season is week 1. Write the names of the people doing the survey in the surveyor's space. The map number is shown on the header information on the map page for each reach and is the same as the reach number; record that number here. Record the reach ID number in the appropriate space (this is the GRTS sample number). Record the air and water temperature in centigrade. Estimate the water visibility in meters with the survey rod as the visible depth to the stream substrate on every survey. Estimate the stream flow at the downstream end of each survey section on every survey and record the stage from the stage gauge, if present. A quick way to estimate the stream flow is to:

(1) measure the wetted width of the channel perpendicular to the flow (in meters) in a run area that lacks surface turbulence, under cut banks, and overhanging vegetation;

(2) measure the depth of the water (in meters) across the channel at three or four points and average these;

(3) multiply the width by the average depth;

(4) hold the wading staff parallel to the stream flow just above the water surface so that you have 1 m in view, drop a leaf at the top end of the 1-m mark on the wading staff while using the second hand of a watch or counting one-one thousand, and so forth, to estimate how long it takes the leaf to float 1 m;

(5) divide the number of seconds it took the leaf to float one meter by the result from step 3 above, which results in flow in cubic meters per second.

Record the drive time as the total drive time to and from the survey site. Record the start and end time of the survey as the time from leaving the vehicle until returning to it. Record the current weather conditions (e.g., sunny and cold, light wind). Note conditions of importance such as landslides, road conditions, or changes in stream visibility.

Redd Data

Record Number

Each redd and fish gets an unique individual record number. The record number is a seven-digit numeric code based on survey date that is linked to the stream and reach information in the header information. The first two numbers are the month (01 would be January and November would be 11). The second two numbers are the day of the month such that the second of the month would be 02 and the 15th would be 15. The following three numbers range from 001 to 999 and each redd and fish gets a consecutively higher number each day. For example, if you see a fish on a redd during a survey of reach 12 in the XYZ River on February 15, the redd number would be 0115001 and the fish number would be 0115002. Write the record number for each redd on the flagging (see Figure 5) and on the data form (see Appendix A). This number should be recorded on the maps and data forms (see Figure 6) for each redd observed.

>   1 ONMY REDD 0115001 CDFG 2003
>   Length = 2.4 Width = 1.4 Re-measure
>   Age 1

FIGURE 5. — Example of properly labeled flagging for a salmonid redd.

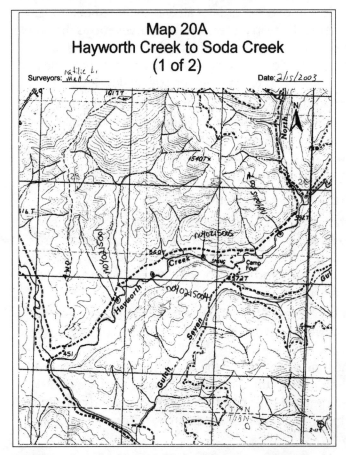

FIGURE 6. — Example of properly filled out map survey page with redd locations and record numbers tied to salmonid redd data form.

Species
Visually identify the species of each redd to the best of your abilities. Use the species code on the data form for fish and redds. Record this on the data form and on the flagging. If a redd is under construction and a fish is on the redd and you clearly identify it, use the species code—the first two letters of the genus and species of the study organisms (Onmy, Onki, Onts, Latr = Pacific lamprey, or Unkn for unknown)—record all data, and write "Remeasure" in the notes and on the flag. If the redd is classified as test or under construction (Fish on = yes), write this on the flagging and record the total length and width (sum of the pot and tail spill length and the maximum width) on the flagging with the word "remeasure". If you come across a flag with that instruction on it, remeasure it. If it is now clearly one species or another, use the species code with the previously used record number on the data form and record all appropriate data on the data form. If it has not changed, do not remeasure it. Do note the redd age (see Redd Age). If you remeasure a test redd and reclassify it, cross out the words "test" and "remeasure" on the flagging and write the species and date on the flag. Leave the original record number on the flag.

Fish on Redd
If you observe a fish on a redd (record "yes"), do your best to identify it to species. If there is not a fish on a redd, write "no".

Redd Age
To determine how long we are able to observe redds and estimate our ability to count all redds present (Gallagher and Gallagher 2005), estimate the redd age and record if and when it was previously measured. Record the redd age as

    1 = new since last survey but still clear,
    2 = still measurable but already measured,
    3 = no longer measurable but still apparent,
    4 = no redd apparent, only a flag, and
    5 = poor conditions; cannot determine if present and measurable or not.

Note that all redds that have not previously been encountered (no flag present) are by definition age 1; if you come across an unflagged redd, it is age 1; if it looks older, write this in the notes column. On a subsequent survey, when a redd is no longer apparent and the flag is still there, record a 4 in this column and write 4 and circle it on the flag. Note that you did this in the notes. During surveys, when you find a redd flag with a circled 4 on it, just keep on going; there is no need to record any further information about this redd. If, however, a new redd has been constructed in this spot, do note the presence of the flag with a 4 on it in the notes and record all information for this new redd.

Remeasured?
This is a yes or no column. Yes if remeasured, no if not.

Distance to Flag
This is the distance in meters from the middle of the redds' tail spill to the flag that identifies this redd. Flagging must be tied to a solid object (see Figure 7). Record

the distance from the tail spill to the flag in this column. Write this information on the flag as well.

Direction to Redd
This is the compass direction from the flag to the middle of the tail spill. Record this information in this column and on the flag (see Figure 5).

FIGURE 7. — **Coho salmon on a flagged redd.**

Notes
Record if the redd is irregularly shaped (not a circle, ellipsis, oval, square, or rhomboid) or if you suspect superimposition. What shape is it? Record the part of the stream where the redd is located (e.g., in the middle, on the side or edge, above or under a log.). Record other pertinent information. Use the back of the page with the record number followed by any additional information.

Page __ of __
The redd data form is page 1 to *n*, depending on how many redd data forms are used for each survey. If there are no redds, do not include the maps; otherwise, the map pages are the next numbers consecutively following the redd data page numbers. Fill out the entire data form for each survey, even if nothing is observed. Staple the sheets together and file them appropriately when you return to the office at the end of the day.

Flagging
For all redds, write the record number, species code, distance flag to redd, direction flag to mid tail spill, year, and redd age on the flag (see Figure 5). Tie the flag securely to the closest solid object directly above and perpendicular to the pot of the redd (see Figure 7). Do not step or walk on redds. Preferably, tie the flag so that it hangs right over the middle of the tail spill. Measure the distance from the middle of the tail spill to the flag location and write this distance on the flag and in the proper column on the data form. Measure the compass angle from the flag to the middle of the tail spill and write this number on the flag and record it on the data form. If the redd is a test redd or under construction (fish on), write REMEASURE on the flagging and in the notes. Examine all flags during each survey (See Redd Age). If the redd was identified as test during previous surveys and it has changed (e.g., is now larger) or is now clearly a redd of one species of another,

record the record number from the flagging on the data form and re-measure the redd. Cross out the words "test" and "remeasure" on the flagging and write the redd species and date on the flag. Leave the record number unchanged. Record all appropriate data. Record the location of the redd on the map and label it with its redd number.

Mapping
Mark the location of all new redds on the field maps. Pay attention to stream and landform features such as left (as you are looking downstream) and right bank tributaries, notable river bends, and other features to keep track of your location so that when you find a redd you can place its location on the map. Draw a dot on the map and connect it with a line to a place on the map away from the stream where you can write the record number for the redd. Do this for all redds observed. If there were no redds or fish observed for a survey, there is no need to include the map in the data packet at the end of the day. Reuse this map on a future survey.

Back at the Office
Put data forms in order and make sure every thing is filled out properly. Staple the forms together and file in a proper spot. Do not leave data forms in the data box or lying about the office, unless the forms are wet and need to dry. Store all equipment in the proper place. All data should be entered into the data base at the end of each day.

Determination of Redd Life
The length of time redds remain visible during spawning ground surveys is termed "redd life" (Smith and Castle 1991). More specifically, redd life reflects the length of time from the postspawning phase to the point that it is no longer discernable (this is a period of days to weeks, depending on stream flow and periphyton accumulation). It is an important aspect in redd counts and has a fundamental bearing on subsequent counts, count expansions, and population estimates. Redd life is variable among species and streams and over years (see Table 1) it is strongly influenced by streamflow, turbidity, periphyton growth, and redd superimposition. In Table 1, we offer summaries of redd life estimates for six species of Pacific salmon. To assess redd longevity, redds should be classified as new, measurable, no longer measurable, or no longer apparent, and recorded on data forms for each redd observed on each survey. Redd longevity (necessary for establishing survey durations and for use with the area-under-the-curve [AUC] population estimates) and observer efficiency (useful for expanding redd counts to account for redds present but not counted during surveys) require that all flagged and newly constructed redds be examined during each survey and that data be recorded regarding redd condition, date first observed, and unique identifier. Use of regional averages developed over a series of years in the AUC, coupled with observer efficiency, may prove reliable for salmon escapement monitoring.

TABLE 1. — Redd life estimates for Pacific salmon. Estimates are from foot surveys unless otherwise noted. Numbers in parentheses are the range of the estimates. (NR = not reported, NA = not applicable.)

Species	Run	Redd Life in Days			Years	River	Location	Source	Comments
		Estimate	Standard error	N					
Chinook Salmon	Fall	~ 42	nr	nr	1948–1992	Columbia	Washington	Dauble and Watson (1997)	Redds visible from air six weeks
		21	nr	nr	nr	Skagit	Washington	Smith and Castle (1991)	
		16 (7–28)	4.4	4	2003	Noyo	California	S. Gallagher (unpublished)	
	Summer and fall	19.1	2.6	nr	1998–2001	Stillaguamish	Washington	P. Hann (unpublished)	Multiyear average
		18.5	0.4	nr	1999–2002	Green	Washington	P. Hann (unpublished)	Multiyear average aerial survey
	Spring	40	nr	nr	nr	Yakima	Washington	Schwatzber and Roger (1986)	
		17.4 (7.3–30.6)	2.5	nr	1998–2001	Suiattle	Washington	P. Hann (unpublished)	Multiyear average
Chum Salmon *	nr	up to 60	nr	nr	2003	Columbia	Washington	Dehart (2004)	
Coho Salmon	na	24.8 (6–36)	0.9	147	2004	Mendocino Coast Streams	California	S. Gallagher (unpublished)	
	na	22.8 (6–84)	2.08	87	2003	Mendocino Coast Streams	California	S. Gallagher (unpublished)	
	na	14.3–25	nr	na	1986–1990	Hoko	Washington	Lestelle and Weller (2002)	Range of means 3 reaches and 4 years
	na	6.4–32.9	nr	na	1986–1989	Skokomish	Washington	Lestelle and Weller (2002)	Range of means 4 reaches and 3 years
Pink salmon*	Fall	< 15	nr	nr	1992–1994	Lake Creek	Alaska	Fukushima and Smoker (1998)	Redd life influenced by stream flow ^
Sockeye salmon*	Fall	< 15	nr	nr	1992–1994	Lake Creek	Alaska	Fukushima and Smoker (1998)	
Steelhead	Winter	40.7 (7–92)	1.39	148	2000	Smith	Oregon	Jacobs et al. (2001)	
	Winter	27.8 (2–88)	2.08	87	2003	Noyo	California	S. Gallagher (unpublished)	
	Winter	20.4 (11–42)	1.59	46	2004	Mendocino Coast Streams	California	S. Gallagher (unpublished)	

* Redd counts may not be applicable for population monitoring of these species due to group spawning behavior and lack of distinct individual redds (J. Haynes, Washington State Department of Fish and Wildlife, pers. comm.).
^ M. Fukushima National Institute for Environmental Studies, Tsukuba, Japan, pers. comm.

Measuring the surface area of redds

While it is not necessary to measure redds for most escapement estimation surveys, in systems with multiple species of salmonids that overlap in spawn timing and area, measurements and other characteristics of redds can aid in determining which species created the redds in question. Redd dimensions are being gathered in some locations to evaluate if hatchery and natural-origin fish make different sizes and shapes of redds (T. Pearsons, personal communication). Other researchers may want to distinguish redds constructed by resident and anadromous forms (e.g., *O. mykiss*) that spawn at similar locations and times.

Redd measurements (consisting of area, substrate, and depth) (see Figures 7–10) can be made to calculate the surface area of each redd for differentiating species, and report redd size, and for using redd areas to estimate escapement

(Gallagher and Gallagher 2005). Pot length (measured parallel to stream flow), pot width (perpendicular to the length axis), and pot depth (the maximum depth of the excavation relative to the undisturbed stream bed) (see Figure 9) should be measured and data recorded. The dominant pot and tail spill substrate should be visually estimated (or otherwise quantified) using a Brusven index (Platts et al. 1983). Tail spill length (longitudinally parallel to stream flow) and tail spill width at one-third and two-thirds from the downstream edge of the pot to the end of the tail spill (perpendicular to the length axis) should be measured. Redd areas can be calculated using the sum of pot and tail spill areas, which can be calculated by treating the pot as a circle or ellipse and the tail spill as square, rectangle, or triangle.

There are other methods for estimating redd surface area, and the specific method used (and precision needed in the data) will depend on study design and goals or the presence of more than one species or life history type. Probably the most precise method in estimating redd surface area was developed by Burner (1951), who estimated redd surface area by creating scale drawings of each redd, measured the maximum width and total length, and used a planimeter to estimate surface area. Estimating redd surface areas by simply multiplying total length by maximum or average width overestimated redd area and was not useful in differentiating chinook and coho salmon and steelhead redds (S. Gallagher, unpublished data). Other habitat-related variables such as gradient, water depth and velocity, distance to cover, stream shade, or channel complexity might also be collected in association with redd counts to address specific questions identified during development of the study plan.

Specific methods for measuring areas of redds in the field
The purpose of measuring redds is to estimate the area of the redd accurately so that these data can be used to differentiate species and estimate escapement. The pot area and tail spill area are calculated from the field measurements, treating the pot as a circle or ellipse and the tail spill as a circle, square, triangle, or rectangle, depending on the individual measurements. In most cases, redds will not conform to this idealized shape, so it is therefore quite important to remember that the focus should be on calculating the total area of the redd.

Pot Dimensions
Pot length (PL) is the total length of the pot parallel to the stream flow in meters to the nearest decimeter (see Figure 8). Measure in meters from the top to bottom edge. When the pot is irregularly shaped, estimate the total length to the best of one's abilities. Record this information on the data form. Pot width (PW) is maximum width of the pot perpendicular to the stream flow or pot length in meters to the nearest decimeter. Measure in meters from one edge to the other. When the pot is irregularly shaped, do estimate the maximum width as best as possible. Record this on the data form (see Figure 9). Pot depth (PD) is the maximum depth of the excavation relative to the undisturbed streambed in meters to the nearest centimeter. Use the staff to measure the depth. Record this on the data form in meters. Pot substrate (PS) is the size of the dominant substrate in the pot. Visually estimate, using the staff gauge to calibrate one's eye, the size of the dominant substrate in the pot in centimeters. The substrate size is the length of the diameter of the smallest axis that will pass through a sieve, in centimeters. Record this on the data form.

Tail Spill Dimensions

Tail spill length (TsL) is the total length of the tail spill parallel to the stream flow in meters to the nearest decimeter. Measure from the top edge of the middle of the pot to bottom edge of the tail spill. When the tail spill is irregularly shaped, do the best to estimate the total length. Record this on the data form.

Tail spill width 1 (TSw1) is the maximum width of the tail spill perpendicular to the stream flow or pot length in meters to the nearest decimeter. Measure from one edge to the other third of the distance down from the top of the tail spill. When the tail spill is irregularly shaped, do the best to estimate the maximum width. Record this on the data form.

Tail spill width 2 (TSw2) is the maximum width of the tail spill perpendicular to the stream flow or pot length in meters to the nearest decimeter. Measure from one edge to the other two-thirds of the distance down from the top of the tail spill. When the tail spill is irregularly shaped, do the best to estimate the maximum width. Record this on the data form.

Tail spill substrate (TS) is the size of the dominant substrate in the tail spill in centimeters. Visually estimate, using the staff gauge to calibrate the eye, the size of the dominant substrate in the tail spill. The substrate size is the length of the diameter of the smallest axis that will pass through a sieve. Record this on the data form.

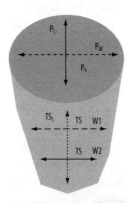

FIGURE 8. — **General measurements for estimating surface area of a salmonid redd. P is pot; TS is tailspill; L is length; W is width; S is substrate (note location in mid-TS for substrate collection not denoted). (Illustration: Andrew Fuller.)**

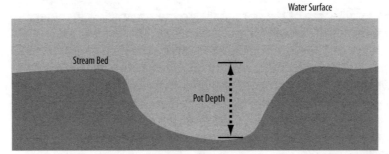

FIGURE 9. — **Cross section of a salmonid redd pot. Pot depth is the distance from the bottom of the pot to the water surface minus the distance from the water surface to the streambed. (Illustration: Andrew Fuller.)**

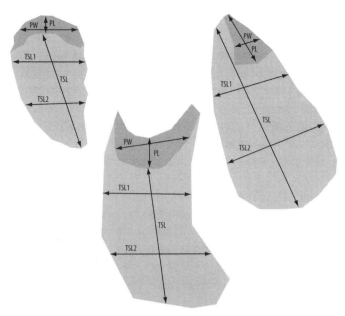

FIGURE 10. — Measurements for unusually shaped redds. (Illustration: Andrew Fuller.)

FIGURE 11. — Measuring a steelhead redd.

Data Handling, Analysis, and Reporting

Metadata procedures

Data common to all redd observations for each survey and survey reach should be linked with a one-to-many relationship to each individual redd observation (see Figure 12). These data should include the date, stream and section name, a unique stream reach identifier such as the latitude-longitude (LLID) number, survey number (1 is the first of the season), surveyors' names, climatic conditions, streamflow, water visibility, and so forth. The subject data discussed earlier could be included in separate data tables and have associated information connected by one-to-many relationships. For instance, the unique stream identifier table might contain the coordinates of the starting and stopping points for the stream reach, its name, general location, driving directions, and sample selection number. The stream section table might include the site description, land use, landowners' names(s) and contact information, and environmental descriptors.

REDD COUNTS

Data specific to each redd would, by virtue of the one-to-many relationships above, contain all the above information needed to form a unique number. For instance, redd number 1 for each survey, coupled with the date the redd was first observed and a unique reach identifier, would serve as a unique number for each observation necessary to track redd longevity, and observer efficiency, and test redd counts. Other data fields specific to each redd should include redd species, fish presence, and specific redd data pertinent to the goals of the study. These fields might include information such as pot and tail spill measurements, substrate measurements, certainty of redd species identification, presence of fish, and general notes.

Quality assurance procedures for the metadata should include site verification, checking stream flow estimates against stream flow gauging stations (where available), and using paper records in conjunction with handheld computers. A series of queries should be designed to test if all redds were observed at least once, and to look for duplicate records and to sort individual redd observations by date to ensure that the date of the first observation exists in the database.

Data Fields

The following list below gives names and types of the data fields associated with redd survey and mapping efforts (adapted from Hahn et al. 1999), and is offered as a starting point for data sheets and database designs:

Field Name in Database	Print Field Name	Type	Size	Definition
River Name	River name			Name of river
River code	River code			
Species	Species			
ReddID	ReddID	I		Unique ID assigned by field biologists: • First 2–4 digits=river and section/reach code month+day when first observed • Next 4 digits=month+day when first observed (MMDD) • last 2–3 digitis=redd number within that section/reach on that date
Sid	Sid	S		Unique, sequential integer for digitizing & maps
Lat/Long	Lat/Long	N		Lat/Long location
ReddNum	RNum	A	2	Last 2 digits ReddID, # of redd found same date & in same river section/reach.
KM	KM	N		River km (or RM=River Mile)
Date1stOb	Date1stOb	D		Date when First Visible
Started	Started	S		Number days redd assumed started before FVDate (1st visible)
FVDate	FVDate	D		Date when judged First Visible (see rules)
FV.Julian	FV.Julian	I		Julian FV date
FV.DOY	FV.DOY	S		Integer day of year for FV date (1–365 —used in plotting by "date")
DateLastObs	DateLastObs	D		Date last observed
DateNotSeen	DateNotSeen	D		Next survey date when redd was no longer visible
LVDate	LVDate	D		Date when assumed Last Visible (see rules)
J.LVJulian	J.LVJulian	I		Julian LV date
SuprImp?	SI	A	1	Was the redd superimposed/overlapped? (Y=yes, P=partial, N=no, blank=unknown)

Field Name in Database	Print Field Name	Type	Size	Definition
SupSPP	SI-spp	A	2	Species which caused redd superimposition
RLife	RL	S		Redd life in days
FieldNote	FieldNote	A		Miscellaneous comments on summary sheets
ReddCat	Rcat	A	2	Code for redd category (A=active, S=start, C=complete...etc)
Comment	Comment	A	10	Comment during visibility assignment
DaysBetween	DaBtwn	S		Number of non-surveyed days between previous & current surveys
Page	Page	S		Page number in field notes
Observer	Obsvr	A	5	Initials of observer(s)

Database design

Figure 12 shows an example of database fields and relationships used for monitoring salmon escapement from spawning ground surveys in coastal California, following the methods of Gallagher and Gallagher (2005) and Gallagher and Knechtle (2005). In this layout, the *daily header* table is information that is included in the header portion of a data sheet, and the specifics for each individual observation are in the redd data and *fish on* tables (see Appendix A). The *fish on* table is for keeping track of fish on redds. The fields in the daily header table are self-explanatory, except for the *daily number* and *start/end mark*. The *daily number* is an automatic number the database uses as the primary key, and the *start/end mark* is for recording the beginning and ending location of the survey. In the redd data table, the first three fields are used to link handheld data recorders to the database. Handheld data recorders limit data transfer errors common when transcribing data from written field records to electronic databases. The *redd fish ID* is an automatic number for linking the *fish on* and redd data tables. The *redd record number* field is the unique number of each redd observation, and *rec date* is the date the redd was first observed. The *field stream marker* is the distance from the start of the reach to the location of the redd. The *fish on* table is the subform for tracking specific information regarding fish observed on redds. The first three data fields are similar to the redd subform, and the rest are self-explanatory. The *location* table is for information on specific reaches. The *field Loc ID* is the location code, the *LLID field* is the geographic coordinates of the starting point of each survey reach, and *HUC* (hydrologic unit code) is the hydrologic unit in which the reach is located.

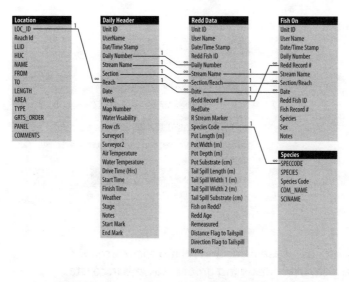

FIGURE 12. — Examples of database relationships for redd count surveys (modified with permission from database created by Dave Gibney, Institute for River Ecosystems, Humboldt State University, Arcata, California).

Data entry

Data should be entered from field data forms into a computer data system upon return to the office after each survey. All data entered from paper records into an electronic database should be rechecked for accuracy and corrected where necessary. A record of data entry errors should be kept and used to identify and alleviate common problems. If handheld data recorders are used, a paper backup should be made periodically in the field to check for errors. A series of queries should be designed to test if all redds were observed at least once, to look for duplicate records, and to sort individual redd observations by date to ensure that a date of first observation exists in the database. To reduce data entry errors, data fields can be given drop-down menus with limited choices such as yes or no, stream name lists, and species lists; some data fields can be set as required (e.g., numbers or text only, limits on decimal places, limits on descriptor length). All original data sheets or data files should be well organized, clearly labeled, and placed in an appropriate long-term storage location.

Data summaries

Gallagher and Gallagher (2005) used logistic regression to differentiate redd species using known redd data (i.e., known species of fish observed on a redd) to develop and check the predictive ability of their model. This technique may be useful in other situations where multiple species or life history forms overlap during spawning. The development of a method to differentiate redd species will likely require a pilot study to determine which, if any, variables might be useful in predicting redd species. Discrimination, principle components analysis, or other multivariate techniques may been useful for differentiating redd species (Fukushima and Smoker 1998; Zimmerman and Reeves 2000).

To assess redd longevity and observer efficiency, all flagged and newly constructed redds should be examined during each survey. To examine redd longevity, redds should be classified as new, measurable, no longer measurable, or no longer apparent. Weekly observer efficiency can be estimated as the percentage of known flagged redds (minus those classified as no longer apparent) observed during each survey (Gallagher and Gallagher 2005). Remember that the goal is to measure observer efficiency in seeing redds—not observer efficiency in seeing flagged redds. Weekly observer efficiency for each species is then averaged for each survey segment in each stream throughout the season to estimate total efficiency for the season (see Table 2). Observer efficiency calculated in this manner can then be used to expand redd counts to account for redds present but not observed. This expanded redd count number, taken along with the associated statistical uncertainty from replicate reaches (see sampling design, pages 203–208), can thereafter be expanded by the length of spawning habitat and presented as the annual redd index for the population of interest.

TABLE 2. — Example of observer efficiency estimate based on marked redds and expanded to account for redds present but not observed during redd surveys

Survey week	Number of redds observed				Observer efficiency			
	New redds	Previously counted	Removed	Known redds	Weekly	Average	Standard	Standard
1	0	0	0	0				
2	0	0	0	0				
3	5	0	0	0				
4	1	5	0	5	1.00	1.00		
5	2	6	0	6	1.00	1.00	0.00	0.00
6	2	6	0	8	0.75	0.92	0.14	0.07
7	1	6	0	10	0.60	0.84	0.14	0.06
8	6	7	0	11	0.64	0.80	0.20	0.08
9	4	7	1	17	0.41	0.73	0.19	0.07
10	6	8	2	20	0.40	0.69	0.25	0.09
11	0	8	0	24	0.33	0.64	0.26	0.09
Total count	27							
Season average	0.64							
Season SE	0.09							
Expanded count	37							
SE	2.43							

Observer efficiency can also be estimated by having several crews (of two people each) follow each other on one survey segment, with each crew recording newly constructed redds. Average field observer efficiency is then calculated by assuming that the largest number of redds observed by any one crew is the "known number," and the totals from each survey crew observing fewer redds can be divided by this number and then averaged. Another method to describe observer error compares "true" redd numbers established by an experienced observer counting redds periodically during the bull trout spawning season to single pass counts made by many observers at the end of the spawning season to model observer error structure and correct redd counts for observer bias (Muhlfeld et al. 2006).

Situations may arise when marking redds for estimating observer efficiency and avoiding double counts may not be practical. In these cases, redd numbers can be estimated from stream surveys using the area-under-the-curve method (Hilborn et al. 1999). This method employs periodic counts during the spawning season and calculates the AUC from the trapezoidal approximation, redd life, and estimates of observer efficiency. The trapezoidal approximation is calculated from the time, usually in days, between surveys and total redd counts on each survey. The trapezoidal approximation in units of redd/days is converted to AUC redd numbers by estimates of redd life (an estimate of the length of time redds remain visible in the stream). The resulting redd numbers are expanded to account for redds present but not counted using estimates of observer efficiency. This method requires that the length of time redds remain visible and observer efficiency estimates are known. Redd longevity is variable among streams (Susac and Jacobs 1999), may be variable among species (see Table 1), and will likely have to be estimated annually for individual streams. Estimates of observer efficiency require multiple passes by survey crews, marking and recounting redds, (Gallagher and Gallagher 2005), or estimates of "true" numbers from experienced observers (Dunham et al. 2001; Muhlfeld et al. 2006). One of the major shortcomings of the AUC is that it lacks a rigorous statistical method for calculating confidence bounds, which, when estimated, require intensive bootstrap computer simulation and independent mark–recapture estimates for their calculation (Korman et al. 2002; Parken et al. 2003).

Converting redd counts into escapement estimates

Number of redds per female

Specific data on the number of redds made by females is still needed. Detailed radio-telemetry efforts may be a useful approach to determine the number of redds a female salmonid constructs. Other options include getting detailed video recordings on the spawning activity of marked females, or by taking genetic samples from emerging fry as they leave the redd (captured via fry emergence traps). If the assumption of one redd per female can be validated, escapement estimates that assuming one redd per female can be made by multiplying the number of redds by the male-to-female ratio observed in each stream or river and summing this with the number of redds. Until the above research is conducted, interim estimates have been proposed. Duffy (2005) suggests multiplying redd counts by 1.2 to expand steelhead redd counts to female escapement. He further suggests that if the female-to-male ratio is known, it can be used to expand female estimates into total escapement. Generally, this approach will require independent escapement estimates to calculate the number of females per redd. There is little evidence to support or refute the idea that estimates of the number of females per redd is consistent among years or between streams. Studies requiring that redd numbers be converted to fish numbers in this manner should evaluate the transferability of this type of data.

Number of fish per redd

In Oregon, steelhead redd counts are significantly correlated with adult escapement, and an estimate of 1.54 females per redd was developed (Susac and Jacobs 2002). In Washington, redd counts are the principle method for monitoring salmonids, and cumulative redd counts are expanded by 2.5 fish per redd to estimate escapement (Boydstun and McDonald 2005). In California,

Gallagher (2005b) found that the number of steelhead and coho salmon per redd differed slightly among streams and years, but the use of one value for all streams to convert redd counts to fish numbers for regional spawning ground surveys appeared reliable. Dunham et al. (2001) found considerable spatial and interannual variation in bull trout spawner-to-redd ratios and attributed it to either strong life history variation among populations or bias and imprecision in redd counts. Al-Chokhachy et al. (2005) estimated an average of 2.68 bull trout spawners per redd from a number of sources and attributed differences between redd counts and mark–recapture estimates to differences in life history forms.

Mark–recapture escapement estimates coupled with redd count surveys, in which the bias in redd counts following improvements to field and laboratory methods suggested in this document, should likely improve estimates of the number of fish per redd for expanding redd counts to escapement. Results of these types of experiments can be used to document redd count-escapement relationships and develop predictive models to estimate escapement from redd counts (Gallagher 2005b). The Bland-Altman method (Glantz 1997) is a useful statistical procedure for determining if two different measures of the same thing are significantly different and may be useful for assessing the transferability of redd counts to index escapement. The transferability of these types of estimates for converting redd abundance to escapement among years and streams needs further evaluation.

Redd area

Another approach to assessing escapement is through the measurement of redd sizes from known species (i.e., the redd area method). Further testing, based on radio-telemetry or other detailed observations of marked female fish, is needed to affirm or refute this approach. Differing water and substrate conditions or the size of the female are among the factors that can affect the size of the redds. Gallagher and Gallagher (2005) estimated salmon escapement using redd areas. Their redd area method assumes that the number of redds a female makes is related to the size of the redd. Coho salmon redd area escapement estimates were based on findings from releases above a counting structure, where it was estimated that females make between one and four redds. Here, redd areas greater than 5.1 m^2 represented one female, redds between 2.1 and 5.0 m^2 represented one-half a female, and redds less than 2.0 m^2 represented one-quarter of a female (Maahs and Gilleard 1993). Female coho redd area escapement estimates were multiplied by the male-per-female ratio observed in each stream and summed with female estimates to estimate total escapement. Observer efficiency estimates were then used to expand the redd area estimates as described earlier for redd counts. To apply this method for estimating steelhead escapement from redd areas, Gallagher and Gallagher (2005) divided the area of the largest known steelhead redd observed over 2 years into quarters and estimated female numbers and escapement in the same manner as for coho salmon. The use of this method for other species will require documenting females making more than one redd and estimating redd surface areas.

Report format

Redd counts are a fundamentally important aspect in salmonid conservation and management, and consistency in the format of data and results will greatly aid our collective efforts. While we recognize slight variations in redd abundance survey reports (due to study plan, objectives, goals, audience, and species of interest), the overall report format should follow the "Introduction, Materials and Methods, Results, and Discussion" structure (Day 1988) and include sufficient details for evaluating the quality of the data for abundance and trend monitoring. The discussion section should include a section on recommendations for future monitoring based on evaluation of annual findings. At the very least, there should be tables of total redd counts, redd densities, observer efficiency, redd abundance, and associated statistical uncertainty for each survey reach and a total for the survey area. This table should also include reach segment lengths, total survey lengths, and total stream length in the sampling area. There should be a map of the survey area showing details of the survey sections and study area. If there is a multiyear data series, figures demonstrating trends with associated statistical evaluation of their significance should be presented. In the Methods section, details of all statistical analysis and hypotheses tested should be thoroughly documented.

Trend analysis

To examine a series of annual redd counts for evaluating population trends, the slope and intercept of the regression line must first be estimated . In some cases, it may be necessary to transform the data so that it fits the assumptions of normality or to use sophisticated multivariate models or nonparametric analysis; however, in general, the simplest method to examine time series data for trends is to use linear regression of adult abundance (e.g., redd counts) versus year to estimate the slope of the trend line. The slope of adult abundance versus year can be graphically examined and statistically tested to determine if it differs from zero (Glantz 1997).

Shea and Mangel (2001) presented models for coho salmon, suggesting that increasing time series and reducing observer uncertainty in juvenile estimates will improve statistical power for detecting trends (changes in long-term abundance) in adult populations from observations of juvenile abundance. To use redd counts for trend detection and population monitoring, Maxell (1999) recommended that errors in redd counts be identified and reduced, levels of significance that adequately balance the risks of committing type I and type II errors should be used, and one-tailed tests for identifying population declines should be used, especially during the first years of a monitoring program. Trend analyses should report the statistical power of tests for trend detection.

Archival procedures

All original data should be well organized, clearly labeled, and archived. Reports should be prepared annually and archived in the files of the primary agency in charge and sent to local or regional natural resource agency libraries. Digital versions of the data sets, as well as hardcopies of reports, should be submitted to international fisheries conservation organizations. Significant findings and new developments in monitoring techniques should be published in the primary literature.

Personnel Requirements and Training

Responsibilities
Redd count survey staff will be responsible for conducting redd counts per the field protocol and training manual. All survey staff will be expected to maintain complete survey field notes per the training manual or annual field protocol. In the field, experienced survey staff will train newly hired survey staff in redd counting techniques.

Qualifications
Redd survey staff should be in such physical shape as to allow for extended and at times strenuous hiking while carrying equipment and personal gear that may weigh 10 kg or more. Survey staff should expect to work extended daily hours as necessary. First-aid certification and swift water rescue training is required and should be provided by the employer during the preseason period. Education requirements for project leaders include a minimum undergraduate degree in fisheries management or related natural resource field or 2-year technical degree with a minimum of two seasons of experience in field survey techniques related to fish management.

Training
A field manual should be made available to all redd count survey staff to promote consistency among survey efforts and to address safety concerns. New hires should be scheduled to go on surveys with experienced redd survey staff and receive training in the field. Safety, aspects of landowner relations, trespassing regulations, and redd count protocol training for all survey crew members should be scheduled and conducted prior to initiating the field season. Safety training for field crews should include first aid, wilderness medicine, swift water rescue training, and wader safety training. Specialized training for using all-terrain vehicles, four-wheel drive vehicles, boats, or other equipment needed for conducting redd surveys should occur during the pre-field season period. Safety and data collection/equipment will require additional specialized training if aircraft, boats, or underwater video are used for redd count surveys. Redd count protocol training should include time for crew members to read and become familiar with the specifics of field procedures, redd identification, and data management.

Operational Requirements

Workload and field schedule
The field schedule should be developed so that all selected reaches can be surveyed less than 14 d apart (Gallagher and Gallagher 2005; ODFW 2005), with one survey occurring just prior to fish entering the stream and continuing until new redds and fish are no longer observed. Crews should work in pairs for safety and have communication devices in case of emergency. Walking upstream, crews can generally cover about 3–5 km in a day depending on redd density and stream complexity. Floating downstream in rafts or kayaks crews can cover 8–10 km per day, depending on redd density, stream size, side channel complexity, and logistics

of access points. Surveying by aircraft can increase the area covered in a day but is limited to midday periods when glare is low and visibility is good.

Equipment needs

Equipment lists for survey vehicles and individual survey crew members should be developed and included in the training manual for redd count surveys (see Table 3). The survey vehicle should contain a first-aid kit (including a snakebite and bee sting kit), fire extinguisher, shovel, flat-tire repair kit, tools, flares, chain saw and safety gear, duct tape, bailing wire, rope, tow strap, WD-40 or other lubricant for stuck locks, rags, toilet paper, and a two-way radio. Cell or satellite phones and citizens' band (CB) radios should be included but may have limited reception in many field locations. Individual staff equipment for walking surveys should include items listed in Table 3. Specialized equipment for rafting or aerial surveys is not included herein.

TABLE 3. — Equipment list for field surveyors conducting redd counts

Equipment
Spawning survey protocol
Data forms
Stream/river reach maps
Pencils, pens, permanent markers
Field notebook
Knife with sheath
Compass
Chest waders and rain gear or dry suit
Wading boots
Hat
Polarized sunglasses
Field vest or backpack
Flagging
Measuring tape (mm)
Watch
Cell or satellite phone
Contact and emergency phone numbers
GPS unit
Swift water safety gear
Machete
Brush axe
Chain saw
Food and water
Extra clothing

Budget considerations

Budget needs will reflect funding for equipment, overtime, travel, training, and administrative overhead. Included in this staff time is an allocation for data management, analysis, and report writing. Field crews should receive salaries equivalent to that of other technical survey crews in the area, a cost of living adjustment for expensive housing markets, medical coverage, and other benefits, including vacation, holiday, and sick-leave pay. The budget should incorporate costs of additional equipment such as all-terrain vehicles or rafts.

Literature Cited

Agrawal, A., R. S. Schick, E. P. Bjorkstedt, R. G. Szerlong, M. N. Goslin, B. C. Spence, T. H. Williams, and K. M. Burnett. 2005. Predicting the potential for historic coho, chinook, and steelhead habitat in northern California. National Oceanic and Atmospheric Administration, NOAA Technical Memorandum NOAA-TM-NMFS-SWFSC-379, Santa Cruz, California.

Al-Chokhachy, R. P. Budy, and H. Schaller. 2005. Understanding the significance of redd counts: a comparison between two methods for estimating the abundance of and monitoring bull trout populations. North American Journal of Fisheries Management 25:1505–1512.

Beard, T. D., and R. F. Carline. 1991. Influence of spawning and other stream habitat features on spatial variability of wild brown trout. Transactions of the American Fisheries Society 120:711–722.

Beland, K. F. 1996. The relationship between redd counts and Atlantic salmon (*Salmo salar*) parr populations in the Dennys River, Maine. Canadian Journal of Fisheries and Aquatic Sciences 53:513–519.

Benson, N. G. 1953. The importance of groundwater to trout populations in the Pigeon River, Michigan. Transactions of the North American Wildlife Conference 18:1–11.

Bjornn, T. C., and D. W. Reiser. 1991. Habitat requirements of salmonids in streams. Pages 83–138 in W. R. Meehan, editor. Influences of forest and rangeland management on salmonid fishes and their habitats. American Fisheries Society, Special Publication 19, Bethesda, Maryland.

Boydstun, L. B., and T. McDonald. 2005. Action plan for monitoring California's coastal salmonids. Final report to NOAA Fisheries, Contract Number WASC-3-1295, Santa Cruz, California.

Briggs, J. C. 1953. The behavior and reproduction of salmonid fishes in a small coastal stream. California Fish and Game, Fisheries Bulletin 94.

Burner, C. J. 1951. Characteristics of spawning nests of Columbia River salmon. U.S. Fish and Wildlife Service, Fisheries Bulletin 61:97–110.

Busby, P., T. Wainwright, G. Bryant, L. Lierheimer, R. Waples, W. Waknitz, and I. Lagomarsino. 1996. Status review of West Coast steelhead from Washington, Idaho, Oregon, and California. National Oceanic and Atmospheric Administration, Springfield, NOAA Technical Memorandum NMFS-NWFSC-27, Virginia.

Cederholm, C. J., D. H. Johnson, R. Bilby, L. Dominguez, A. Garrett, W. Graeber, E.L. Greda, M. Kunze, J. Palmisano, R. Plotnikoff, B. Pearcy, C. Simenstad, and P. Trotter. 2001. Pacific salmon and wildlife—ecological contexts, relationships, and implications for management. In D. H. Johnson and T. A. O'Neil, managing directors. Wildlife-habitat relationships in Oregon and Washington. Oregon State University Press, Corvallis.

Chilcote, M. W. 2001. Conservation assessment of steelhead populations in Oregon. Oregon Department of Fish and Wildlife, Portland.

Crisp, D. T., and P. A. Carling. 1989. Observations on sitting, dimensions, and structure of salmonid redds. Journal of Fish Biology 34:119–134.

Dauble, D. D., and D. G. Watson. 1997. Status of fall chinook salmon populations in the Mid-Columbia River, 1948-1992. North American Journal of Fisheries Management 17:283–300.

Day, R. A. 1988. How to write and publish a scientific paper, 3rd edition. Oryx Press, New York.

DeVries, P. 1997. Riverine salmonid egg burial depths: review of published data and implications for scour studies. Canadian Journal of Fisheries and Aquatic Sciences 54:1685–1698.

Duffy, W. G. 2005. Protocols for monitoring the response of anadromous salmon and steelhead to watershed restoration in California. California Cooperative Fish Research Unit, Humboldt State University, Arcata. Prepared for California State Department of Fish and Game's salmon and steelhead trout restoration account agreement no. P0210565.

Dunham, J., B. Rieman, and K. Davis. 2001. Sources and magnitude of sampling error in redd counts for Bull Trout. North American Journal of Fisheries Management 21:343–352.

Eddy, S., and J. C. Underhill. 1978. How to Know the Freshwater Fishes, 3rd edition. Wm. C. Brown Company Publishers, Dubuque, Iowa.

Erman, D., and V. Hawthorne. 1976. The quantitative Importance of an intermittent stream in the spawning of rainbow trout. Transactions of the American Fisheries Society 6:675–681.

Everest, F. H. 1973. Ecology and management of summer steelhead in the Rogue River, Oregon. State Game Commission, Corvallis, Oregon.

Firman, J. C., and S. T. Jacobs. 1999. A survey design for integrated monitoring of salmonids. Oregon Department of Fish and Wildlife, Corvallis.

Fukushima, M., and W. M. Smoker. 1998. Spawning habitat segregation of sympatric sockeye and pink salmon. Transactions of the American Fisheries Society 127:253–260.

Gallagher, S. P. 2005a. Annual coho salmon (*Oncorhynchus kisutch*) and steelhead (*O. mykiss*) spawning ground escapement estimates 2000 to 2004 in several coastal Mendocino County, California streams and recommendations for long term monitoring of coastal salmonids. California Department of Fish & Game, Arcata, California.

Gallagher, S. P. 2005b. Evaluation of coho salmon (*Oncorhynchus kisutch*) and steelhead (*O. mykiss*) spawning ground escapement estimates for monitoring status and trends of California coastal salmonids: 2000 to 2005 escapement estimates for several Mendocino County, coastal streams. California Department of Fish & Game, Arcata, California.

Gallagher, S. P., and C. M. Gallagher. 2005. Discrimination of chinook and coho salmon and steelhead redds and evaluation of the use of redd data for estimating escapement in several unregulated streams in northern California. North American Journal of Fisheries Management 25:284–300.

Gallagher, S. P., and M. K. Knechtle. 2005. Coastal northern California salmon spawning survey protocol. California State Department of Fish and Game, Fort Bragg, California.

Geist, D. R., T. P. Hanrahan, E. V. Arntzen, G. A. McMichael, C. J. Murray, and Y. Chien. 2002. Physicochemical characteristics of the hyporheic zone affect redd site selection by chum salmon and fall chinook salmon in the Columbia River. North American Journal of Fisheries Management 22:1077–1085.

Glantz, S. A. 1997. Primer of biostatistics, 4th edition. McGraw-Hill, New York.

Hay, D. W. 1984. The relationship between redd counts and the numbers of spawning salmon in the Girnock Burn (Scotland). International Council for the Exploration of the Sea, ICES/C.W. No. M:22, Charlottelund, Denmark.

Heggberget, T. G., T. Haukebo, and B. Veie-Rosvoll. 1986. An aerial method of assessing spawning activity of Atlantic salmon, *Salmo Salar L.*, and brown trout, *Salmo trutta L.*, in Norwegian streams. Journal of Fish Biology 28:335–342.

Hilborn, R., B. Bue, and S. Sharr. 1999. Estimating spawning escapements from periodic counts: a comparison of methods. Canadian Journal of Fisheries and Aquatic Sciences 56:888–896.

Hobbs, D. F. 1937. Natural reproduction of quinnat salmon, brown and rainbow trout in certain New Zealand waters. New Zealand Marine Department Fisheries Research Division Bulletin 6.

Irvine, J. R., R. C. Bocking, K. K. English, and M. Labelle. 1992. Estimating coho (*Oncorhynchus kisutch*) spawning escapements by conducting visual surveys in areas selected using stratified random and stratified index sampling designs. Canadian Journal of Fisheries and Aquatic Sciences 49:1972–1981.

Isaak, D. J., and R. F. Thurow. 2006. Network-scale spatial and temporal variation in chinook salmon (*Oncorhynchus tshawytscha*) redd distributions: patterns inferred from spatially continuous replicate surveys. Canadian Journal of Fisheries and Aquatic Sciences 63:285–296.

Isaak, D. J., F. R. F. Thurow, B. E. Rieman, and J. B. Dunham. 2003. Temporal variation in synchrony among chinook salmon (*Oncorhynchus tshawytscha*) redd counts from a wilderness area in central Idaho. Canadian Journal of Fisheries and Aquatic Sciences 60:840–848.

Jacobs, S., J. Firman, and G. Susac. 2001. Status of Oregon coastal stocks of anadromous salmonids, 1999-2000. Oregon Department of Fish and Wildlife, Monitoring program report number OPSW-ODFW-2001-3, Portland.

Jones, E. L., T. J. Quinn, and B. W. Van Allen. 1998. Observer accuracy and precision in aerial and foot survey counts of pink salmon in a southeast Alaska stream. North American Journal of Fisheries Management 18:832–846.

Korman, J., R. M. N. Ahrens, P. S. Higgins, and C. J. Walters. 2002. Effects of observer efficiency, arrival timing, and survey life on estimates of escapement for steelhead trout (*Oncorhynchus mykiss*) derived from repeat mark–recapture experiments. Canadian Journal of Fisheries and Aquatic Sciences 59:1116–1131.

Krebs, C. J. 1989. Ecological methodology. Harper & Row, New York.

Kuligowski, D. R., M. J. Ford, and B. A. Berejikian. 2005. Breeding structure of steelhead inferred from patterns of genetic relatedness among nests. Transactions of the American Fisheries Society 134:1202–1212.

Lestelle, L. C., and C. Weller. 2002. Summary report: Hoko and Skokomish river coho salmon spawning escapement evaluation studies 1986-1990. Point No Point Treaty Council, Kingston, Washington.

Maahs, M., and J. Gilleard. 1993. Anadromous salmonid resources of Mendocino coastal and inland rivers 1990-91 through 1991-92: an evaluation of rehabilitation efforts based on carcass recovery and spawning activity. California Department of Fish and Game, Fisheries Division, Fisheries Restoration Program, Contract Number FG-9364, Saacramento.

Madden, G. A., P. J. Cronin, B. L. Dubee, and P. D. Seymour. 1993. Using artificial barriers to protect and enhance Atlantic salmon. New Brunswick Department of Natural Resources and Energy, Technical report, Fredericton.

Maxell, B. A. 1999. A power analysis on the monitoring of bull trout stocks using redd counts. North American Journal of Fisheries Management 19:860–866.

McElhany, P., M. Ruckelshaus, M. Ford, T. Wainwright, and E. Bjorkstedt. 2000. Viable salmonid populations and the recovery of evolutionarily significant units. U.S. Department of Commerce, NOAA Technical Memorandum NMFS-NWFSC-42, Seattle.

Meffe, G. K. 1986. Conservation genetics and the management of endangered fishes. Fisheries 11:14–23.

Minta, S., and M. Mangel. 1989. A simple population estimate based on simulation for capture–recapture and mark–recapture data. Ecology 70:1738–1751.

Muhlfeld, C. C., M. L. Taper, and D. F. Staples. 2006. Observer error structure in bull trout redd counts in Montana streams: implications for inference on true redd numbers. Transactions of the American Fisheries Society 135:643–654.

Neilson, J. D., and G. H. Geen. 1981. Enumeration of spawning salmon from spawner residence time and aerial counts. Transactions of the American Fisheries Society 110:554–556.

ODFW (Oregon Department of Fish and Wildlife). 2004. Coastal steelhead spawning survey procedures manual (2004). Coastal salmonid inventory project. Oregon Department of Fish and Wildlife, Corvallis.

ODFW (Oregon Department of Fish and Wildlife). 2005. Coastal salmon spawning survey procedures manual (2005). Oregon adult salmon inventory and sampling project. Oregon Department of Fish and Wildlife, Corvallis.

Orcutt, D. R., B. R. Pulliam, and A. Arp. 1968. Characteristics of steelhead trout redds in Idaho streams. Transactions of the American Fisheries Society 97:42–45.

Parken, C. K., R. E. Bailey, and J. R. Irvine. 2003. Incorporating uncertainty into area-under-the-curve and peak count salmon escapement estimation. North American Journal of Fisheries Management 23:78–90.

Platts, W. S., W. F. Meghan, and G. W. Minshall. 1983. Methods for evaluating stream, riparian, and biotic conditions. U.S. Forest Service, Intermountain Forest and Range Experiment Station, General Technical Report INT-138, Ogden, Utah.

Reingold, M. 1965. A steelhead spawning study. Idaho Wildlife Review 17:8–10.

Rieman, B. E., and J. D. McIntyre. 1996. Spatial and temporal variability in bull trout redd counts. North American Journal of Fisheries Management 16:132–141.

Rieman, B. E., and D. L. Myers. 1997. Use of redd counts to detect trends in bull trout (*Salvelinus confluentus*) populations. Conservation Biology 11:1015–1018.

Schwartzberg, M., and P. B. Roger. 1986. Observations on the accuracy of redd counting techniques used in the Columbia River basin. Columbia River Inter-Tribal Fisheries Commission, Technical Report 82-6, Portland, Oregon.

Semple, J. R., P. J. Zamora, and R. Rutheford. 1994. Effects of dredging on egg to fry emergence, survival, and juvenile Atlantic salmon abundance, Debert River, Nova Scotia. Canadian Technical Report of Fisheries and Aquatic Sciences 2023.

Shea, K., and M. Mangel. 2001. Detection of population trends in threatened coho salmon (*Oncorhynchus kisutch*). Canadian Journal of Fisheries and Aquatic Sciences 58:375–385.

Smith, C. J., and P. Castle. 1991. Northwest Fisheries Resource Bulletin: Pudget Sound chinook salmon (o.t.) escapement estimates and methods. Washington Department of Fish and Wildlife project report series, Northwest Indian Fisheries Commission, Olympia.

Steel, E. A., B. E. Feist, D. W. Jensen, G. R. Pess., M. B. Sheer, J. B. Brauner, and R. Bilby. 2004. Landscape models to understand steelhead (*Oncorhynchus mykiss*) distribution and help prioritize barrier removals in the Willamette basin, Oregon, U.S.A. Canadian Journal of Fisheries and Aquatic Sciences 61:999–1011.

Stevens, D. L. 2002. Sampling design and statistical methods for the integrated biological and physical monitoring of Oregon streams. Oregon State University, Department of Statistics, Corvallis, Oregon, and EPA National Health and Environmental Effects Research Laboratory, Western Ecology Division, Corvallis, Oregon.

Susac, G. L., and S. E. Jacobs. 1999. Evaluation of spawning ground surveys for indexing the abundance of winter steelhead in Oregon coastal basins. Annual progress report 1 July 1997 to 30 June 1998. Oregon Department of Fish and Wildlife, Corvallis.

Susac, G. L., and S. E. Jacobs. 2002. Assessment of the Nestucca and Alsea rivers winter steelhead, 2002. Oregon Department of Fish and Wildlife, Portland.

Thurow, R. F., and J. G. King. 1994. Attributes of Yellowstone cutthroat trout redds in a tributary of the Snake River, Idaho. Transactions of the American Fisheries Society 123:37–50.

Zimmerman, C. E., and G. H. Reeves. 2000. Population structure of sympatric anadromous and non anadromous *Oncorhynchus mykiss*: evidence from spawning surveys and otolith microchemistry. Canadian Journal of Fisheries and Aquatic Sciences 57:2152–2162.

Appendix A: Example redd count data sheet

Fill out even if no redds were observed

Page 1 Of 3

Species Codes: Onki = Coho Salmon
Onmy = Steelhead
Onts = Chinook Salmon
Latr = Pacific Lamprey
Unkn = Unknown Species

Stream Name:	Caspar Creek	Week:		Water Temp.	52	8
Section Name:	Below Forks	Map #	CAS 1	Air Temp.		11
Reach ID:	CAS 1	Water Visibility (m):	1.5	Start Time:		12:50
Date:	12/12/2005	Stream Flow (m^3/s):	0.25	End Time:		13:50
Surveyors:	Mike Morrison, Scott Harris	Stage:	99.25	Drive Time:		0.75
Weather:	Ligth Rain					
Notes:	Steam Flow up at end of survey to 1.45 Visibility down to 1.45					

Directions: Visually identify each redd to species. Record the record number. Mark redd location with record number on the map. If a fish is on a redd record this on both fish and redd data form s, write fish number of redd form and redd number of fish form. If you can not measure a redd visually estimate the dimensions, record in appropriate spaces on data form, and write in the notes "estimated measurements". Flag all redds and write record number, species, year, and total length and width of redd on the flag. If test or under construction write "REMEASURE" and total length and maximum width on the flag. Remeasure all test and under construction redds on next survey. Redd age: 1 = New since last survey, 2 = Still measurable, 3 = Not measurable but still apparent, 4 = No longer present, 5 = Pre-condition - can not measure or determine age.

Record Number	Species Code	Pot Length (m)	Pot Width (m)	Pot Depth (m)	Pot Substrate	Tail Spill Length	Spill Width 1 (m)	Width 2 (m)	Tail Spill Substrate	Fish on? Yes/No	Fish #	Redd Age	Remeasured Yes/No	Distance Spill to Tail	Flag Direction Spill to Tail	Flag #	Notes
121200	Onmy	1.2	2.2	0.15	4	1.5	1.2	1.1		5	yes		1212002	1	No	1.35m	230 degre	fish spooked off flag above not was test now is
121200	Onki	2.3	3.2	0.2	6	3.3	2.5	1.5		2	yes		1212003	1	no	0.25 m	145 deg	clearly onmy
12021	test	1	1.2	0.1	5	1.6	1.4	1.2		4	no	na		2	yes			remeasure again fish still working
12020	Onki	2.5	2.3	0.15	6	2.5	2.2	2		3	yes		1212005	2	yes			redd perpendicular to stream flow
121200	Unkn	2.8	1.5	0.2	3	1.1	0.9	0.8		2	no	na		1	no	1.75 m	355 deg	flag syas remeasu too turbid to
12020	Onki										no			5	no			no longer presen
12020	Unkn										no			4	no			4 on flag

Appendix B: Redd Counts Data Sheet

Stream Name:		Week:		Page ___ Of ___
Section Name:		Water Temp.		**Species Codes:** Onki = Coho Salmon
Reach ID:	Map #	Air Temp.		Onmy = Steelhead
Date:	Water Visibility (m):	Start Time:		Onts = Chinook Salmon
Surveyors:	Stream Flow (m³/s)	End Time:		Latr = Pacific Lamprey
	Stage:	Drive Time:		Unkn = Unknown Species

Weather:

Notes:

Directions: Visually identify each redd to species. Record the record number. Mark redd location with a fish on the map. If a fish is on a redd, record this on both fish and redd data form, write fish number of redd form and redd number of fish form. If you cannot measure a redd visually estimate the dimensions, record in appropriate spaces on data form, and write in the notes "estimated measurement". Fill all dimension number spaces by species, year, and total length and width of redd on the fin. If tester under construction write "REMEASURE" and total length and maximum width on the flag. Remeasure all test and under construction redds on next survey. Redd age: 1 = New since last survey. 2 = Still measurable. 3 = Not measurable but still apparent. 4 = No longer there. 5 = poor conditions - can not measure or determine age.

Record Number	Species Code	Pot Length (m)	Pot Width (m)	Pot Depth (m)	Tail Spill Length (m)	Tail Spill Width 1 (m)	Tail Spill Width 2 (m)	Tail Spill Substrate	Pot Substrate	Fish on?	Fish #	Redd Age	Remeasured? Yes/No	Distance to Tail Spill	Flag Direction Tail to Tail Spill	Flag Tail Spill	Notes

Rotary Screw Traps and Inclined Plane Screen Traps

Gregory C. Volkhardt, Steven L. Johnson, Bruce A. Miller, Thomas E. Nickelson, and David E. Seiler

Background and Objectives

Inclined plane screen traps and rotary screw traps have long been used by biologists to capture downstream migrating juvenile anadromous salmonids from medium- and large-sized streams (Schoeneman et al. 1961; Seiler et al. 1981; Kennen et al. 1994) and from small tributary streams (Solazzi et al. 2000). In its original fixed screen design, the floating inclined plane screen (scoop) trap has been used to capture juvenile migrants for more than 40 years (Schoeneman et al. 1961). William Humphreys replaced the fixed screen with a traveling screen powered by a paddle wheel and added a debris drum at the back of the live well (Humphreys trap) in 1966 (McLemore et al. 1989). The rotary screw (screw) trap was developed and patented by two biologists from the Oregon Department of Fish and Wildlife (ODFW) in the late 1980s. All these traps are anchored at a fixed point in the stream channel and intercept a portion of the juvenile salmonids or smolts migrating downstream.

Traditionally, fishery managers have relied on escapement estimates to monitor anadromous salmonid population status and management effectiveness (Ames and Phinney 1977; Beidler and Nickelson 1980; Hilborn et al. 1999); however, estimation of population abundances at earlier life stages enables partitioning survival among life stages and developing hypotheses for restoration actions (Moussalli and Hilborn 1986; Mobrand et al. 1997). Juvenile fish traps have often been used to estimate the abundance (Tsumura and Hume 1986; Baranski 1989; Orciari et al. 1994; Thedinga et al. 1994; Letcher et al. 2002; Johnson et al. 2005), timing (Wagner et al. 1963; Hartman et al. 1982), size (Orciari et al. 1994; Olsson et al. 2001), survival (Schoeneman et al. 1961; Wagner et al. 1963; Tsumura and Hume 1986; Olsson et al. 2001; Letcher et al. 2002), and behavior (Brown and Hartman 1988; Roper and Scarnecchia 1996) of downstream migrant anadromous salmonids. In many salmon-bearing systems, population abundance is only monitored during the adult (spawner) stage. Additional monitoring of smolt abundance is a particularly powerful tool because it enables partitioning mortality between the freshwater life stages (egg-to-smolt) and marine life stages (smolt-to-adult).

While estimating smolt abundance is the most common reason for operating an inclined plane screen trap or screw trap, the capture of downstream migrants has wide utility. Traps can be used to monitor the effects of river management on wild stocks, such as the effectiveness of diversion, lock, and dam management. They are powerful tools for validating assumptions regarding the effects of watershed restoration programs and land-use policies on fish populations (Solazzi et al. 2000; IMW SOC 2004; Johnson et al. 2005). They can also be used to assess survival between life stages, such as egg-to-smolt survival or parr-to-smolt overwinter survival (Solazzi et al. 2000; Seiler et al. 2003; Johnson et al. 2005). Smolt-to-adult survival estimates can be developed for wild populations by coded wire tagging smolts that are captured in inclined plane screen traps and screw traps and estimating the escapement and fishery impacts on the tagged population.

In addition to serving as a tool to monitor wild populations, inclined plane screen traps and screw traps are useful for evaluating hatchery programs and hatchery/wild fish interactions. Such studies may include evaluating the instream survival of hatchery production following release and evaluating treatments such as rearing strategy, release timing, release location, and flow manipulation on groups of hatchery fish. These latter uses can be applied to evaluate a variety of projects or actions, ranging from hatchery supplementation strategy to avoidance of hatchery and wild fish interactions. In addition to abundance estimates, investigators use inclined plane screen traps and screw traps to collect samples of downstream migrants for purposes such as genetics sampling, fish disease research, predation (gut content) evaluations, and wild stock marking and tagging projects.

Operating a downstream migrant trap allows the investigator to sample wild salmonids produced in a watershed or tributary over time. The sample in itself is valuable because it documents the presence/absence of migrating juveniles and enables determination of age and size at migration, condition, timing, species, and genetic characteristics. Furthermore, catch of a given species or catch-per-unit-effort (CPUE) can be used as an index of downstream migrant production if the location of the trap, its placement, and hours of operation are sufficient and held reasonably constant from year to year.

More importantly, trapping information can also be used to create estimates of total freshwater production by use of simple mark–recapture methods to estimate abundance. The rationale is simply that the proportion of marked fish appearing in a random sample is an estimate of the marked proportion in the total population. The proportion captured (trap efficiency) is estimated by conducting a series of trap efficiency experiments throughout the trapping season.

On the west coast of the United States and Canada, juvenile fish traps have been used primarily to estimate the natural production of juvenile coho *Oncorhynchus kisutch,* sockeye *O. nerka,* and steelhead *O. mykiss* from fifth-order and smaller basins (Nickelson 1998). Nevertheless, with careful planning, reasonably accurate production estimates have been obtained when sixth-order and larger systems have been trapped (Schoeneman et al. 1961; Thedinga et al. 1994). For example, side-by-side scoop and screw traps have been successfully used to make estimates of yearling coho and sub-yearling chinook *O. tshawytscha* migrants since 1990 in the Skagit River (a seventh-order basin) (Seiler et al. 2003) (see Figure 1).

This protocol describes methods to estimate wild downstream migrant salmonid production using either an inclined plane screen trap or a rotary screw trap. Because the traps strain the upper portion of the water column, they are generally not very useful for capturing species that migrate along the bottom of the river (e.g., lamprey). The traps can be scaled to operate in various-sized streams but are most commonly used in streams that are too large or powerful to employ a fence weir (i.e., ~10–15-m or larger channels).

FIGURE 1. — Skagit River screw and scoop traps.

Inclined plane screen trap

The design of inclined plane screen traps permits trapping a range of stream velocities and depths. There may be any number of derivations from the basic scoop trap design, which is simply a wedge-shaped screened rectangular tube suspended in the water column from a pontoon barge. The screen section is typically constructed of galvanized woven-wire mesh (hardware cloth) or perforated plate aluminum sheet metal riveted to a frame. All seams are coated with a sealant so that no sharp metal edges are exposed that can injure fish. The scoop trap is typically suspended inside a pontoon barge from support winches at the corners of the fore and aft decks (see Figure 2). The trap position is fixed using anchor lines that extend from each pontoon to shore, a fixed in-stream structure (e.g., bridge), or a high lead that extends across the river.

FIGURE 2. — Scoop trap mounted on a pontoon barge, shown in the nonfishing position (Clearwater River, Washington).

Other inclined plane screen trap designs been developed to reduce debris buildup on the trap and to adapt to specific site characteristics. The traveling screen trap or Humphreys trap, originally designed by William Humphreys (ODFW), uses a traveling screen instead of a fixed screen along with a trash drum at the back of the livewell to reduce debris buildup on the trap (McLemore et al. 1989). The basic Humphreys trap uses a paddle wheel and gear assembly attached to one or both of the pontoons supporting the trap to power the traveling screen and trash drum. A similar design, the motorized Humphreys trap, uses a 12-V DC motor instead of paddle wheels to power the traveling screen (see Figure 3). This design is best suited to smaller streams that lack the hydraulic power to drive the Humphreys trap or a rotary screw trap. In another variation of the fixed screen design, the upstream end of the inclined plane screen is attached to a low-head dam or weir and collects fish passing over the structure (e.g., DuBois et al. 1991). A lightweight inclined plane trap for sampling salmon smolts has also been used in Alaska (Todd 1994).

FIGURE 3. — Motorized incline plane trap used in small Oregon coastal streams to monitor downstream migration of juvenile salmonids.

FIGURE 4. — Fishing scoop trap showing the fore and aft winches used to raise and lower the trap, the trap apex, live well, and catch processing station (Chehalis River, Washington).

FIGURE 5. — Close-up of a motorized incline plane trap's moving screen with attached cups that help keep juvenile fish from escaping off the screen.

When the inclined plane screen trap is lowered into the current, water is strained through the screens and downstream migrants are swept up the screen incline and deposited into a protected, solid-sided and floored live box at the back (see Figure 4). To capture and retain migrants in a scoop trap, water velocity through the trap must exceed swim speed. As swimming ability is directly related to body length, higher velocities are required to trap large migrants. Fry (less than 50 mm fork length) may be captured at relatively low velocities, whereas trapping the larger migrants, such as steelhead smolts (up to 250 mm), requires velocities greater than 2 m/s (mps). At less-than-optimal velocities, larger migrants may avoid or swim out of the trap. Velocity requirements may be partially mitigated with a traveling screen trap since the screen can be fitted with baffles or perforated L-shaped cups to help carry fish to the livewell and reduce the chance of escape (see Figure 5). As velocity increases, the volume of water and suspended debris passing through the trap also increases, requiring more frequent inspection and cleaning of the trap and live box.

Flow into the trap is regulated by positioning the trap (laterally and longitudinally) in the stream and by adjusting the level and angle of the inclined screen through its four support winches. Proper adjustment of a scoop trap is indicated by a smooth flow over the apex of the incline into the holding chamber, with a water depth over the apex of 1.5 to 2 cm. As the screen accumulates debris, its ability to pass water decreases and the depth and velocity over the incline increases, causing turbulence in the holding chamber. Debris load is affected by streambank vegetation, weather (rain and wind transport debris into the river) and, most importantly, river discharge. Trap operation through a freshet requires that the screens are carefully monitored and regularly cleaned and that the catch is frequently removed from the live box and processed.

Traveling screen traps fitted with baffles may be adjusted so that the top of the screen extends slightly out of the water, since the movement of the screen and baffles carries the fish to the live well (see Figure 5).

Rotary screw trap

The screw trap consists of a cone covered in perforated plate that is mounted on a pontoon barge (see Figure 6). Within the cone are two tapered flights that are wrapped 360 degrees around a center shaft. The trap cone is oriented with the wide end facing upstream and uses the force of the river acting on the tapered flights to rotate the cone about its axis, similar to an Archimedes screw (see Figure 7). Downstream migrating fish are swept into the wide end of the cone (typically either 1.5 m or 2.5 m in diameter) and are gently augured into a live box at the rear of the trap (see Figure 8). A winch is used to adjust the forward elevation of the screw, and an additional winch may be used to raise and lower the aft end of the screw if desired. A small drum screen, powered by the rotating cone or a paddle wheel, may be located at the rear of the live box to remove organic debris.

FIGURE 6. — Screw trap mounted on a pontoon barge in the nonfishing position (Puyallup River, Washington).

FIGURE 7. — Two rotary screw traps in the nonfishing position suspended from a three pontoon barge (Wenatchee River, Washington).

When positioned in the river, both inclined plane screen traps and screw traps are navigation hazards to boaters, float tubers, and swimmers. Signage should be positioned upstream to instruct river users how to avoid the trap safely. Other protective measures may include installing flashing lights to improve the visibility of the trap (see Figure 8) and deflectors to help prevent water users and large woody debris from entering the trap (see Figure 9).

FIGURE 8. — Back end of the screw trap showing the auger cone, live well (covered), and trash drum (Green River, Washington).

FIGURE 9. — Screw trap with deflector (Green River, Washington).

Rationale

While scoop traps and other inclined plane screen trap designs have been used to capture downstream migrants for more than 40 years, screw traps have only been in use since the early 1990s. Screw traps and traveling screen traps incorporate a number of improvements over the older scoop trap. For example, since the screened surface of a screw trap rotates about an axis in and out of the water, small debris falls off of the screen and is flushed into the live well. The frequency of trap cleaning is greatly reduced for screw and traveling screen traps compared to the scoop trap, which more readily accumulates debris on its screened surfaces. Another shortcoming of scoop traps is that they are effective only where water velocity exceeds the burst swimming speed of the target species. This problem is most apparent for strong swimmers such as steelhead and cutthroat smolts, which often can swim out of a scoop trap. As screw traps rotate about their axes, the tapered flights block captured fish from swimming back out of the trap; therefore, screw traps can be more effective for capturing larger migrants. Traveling screen traps employing baffles or other capture aids are typically more effective in retaining larger migrants than scoop traps but are less effective than screw traps.

Despite these deficiencies, scoop traps are attractive for their simplicity. The lack of moving parts makes them very reliable. Scoop traps are generally more effective than are screw traps at capturing smaller migrants (i.e., <80 mm for salmonids). For example, the Washington Department of Fish and Wildlife (WDFW) has operated a scoop trap and a screw trap side by side to evaluate anadromous salmonid production in the Skagit River, Washington. Although these traps strain nearly equal volumes of water, the scoop trap consistently traps a higher proportion of subyearling migrants, and the screw trap consistently traps a higher proportion of larger, older age migrants, such as steelhead and cutthroat trout trout *O. clarkii*. Yearling migrants such as coho are captured at about equal rates. A potential explanation is that downstream migrants are more likely to avoid the screw trap because of the noise it generates; therefore, the quieter scoop trap probably has a higher initial capture rate. Nevertheless, because larger migrants are not retained as well in the scoop trap, the screw trap is less size-selective.

Site characteristics are another important consideration for trap selection. In general, screw trap operation is better accommodated in larger streams, where sufficient water depth and velocity are available to accommodate and power the screw. E. G. Solutions, the patent-holding company of the screw trap design, currently produces only 1.5- and 2.5-m-diameter traps, which limit the size of stream in which these traps can be operated. Nevertheless, screw traps can be operated in streams of variable size by scaling the size of the pontoon barge that supports the trap. Whereas pontoon barges supporting scoop and screw traps on large rivers may be 9–13 m in length and fabricated from steel, those used on small streams are fabricated from aluminum and are much smaller (e.g., 4 m in length). For example, ODFW has successfully employed screw traps in streams with catchments as small as 14 km^2. Inclined plane screen traps are often built by the investigator and can be scaled to operate in very small streams. Because of their adjustable screen depth, traveling screen traps and scoop traps are less constrained by shallow water depths (i.e., smaller streams) than are screw traps.

Sampling Design

The sampling design will depend on the objectives of the study. Since considerable effort is required to install and operate inclined plane screen traps and screw traps, most investigators will use these gear types only if sampling over many days is desired. Although there are many potential uses, most investigators will use these traps either to estimate the total freshwater production of wild salmonids or, where conditions preclude this, to develop an index of production. The following discussion is oriented towards estimating freshwater production; however, in most cases, investigators will be able to adapt these methods to meet the objectives of their studies.

Site Selection

Selection of trapping sites should be viewed from a variety of scales. If the natural production of salmon is to be monitored at the watershed scale, no hatchery fish should be present in the river or stream; if they are, all hatchery fish should be identifiable so that wild fish may be enumerated. Precision of the estimates increases with higher trap efficiency (i.e., proportion of migrants captured);

therefore, it is generally better to select sites where a higher proportion of the total flow can be screened through the trap. This becomes a trade-off, however, if the trap is placed below a hatchery release site, since higher trap efficiencies can result in very large numbers of hatchery fish entering the trap following release. When this occurs, good communication between trap operators and hatchery staff must be maintained to avoid a fish kill. In general, it is best to avoid these situations when choosing a trap site.

Another consideration when selecting watersheds (or catchment basins) is the stream hydrograph. Flow is dependent on such variables as landform, geology, land cover, climate, and precipitation patterns, which of course cannot be controlled. The effect of these factors on the stream discharge needs to be considered when attempting to estimate total freshwater production. Streams and rivers exhibiting a flashy hydrograph are very difficult to trap due to high fluctuations in flow conditions and debris loads. Because trap efficiency and migration rates often change dramatically with discharge, it is very difficult to estimate migration accurately. Furthermore, traps may become difficult to access safely without prior planning and preparation.

If the monitoring objective is to measure total abundance within a watershed, the trap should be placed as low in the watershed as is practicable. It is vital to take into account the life history and in-river migration patterns of your target species. Species exhibiting a stream-type life history pattern, such as coho salmon and steelhead, often migrate within basin and rear away from their natal streams; therefore, the smolt production measured from a tributary trap may represent a variable proportion of the progeny from the adults that spawned upstream of the trap. Species with an ocean-type life history pattern (e.g., pink salmon *O. gorbuscha*) often spawn lower in the watershed, so it is prudent to place traps as low in the system as possible in order to estimate production. Tributary traps are often used in a before-after monitoring design to evaluate differences in abundance resulting from changes in management or restoration. Care must be taken in interpreting the results, however, since improved smolt production could be the result of parr movement into the enhanced tributary for rearing rather than increased egg-to-smolt survival.

At the site scale water velocity, depth, and proportion of the flow screened are also important considerations for trap placement. Velocity is an especially relevant consideration if trapping strong swimming species such as steelhead, and becomes less so when trapping newly emerged fry. For most species, water velocities of at least 1 m/s are desirable for scoop trap operation; for most steelhead smolts, velocities greater than 2 m/s may be required for capture and retention. For Humphreys traps, McLemore et al. (1989) suggest a minimum water velocity of 0.9 m/s, with most efficient operation at 1.5–2 m/s. We have found that water velocities of 0.8–2 m/s work well for screw traps operated in Oregon coastal streams. Ideally, screw trap sites should have sufficient velocity to conduct at least 5–6 trap rotations per minute (rpm) for capturing larger smolts. Velocity does have its limits: 2.5-m diameter screw traps can be damaged at rotation speeds greater than 14–15 rpm. Trap avoidance is minimized in cases where higher velocities occur; nonetheless, fish can certainly be captured at lower velocities. For example, researchers in California have found that successful capture and retention of steelhead smolts occurs at speeds as slow as 1.67 rpm.

Care must be taken that the water depth under the trap and live well will be sufficient over all flow conditions that are expected during the outmigration period. To achieve the highest possible trap efficiency, it is usually best to select a site where a relatively high proportion of the total flow can be screened through the trap. The requirement for adequate velocity, depth, and trap efficiency usually argues for placing the trap in the thalweg of the channel. Consideration must be given, however, to the number and behavior of migrants captured. The investigator may choose to operate the trap in a slightly less advantageous position to avoid causing stress or predation in the live well by capturing and holding too many migrants. In addition, a substantial proportion of the migrants from some species/age-classes may migrate along the channel margins. If these fish are targeted, placing the trap nearer to margin habitats and using weir panels to lead these fish to the trap entrance may achieve higher capture rates.

Stream flow should be moving in a straight line as it enters the trap. Pools with sharp changes in direction that result in large back-eddy currents should generally be avoided. Streamflow in smaller streams may diminish to the point that water velocity is not sufficient to turn a rotary screw trap. Operating a trap in low-flow situations presents its own challenges and requires prior planning for successful trapping. Small boulders, sandbags, or screened weir panels can be used to improve the hydrodynamics of a site. Where bedrock is the predominant substrate upstream from the trapping location, tripods can be bolted directly to the bedrock, providing a firm foundation for attaching screen panels to direct flow towards the trap (see Figure 10). Tripods and screen panels can also be erected in gravel-dominated streams to improve trapping conditions, but these sites usually require substantial reinforcement with sandbags in front of and behind the panels to minimize the stream's ability to undercut the panels in high-flow events. These types of channel modifications or treatments can also be used to increase the functionality of locations with poor site characteristics; however, the researcher should evaluate whether these actions are subject to regulatory authority.

FIGURE 10. — Tripods and screen panels bolted to bedrock substrate to improve screw trap operation in a low-gradient coastal stream.

Additional consideration needs to be given to site selection for screw traps because of the noise they generate. Migrants will avoid the trap if they are aware of its presence; therefore, it is best to select a site where the trap noise can be masked to maintain higher trap efficiency. Fortunately, higher velocity reaches are generally noisy reaches. In smaller rivers, these conditions occur at the head end of a pool or chute where water velocities over an elevation drop (e.g., riffles, cascades, falls) can be directed into the trap. In larger rivers, channel constrictions may afford the best sites.

In addition to the aforementioned criteria, consideration must be given to anchoring the trap in the stream. Scoop and screw traps can be anchored by cables to the base of stout trees on each bank; to anchors affixed to bridge abutments, retaining walls, or bedrock; or to a high lead suspended across the river. Where anchoring structures are unavailable, some researchers have created anchors by burying 4–6 fence posts tied together with a steel cable or by driving a series of 6 fence posts into the substrate in a triangular arrangement. In the early 1960s the mainstem Columbia River was trapped using a series of scoop traps cabled to large concrete blocks submerged in the river (Schoeneman et al. 1961).

Finally, investigators need to consider access and security when selecting trapping locations. Traps anchored in the river are a curiosity, which can draw theft or vandalism when not attended. Ideally, the trap site will be located near a launch/recovery site to ease trap installation and removal.

Period of operation

The time frame for trap operation varies with the target species and trapping location. Table 1 provides general migration timing for anadromous salmonids in Oregon and Washington rivers. Downstream migration timing in specific rivers can vary from these general guidelines. Timing may need to be investigated during the first year of monitoring where it is not well known.

TABLE 1.—Generalized migration timing for anadromous salmonids in Oregon and Washington.

Species	Age	Migration period
Chinook	0, 1	January–July/August
Coho	1	March/April–June
Sockeye	0	January–May
Chum	0	February–April
Pink	0	January–May
Steelhead	2	March–May
Cutthroat	0, 1, 2	January–December*

* Migration timing for cutthroat varies widely.

To estimate production, traps should be operated throughout the migration period for the target species. For most species, migration rates are often highest at night; yet daytime migration rates can also be high on some streams, particularly where turbidity levels are high. At a minimum, the investigator should stratify trapping periods to reflect different migration/capture rates. This often means checking the trap and processing the catch at dawn and at dusk to measure day and night catch rates. These are not, however, the only times to check the trap; the frequency of catch processing and trap maintenance should be determined by catch rates and debris loads. Stratification facilitates subsampling and estimating catches during periods when trapping is suspended.

Field/Office Methods

Before trapping can begin, all equipment and supplies must be assembled to accomplish project objectives. At a minimum, these include the trap/pontoon structure and anchor cables, a means to get to the trap (e.g., boat, gangplank), dip nets for removing and handling fish, data forms, fish anesthetic, a marking device (e.g., scissors, dye), tanks or buckets for working up captured fish, a trap cleaning device (e.g., brooms, water pump, nozzle), and lights for night work. Permits may also need to be secured for placement of the trap and/or handling fish from various jurisdictions. Sufficient time must be allotted during the planning period to secure permits.

The approach for trap installation depends on the size and weight of the trap used. Small inclined plane screen traps and screw traps that use lightweight aluminum pontoons can be transported in pickup beds and assembled on-site. Components of larger, heavier traps, such as the scoop trap shown in Figure 2, can be trucked to the site using a low-boy trailer. In this case, onsite assembly requires the use of a loader or other heavy equipment to move the components into place. A third option is to truck an assembled trap to the site and position it in the water using a boom truck or crane.

Once in the water, the trap is ready to be positioned in its fishing location. The approach used to accomplish this will depend on the size of the trap and stream and the distance from the launch point to the fishing site. Small traps operating on small streams can be moved into position by hand. Bow-mounted cables or ropes can be attached to trees or other anchoring structures on the banks. Movement of the trap into its final position can be accomplished using hand winches or chainfalls. If the trap is anchored to trees, the load needs to be distributed over the trunk to prevent girdling. Fabric straps make useful attachments.

If the launch point is some distance from the fishing site, the trap can be "walked" into position by alternating port and starboard attachment points either upstream or downstream and tightening or loosening the bow cables as necessary using winches. In navigable waters, a boat can be used to push the trap to a site where one of the methods described can be used to secure the trap to its fishing position.

Larger traps may use bow winches mounted port and starboard to store attachment cables (see Figure 2). The most direct approach is to run the cables out to the attachment points and pull the trap into position using the winches. Another approach is to attach cabling directly from the trap to a highline that has been strung over the river (see Figure 11). The use of bow-mounted winches is the preferred approach, since it makes repositioning the trap much easier.

FIGURE 11.—Example of a highline anchoring system for two screw traps operating side by side (Wenatchee River, Washington).

Traps on larger streams and rivers are accessed primarily by boat. On large rivers an aluminum flat-bottomed skiff powered by a motor outfitted with a jet pump can be used to reach traps. To reach traps on smaller rivers, WDFW uses a pulley that attaches the bow of an aluminum skiff to a cable extending from the bank to the trap. Also, extending from the trap to shore is a rope, which is suspended over the skiff and used by the operator to ferry the boat back and forth along the cable from shore to trap. When traps are located 3 m or less from a steep bank, a portable gangplank bridge can sometimes be constructed and used for access (see Figure 9).

Although traps on small streams are generally accessed by wading, a cabling configuration may also used to bring the trap to shore when high flows preclude wading access. For example, researchers often use a highline system for anchoring the trap (see Figure 11). A bridle is attached to the front of the trap, and a main line from the bridle is routed through a pulley on the highline crossing the stream upstream of the trap. The main line then runs through another pulley attached to a tree or anchor on the bank. An additional rope-and-pulley setup is also attached to the front of the trap and brought directly to the bank, allowing the trap to be positioned from side to side. This method exerts tremendous tension on the highline that spans the stream, so the cable size and anchors need to be sized accordingly. Alternatively, single lines may be led from each pontoon through blocks mounted on each shore. With this configuration, downstream force is spread between two lines instead of one main cable. The line attached to the pontoon opposite the access side of the river should be led back across the river so that both lines can be controlled from the same bank. Depending on the site, additional pulleys can be added on each line so that one person can manipulate the position of the trap. Using these rope riggings, operators can safely manipulate the trap's position during high-flow events from the bank of the stream. A safety cable should also be attached to the downstream end of the trap and to the pontoon farthest from the bank to which the cable is attached. With this configuration, if the trap breaks free from the main front cable, the trap should rotate and face downstream with the trap held against the bank by the safety cable.

Trap operation

Inclined plane screen trap

Once the pontoon barge is in position, trap operation begins by lowering the trap into the water using the cables or rope attached to each of the four corners of the trap. On larger traps, winches are used to raise and lower the corners (see Figure 4). For scoop traps, the trap's apex—the point where fish riding up the scoop screen drop over the top and into the live box—should have about 1.5–2 cm water depth. The two forward corners should be level and straining the top meter or less of water, depending on the size of the trap. The two aft corners should be laterally level, and the live well should have sufficient water to maintain the captured fish. Water in the live well should be relatively quiet. Most of the streamflow entering the scoop will pass through the trap screens, but sufficient sweeping flow must be available to carry fish up the incline and into the live well. These conditions are usually met when the trap is level fore to aft or slight higher at the forward end (see Figure 4). When the trap is properly set for fishing, marking the leading edge of the scoop at the water's edge so that the trap can be reset to fish at the same depth each time can reduce variation in trap efficiency.

When checking the trap, the forward end of the trap is raised until the bottom of the scoop is above the water level and no longer fishing. Then, the aft end is raised as needed to remove the catch for processing. Brooms or a water pump with nozzle are used to clean the trap screens before lowering the trap back into its fishing position. The date and time that the trapping is suspended and resumed are recorded. The catch is enumerated by species, and other data and samples are taken as appropriate to the study.

Operation of the traveling screen traps is similar to the scoop trap. Because these traps must operate during high streamflows, there is always a risk that the traps will become jammed with debris. At these times, traps require constant or frequent attention to minimize potential mortalities and to ensure that traps are functioning properly. Still, there may be times when traveling screen traps stop rotating while no one is on-site. To determine the length of time a motorized Humphreys trap actually runs, a 12-V clock is connected to a fuse in line with the trap motor. If the trap is jammed with debris, the fuse is blown, stopping the clock. By recording the starting and ending time on the clock, the length of time the trap was fishing before becoming jammed can be determined. The fuse has the additional benefit of preventing motor burnout by cutting off power to the motor when the trap becomes jammed.

Screw trap

The screw trap is lowered into its fishing position by cables attached to the forward and/or aft ends of the trap structure. Typically, one or two hand winches or chainfalls are used to raise and lower the forward end (see Figure 12) or both ends (see Figure 8), depending on trap design. The forward end of the screw should be lowered until the axle is at the water's surface (see Figure 13). The aft end is lowered so that fish can swim from the aft screw chamber into the live well, but not so low that they can ride the debris drum (if there is one) over the back of the trap.

FIGURE 12.—Rotary screw trap outfitted with a single forward winch in the nonfishing position.

FIGURE 13.—Rotary screw trap in fishing position.

Since the screw is constantly rotating, relatively little debris builds up on the screw's outer screen. As the debris drum removes much of the debris entering the trap, this gear requires less cleaning than a scoop trap. During each trap check, debris remaining in the live well is removed and captured fish are dipnetted out. The trap can usually remain in operation during this procedure. The date and time of the trap check is recorded. Catch is enumerated by species and other data/samples are taken as required by the study.

Debris can also prevent operation of rotary screw traps. To estimate when rotation of the trap ceased, we recommend using a trucking industry hub-odometer. The hub-odometer is placed on the front or rear of the central shaft of the trap (see Figure 14) and records the "distance" that the screw turns between trap checks. By monitoring the number of revolutions per minute, and knowing the distance that the screw turned, the length of time the trap fished before jamming can be estimated. The hub-odometer that we use records 1.6 km for every 500 revolutions of the trap shaft. Thus, the length of time the trap fished can

be estimated by

$$\text{hours trapped} = [\text{total revolutions}/(\text{measured revolutions/minute})] \div 60,$$

where

$$\text{total revolutions} = (\text{ending odometer reading} - \text{beginning odometer reading}) \times 500$$

FIGURE 14.—Hub-odometer attached to shaft of rotary screw trap. Hub-odometer readings are used to determine the number of hours the trap fished if trap is stopped by debris when staff is not present.

As stream discharge diminishes in the Spring, screw trap operation can become increasingly difficult in small streams or in stream reaches with marginal site characteristics (e.g., pool depth, velocity). These reduced flows may create a situation where there is insufficient velocity to turn the trap screw. Decreasing stream discharge may also reduce pool depth at the trapping site so that the screw cone encounters the bottom, either stopping the cone from turning or damaging the cone as it scrapes on the bottom. Considerable effort should be made to find sites that enable trap operation over an anticipated range of streamflows. If such a site cannot be found, a trap design more conducive to site characteristics (e.g., smolt fence) should be considered. Nevertheless, if screw trap operation is necessary at a marginal site, there are a number of actions that can be attempted until conditions improve.

For example, adjustable legs can be added to the trap to prevent the trap cone from contacting the streambed during low-flow periods (see Figure 15). Screen panels can also be extended to force the entire streamflow in front of the trap to generate enough water velocity to turn the trap screw (see Figure 10). Loosening or disconnecting the mechanism that drives the drum screen will reduce friction on the cone and help the cone spin in low-flow situations. Additional "wings" or "flights" can also be fabricated and attached inside the cone between the permanent wings to increase the surface area contacting the water entering the trap, thus helping to keep the cone rotating.

FIGURE 15.—Rotary screw trap in operation during low stream flows with the pontoon barge supported above the water on adjustable legs that allow the trap to spin during periods when the trap cone would normally contact the bottom of the stream.

If these measures fail to adequately turn the screw, rotation can be achieved using equipment similar to that which runs the motorized Humphreys traps described by McLemore et al. (1989). The upstream end of the shaft provides an attachment point for a sprocket gear that can be driven by a motor assembly, applying rotational force to the drum. The sprocket gear mounted to the drum shaft is driven by a riveted roller chain leading to a smaller sprocket gear mounted on a motor and speed-reducer assembly, which in turn is mounted on the front cross-member (see Figure 16). The 92:1 speed reducer allows the power transfer between the high rpm of the motor and the low rpm of the trap drum. The sprocket gear mounted to the front drive shaft is 34.3 cm in diameter with 84 1.27-cm pitch teeth. A section of pipe of a diameter that will fit either inside or over the trap drum shaft is centered and welded square to the sprocket gear. Once fitted, a hole is bored through the shaft and the pipe/gear is held in place with a 10-mm stainless steel bolt. This combination of gears, speed reducer, and a standard 1,750-rpm motor will turn the drum at a calculated speed of 3.2 rpm, but with some force provided by the stream flow, the actual rpm is generally closer to 3.5–4 rpm.

FIGURE 16.—Rotary screw trap with 12-V motor turning the drum when stream flows are insufficient.

For remote sites where high-voltage power is not available, the trap may be motorized using 12-V equipment. A 1/14 hp, 12-V motor that draws 6.9 A s at full load may be powered by several high-amperage deep-cycle batteries wired in parallel. The amperage capacity must be matched to the anticipated run time with no more than 50% discharge on the battery pack. For 24-h operation, a minimum capacity of 330 amp-hours is needed. Alternatively, limiting trap operation to sample only at night, if previous sampling has indicated little migration occurs during daylight hours, could reduce capacity requirements. The simplest method to achieve this is to wire in a 12-V photo switch that closes at low light, turning the trap on at dusk, and then opens the circuit in light, turning the trap off at dawn. Another option would be to use a 12-V programmable timer switch. This type of switch can be programmed for several on-off operations in a 24-h period. For any type of motorized configuration, the circuit should contain an in-line fuse to protect the motor should debris cause the drum to jam. A 10-A thermal fuse should be wired between the motor or photo switch and the battery pack.

Use of a motor to provide driving force to the drum also permits deployment of a rotary screw trap in tidal reaches of a stream, where gradient is low and streamflows are generally inadequate to drive the trap drum. In this circumstance, motorizing the drum during the tidal flood is inefficient because fish are unlikely to migrate downstream against the current. A motorized trap deployed in a tidal channel can be controlled by installing a float switch in the motor circuit that turns the drum on during ebb and low tide and off during flood and high tide. When deployed in a tidal channel, it is usually necessary to construct a plywood or screen weir in front of the trap to focus flow and migrating fish towards the trap drum. This weir provides a location to mount a float switch. The base of the switch is attached to either the weir or a freestanding mount next to the trap. A pole is mounted horizontally to the switch base with a pivot bolt near one end of the pole. A foam crab-pot buoy is mounted to the pole at the end farthest from the pivot bolt, and a copper contact plate is mounted to the end closest to the pivot bolt. A second copper contact plate is attached to the base, forming a switch. The base is mounted such that when the pole is near horizontal, the switch plates are in contact, completing the circuit and permitting the drum to be powered. When the flood tide lifts the crab float and pole, the circuit breaks and the drum stops.

The wires leading to the switch need to be long enough to remain slack during the highest spring-series tides. A float switch, in conjunction with a 12-V photo switch, limits hours of trap operation to ebb and low tide only at night and makes most efficient use of batteries.

Fish handling

Traps are checked as often as necessary to provide for the safe holding and handling of captured fish and to maintain the efficient operation of the gear. At a minimum, the trap should be checked at dawn and at dusk to evaluate day versus night capture rates. Where subyearlings are captured, holding them in close proximity to larger piscivorous fish such as cutthroat smolts and sculpins increases the likelihood that catch counts on the subyearlings will be biased low due to live-box predation.

Some investigators have placed tree branches or other debris in the live well to provide a refuge for small fish. Care must be taken when using this approach since the debris may cause descaling as turbulence in the live well increases. The safest approach for maintaining fish health and minimizing predation is to frequently check and remove fish from the trap.

Creating a workstation for employees to collect data on the catch will alleviate stress, both on the fish and the sampler. The workstation can be built on the pontoon deck of larger traps (see Figure 7) or built to stand in the stream or along the shoreline when processing fish from small traps (see Figure 17). Although this equipment will vary depending on the site and trap design, constructing a sampling table with adjustable legs and racks to hold screened buckets will often allow the sampler to process the catch while remaining in the stream. This will reduce the need for maintaining adequate oxygen and water temperature for the fish as they are being processed. Workstations mounted on larger traps can be supplied with water from generator or battery-powered sump pumps. Since capture and handling often occur at night, workstations should be provided with artificial lighting.

FIGURE 17.—A workstation at a trapping site in an Oregon coastal stream is used to process screw trap catches and record data.

Generally, all fish are anesthetized prior to processing (e.g., marking, mark recovery, length/weight measurement). Tricaine Methanesulfonate (MS222), CO_2, and clove oil are the most common anesthetics used. When anesthetizing fish, it is important to remember that water temperature, anesthetic concentration, and fish density and size can all increase the stress load on the fish. Care needs to be taken so that no more fish are anesthetized at one time than can be safely processed. This will vary with the experience of the sampler and the amount of information being collected. Fewer fish should be anesthetized at one time as water temperature increases, since higher temperatures generally increase the effectiveness of the anesthetic as well as handling stress on the fish. Anesthetic water should be regularly changed to keep it cool and well oxygenated. Once all fish are processed, the recaptured marked fish and fish not needed for trap efficiency experiments are released far enough downstream to minimize the potential for recapture.

A convenient approach for anesthetizing fish with MS222 involves employing a premixed concentrated solution. For example, in a dark plastic bottle, mix 5 g of powdered MS222 in 500 mL of water. The bottles we use are 500 mL-squeeze-and-pour dispenser bottles. These bottles incorporate a graduated cup on the cap to measure out the concentrated anesthesia. To mix anesthesia for field use, combine 25 mL solution from the bottle with 5 L of river water (50-mg/L MS222) in a dishpan (premarked at 5 L). We have found that this concentration does not fully anesthetize the juvenile fish but partially sedates them and enables rapidly mixing anesthetic to consistent concentrations.

Data Handling, Analysis, and Reporting

Trap efficiency testing

Procedures
Trap efficiency is measured by the rate that marked fish released above the trap are recaptured. A variety of techniques can be used to mark fish for trap efficiency testing. Probably the simplest approach is to anesthetize the fish and apply a partial fish clip (e.g., upper or lower caudal lobe, posterior/anterior anal lobe, various caudal punches). Other approaches include dyeing, freeze branding, panjet marking, and tagging. Fish should be fully recovered from the anesthetic and handling prior to release.

Mark groups can be composed of hatchery fish or fish that have been previously captured in the trap. Using hatchery fish complicates the study since one must assume their probability of capture is the same as for naturally reared fish. Groups of marked fish representing each targeted species and age-/size-class are released upstream of the trap over the period of their migration. The release point selected should be far enough upstream to provide for a similar distribution across the channel compared to unmarked fish (at least 2 pool/riffle sequences), but not so far upstream that predation on marked fish is substantial. Each group of marked fish should be released evenly across the river to avoid biasing their lateral distribution. To reduce predation subsequent to recapture, marked fish should be released during the time strata that they migrate.

In small streams, migration primarily occurs at night and smolt traps are typically checked in the morning. Instead of making additional trips back to the trap at night to release mark groups held during the day, some researchers have used a timer-activated, self-releasing live box to release the marked fish automatically at dusk (Miller et al. 2000). This device consists of three recovery chambers. Dark-colored 5-gal buckets work well for fish less than 120 mm. When larger steelhead and cutthroat trout migrants are being held for trap efficiency tests, fabricating larger recovery chambers from perforated plate is more appropriate. The recovery chambers are suspended between two small floating pontoons. A spring-wound timer is connected to a 12-V automobile door lock actuator. At the appropriate time, the timer energizes the door lock actuator, which pulls a pin to release the recovery chambers. The recovery chambers pivot on a pipe inserted through holes in their base, turn upside down, and release the fish. To avoid predation, fish are separated by size into the three recovery chambers, which are set to release the fish at different time intervals. In trapping small streams, marked fish are typically released at least 2 pool/riffle units, but no more than 300 m, above the trap.

Factors affecting trapping efficiency
Flow is the dominant factor affecting downstream migrant trapping operations in any system. It affects trapping efficiency and migration rates since high flows often stimulate fish to migrate; therefore, minimal trap efficiencies may occur at the same time that peak flow events are causing migration rates to increase.

Visibility, fish size, and noise are other factors that affect trapping efficiency. Larger downstream migrants, especially steelhead and cutthroat trout, may be able to avoid capture when the trap is visible by swimming around the trap or back out the mouth of the trap, especially when velocities are low. Some portion of ocean-type chinook and chum salmon may rear upstream for a short period of time and grow prior to migration; therefore, efficiency for a species may change over time. Behavior may also be important. Some species may primarily migrate down the thalweg of the channel, whereas a higher proportion of others may use the channel margins. Noise created by the trap causes an avoidance response. This is mitigated through proper site selection, as previously discussed.

Of course, human actions also affect trap efficiency. On larger streams and rivers, researchers may be forced to move the trap away from the thalweg during high-flow events to avoid debris entrainment and subsequent trap damage as well as for safety concerns. On small streams, temporary hydraulic modifications (e.g., screen panels) are often erected over the course of the season to direct flow to the trap. This is often necessary to keep a trap turning, and it obviously influences the trap efficiency.

These factors indicate that efficiency tests should, if possible, be conducted over the entire migration period, over a range of flows and turbidity levels, and for each species whose production is to be estimated. Human actions such as trap repositioning and installation of hydraulic modifications call for stratification of tra- efficiency tests. If possible, treatments should be done consistently each time to minimize the number of efficiency strata created.

Selection of calibration test fish

Fish marked and released for trap efficiency trials should be representative of the entire target population. Care should be taken to minimize bias relative to such factors as size and origin. For example, although hatchery fish used for calibration may be of the same species and age as their wild counterparts, they may be larger or behave differently and consequently may be captured at different rates than wild fish. Rates of in-stream predation and residualism are likely higher for hatchery fish. For these reasons, trap efficiency estimates resulting from release groups using hatchery fish may be biased low.

The importance of expanding trap counts by appropriate measures of trap efficiency is illustrated for a small Oregon coastal stream in tables 2 and 3. In the examples shown, inferences of abundance would be incorrect if trap efficiency estimates based on species and size-class characteristics had not been used to adjust the estimate of downstream migrants. These effects would be greatly magnified in larger streams and rivers where fewer than 10% of the downstream migrants are typically captured.

TABLE 2. — Comparison of unadjusted trap catches and adjusted estimates of total migrants for three species of salmonids in Tenmile Creek, Oregon, Spring 1993.

Species	Catch	Seasonal trap efficiency	Migration
Coho (age 1+)	2,429	48%	5,050
Steelhead (>120 mm)	1,298	17%	7,591
Chinook (fry)	242	63%	387

TABLE 3. — Comparison of unadjusted trap catches and adjusted estimates of total migrants for three size groups of juvenile steelhead in Tenmile Creek, Oregon, Spring 1993.

Length strata	Catch	Seasonal trap efficiency	Migration
60–89 mm	952	26%	3,719
90–119 mm	546	22%	2,516
120–200 mm	637	13%	4,977

Estimating total migration

Estimating migration for any period, whether a short time interval or an entire season, involves mark–recapture experimentation that requires a catch and an estimate of trap efficiency. A number of approaches are available to estimate population size by use of mark–recapture techniques. The simplest approach is a Petersen equation written as follows:

$$N_i = \frac{\hat{n}_i M_i}{m_i} = n_i \hat{e}_i^{-1} \qquad \text{(eq 1)}$$

where

$$\hat{e}_i = \frac{m_i}{M_i} \qquad \text{(eq 2)}$$

and where
$\hat{N}_i$ = Estimated number of downstream migrants during period i
M_i = Number of fish marked and released during period i

n_i = Number of fish captured during period *i*
m_i = Number of marked fish captured during period *i*
$\hat{e}_i$ = Estimated trap efficiency during period *i*

The six basic assumptions for the Petersen estimate are:
1) The population is closed;
2) All fish have an equal probability of capture in the first period;
3) Marking does not affect catchability;
4) All fish (marked and unmarked) have an equal probability of being caught in the second sample;
5) The fish do not lose their marks; and
6) All recovered marks are reported.

Seber (1982) discusses these assumptions in detail and provides tests for validating the assumptions. Results from these tests will help determine the best approach for data analysis. In many cases, the most appropriate approach will not become apparent until after all the fieldwork has been completed and the data examined. The biologist always needs to temper his/her decision on the approach with knowledge of the behavior of the targeted species. A plausible rationale should be developed to explain and support these decisions. Four general approaches are outlined in this section.

1. Stratified mark–recovery approaches
This approach estimates migration over a season by stratifying the mark and recovery data into a number of discreet time periods. Time strata can be a day, a week, or longer in some cases. Three conditions are described:
a. Two partial capture traps are used: an upstream trap for marking and a downstream trap for recovery of marked migrants;
b. One total capture (upstream) trap for marking and one partial capture (downstream) trap for recovery of marked migrants are used; and
c. A single partial capture trap is used where a portion of the catch is marked and released upstream for efficiency trials or marked hatchery fish are used for the efficiency trials.

1a. Use of two partial capture traps
This approach employs an upstream partial capture trap such as an inclined plane screen trap, screw trap, or fyke trap to capture and mark or tag downstream migrants and a downstream partial capture (inclined plane screen or screw) trap to recover the marked or tagged fish. Migration over the discreet period, $\hat{N}_i$, is estimated using the Petersen equation (equation 1). Chapman (1951) found this estimate to be biased and suggested the following modification:

$$\hat{N}_i = \frac{(M_i + 1)(n_i + 1)}{(m_i + 1)} - 1 \qquad \text{(eq 3)}$$

This estimate is exactly unbiased when $(M_i + n_i) \geq \hat{N}_i$, and approximately unbiased when $(M_i + n_i) < \hat{N}_i$. An unbiased estimate of the variance, , was developed by Seber (1970):

$$V(\hat{N}_i) = \frac{(M_i + 1)(n_i + 1)(M_i - m_i)(n_i - m_i)}{(m_i + 1)^2(m_i + 2)} \qquad \text{(eq 4)}$$

Regular discreet time periods (e.g., 1, 2, 3 ... 7 days) should be established a priori. The length of the time period is dictated by assumption 2 on page 254 (constant probability of capture in the first sample). Since trap efficiency changes with stream discharge and other factors, it becomes difficult to meet this assumption with longer time periods. Marks used in efficiency trials (e.g., partial fin clip) must be changed between time periods. Although most recaptures typically occur soon after release, some recaptures may occur along with those from subsequent mark groups. Because strata are nonoverlapping and independent, estimated total juvenile production, , is calculated by the sum of n migration period estimates as follows:

$$\hat{N} = \sum_{i=1}^{n} \hat{N}_i \quad \text{(eq 5)}$$

with associated variance

$$V(\hat{N}) = \sum_{i=1}^{n} V(\hat{N}_i) \quad \text{(eq 6)}$$

The 95% confidence interval (CI) is then estimated using

$$\hat{N} \pm 1.96 \sqrt{V(\hat{N})} \quad \text{(eq 7)}$$

1b. Use of a total capture trap and a partial capture trap

Assumption 2 (constant probability of capture in the first [upstream] trap), can be difficult to meet as streamflow conditions change using a partial capture trap; however, effect of streamflow on trap efficiency is negated if a total capture trap (e.g., smolt fence) is used as the upstream trap. In this case, one smolt fence (or more) is installed in tributaries upstream of the scoop or screw trap. Downstream migrants captured in the tributary traps are marked and released downstream, where a portion are recovered in the scoop or screw trap. The estimated migration and variance are calculated using equations 3 and 4, respectively. Using this approach, temporal stratification of the mark–recapture data is not required. A single time period (i.e., the entire out-migration period) can be used, since marking rates at the upstream trap integrate the effects of changing environmental conditions and migration timing.

This approach is most useful where the target species originates in streams where a total capture trap can be used (e.g., coho salmon). Technically, assumption 2 is still not met, since every fish will not have an equal chance of being marked (not all tributaries upstream of the partial capture trap will be trapped); however, if one assumes that the migration timing from the trapped tributaries is the same as for the untrapped tributaries, then a constant proportion of the total population is marked throughout the migration period and the desired outcome addressed by assumption 2 is achieved.

1c. Use of a single partial capture trap

In most cases, researchers will use a single trap and will conduct trap efficiency experiments by marking and transporting/releasing part of the catch or hatchery fish upstream of the trap. In this case, the marked fish captured in the trap must not be included in the population estimate since they were either counted as unmarked fish before being marked or were of hatchery origin and not part of the migration estimate. Carlson et al. (1998) advocate the following in lieu of equations 3 and 4:

$$\hat{U}_i = \frac{u_i(M_i + 1)}{m_i + 1} \quad \text{(eq 8)}$$

$$V(\hat{U}_i) = \frac{(M_i + 1)(u_i + m_i + 1)(M_i - m_i)u_i}{(m_i + 1)^2(m_i + 2)} \quad \text{(eq 9)}$$

where

U_i = Number of unmarked fish migrating during discreet period i
u_i = Number of unmarked fish captured during discreet period i

Total juvenile production $\hat{U}$ and associated variance $V(\hat{U}_i)$ are estimated by Equations 5 and 6, respectively, substituting $\hat{U}, \hat{U}_i, V(\hat{U}),$ and $V(\hat{U}_i)$, for $\hat{N}, \hat{N}_i, V(\hat{N})$, and $V(\hat{N}i)$ in the equations. The 95% CI is estimated by

$$\hat{U} \pm 1.96 \sqrt{V(\hat{U})} \quad \text{(eq 10)}$$

Alternatively, variance and confidence intervals can be estimated using bootstrap methodology (Thedinga et al. 1994).

Other considerations

Stratified mark–recapture approaches assume that each estimate of trap efficiency is an accurate measure of the proportion of downstream migrants caught in the trap. Since each test actually represents a single measure, it would be expected to include error. Assuming that error is normally distributed with zero mean, this approach argues for estimating discrete periods of short duration (e.g., 1 d) since the expected error over many samples should approach zero. Conversely, small sample sizes (mi) can greatly bias trap efficiency estimates, which argues for marking more fish, if available, or for strata of longer duration so that larger numbers of fish can be marked and recaptured. Researchers often balance these opposing elements by setting the duration of strata from 3 d to a week, depending on trap efficiency and the number of fish available for marking. A rule of thumb from mark–recapture studies is that at least five recaptures should occur for each stratum to minimize bias (Schwarz and Taylor 1998).

During some strata, fewer than five marked fish may be recovered. On small streams, this is most likely early and late in the migration period, when few fish are captured and marked; however, this can also happen during other times when trap efficiency is low. To avoid biasing the estimate, adjacent strata can be pooled to achieve at least five recaptures. Yet this approach should not be used when dissimilar recapture rates are likely to have occurred between the adjacent strata (e.g., dissimilar streamflows). If pooling is not appropriate, the researcher should consider using the estimated trap efficiency for the stratum and accepting the bias or dropping the efficiency test and using another approach for estimating efficiency for the stratum (e.g., mean of all tests, alternative stratification [see approach 3 on page 261]). Every effort should be made to avoid the situation of low sample size in the stratum.

2. Modeling trap efficiency

This approach estimates trap efficiency from an independent variable, typically streamflow. A series of trap efficiency tests are conducted over a range of flows and analyzed to determine if a significant relationship can be established (see Figure 18). When using regression analysis, it has been suggested that the observed F statistic should exceed the chosen test statistic by a factor of four or more for the relationship to be considered of value for predictive purposes (Draper and Smith 1998). Furthermore, a wide range of flows reflecting conditions across the entire season would maximize precision.

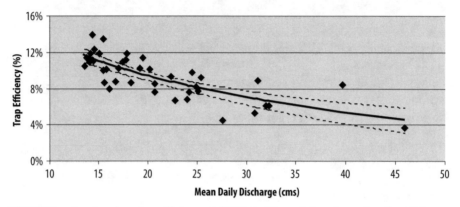

FIGURE 18. — Age-0 sockeye trap efficiency and 95% confidence intervals as a function of stream discharge. (Cedar River, Washington.)

Using this approach, migration on day i is calculated using equation 1, and its variance, $V(N_i)$, is estimated by;

$$V(\hat{N}_i) = V(\hat{e}_i)\left(\frac{\hat{n}_i}{\hat{e}_i^2}\right)^2 + \frac{Var(\hat{n}_i)}{\hat{e}_i^2} \qquad (eq\ 11)$$

If linear regression is used to estimate trap efficiency, the variance is estimated as follows:

$$V(\hat{e}_i) = MSE\left(1 + \frac{1}{n} + \frac{(X_i - \bar{X})^2}{\sum_{i=1}^{k}(X_i - \bar{X})^2}\right) \qquad (eq\ 12)$$

where:
$\hat{e}_i$ = The trap efficiency predicted on day i by the regression equation, $f(X_i)$
MSE = The mean square error of the regression
k = The number of trap efficiency tests used in the regression
X_i = The independent variable on day i

If catch n_i is not estimated, the second part of equation 11 reduces to zero and is not part of the calculation.

Estimating catch

Unlike the previous equations, equation 11 introduces variance in the unmarked catch estimate. Up to this point, we have treated catch as a count, but there are usually times during the season when the trap is not operated (e.g., debris stops the trap). Catch must be estimated during these periods—with the exception that approaches 1a, 1b, and 4 (see page 259–262) integrate unfished periods into the population estimate so that estimates of missed catch are not needed. Catch expansion may or may not be required using approach 1c, depending on whether the efficiency trials encompass nonfishing periods.

Generally catch can be estimated by interpolating catch rates from the previous and following fishing periods. Complications occur, however, when the unfished period extends through periods of rapidly changing catch rates (e.g., from night to day periods). When this occurs, the researcher may need to evaluate the magnitude of the catch rate change in the interpolation. Catch may also need to be extrapolated to account for migration before and/or after the period of trap operation. The variance of these catch estimates can have a substantial effect on the variance of the population estimate when estimated missed catches are high.

3. Stratifying trap efficiency

Like approach 2, this approach predicts trap efficiency using an independent variable or condition class. In this case, efficiencies are fairly constant over some range of the independent variable or condition class. As the independent variable passes some threshold or another condition class occurs, efficiencies change or "step" to a new level. For example, if the trap is placed in a U-shaped channel adjacent to a wide gravel bar, trap efficiencies may be at one level when flows are contained in the channel and at another when higher discharge causes a substantial portion of the flow to spread out across the gravel bar. Fish size may change over the trapping season, causing changes in trap efficiency by time strata. Turbidity levels may cause changes in efficiencies as well. In some locations, fish are better able to avoid traps during day fishing periods. In this case, efficiency data would be stratified by diel period (see Figure 19).

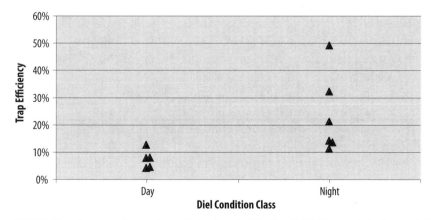

FIGURE 19.— Range and mean trap efficiencies stratified by diel fishing periods (Issaquah Creek, Washington)

Another cause for stratification is when human actions over the course of the season affect efficiency. Obvious examples include adding screen panels upstream of the trap to increase flow and direct more fish into the trap, and repositioning the trap, whether to increase or reduce catch rates or avoid damage during freshets.

Mean trap efficiency is calculated for each stratum, $\bar{e}_j$, and the means are tested for significant differences. The investigator may consider pooling strata where the means are not significantly different. Migration is estimated for discreet periods, when the independent variable is within a defined stratum, $\hat{N}_j$, by dividing the sum of the catch, n_j, or estimated catch, $\hat{n}_j$, by the mean trap efficiency for the stratum, under the assumption of homogeneous trap efficiencies. The variance of the estimate is estimated using the delta method (Goodman 1960) by

$$Var(\hat{N}_j) = \hat{N}_j^2 \left(\frac{Var(\bar{e}_j)}{\bar{e}_j^2} + \frac{Var(\hat{n}_j)}{\hat{n}_j^2} \right) \qquad (eq\ 13)$$

4. Back-calculating production

Using this approach, fish captured in the scoop or screw trap are marked or tagged and released downstream. Recapture occurs at another life stage, and a Petersen estimate of production is made. Typically, recapture occurs when the returning adults are sampled in a fishery or upon the spawning grounds. The term "back-calculating production" was coined as a result of the length of time between trapping and when the migration estimate was made (Seiler et al. 1994).

Production and variance are estimated using equations 3 and 4, respectively. The analysis is similar to approach 1b, since the data are not stratified. This approach is most useful where trap efficiency estimates are difficult to make. This method is more easily applied where nearly the entire cohort returns in a single year (e.g., coho). Age sampling would be required for this approach to work for species that return to spawn in multiple year-classes.

Other tools

This section focuses on the basic analysis of downstream migrant mark–recapture data for the estimation of smolt abundance using the Petersen equation and its derivations. Software is readily available on the Internet for analyzing stratified mark–recapture data. For example, Bjorkstedt (2005) describes the use of DARR 2.0, a software application that develops smolt abundance estimates using the Darroch (1961)-stratified Petersen estimator discussed in approach 1. This document and software are from the NOAA Fisheries Southwest Regional Fisheries Science Center–Santa Cruz Web site, <http://santacruz.nmfs.noaa.gov/publications/date_2000.php>.

Another application available online is the Stratified Population Analysis System (SPAS) described by Arnason et al. (1996). This program can analyze mark–recapture data using the Darroch moment estimate (Chapman and Junge 1956), a maximum likelihood method for the Darroch estimate (Plante 1990), the Schaefer estimate (Schaefer 1951; Warren and Dempson 1995), and the pooled Petersen estimate described by Seber (1982). It also includes a simulation capability that can be used for planning experiments and assessing the properties of the estimates. Arnason et al. (1996) and SPAS are available from the University of Manitoba Population Analysis Software Group Web site, <www.cs.umanitoba.ca/~popan>.

Additional information on the analysis of smolt trapping data is provided in Schwarz and Dempson (1994), MacDonald and Smith (1980), Mäntyniemi and Romankkaniemi (2002), Carlson et al. (1998), and Thedinga et al. (1994).

Personnel Requirements and Training

A successful trapping operation requires a team of professional and technical staff that is dedicated to the success of the project. This statement is easily set aside since, after all, "Aren't we all dedicated to the project's success?" Nevertheless, when put to the test, many projects have failed to develop precise estimates of smolt production as a result of poor decision making or lack of tenacity during critical periods in the trapping season. Spring storm events increase river discharge and debris loads, making trap operation more difficult, dangerous, and time consuming. Like river discharge, catch rates can also steeply increase and decrease over a relatively short period of time. Trap operation during these periods often requires working extremely long and physically taxing periods.

The number of personnel required to operate a trap over a smolt outmigration period depends on the objectives of the study and the size of the river. On small systems where only an index of production (e.g., catch-per-unit-effort) is desired, a field crew of one may be all that is necessary. When operating a trap continuously or counting large numbers of fish, a larger crew is required. Three people make up a good-sized crew during the peak of the migration, when the workload is high. Each person can be put on an alternating 6-d-on/3-d-off schedule so that there are always two people available to work the gear. Later in the season when fewer fish migrate and flows subside, the crew can be reduced to two.

Crew

The crew is responsible for day-to-day trap operations, deciding when and how often to check the trap and process the catch, record data, and evaluate/maintain the gear. The crew also conducts trap-efficiency tests to measure the proportion of downstream migrants captured.

Project Leader

The project leader, typically a fish biologist, supervises the field crew. The project leader oversees the project, schedules the crew, and maintains communication with crew members over the trapping season. The project leader may also help with the fieldwork as needed.

Protocols

The protocols discussed in this section, although not exhaustive, should provide researchers with a basic understanding of how production is assessed using floating inclined plane screen traps or rotary screw traps; however, no matter how complete the protocols are, much of the knowledge for successfully operating this gear can only be gained through experience. Inclined plane screen traps and screw traps are operated by nearly all agencies that manage anadromous salmonid populations. Project leaders are well advised to receive mentoring from experienced investigators or gain experience from successful programs before starting projects.

Acknowledgments

Reviews and comments from D. Rawding, K. Ryding, P. Wagner, M. Sparkman, F. McCormick, and B. Allen greatly improved this manuscript. K. Kenaston, Oregon Department of Fish and Wildlife, provided information on the origin of the rotary screw trap. Special thanks go to M. Ackley, T. Miller, S. Neuhauser, and P. Topping, who provided the photographs used in this chapter.

Literature Cited

Ames, J., and D. E. Phinney. 1977. Puget Sound summer-fall chinook methodology; escapement goals, run size forecasts, and in-season run size updates. Washington Department of Fisheries Technical Report 29, Olympia.

Arnason, A. N., C. W. Kirby, C. J. Schwarz, and J. R. Irvine. 1996. Computer analysis of data from stratified mark-recovery experiments for estimation of salmon escapements and other populations. Canadian Technical Report Fisheries and Aquatic Sciences 2106.

Baranski, C. 1989. Coho smolt production in ten Puget Sound streams. State of Washington Department of Fisheries, Technical Report 99, Olympia.

Beidler, W. M., and T. E. Nickelson. 1980. An evaluation of the Oregon Department of Fish and Wildlife standard spawning fish survey system for coho salmon. Oregon Department of Fish and Wildlife Information Report Series 80-9, Portland.

Bjorkstedt, E. 2005. DARR 2.0: updated software for estimating abundance from stratified mark–recapture data. NOAA-TM-NMFS-SWFSC 68. 21pp. Available: http://santacruz.nmfs.noaa.gov/files/pubs/00439.pdf (August 2005).

Brown, T. G., and G. F. Hartman. 1988. Contribution of seasonally flooded lands and minor tributaries to the production of coho salmon in Carnation Creek, British Columbia. Transactions of the American Fisheries Society 117:546–551.

Carlson, S. R., L. Coggins, and C. O. Swanton. 1998. A simple stratified design for mark–recapture of salmon smolt abundance. Alaska Fisheries Research Bulletin 5(2): 88–102. Available: www.adfg.state.ak.us/pubs/afrb/vol5_n2/carlv5n2.pdf (August 2005).

Chapman, D. G. 1951. Some properties of the hypergeometric distribution with applications to zoological sample censuses. University of California Publications in Statistics 1:131–160.

Chapman, D. G., and C. O. Junge. 1956. The estimation of the size of a stratified population. The Annals of Mathematical Statistics 27:375–389.

Darroch, J. N. 1961. The two-sampled capture–recapture census when tagging and sampling are stratified. Biometrika 48:241–260.

Draper, N. R., and H. Smith. 1998. Applied regression analysis, 3rd edition. John Wiley & Sons, Inc., New York.

DuBois, R. B., J. E. Miller, and S. D. Plaster. 1991. An inclined-screen smolt trap with adjustable screen for highly variable flows. North American Journal of Fisheries Management 11:155–159.

Goodman, L. A. 1960. On the exact variance of products. Journal of the American Statistical Association 55:708–713.

Hartman, G. F., B. C. Andersen, and J. C. Scrivener. 1982. Seaward movement of coho salmon (*Oncorhynchus kisutch*) fry in Carnation Creek, an unstable coastal stream in British Columbia. Canadian Journal of Fisheries and Aquatic Sciences 39:588–597.

Hilborn, R., B. G. Bue, and S. Sharr. 1999. Estimating spawning escapements from periodic counts: a comparison of methods. Canadian Journal of Fisheries and Aquatic Sciences 56:888–896.

IMW SOC (Intensively Monitored Watersheds Scientific Oversight Committee). 2004. Evaluating watershed response to land management and restoration actions: intensively monitored watersheds (IMW) progress report. Salmon Recovery Funding Board. Available: www.iac.wa.gov/Documents/SRFB/Monitoring/IMW_progress_rpt.pdf (August 2005).

Johnson, S. L., J. D. Rodgers, M. F. Solazzi, and T. E. Nickelson. 2005. Effects of an increase in large wood on abundance and survival of juvenile salmonids (Oncorhynchus spp.) in an Oregon coastal stream. Canadian Journal of Fisheries and Aquatic Sciences 62:412–424.

Kennen, J. G., S. J. Wisniewski, N. H. Ringler, and H. M. Hawkins. 1994. Application and modification of an auger trap to quantify emigrating fishes in Lake Ontario tributaries. North American Journal of Fisheries Management 14:828–836.

Letcher, B. H., G. Gries, and F. Juanes. 2002. Survival of stream-dwelling Atlantic salmon: effects of life history variation, season, and age. Transactions of the American Fisheries Society 131:838–854.

MacDonald, P .D. M., and H. D. Smith. 1980. Mark–recapture estimation of salmon smolt runs. Biometrics 36:401–417.

Mäntyniemi, S., and A. Romakkaniemi. 2002. Bayesian mark–recapture estimation with an application to a salmonid smolt population. Canadian Journal of Fisheries and Aquatic Sciences 59:1748–1758.

McLemore, C. E., F. H. Everest, W. R. Humphreys, and M. S. Solazzi. 1989. A floating trap for sampling downstream migrant fishes. United States Forest Service, Pacific Northwest Research Station, Research note PNW-RN-490, Corvallis, Oregon.

Miller, B. A., J. D. Rodgers, and M. S. Solazzi. 2000. An automated device to release marked juvenile fish for measuring trap efficiency. North American Journal of Fisheries Management 20:284–287.

Mobrand, L. E., J. A. Lichatowich, L. C. Lestelle, and T. S. Vogel. 1997. An approach to describing ecosystems performance "through the eyes of salmon." Canadian Journal of Fisheries and Aquatic Sciences 54:2964–2973.

Moussalli, E., and R. Hilborn. 1986. Optimal stock size and harvest rate in multistage life history models. Canadian Journal of Fisheries and Aquatic Sciences 43:135–141.

Nickelson, T. E. 1998. ODFW coastal salmonid population and habitat monitoring program. Oregon Department of Fish and Wildlife, Salem.

Olsson, I. C., L. A. Greenberg, and A. G. Eklov. 2001. Effect of an artificial pond on migrating brown trout smolts. North American Journal of Fisheries Management 21:498–506.

Orciari, R. D., G. H. Leonard, D. J. Mysling, and E. C. Schluntz. 1994. Survival, growth, and smolt production of Atlantic salmon stocked as fry in a southern New England stream. North American Journal of Fisheries Management 14:588–606.

Plante, N. 1990. Estimation de la taille d'une population animale à l'aide d'un modèle de capture–recapture avec stratification. M.Sc. thesis. Université Laval, Quebec.

Roper, B., and D. L. Scarnecchia. 1996. A comparison of trap efficiencies for wild and hatchery age-0 chinook salmon. North American Journal of Fisheries Management 16:214–217.

Schaefer, M. B. 1951. Estimation of the size of animal populations by marking experiments. U.S. Fish and Wildlife Service Fisheries Bulletin 69:191–203.

Schoeneman, D. E., R. T. Pressey, and C. O. Junge, Jr. 1961. Mortalities of downstream migrant salmon at McNary Dam. Transactions of the American Fisheries Society 90:58–72.

Schwarz, C. J., and B. D. Dempson. 1994. Mark–recapture estimation of a salmon smolt population. Biometrics 50:98–108.

Schwarz, C. J., and C. G. Taylor. 1998. Use of the stratified-Petersen estimator in fisheries management: estimating the number of pink salmon *(Oncorhynchus gorbuscha)* spawners in the Fraser River. Canadian Journal of Fisheries and Aquatic Sciences 55:281–296.

Seber, G. A. F. 1970. The effects of tag response on tag-recapture estimates. Biometrika 26:12–22.

Seber, G. A. F. 1982. The estimatioin of animal abundance and related parameters, 2nd edition. Charles Griffin and Company, London.

Seiler, D., S. Neuhauser, and M. Ackley. 1981. Upstream/downstream salmonid trapping project, 1977–1980, progress report #144. State of Washington Department of Fisheries, Olympia.

Seiler, D., P. Hanratty, S. Neuhauser, P. Topping, M. Ackley, and L. E. Kishimoto. 1994. Annual performance report October 1992 – September 1993: wild salmon production and survival evaluation. Washington Department of Fish and Wildlife, Olympia.

Seiler, D., S. Neuhauser, and L. Kishimoto. 2003. 2002 Skagit River wild 0+ chinook production evaluation annual report. Washington Department of Fish and Wildlife, FPA 03-11, Olympia.

Solazzi, M. F., T. E. Nickelson, S. L. Johnson, and J. D. Rodgers. 2000. Effects of increasing winter habitat on abundance of salmonids in two coastal Oregon streams. Canadian Journal of Fisheries and Aquatic Sciences 57:906–914.

Thedinga J. F., M. L. Murphy, S. W. Johnson, J. M. Lorenz, and K. V. Koski. 1994. Determination of salmonid smolt yield with rotary-screw traps in the Situk River, Alaska, to predict effects of glacial flooding. North American Journal of Fisheries Management 14:837–851.

Todd. G. L. 1994. A lightweight, inclined-plane trap for sampling salmon smolts in rivers. Alaska Fishery Research Bulletin 1(2):168–175.

Tsumura, K., and J. M. B. Hume. 1986. Two variations of a salmonid smolt trap for small rivers. North American Journal of Fisheries Management 6:272–276.

Wagner, H. H., R. L. Wallace, and H. J. Campbell. 1963. The seaward migration and return of hatchery-reared steelhead trout, *Salmo gairdneri* Richardson, in the Alsea River, Oregon. Transactions of the American Fisheries Society 92:202–210.

Warren, W. G., and J. B. Dempson. 1995. Does temporal stratification improve the accuracy of mark–recapture estimates of smolt production: a case study based on the Conne River, Newfoundland. North American Journal of Fisheries Management 15:126–136.

Beach Seining

Peter K. J. Hahn, Richard E. Bailey, and Annalissa Ritchie

Background and Objectives

Background

Seining is a fishing technique traditionally done in areas with large schools or groups of fish. The earliest form of seining was dragnetting (also called beach seining). There is evidence of seine nets used in artisanal fisheries several thousands of years ago and on every continent (von Brandt 1984), including North America, where native peoples used them to catch salmon in the Columbia River (Craig and Hacker 1940) (see Appendix B) and elsewhere. Nets ranged in size from very small, single-person "stick seines" to seines in New Zealand that measured more than 1,600 m long and employed hundreds of people to retrieve.

The typical modern seine net has weights on the bottom (lead line) and buoys on the top (float or cork line) to keep the net vertical when pulled through the water to entrap fish. Some seine nets are designed to sink or to float, but most remain in constant contact with both the bottom and the surface and thus are best suited for shallow waters. A beach seine is often set from shore to encircle a school of fish and is then closed off to trap them against the shore. One variation is to set a seine net parallel to and some distance from shore and then pull it to the beach. Another variation is to encircle fish some distance from shore but still in shallow water and pull the net onto boats. This latter method evolved into the purse seine, which has rings along the lead line through which a rope is pulled to "purse" or tighten the bottom of the net together before the net is gathered to the side of a boat; purse seines, however, are not limited to shallow waters for their effectiveness. Between a beach and purse seine is the lampara net, which is fished at the surface in deep water. It has a lead line much shorter than the float line, which shapes the seine much like a dust pan and prevents epipelagic fish from diving and escaping (von Brandt 1984; Hayes et al 1996). Some seines are even fished through holes cut in ice-covered lakes to capture semitorpid aggregations of fish.

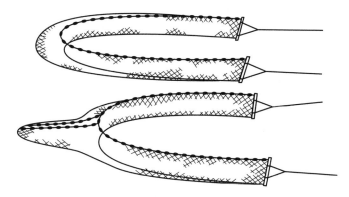

FIGURE 1.—Beach seine diagrams (FAO technical paper). The lower net has a "bag" in the "bunt" (middle) section in order to hold more fish and to prevent fish from escaping. Bunt mesh size is usually smaller than for "wing" sections. Typically "tow lines" are attached to each end to allow pulling the net in from a distance.

In the early years of the nonnative commercial salmon fisheries, beginning before 1880 in the Columbia River (Craig and Hacker 1940), beach seines were set from large rowboats and the nets would be hauled in by hand or with horses. The fishermen rowed their boats to the fishing grounds or hitched a tow from a steam-powered cannery tender. Gradually, the fisheries became more mechanized, utilizing power boats and motor winches. By 1908, some seine nets were nearly 800 m long (see Appendix B) and highly effective. In 1917, purse seining and again in 1934–1935, beach seining, trap nets, and fish wheels were outlawed on the Columbia River.

Today, beach seining is generally not permitted for commercial purposes in any North American rivers, except in some areas of the far north; however, research seines can be employed in wadable and nonwadable systems across a variety of habitat types to capture both juvenile and adult salmonids. In these habitats, seines can be deployed by wading or from drift boats or powerboats. Hayes et al. (1996) described generic applications of seining for fish capture. Specific seine applications include capturing fish to estimate total abundance (usually by mark–recapture studies), estimate relative abundance over time and space, describe fish population diversity and distribution, capture broodstock, monitor effectiveness of habitat alterations, and mark fish and collect biosamples (Dawley et al. 1981; Farwell et al. 1998; Brandes and McLain 2001; Rawding and Hillson 2002; Fryer 2003; Hahn et al. 2003; Kagley et al. 2005).

Rationale

Beach and pole seining is an efficient method to capture salmonids and some nonsalmonid fishes in a wide variety of habitats (see Appendix B), including rivers, estuarine, and nearshore lake, reservoir, and marine habitats (Pierce et al. 1990). It is most effective when used in relatively shallow water with few obstructions, where fish are in high concentrations, and for species that are less likely to outswim the net; however, in some circumstances seining can capture highly mobile species such as adult salmon. Seining permits the sampling of relatively large areas in short periods of time as well as the capture and release of fish without significant stress or harm, as long as the bunt of the seine is kept in water and the fish are not too crowded (or fish are quickly moved to a holding container). Cost of gear, boats, and personnel range from relatively inexpensive to modest, depending on the scope and frequency of sampling. Purse and lampara net seining can also be effectively conducted in deeper water. These two techniques often require larger boats and nets (and thus greater financial investment) and are mentioned only briefly in this chapter.

Seining is a useful technique for objectives such as collecting fish for biological samples, sampling fish diversity within a given habitat (low-precision requirements), and estimating relative abundance (with modest precision) or population abundance with high accuracy and precision (via mark–recapture). Seining is frequently used for capturing small juvenile salmonids, where a measure of relative abundance or catch-per-unit-effort (CPUE, as fish/set, fish/area, or fish/volume sampled) is needed. By using standardized nets and deployment methods, scientists have attempted to characterize abundance over time and space, either within or across years. Other capture methods, such as midwater trawls, can contribute results with similar units (Brandes and McLain 2001). Beach seines allow the selective capture and subsequent release of a wide range of salmonid fish sizes. This characteristic makes beach seining a useful capture method for many mark–recapture based salmonid assessments, in which marking more fish allows for greater precision of the population estimate.

Objectives

Selecting seines as a method of fish capture should depend on requirements for data and specific study objectives. It should not depend merely on the ease of deployment or historical efforts, or because of limited exposure or training in the various gears available (Rozas and Minello 1997). Specific objectives will also determine the size and species of fish to be targeted for collection and what habitats will be sampled (see Appendix A). These elements (purpose for collecting fish, target fish size, and habitat conditions where sampling is proposed) drive the selection of gear type and then seine type, length, depth, mesh size, and the method of setting and retrieving the seine.

We categorized objectives for seining into six types or purposes: (1) relative abundance estimation, (2) absolute abundance estimation via indirect measures, (3) relative survival estimation, (4) biological sampling, (5) estimating species diversity or presence, and (6) absolute abundance estimation via direct measures. There may be some studies in which two or more of these are applicable, but the most important objective should determine the capture method(s) of choice, which may include methods other than seining.

1. Relative abundance estimation

To obtain estimates of relative abundance, seining can be conducted with prescribed methods and nets such that the area sampled is nearly constant or where the area (and depth) swept can be measured and estimated. Results are often commonly reported as CPUE: fish/set, fish/area, or fish/volume. Results are compared across sites within a defined region (e.g., one estuary complex, shorelines within marine bays, straits, sounds) by time of year or across years or by time of day and night (Miller et al. 1977; Dawley et al. 1981; Nelson et al. 2004; Kagley et al. 2005; Nobriga et al. 2005). Seine data may be combined with CPUE results from other gear such as midwater trawls or tow nets (Brandes and McLain 2001). Annual results may be related to parent spawner population size or environmental variants. Estimates of residence or migration time can also be made using marked fish releases (Duffy et al. 2005).

Typical target fish in North America are age-0 chinook salmon *Oncorhynchus tshawytscha*, chum salmon *O. keta*, and pink salmon *O. gorbuscha* for beach seines and yearling chinook, coho, sockeye, and steelhead for purse seines. These are abundant migratory species that pass through specific habitats at specific times. There are also situations where eggs, larvae, and fry can be caught by fine-mesh purse seines (Bagenal and Nellan 1980). The capture of nontarget species is often not a goal for studies, and therefore, variation in capture efficiency or selectivity by species may not be an important issue. Often the same suite of sites is repeatedly sampled over time so that variation in efficiency caused by different habitats is not an issue again unless habitat is changing over time.

If the goal is to sample a variety of habitats to see which are utilized by the target species or the full suite of species, then seining efficiency is an issue to consider (Parsley et al. 1989; Rozas and Minello 1997). Techniques may have to be developed to estimate efficiency, and other capture techniques should be considered as alternates or complements.

2. Absolute abundance estimation via indirect measures

The primary method for estimating population size indirectly is the mark–recapture estimate. Seining becomes a tool to capture fish for marking and release or for recapture (Farwell et al. 1998, 2006; Hahn et al. 2004; Rawding and Hillson 2003). Estimating CPUE or the true proportion of the population sampled is not relevant. Maximizing catches (to reduce cost) and representatively sampling the population (low bias) is the goal. Generally, multiple capture techniques are desirable to overcome biases that may be inherent in a single method. Caution is warranted when the same technique is used for both initial capture and recapture (potential gear bias), particularly if the elapsed time is short between events (due to learned net avoidance by the target fish). Seines can be highly effective for capturing a wide range of sizes of adult salmon and can also be used to capture juvenile salmon, particularly in riverine situations. Species that are migratory and abundant and have constrained migration timing are well suited to capture by seines. Adult chinook, pink, chum, coho, and sockeye salmon are well suited, summer steelhead sometimes are, and other species may have populations that are vulnerable to seines. Salmon smolts might be captured and marked via smolt trap, and then seines may be used in the lower river to sample for recaptures.

3. Relative survival estimation

When two or more groups of marked fish of the same species are released in known numbers, later sampling can allow estimates of the relative survival for each group. This could be considered a subset of relative abundance estimation, but estimation of CPUE is not needed. Estimating relative survival is especially important for migratory salmonids that travel by various routes through or around dams (see Dawley et al. 1981 for the Columbia River, where both beach and purse seines were used). It can also be used to test experimental versus control hatchery rearing or time-of-release groups, or survival by various size groups (as long the seine has the same or known capture efficiency for all size groups). Seining could also be used in lakes and ponds or small streams. Comparing survival across species should be done cautiously because capture selectivity and efficiency may differ substantially by species.

4. Biological sampling

Some studies merely require that individual fish be captured (e.g., gut contents, tissue samples for analyzing DNA, electrophoresis, or contaminants) or that individuals of various sizes be caught (e.g., for scale or otolith sampling for length-by-age analysis). For these studies, seines can be a very effective means of collecting fish specimens. Seines may or may not function well to describe abundance across a range of fish sizes if capture efficiency varies greatly by size. In the Green River in Washington, mature chinook salmon 25–115 cm fork length seemed to be caught equally well, based on underwater observation of beach seines in action (P. Hahn, Washington Department of Fish and Wildlife, personal communication).

5. Estimating species diversity or presence

When estimating species diversity or documenting presence, several capture methods may be used, including seines (Klemm et al. 1993; Meador et al. 1993; British Columbia Ministry of Environment, Lands and Parks 1997; Lazorchak et al.

1998; Moulton II et al. 2002). Diversity estimates that incorporate abundance may require efficiency (calibration) tests because seines are known to have variable and species-specific efficiencies and are especially poor for epibenthic species in complex habitats. Other applications include documenting if and when certain habitats are used by certain species (e.g., electrochemical impedance spectroscopy studies, habitat restoration monitoring). Diurnal-nocturnal and tide stage effects must be considered.

6. Absolute abundance estimation

Studies that attempt to directly measure the absolute abundance per unit area or volume are difficult to realize with the use of seines alone (Rozas and Minello 1997). Varying habitats (e.g., substrate, vegetation, clarity, currents) can allow fish to escape capture by seines, and each species and sometimes even each size group may have different catchability (Bayley and Herendeen 2000). Tests may be needed in each habitat to measure the selectivity and efficiency for each type of gear. Underwater observation may be needed to document the behavior of the gear and the fish while the seine is being deployed and retrieved.

Effectiveness

Some objectives (e.g., biological sampling of individual fish) can be met very well by using a single haul with a seine. Seining in multiple sites over time can be an excellent way to capture fish for marking and release or to recapturing fish for survival studies. Repetitive seining over time with standardized nets and standardized deployment in relatively similar habitat can be an effective way to quantify the relative abundance of certain species over time and space, especially for small juvenile migrating salmon. Species richness (diversity index), species rank, and the size distribution within species can sometimes be estimated using a single seine haul (Allen et al. 1992, cited in Hayes et al. 1996). Knowing when this is true is problematic because each species and size often has its own selectivity or capture efficiency. These factors can vary greatly if nonsalmonid species are included, if habitat varies substantially from site to site, or if some habitats cannot be sampled. In Allen et al. (1992), the six dominant taxa had capture efficiencies (CE) that ranged from 7% to 91%, and the highest average CE was 52% (this was for a tidal pool isolated by low tide and block nets). Rare taxa were not well represented by a single haul; these were better assessed using multiple seine hauls and/or multiple gear types. Estimating absolute abundance or biomass by direct measure is not easily accomplished by the use of seines alone, especially by single seine hauls, without calibration studies (see pages 279–280). Thus, effectiveness of beach seining depends on the species sampled, the population of interest, the habitat that is seined, and the overall goal of the sampling effort. Refer to Factors that affect capture efficiency and selectivity on pages 274–279.

Sampling Design

General site selection

Seining may be carried out in a variety of habitat types, depending on the population and life stage of the targeted species (see Appendix A). Sites with firm sloping beaches are favorable but not required. Adult salmonids are seined from pool (holding water) and nearshore lake habitats adjacent to and on the migratory routes to spawning grounds. Juvenile salmon can be seined from streams, estuaries, or nearshore lake and reservoir areas. Sites with irregular bottom topography, significant accumulations of debris or larger rocks, or dense stands of aquatic vegetation may not be suitable for seining due to net snagging or lifting and reduced fish retention. Current velocity and depth influence site selection and choice of net design. For a general guide to seining methods that can be used in varying habitats, see Appendix A.

Gear and method selection

Seines vary greatly in size depending on the target species, water depth, habitat, currents, and purpose (see Appendix B). For estuary and intertidal habitats, smaller nets can be used for sampling the shallow, intertidal shoreline areas (less than 1.2 m deep) with relatively homogenous water depth, velocity, vegetation, and substrate. Larger nets can be used for the intertidal-subtidal fringe with depths ranging from 1.8 to 4.6 m or deeper. Faster current generally requires a larger mesh size, at least in net wings, to reduce drag while deeper waters require wider nets to reach from the surface to the substrate. Often the wings are tapered to reduce overall net mass and because the ends of the seine are usually in shallower water than the midsections (e.g., see figures 2 and 3). Larger mesh in the wings than in the bunt and bag reduces drag in the initial stages of retrieval. For wadable waters and juvenile salmonids, small net beach or pole seine methods are appropriate (see Appendix B). Nets ranged from 10 to 24 m in length with 3.2–6.4-mm knotless-mesh nylon netting. For juvenile sampling in nonwadable waters, larger nets are used (37–95 m long, with mesh 3.2–9.5 mm). Typical net constructions for capture of adult salmon in nonwadable rivers vary between 45 and 70 m in length (historically, nets up to 777 m long were used for commercial harvest) and 5 m and 9 m in depth with lead lines varying between 0.5 and 1.75 kg/m (Appendix B). Floats are typically installed at 30–50-cm intervals on the cork lines. Stretched mesh sizes vary, but 5–6.3-cm (2–2.5-in) mesh is commonly employed. Twines used in mesh construction range in gauge from 48 to 96, with 96 being heavy and 48 being light. Twines are sometimes tarred to increase durability.

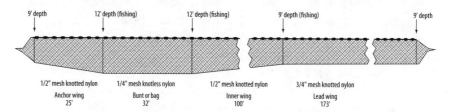

FIGURE 2. — **Example of beach seine dimensions with unequal tapered wings (from Sims and Johnsen 1974) used in the Columbia River estuary, Washington. (Illustration: Andrew Fuller, from Sims and Johnsen.)**

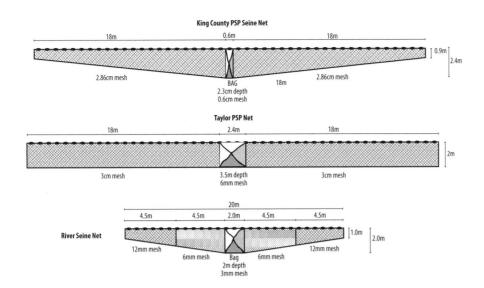

FIGURE 3.—Example of three beach seines used in Puget Sound, Washington. (Illustration: Andrew Fuller from Nelson et al. 2004, King County Department of Natural Resources.)

When to sample and sampling frequency

For purposes of mark–recapture estimates, seining for adult salmon should occur throughout the return run, such that returning fish are captured in a manner proportional to their abundance within the river (Farwell et al. 1998; Hahn et al. 2002; Rawding and Hillson 2002). For species such as sockeye, this may involve deployment of seining gear at passage sites downstream from spawning areas. For other adult salmon species, this may involve fishing at a variety of pool and glide locations throughout the spawning range (R. E. Bailey, Canadian Department of Fisheries and Oceans, unpublished data). Lazorchak et al. (1998) provide sampling schedules for use of beach seines in habitat assessments associated with juvenile salmonids as well as nonsalmonid species. Beach seining for juvenile salmonids is often conducted during the period of smolt out-migration in rivers and estuaries. Peaks in smolt migration vary by location and should be investigated as part of the planning process for a full-scale beach seining sampling effort.

To determine assemblage diversity or provide fish for biosampling, one or many sets may be employed, either on one day or throughout a longer period. To establish an abundance estimate, a single set per site may be adequate if the seine has been calibrated for efficiency. Alternatively, an area can be blocked off and seined until no more fish are caught, or a mark–recapture approach can be used for resident population estimate of small fish. A minimum of three sets per site is recommended (SSC 2003) if characterization of each site is an objective. If only the larger area needs to be characterized, single samples may be scattered randomly (but perhaps stratified into habitat types). If trend data is the objective, then nonrandomly selected permanent sites may be adequate (Brandes and McLain 2001). For capturing adult salmon for mark–recapture estimates, seining once per day or every other day at several sites may provide an adequate sample size. Some sites may allow two or more samples per day if migration is active. In small, clear rivers, like the Green River in Washington (Hahn et al. 2002), much of the chinook salmon migration may occur at night, so multiple sets per site per day are not fruitful. For coho salmon in the same river, however, active migration was noted during daylight (Hahn, personal communication).

Bias, selectivity, and efficiency

Factors that affect capture efficiency and selectivity

Gear selectivity is a quantification of the varying probability of capture for different sizes and/or species of fish (Backiel 1980). Capture efficiency (CE, sometimes called catch efficiency or catchability) is similar to selectivity but can be defined as the percentage removal or rate of exploitation for different sizes and/or species of fish in the area fished by the seine. Both terms may be understood for (a) the entire population in the body of water being sampled, (b) the subpopulation in the habitat area being sampled, or (c) that subpopulation of fish within the swept area of the seine. For this protocol, we follow Bayely and Herendeen (2000) and define capture efficiency as the product of encirclement efficiency when laying the net and retention efficiency while hauling in the net. Rozas and Minello (1997) suggest adding recovery efficiency as another component (see equation 1), defined as those fish that were retained within the net when pursed and were observed and counted; however, the individual efficiencies are difficult to estimate separately and no author has quantified all three. In some habitats (i.e., those without much vegetation) and for larger fish such as salmonids juveniles or adults, recovery efficiency is essentially 100%. CE for a beach seine is often in the range of 20–80%.

$$CE = \text{capture efficiency} = (\text{encirclement E})(\text{retention E})(\text{recovery E}) \quad (\text{eq 1})$$

We will define selectivity more broadly to acknowledge that the area that can be seined may contain only part of the total population of interest and species and fish of varying sizes are probably not randomly distributed over large areas. As an example, Duffy et al. (2005) sampled beaches in northern Puget Sound, Washington within single 24-h periods and found that size of chinook, coho, and chum salmon juveniles were significantly and often substantially smaller in daylight samples compared to crepuscular (dawn/dusk) or nocturnal samples. For southern Puget Sound beaches, differences were mostly insignificant and small. In the Columbia River estuary, purse seines consistently caught larger juvenile salmonids than did beach seines fished nearby (Johnsen and Sims 1973; Sims and Johnsen 1974; Dawley et al. 1986).

To help understand what affects selectivity and capture efficiency—and therefore, effectiveness of seining—we considered the following categories: habitat, water, fish, time of year and day, net, and method. Considering these factors in your study design before you sample and after you begin sampling (by underwater observation and calibration studies) will help you to decide when seining may be a good technique to meet your objectives; it will also help improve the analysis of data collected. We relied on our own experience, that of our peers, and the following authors: Allen et al. (1992), Backiel and Welcomme (1980), Bayley and Herendeen (2000), Dewey et al. (1989), Lyons (1986), Parsley et al (1989), Pierce et al. (1990), Rozas and Minello (1997), and others.

Habitat

- Substrate (e.g., roughness, softness)
- Vegetation (e.g., submergent, emergent, compressibility)
- Wood (e.g., trees, brush, and parts thereof)—underwater and above water and at shoreline

- Manmade objects (e.g., pilings, docks, junk, riprap)
- Small debris
- Seine rolling

Any object that snags a seine or causes it to lift off the bottom can allow fish to escape. Uniform, small substrates form a better seal with the lead line of the net than do larger cobbled or uneven bottoms. Aquatic vegetation weakens the seal and provides hiding places for small fish underneath or within clumps. Additional leads or adding a heavy chain (Penczak and O'Hara 1983) can reduce fish loss under the seine. In addition to substrate unevenness, snagging on logs, rocks, and other debris will slow down seining and decrease seining efficiency by allowing fish to escape prior to (delayed) net closure or underneath the net. Some soft substrates such as sand and silt will allow the lead line to sink and act like a dredge, which requires a reduction in weights. This dredging effect can also happen on gravel and pebble substrates and can fill the bag with unwieldy weight. One solution is to start the bag of the seine several inches above the lead line (J. Fryer, Columbia River Inter-Tribal Fish Commission, personal communication). Algae, leaves, small wood, and other debris also can clog seines, slowing retrieval and possibly lifting the lead line. Any inanimate matter in the bag or bunt can result in fish injury or death. Seine rolling (Pierce et al. 1990) may occur in dense vegetation where up to one-third or more of the width of a seine may roll upon itself. This net behavior increases the probability of fish escape. Additional tow lines clipped onto the lead line at intervals and pulled after initial setting is complete may reduce rolling. More lead on the lead line, by adding trailer sticks (Threinen 1956), may also help. For efficiency calibration studies, consider hand-pulling all the vegetation inside the deployed seine before retrieving if overall habitat impact is small.

Modifying habitat in seining sites with devices that lift the seine over obstructions (see Site testing and modification, page 283) can make seining possible, but it may result in temporary opportunities for fish to escape. It can allow seining when quite large obstructions, such as remnant vertical pilings up to 0.6 m tall, are present. It works best during the early phase of seine retrieval (before fish are concentrated) for large target fish (Hahn, personal communication) and when direct abundance estimates are not needed. Underwater observation should be used to confirm intended seine and fish action, and tending the net with snorkelers may be needed to keep fish from escaping.

Water

- Temperature
- Clarity (turbidity)
- Depth
- Currents
- Wave action
- Tide stage
- Ice depth

Within the preferred or tolerated range for each species, warmer water enhances fish swimming speed and thus increases the likelihood of escape. Compared to turbid water, clear water allows fish to see an oncoming net, boat, or tow line and initiate escape behavior earlier. Noise, vibrations, and changes in water pressure are induced by moving seines and perceived by fish. Some of these factors can drive fish ahead of the net and into the area it encircles, but it may also give them time to escape. Vibrating the tow lines may keep fish from darting out of the path of the seine. Water current either enhances or hinders the speed of seine retrieval. Currents and wave action affect the shape of the seine and can temporarily lift lead lines or submerge float lines. River currents, in particular, commonly can billow or push a net so that the lead line lags behind and allows fish to dive and escape under it. The tidal stage affects which substrate habitats are available to be seined and can affect where fish are prior to seining. Tidal channels are invaded with incoming tides and vacated on outgoing tides. For those who attempt to seine through ice, ice depth (via hole drilling speed) affects how rapidly a seine can be deployed to encircle schooled fish.

Fish

- Species
- Size
- Swimming speed
- Water column orientation (e.g., epibenthic, pelagic, epipelagic)
- Macro- and microhabitat association (e.g., nearshore/offshore, structure)
- Jumping ability and proclivity

Each species of fish has a suite of behaviors that characteristically cause fish to distribute vertically and horizontally in a water column in response to light, habitat, prey, other fish, and disturbance. This distribution may change with fish size (age) and time of year. Fish species that associate with the bottom (epibenthic) and with complex habitat (vegetation, rocks, wood) are generally more difficult to capture with seines than are fish that are pelagic or epipelagic (Murphy and Willis 1996). Sinking seines are sometimes used for fish that are epibenthic, and floating seines are used for fish that are (epi)pelagic. These two types of seining allow a greater reach from shore and thus a greater swept area. Usually, the sinking or floating occurs only during the first half of retrieval (typically a parallel set is used; see later section).

Fish that live in or migrate through nearshore waters are more often suitable targets for beach seines, whereas offshore fish can often be caught with purse seines or lampara seines if they are sufficiently close to the surface. Larger salmonid smolts tend to stay farther from shore, at least during the day. Johnsen and Sims (1973) compared catches of juvenile chinook salmon made in the same area and time of day in the Columbia River estuary: the mean length was 11.9 cm, versus 7.9 cm for fish caught by purse and beach seines.

If particular species or sizes of fish can outswim the net that is being set, encirclement efficiency is reduced and the accuracy of the CPUE estimate is likely to be low. In general, encirclement efficiency decreases and retention efficiency increases with bigger fish because swimming speed increases but ability to penetrate through net meshes decreases. Catches are higher for fish species that swim ahead of nets than for species that dive under or jump over nets. Mullet are notorious for jumping when startled or cornered. Scientists have seen chinook salmon turn on their sides, push their snouts under the lead line, and wiggle to escape under a seine (Hahn, personal observation). This behavior was avoided when the lead line remained ahead of the cork line; in those cases, fish tended to dart into the outward bulging main body of the net. Longer nets, more rapid deployment, and faster retrieval are needed as fish size increases. Herding or scaring fish toward the net may increase capture efficiency. Sometimes the haul lines (end ropes) create enough disturbance when retrieving the net so as to keep fish from passing under them. This can be enhanced by purposely slapping and splashing the water with these lines.

Time of year and day
- Time of year
- Time of day (which is interrelated with amount of light)

Effectiveness of seining is increased when fish are present! Therefore, learn about the behavior of your target fish species and design capture strategies to take advantage of their distribution and behavior. Time of year and day are very connected with the life history and behavior of the fish you wish to collect. Some species of fish migrate at certain times, places, or sizes. Resident fish gain size throughout the year and may seek different habitat as they age.

Within each day, fish often move in response to the amount of light. Larger fish often come into shallow water at night. The amount of light affects the ability of fish to detect and to avoid the sample gear; therefore, sampling at night can increase effectiveness of seining. However, nighttime seining also affects the ability of samplers to see what they are doing and may reduce safety.

Net (gear)
- Construction (how webbing is connected to float and lead lines)
- Mesh size (relative to fish size and different for wings, bunt, and bag)
- Mesh shape when retrieved (under tension)
- Twine size, knotless versus knotted
- Length of net
- Depth of net
- Float line (size of floats—does net sink or float when water depth is greater than net depth?)
- Lead line (amount of weight, lead-core versus weights, sinking speed)
- Wing and bunt design (with or without bag)
- Wear

Beach, pole, and purse seines need to have a tight attachment of webbing to the cork line and around floats so that openings are minimized. Beach and pole seines need to have a similarly tight attachment to the lead line. Worn nets may tear and allow fish to escape. To reduce wear, additional webbing is sometimes wrapped and sewn around the lead line and lead weights.

Larger mesh size and smaller twine diameter have reduced water resistance and allow quicker retrieval of seines. The trade-offs are greater encirclement efficiency from the gain in speed but possible lower retention efficiency. A weaker net may tear or rip more easily. Mesh size in the wings can often be larger than the cross-sectional dimensions of the target fish. Under tension, rectangular webbing assumes a diamond shape (i.e., the cords that form the sides of the mesh openings in a seine net run at 45° angles to horizontal and vertical), which reduces the space through which fish wiggle. Also, the early part of the seine retrieval often drives fish ahead of the net, so they do not usually attempt to wiggle through a net even when they could successfully do so. When fish finally sense that they are confined, they may panic and dive headlong into the net, but by then, the wings should be on the beach or in the boat; therefore, smaller mesh size is used in the bunt (center) section so that fish of all target sizes are unable to penetrate through individual meshes. Knotless twine tends to be gentler on fish (minimizes descaling) and is best for the bunt and bag sections, if not the whole seine. A bag helps concentrate fish away from cork and lead lines and reduces possibility of escape.

A long net can be effective in surrounding and trapping large and fast fish, but it is bulkier and harder to retrieve (requiring more people). Short nets are most appropriate for small fish capture in shallow water; however, shorter length reduces the swept area and thus the total capture. The width (depth) of the seine must exceed the maximum depth of the water to be fished, unless floating or sinking seine retrievals are intended. When relaxed, a seine can reach deeper than its width, but under tension and billowed with the current, its width is reduced by as much as 25% or more.

Floating seines have buoyancy that exceeds the weight of the lead line and can be used to capture fish that are epipelagic. Sinking seines have lead weight that exceeds buoyancy and can be used to capture demersal (epibenthic) fish. In both cases, the cork and lead line come in contact with the surface and substrate during the last half of retrieval to prevent fish from surfacing or diving to escape. Lead core line may offer less opportunity than external weights to snag in rocky substrate; however, external weights can be added or removed to customize a single seine for different conditions. Too little weight will allow the lead line to lift when the seine is pulled vigorously, allowing fish to escape. More lead means the seine sinks more quickly to reach the bottom, cutting off the escape of diving fish. Underwater observation should be used to confirm the intended action of the seine and fish.

Method (gear deployment)
- Speed of net deployment
- Boat motion and engine noise
- Speed of retrieval
- Shape of set (and effect of current)
- Tow line action

Speed is an essential part of successful seining. Fish should be given the least amount of time to flee and attempt escape. The size and swiftness of the target fish should influence both the length of the seine used and the speed at which it is deployed and retrieved. More power (additional personnel or winches) can increase the speed of retrieval. Boat and motor size affect the size of seine and the speed at which it can be deployed. Larger, faster boats create stronger vibrations and greater visual disturbance. Underwater observation can provide guidance on the adequacy of your seine size (sections can be added) and deployment speed. Pierce et al. (1990) wrote, "Fish [in lakes] generally seemed to ignore the [small] boat until it came within 2 or 3 m, so we assumed that evasion or other movements into or out of the enclosed area ... were negligible." Bayley and Herendeen (2000) also noted that noise and disturbance effects attenuated quickly. Hahn et al. (2003) depended on jetboat noise and motion to scare, herd, and concentrate adult salmon downstream from the heads of pools and towards the tailouts in the Green River, whereas Farwell et al. (2006) used a second boat to chase fish upstream towards the seine boat.

Too great a retrieval speed will allow the lead line to lift (or the cork line to dive) when the seine is pulled vigorously, allowing fish to escape. Additional weight or floats may need to be added. It is best to always attempt to retrieve or pull the lead line even with or ahead of the cork line; otherwise, fish tend to be directed down the netting to the lead line, where they might be tempted to dive and wiggle under. The shape of the seine as set, and its subsequent shape when towed, can affect the outcome of attempted escape behavior by the target fish. A curved net probably reduces the opportunity for lateral escape, forcing the fish to attempt to outswim the net. For pole seines pulled along the shore, large and swift species can probably escape. For seines set rapidly in an arc away from and then back to shore—as compared to seines set parallel to shore—escapement may be reduced, but this depends on the size of the target species. Underwater observation should be used to verify seine action and fish behavior. This should be done during all stages of seining. In the final stages of retrieval and confinement prior to processing the fish, the lead line should be brought on shore (or into the boat) and the cork line elevated if fish are inclined to jump.

Calibration: Measuring seine efficiency

Although salmonids were not involved, a paper by Rozas and Minello (1997) provides one of the best overviews of the various issues involved with seining and elaborates on the need for knowing capture efficiency for certain uses of sampling data. They reviewed many studies, mostly related to sampling estuarine areas in the Gulf of Mexico. They outlined catch efficiency as having two components: gear capture efficiency (which describes the proportion of fish in the path of the seine that are caught) and recovery efficiency (which describes the proportion of the caught fish that are actually found and observed by the samplers; this is also called retention efficiency). Some of their conclusions and recommendations were as follows:

(1) Seines (and trawls) had low and variable catch efficiency and were particularly affected by aquatic vegetation;

(2) Enclosure samplers might be better than seines in estimating the density of small nekton in estuaries;

(3) Tide stage must be considered in designing and analyzing results in estuaries;

(4) Catch efficiencies can be estimated by mark–recapture within block nets;

(5) Observing and measuring gear avoidance by direct observation (diving and/or underwater cameras) was worthwhile; and

(6) Clumped distribution of fish (either due to habitat relationship or schooling behavior) requires an increase in the number of samples or a greater sample area per unit of effort (seines functioned well in this regard).

The most practical method of measuring the CE for a seine net (calibrating the net to particular habitats) is to use block nets to trap a representative group of fish within an enclosed space, introduce marked fish, conduct one or more seine sets within the enclosure so that perhaps 40–70% of the enclosed space is swept, retrieve the block nets carefully, and then calculate capture efficiency (Wiley and Tsai 1983; Parsley et al 1989; Bayley and Herendeen 2000). The first seine set is used to estimate CE for a single set (as in Bayley and Herendeen 2000), or the entire series of sets could be used in aggregate (as in Weinstein and Davis 1980; Wiley and Tsai 1983). This allows as natural a situation as possible in which fish can choose all options for escape. This is illustrated in equation 1, where CE is a product of three probabilities. Some authors chose to set tightly within a block net enclosure so that close to 99% of the area was swept (Penczak and O'Hara 1983; Lyons 1986; Pierce et al. 1990; Holland-Bartels and Dewey 1997). However, this technique does not allow fish much opportunity to escape the ends of the seine, thus increasing CE above its true value. A few authors chose to seine a naturally blocked population, such as a tidal estuary pool, and compared the first seine haul to the total population removed by repeated hauls and/or following rotenone application (Weinstein and Davis 1980; Allen et al. 1992). Calibration trials should be repeated in several locations to represent as best as possible the habitat sampled and also to compensate for erratic individual CE values caused by schooling fishes (wherein the school might or might not be caught within any one seine set).

When using a seine within a block net method, CE is calculated for each species (or species-size group) by dividing the number of fish caught by the total estimated population in the enclosure. Using marked fish allows for an estimation of the number of fish not recovered when the block net is retrieved (mark–recapture estimate), as summarized in equation 2. Some authors have used the removal method and repeated seining to estimate the total population (Lyons 1986) and others have used rotenone and marked fish (Weinstein and Davis 1980).

$$N_i = \text{total population species } i = \text{(recovered by seine)} + \text{(recovered by block net)} + \text{(estimated not recovered)} \quad \text{(eq 2)}$$

Fish for marking can be obtained nearby by seining outside the block net enclosure. This ensures that they are accustomed to the habitat enclosed and behave similarly to fish already inside the enclosure; however, there could be some learned net avoidance behavior (but this should not be much of a factor for slow and careful block net retrieval). Marked fish should be introduced gently into the enclosure (by using long-handled dip nets) and distributed evenly (Bayley and Herendeen 2000).

Calibration methods are only needed when seine catches must be expanded to estimate the total populations present in the selected habitat or when calculating species diversity indices that incorporate abundance. As mentioned before, other applications of seining simply capture fish for marking or mark–recapture, biological sampling, or estimating relative CPUE or survivals.

Data extrapolation within and among locations and times
Abundance estimates by seining, without using mark–recapture or efficiency estimates, would not be very comparable among different systems or habitats. Data could be somewhat comparable if most factors were the same, which is most likely to occur within a system being sampled by the same crews during the same sampling period. For example, comparing the catch rate (i.e., number per square meter sampled) may be somewhat comparable if species, species size, habitat characteristics (e.g., river gravel bar, of medium velocity, similar water clarity, time of season), sampling gear (e.g., seine dimensions, mesh size), and deployment methods were very similar.

Field/Office Methods

Planning and setup
Prior to the start of the sampling season, preseason preparation is needed to ensure a smooth implementation of a seining program. At this point, we assume that you have formally developed the purposes and objectives for your research. Now is the time to add detail to your written study proposal. Your general methods will now need evaluation and revision to accommodate the realities of the sampling sites.

You will face a number of questions: Where exactly do you want to sample? What are the habitat conditions there (e.g., lotic versus lentic, fast versus slow currents, substrate type, beach angle, depth)? What are your target species/life stage(s)/size range(s)? Answers to these questions will help determine the size of nets, boats, other gear, personnel, and permitting required for the project.

Site selection and inspection
If you are not personally familiar with the body of water and fish populations for which you believe seining may be a useful capture method, then first consult with local experts, who may be fishery biologists, fishermen, guides, or local residents. (See page 282, Preseason activities for seining.) Find out what research has already been conducted in the area of interest. Do a literature search and contact agencies, Indian tribes, and universities to see what monitoring activities may already have been conducted. Their reports and publications will describe sampling sites and when the target fish are likely to be present. Next, consult maps and aerial photos to look for likely new sites and to find potential access. Obtain a geographical information system-produced map that shows generic property ownership. Visit a local government office or Web site to look for specific property owners. Remember to contact the owners to inform and obtain consent before crossing private property. Make a list of the characteristics of seinable sites for your project. For example, to collect adult salmon in a river for a mark–recapture study, you might want to note: places where target fish aggregate in sufficient quantities to make their collection worthwhile with a 5–7-person crew, pools with fast and shallow water above and below, smooth substrate or only a modest amount of snagging objects that can be "fixed," and lateral sand or gravel bars where a seine can be hauled near the shore.

Use a boat to explore likely stretches of river, bay, or lake. Use a digital camera, global positioning system (GPS) unit, map notes, and sketches to document potential sites. Every potential site must have a suitable haul-out beach or shallow water with modest or no current where the fish can be held until processing. This must be located properly with respect to the anticipated seine deployment and retrieval location. For marine waters and estuaries, obtain a tide schedule and visit the sites at various tide stages to understand currents, substrate exposure, and access issues. For rivers, find the nearest flow gauge, record flows for every site visit, and compare them to help determine what is expected on the dates when seining needs to occur.

On the initial or subsequent inspection trip, for waters that are sufficiently clear, employ an experienced snorkeler or agency-certified SCUBA diver to look at the sites underwater. On a map, record locations of any objects that might snag a seine, the substrate composition, and the direction of currents.

> **Box 1: Preseason activities for seining**
> (1) Develop objectives; write a sampling and safety plan.
> (2) Gather information on tides, currents, flow volumes, and water clarity for the proposed sampling time period and locations. Get aerial photos or visit Internet sites that have digital orthophotos as background.
> (3) Collect local knowledge (contact expert biologists, anglers, tribal members, commercial fishers, and/or property owners.
> (4) Inspect seining sites, access points, and routes for boat, vehicle, and foot travel where required. Use snorkel surveys to check sites. Use a GPS device to store coordinates of all sites and access points. Take pictures and make hand-drawn maps (including maps of underwater habitat features).
> (5) Contact private landowners to ask whether boats, nets, and equipment may be stored securely on their property or if they will allow access.
> (6) If needed, conduct a pilot study to demonstrate feasibility.
> (7) Choose appropriate net designs and deployment vessels. Estimate requirements for winches or other mechanical aids to recover nets.
> (8) Order nets and discuss needs with net company experts.
> (9) Inspect nets to ensure that mesh, lines, floats, and weights are all secure and functional.
> (10) Determine required crew size. List all equipment needed and gather or purchase items.
> (11) Prepare data sheets and develop a data entry plan.
> (12) Buy and prepare holding boxes and tags to apply to fish (if needed).
> (13) Apply for permits (where required) for access or for take or handling of fish (that may be listed as threatened or endangered).
> (14) Prepare vessels, vehicles, and trailers, including safety equipment.
> (15) Arrange for and provide safety training.
> (16) Prepare sites for seining and test the seines.

Suitable: shallow silt, sand, gravel, cobble, hard clay (Note: deep silt can "suck in" a lead line and can be impossible for humans to wade through).

Potentially "fixable" objects: rocks, small and scattered boulders, occasional short pilings (less than 1 m), loose wood, occasional imbedded wood (e.g., trunks, branches), very small rootwads, fishing lures and line, metal (e.g., wire, poles, shopping carts, bicycles). (Note: large rootwads or log jams, if they lie to the side of the main seining path, sometimes can be "fixed" or marked to prevent snagging a net that has been set too close.)

Unfixable objects: medium to large rootwads, tall or abundant pilings, abundant boulders and embedded wood, long sweeper branches, entire trees, car bodies (unless they can be hauled out).

Estimate likely net sizes that might accomplish fish capture in the identified sites. Is more than one net needed?

Site testing and modification (preparing for seining)

Testing

Once sites have been visually evaluated and inspected underwater, either plan for habitat modifications (see below) or use a bare lead line and conduct trial sweeps. Even seemingly small objects can be serious snags. This activity should occur well before the date when seining needs to occur. If a site seems to be devoid of snags or if water turbidity precludes visual underwater inspection, first use a bare leadline to conduct trial sweeps. Mark snag locations with a float and line tied to a heavy weight. After assurance that major snags are absent, use your seine for additional trials. All this will require advance planning and sufficient personnel (see later sections, including safety). If the sites are clean and usable as is (note that occasional snags may remain marked and avoided), then skip the modification steps. If the site seems to have a number of snags but is desirable for sampling, then read the next section and decide whether to abandon the site or modify it. After site modifications, testing with a seine should occur.

Modification

The following techniques are useful only when desirable seining sites have a modest amount of structure that can be overcome and when sufficient alternative sites do not exist. Our focus is on river sites, but there may be applicability to other environments. The amount of effort to invest in "fixing" a site is proportional to the belief and/or knowledge that target fish inhabit and can be caught by seines in the site, and the number of times sampling is planned. The answer to the question "Does it appear that seining can successfully occur if underwater obstructions were overcome?" should be yes before planning to modify a site. As always, safety is the first priority and good judgment is necessary in all steps in this process.

Remember that underwater objects are habitat for fishes. Often, the better the habitat, the more difficult it will be to seine. The goal of site modification is to find creative, temporary ways to allow a seine to slide over and around objects that normally might snag. Only as a last resort should occasional, small, protruding objects be cut. Unnatural objects (e.g., trash) may be removed. After your research is complete, you may remove the devices you had installed, and the habitat should be nearly identical to what it was before.

The devices that have been used to facilitate seining are pitchfork, J-bar, U-bar (or bridge), straight-bar, sandbags or gravel bags and marker buoys (P. Hahn, Washington Department of Fish and Wildlife, personal communication) (see Figure 4). The simplest are sandbags, which should either be made of biodegradable burlap (if left in place) or affixed with a short looped cord to facilitate later removal. These bags can be piled on and around rough cobble and small boulders or small protrusions from the substrate. They act to remove crevices into which lead lines may slide. The other devices all consist of 0.95-mm (3/8-in) iron rebar used in construction to strengthen concrete. This size can be bent underwater by a diver; thicker rebar is more difficult to work with. The straight bar can be up to 6.1 m long and is worked underwater to bend over and around large objects, often in a crosshatch fashion. The upstream ends must be embedded in the substrate. The U-bar has a washer welded on each end and is useful if there are large tree trunks underwater (generally oriented parallel to the current) that have snapped off limbs or jagged cavities. Usually, these trunks lie on either side of the seining path; the U-bar helps avoid snagging by the mesh, which might billow out or be set too close. The washer in this and the other devices allows the use of nails to fix the device to woody objects. The J-bar devices come with and without washers welded to or near the curved end. The long end is jammed into the substrate and the remainder angled downstream to cover boulders and woody snags. The pitchfork devices require the most welding and are constructed of two curved pieces of rebar with cross bracing (see Figure 4). They can be from 1 m to more than 3 m long. The tines are pointed upstream and jammed into the substrate, with the curved end over the obstacle. Several can be placed in overlapping fashion. The most common use is as a net-lifter, but they can also be placed around the sides of rootwads to deflect a seine laterally. They can successfully lift a seine up to 1 m over an object such as a piling. However, be aware that any lifting of the lead line allows the possibility of fish escape.

There are other habitat modification or avoidance techniques. Selective trimming of occasional, small sweeper branches may be necessary, but should be minimized to avoid permanently changing the habitat. Sometimes lone boulders (such as from highway shoulder riprap) can be moved back to the bank using tire chains, steel cable, and winches. Finding and removing tangles of fishing line, hooks, and lures is an important task of the snorkel team. Just one treble hook tied to stout line can stop a seine, be dangerous to remove underwater, or injure the seining crew. Last, buoys can be tied to underwater obstacles so that the boat operator can see and avoid them during seine deployment. If there is a fish escape route between the obstacle and the far shore, a snorkeler can splash and keep fish out of such an area until the seine has been deployed downriver.

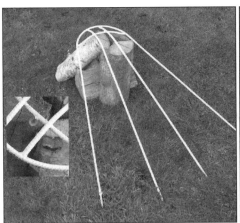 

FIGURE 4.—Devices built out of rebar to bridge over rocks, boulders, snags, large woody debris, and so forth, to allow beach seining for capturing adult chinook salmon in the Green River in Washington in 2000–2002: a pitchfork device (left); two J-bar and a bridge (U-bar) devices (right). The insets show placement of welded washers that allow nailing the devices to wood objects. (Photo: Peter Hahn, Washington Department of Fish and Wildlife.)

Freeing a snagged seine

Depending on bottom topography and the presence of debris or large rocks, nets may become trapped during retrieval. Slacking tension on lead lines and pulling up on the cork line and webbing (into any current) from a boat may work in deep water. A gaff hook on a long pole can be of assistance. Having snorkelers in the water can be very helpful (see the snorkel safety section, page 311, for detailed procedures and cautions); they can attach a pulling line to the lead line. In shallow water, a snorkeler simply wades to the snagged location and frees the net. Note that while a net is being freed, fish may escape.

Seining methods and events sequence

General

A variety of net types and sizes and usage methods have been used by scientists. In nonwadable lotic habitat, seines are typically set from unpowered crafts, such as drift boats or rowboats, or from jet or propeller-driven boats. For maximum efficiency of setting, a seine table (a flat deck free from obstructions or cleats) is desirable. For powered boats, when setting from the stern, a centrally mounted tow post equipped with a quick-release device is desirable, as is a cowling around outboard motors to reduce the chance of nets tangling. In wadable systems, smaller nets are used and deployed by hand with one end of the net anchored to the shore and the other end extended out from shore and then looped around to encircle the fish as the ends are pulled in against the beach. Alternatively, both seine ends may be fixed to poles held by people who walk and push or pull the net. Each pole is held vertically or slightly angled (bottom forward) to keep the lead line against the substrate. In the latter case, the net is generally shorter and may be brought up against the shore or swooped upwards into the air midchannel to trap the fish.

With most seine sets, lead and cork lines should be withdrawn at approximately equivalent rates until close to shore. Once the lead line approaches the shore, it should be withdrawn more than the cork line until a secure pond or corral is formed in the bag of the net and the lead line is on the beach. For circle sets, lampara sets, and purse seine sets, the lead line is retrieved first, followed by the remainder of the cork line and net. Fish may then be allowed to rest within the bag until they are withdrawn for sampling, tagging, or transport to hatchery for use as broodstock. For some methods (e.g., circle set), vegetation may need to be removed methodically and inspected for fish before the seine can be pursed.

Once all fish have been withdrawn from the net, the net is cleaned of all leaf litter, sticks, rocks, and other debris; checked for damage; and reloaded for the next set. Damage to seines can be repaired following instructions in Gebhards (in Murphy and Willis 1996).

Method details

We categorized seining methods into the following groupings. Methods and variations are described and illustrated below.

Pulled linear sets
    1. Parallel set
    2. Perpendicular set
    3. Perpendicular quarter-arc set
    4. Wandering pole seine
    5. Lampara set

Arc sets
    6. Simple arc set (and fast pursuit sets, including double-arc single net option)
    7. Double-seine simple arc set
    8. Beach-lay elliptical arc set
    9. Rectangular arc set
    10. Circle set

Trap sets
    11. Cable-L trap set
    12. Block net sets
    13. Enclosure net tide set
    14. Channel trap tide set

Purse-seine sets
    15. Purse seine set

Pulled Linear Sets

General characteristics: A seine is fully deployed in a straight line, and then both ends are kept apart and pulled through the water some distance until brought to shore or pursed offshore. Usually used to capture small or "slow" fish, where extra speed, stealth, and/or long nets are not needed. (Here "slow" is relative to the length of net deployed; the longer the net, the faster a fish can be and still not escape.)

1. Seine method: Parallel set

Citations: Schreiner 1977; Fresh et al. 1979; Bax 1983; Simenstad et al. 1991; Brandes and McLain 2001; Toft et al. 2004 (see Appendix B).

Procedure: The seine net (see example in Appendix B) is fully deployed in a straight line (AB in Figure 5) by boat at a predetermined distance from shore and parallel to the shoreline. The pulling line AD has a marker a set distance from the seine (30 m for the Puget Sound protocol) (Simenstad et al. 1991 and others) so that the shore tender can signal the boat operator when to begin laying out the seine. When the boat reaches B, the end of pulling line BC is brought to shore without pulling the net. At a signal, people at D and C rapidly begin pulling the seine to shore. In the Puget Sound protocol, when a second marker 10 m from the net is reached, both groups of pullers begin to run towards each other (C–K and D–L) and finish pulling in the seine at HJ. The lead line (JNH) is pulled more rapidly than the cork line near the end of the retrieval.

Analysis: May be used for fish/set, fish/area, or fish/volume analysis. This procedure allows a fixed length seine to be fished in a consistent manner. Length and width of the swept area must be known, and maximum depth must also be known for volume estimates. Caution: the distance between ends of the seine decreases once pulling to shore commences (because the seine assumes an arc shape); thus area and volume should be based on the actual travel path (see approximate polygon LGABEK in see Figure 5).

Where: Lakes or protected marine shoreline with no or very little current.

Variations:

A. Clip-on/removable floats allow the seine to be fished initially as a sinking or a floating net. When sufficiently shallow water is reached during the retrieval, the cork and lead lines keep the seine stretched from surface to substrate.

B. Kubecka and Bohm (1991) and Kubecka (1988) oriented the two pulling lines about 45° away from the seine, apparently in an attempt to keep the seine stretched as much as possible and maximizing the swept area.

C. Brandes and McLain (2001) began a set with the seine piled on shore. One person pulled one end straight out from shore until maximum safe depth was reached and then turned and moved parallel to the beach. The second person followed the path of the first person until the seine was stretched parallel to shore. Both persons then pulled straight to shore. The length of the seine was used as the width of the swept area.

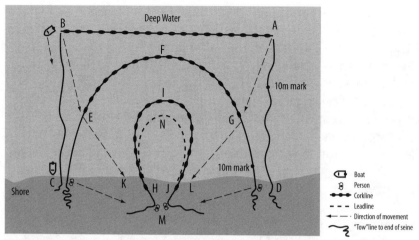

FIGURE 5.—Parallel set deployment and retrieval of a beach seine as used in Puget Sound sampling protocols (not to scale). (Illustration: Andrew Fuller from design by Peter Hahn.)

2. Seine method: Perpendicular set

Citations: Hayes et al. 1996; Fryer 2003 (see Appendix B).

Procedure: Two persons start on shore (at A, Figure 6); each one holds a pole (or stick) fastened to opposite ends of the seine. The seine is stacked on the beach in a looped fashion—lead lines under the cork lines—ready to pull out. One person pulls one end of the seine straight out from shore until the end of the net or the deepest safe water (1.2 m in Brandes and McLain 2001) is encountered (at C). The cork line is marked in 1-m increments so that the distance from shore can easily be noted (in case only part of the net is stretched out). Each pole is held to keep the cork and lead lines spread apart, and the bottom end of each pole keeps the lead line in constant contact with the substrate. Each pole is angled so that the lead line is ahead of the cork line. The nearshore end of the seine is kept in very shallow water or slightly on shore. Both persons then pull the seine along the shore some variable or set distance (CEF and ADI), whereupon both ends are brought onto shore (at G–I). Pole seines can be designed to be operated by one (very short net) or two persons (long net).

Analysis: May be used for fish/area or fish/volume analysis (length, width, and depth must be measured), but fish/set is appropriate only if the distance pulled is consistent. The poles may be marked with 0.1-m-depth increments to facilitate measuring the water depth at the offshore end of the seine (note: take an average of two or more depths). To ensure consistent width of swept area, a rope of known length can be attached to the upper end of each pole (DE in Figure 6) and stretched tight during seining.

Where: Large to small rivers, creeks, lakes, and protected marine shoreline. Currents must be modest or water so shallow that current is not a safety factor.

Variations:

A. A motorized boat or rowboat may be used on the deepwater end of the seine, which allows a greater reach from shore and thus greater swept area (see Figure 7). This variation causes greater difficulty in measuring the offshore depth or the distance between ends of the net; therefore, it is most appropriate for collecting fish for marking or biological measurement. The substrate must be known to be free of snags (note: test with bare lead line first).

B. The seine may be walked quickly through areas to be sampled, not necessarily tight to the shoreline (see Figure 8). The net is lifted to finish a set or pursed to shore. This variation is for collecting fish for marking or biological measurement, or for qualitative rather than quantitative analyses. (See Wandering Pole Seine method on pages 290–291.)

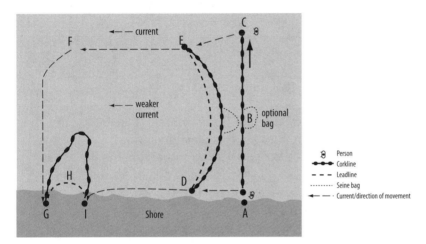

FIGURE 6. — Perpendicular set deployment and retrieval of a beach seine (not to scale). (Note: the distance A–I may be very long—perhaps several hundred meters.) (Illustration: Andrew Fuller from design by Peter Hahn.)

FIGURE 7. — Perpendicular set beach seine operated with a jet boat on the offshore end, Columbia River, Hanford Reach juvenile chinook salmon tagging study. (Photo: Jeff Fryer, Columbia River Inter-Tribal Fish Commission.)

FIGURE 8.—Perpendicular set two-person pole seine pulled along shore to catch juvenile chinook salmon, Columbia River, Hanford Reach juvenile chinook salmon tagging study. (Photo: Jeff Fryer, Columbia River Inter-Tribal Fish Commission.)

3. Seine method: Perpendicular quarter-arc set

Citations: Levings et al. 1986; Parsley et al. 1989 (see Appendix B).

Procedure: One person sets the seine straight out from shore until the end of the net or the deepest safe water is encountered. The end on shore is fixed, and the end away from shore is then pulled in a semicircle back to shore, keeping the net as elongated as possible. By using a fixed length of net and pulling the offshore end in a consistent manner, a consistent swept area can be attained. In the one citation where this method was used, the seine was set inside a rectangular block net, and the swept area was calculated as 64% of the area inside the rectangle.

Analysis: Parsley et al. (1989) used this method within a rectangular set block net to calibrate efficiency. May be used for fish/area or fish/volume analysis (note: length, width, and depth must be measured). The poles may be marked with depth increments to facilitate measuring the water depth at the offshore end of the seine. To ensure consistent width of swept area, a rope of fixed length can be attached to the upper end of each pole.

Where: Lakes, reservoirs, or protected marine shoreline with no current. This set may also be used in large rivers, but currents must be nil and water sufficiently shallow. (Currents would cause billowing of the seine and reduce the efficiency of the set.)

Variations: None.

4. Seine method: Wandering pole seine

Citations: Allen et al. 1992; Flotemersch and Cormier 2001; Toft et al. 2004.

Procedure: A pole seine is stretched out between two persons and pulled through the water, linearly or in a meandering path; neither end stays at the shoreline (see Figure 9). At the end of the set, the lead line is often simply scooped into the air, trapping the catch in the bunt of the net (see Figure 10); however, this only works with short seines.

Analysis: Generally qualitative; used to capture fish for tagging or sampling. Can be used for species presence documentation, but selectivity will be unknown.

Where: Any body of water (e.g., lakes, rivers, ocean) where two persons can safely walk and pull a seine.

Variations: Somewhat similar to perpendicular set when one end is not kept onshore. (A) Two seines can be used simultaneously by three persons (but it is better to prepare in advance and purchase a longer seine), or (B) two seines can be fished next to each other to take advantage of a broader uniform habitat.

FIGURE 9.—Wandering pole seine set in Whatcom Creek, Washington. (Photo: Charmane Ashbrook, Washington Department of Fish and Wildlife).

FIGURE 10.—Scooping the seine at the end of a wandering pole seine set, in order to capture fish entrained in the bunt section. The seine could also be brought to shore. (Photo: Washington Department of Fish and Wildlife).

5. Seine method: Lampara set

Citations: Hayes et al. 1996; Bayley and Herendeen 2000.

Procedure: This specialized net is made with a lead line that is much shorter than the cork line and is much wider in the middle than in the wings. It is generally set from one boat to a second boat so that the seine becomes stretched between them. Both boats then move in parallel, towing the seine some distance before coming together to purse the net.

Analysis: May be used for fish/set, fish/area, or fish/volume analysis (note: length of tow, width of opening between ends, and depth to lead line must be measured). Only pelagic and epipelagic fish will be caught. Acts much like a trawl, with the similar issues involving efficiency (fish avoidance).

Where: Used in large bodies of water that are too deep to wade or where the lead line cannot easily reach the substrate and with no or modest currents.

Variations: Motor-powered or hand-paddled boats may set and pull the nets.

Arc Sets

6. Seine method: Simple arc set (and fast-pursuit sets, including double-arc single net option)

Citations: Sims and Johnsen 1974; Healey 1980; Dawley et al. 1981, 1986; Levings et al. 1986; Pierce et al. 1990; Hayes et al. 1996; Bayley and Herendeen 2000; Hahn et al. 2003; SSC 2003; Kagley et al. 2005; Farwell et al. 2006 (see Appendix B).

Procedure: In its most simple and perfect form, a seine is laid in a half-circle by starting with one end on shore (A in Figure 11; see also Figures 13 and 14); a boat or people then carry and lay out the net into deeper water and then arc back to shore (at D). The lead line should be stacked in the direction of the arc (i.e., on the downstream side if making arc downstream). The speed at which this is done and the length of the seine depends on the target fish species and size.

Pursing the seine generally commences once the setting end of the seine reaches shore. The lead line is pulled to keep it in front of or even with the cork line. Once the lead line is on shore, the remaining part of the seine may be kept in shallow water until fish are processed (if intended for live release) or it is pulled and shaken to concentrate the fish into the bag or bunt section. Fish can then be emptied into a container. Aquatic vegetation may slow the retrieval considerably, especially if vegetation is pulled and carefully inspected to recover all fish. If a repetitive half-circle shape is not needed, the seine may be laid in an oblong or irregular shape that emphasizes sampling in a particular area (but area and volume calculations become impossible).

Analysis: May be used for fish/set, fish/area, or fish/volume analysis. This procedure allows a fixed length seine to be fished in a consistent manner, given that the boat operators or seine setters are experienced. If the seine can be consistently set in a semicircle (as in Figure 11), then it will be fairly straightforward to calculate area if the radius (GH) is known. Likewise, the volume can be calculated (Box 2) if the depth at the apex B in Figure 11 is measured. (After the set, a range finder or a rope equal to the length of radius could be used to find the correct distance offshore.) If the net is set in an elliptical or other shape, the user will need to determine how to estimate the volume. The fast pursuit options are used mainly to capture the maximum number of fish for tagging and release and not for quantitative analysis of CPUE.

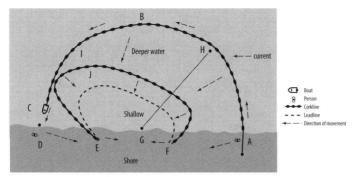

FIGURE 11.—Simple arc set for a beach seine using a motorboat. The seine is initially set from A to B to C. Note that a perfect semicircle is rarely attained (see radii GA, GH, GB, GI and GD in this example) so that area and volume calculations will be approximate. One possible position of the net is shown after half-retrieval (FJE), affected by a current coming from the right and the persons at A and D moving towards each other (note that the lead line is in advance of the cork line). (Illustration: Andrew Fuller from design by Peter Hahn.)

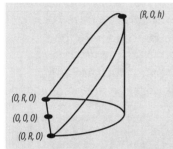

Box 2. Volume of a cylinder intersected by a plane.

A special case of the cylindrical wedge, also called a cylindrical hoof, is a wedge passing through a diameter of the cylinder base. Let the height of the wedge be h and the radius of the cylinder from which it is cut be R. Turn the above image upside down, and imagine the flat surface of the elliptical plane to be the bottom of a lake or river, and the half circle to be the seine corkline on the water surface. The perimeter of the ellipse is defined by the leadline. The water depth at the apex of the net is then h. The equation for the volume is: (Illustration: Andrew Fuller from design by Peter Hahn.)

$$V = \frac{2}{3} R^1 h$$

Where: Large-medium rivers, lakes, ponds, protected marine shoreline. Currents must be modest or absent. Small rivers, or any area with complex habitat, generally have space constraints so that only a small seine can be set in a consistent manner.

Variations:

- A. The seine may be set by hand from a floating tub or small boat in shallow water (see Figure 13, parts B and C) (E. Beamer, Skagit System Cooperative, personal communication). (See also Rectangular Arc Set.)
- B. The seine may be piled on the shore and pulled out by a boat and then around in an arc back to shore. This tends to be like a perpendicular quarter-arc set (see Levings et al. 1986).
- C. Hold-open option: A boat starts by setting half the seine from shore into

the current, holding the billowed seine against the current for specified time (e.g., 4 min), then completing deployment of the seine in an arc back to shore (Figure 12, part B). The intention is to allow fish to move down current into the seine. Recent evidence suggests that this strategy does not increase the catch and may actually decrease it due to avoidance by larger fish (Beamer, personal communication). A powerful engine is needed to counteract drag.

D. Double-arc single net option: Two boats, each with half the seine stacked on them, travel together to a point offshore. At a signal, each proceeds away from each other, each completing a quarter-arc back to shore. Using two boats increases the speed at which the entire net is set. This may be best used where an extremely long seine is to be set in relatively calm water. It could be considered a fast pursuit option.

E. Fast-pursuit, forward-setting option: (See Figures 15 and 16) (R. E. Bailey, Canadian Department of Fisheries and Oceans, personal communication) Although a seine may be deployed from the stern (forward travel) or bow (reverse travel) of a boat, where speed is required, setting from the stern is preferred. A fast, forward-moving boat can minimize the set time and maximize surprise and capture efficiency for large fish such as Pacific salmon. With this variation, the seine typically is not set entirely across the river but in something more like a half-circle arc.

Background: The lower Shuswap River in British Columbia runs about 23–34 m^3/s (800–1,200 ft^3/s) when this seining technique has been used, mainly in deeper glides and holding pools. In this river, 61 × 9 m and 61 × 12 m seines were used. In the wider Harrison River, a 73 × 12 m seine was used. The sites must be free of snags or prepared in advance for seining. Snorkeler assistance was not appropriate due to needed stealth and safety hazard from rapid boat motion. More than 600 adult chinook salmon have been caught in a single set.

Procedure: During daily operations, crews arrive at the seining site by boat and proceed to organize any sampling equipment prior to deploying the net. Once equipment is organized and the crew is ready to proceed, the net is loaded onto a custom seine deck (at the stern, note that a protective cowling is needed around the engine well). The tow line, connecting the upper (cork line) and lower (lead line) bridles, is attached to the quick release mechanism on the tow post. If needed, extra lengths of rope may be added to the tow line to facilitate passing the line to the crew on shore prior to release from the post. The net is stacked back and forth on the seine table, corks forward and lead line to the rear, more to the side of the vessel from which the line will exit. Sets are made by deploying the net in either a downstream or an upstream arc. For downstream sets from the left bank, the seine should exit on the port side of the vessel; for sets from the right bank, the seine should be arranged to exit from the starboard side of the vessel. In Figure 14, the crew is preparing for a set from the left bank, and thus the line is arranged to pull net from the port side of the stern. When the net has been fully stacked, the beach line from the head end should be tied off to a solid attachment point, preferably close to the waterline. Sets are typically made in a downstream arc at speeds up to 20 km/h. The boat operator should attempt to run out of net just as the boat reaches the beach, closing the set.

On very large rivers (more than 100 m wide), a beach line 50 m long is used and tied to a hydraulic (motorized) winch. After the boat takes off from shore, the net first starts to spill when the end of the line is reached. The winch then begins retrieving the seine head back to shore at about 1 m/s. The head should reach the shore just as the boat completes setting the rest of the net. The tail end of the net is then pulled in using a 4 × 4 pickup truck traveling perpendicular to the river. These modifications maximize retrieval speed and allow a fixed length of net to effectively reach farther but still cut off upstream escape.

- F. Two-boat herding for the fast-pursuit arc set: When a signal is made to a waiting chase boat downstream, the chase boat proceeds to zigzag noisily upstream towards the seine boat (waiting on shore), herding fish before it. Before the chase boat arrives, the seine boat begins to set the seine rapidly in a downstream arc, timing movement to complete the set just after the chase boat arrives. This option can be used in rivers where fish escape routes are not limited by shallow water. (Bailey, personal communication.)
- G. Fast-pursuit reverse set with snorkelers and site preparation:

This suite of modifications was developed by Peter Hahn (Hahn et al. 2003, 2004) in the Green-Duwamish River in Washington to capture adult chinook salmon. These techniques can be used in any small-to-medium river (less than 20 m^3/s, ~700 ft^3/s) to allow seining that might otherwise be precluded due to abundant snags. The key elements include

(1) site location and evaluation by snorkeling;

(2) site preparation to shield the seine from snags;

(3) use of a jet sleds, snorkelers, and human "beaters" to herd and keep fish from escaping (a jet sled is a flat-bottom, wide, stable boat with a blunt or squared bow, and with a jet-drive outboard on the stern; it could also be powered by an inboard engine);

(4) setting the seine bank to bank before turning downstream; and

(5) using snorkelers to help manage the seine during and after deployment and to keep fish from escaping.

Background: The Green-Duwamish is a small, clear river in August and September (7–14 m^3/s, ~250–500 ft^3/s) that has abundant natural snags (e.g., tree roots, trunks, and limbs), remnants of submerged pilings (from the historic log rafting period), and riprap boulder banks from extensive lateral dikes. Just prior to the onset of spawning in mid-September, large numbers of chinook began moving upstream and holding briefly in pools during the day. Usually, they congregate in the upper half of a 2–6-m-deep pool under a turbulent surface or in deep water (between B and S2 in Figure 16). Up to 400 adult chinook salmon have been caught in a single set.

BEACH SEINING

FIGURE 12. —Large beach seine methodology. (a) Design of net (not to scale); (b) setting and towing net; (c) hauling or pursing the net to shore. (Credits: A—Eric Beamer, Skagit System Cooperative, 2003; B and C—Richard A. Henderson).

FIGURE 13.—Small net beach seine methodology. (A) Design of net (not to scale); (B) setting the net on a shallow beach; (C) beginning to haul (pursing) the net. (Photos: A—Eric Beamer, Skagit System Cooperative, 2003; B—Karen E. Wolf; C—Richard A. Henderson).

FIGURE 14. — Fisheries and Oceans Canada stock assessment crew and beach seine boat, ready to make a simple arc, fast pursuit set on the Lower Shuswap River, British Columbia, October 2001. The crew is waiting for a second boat to begin herding fish upstream. Note the protective cowling around the motor, the seine stacked on the stern, and helmets worn by the crew. (Photo: Richard E. Bailey, Canada Department of Fisheries and Oceans)

FIGURE 15. — Making a simple arc, fast pursuit set in a downstream arc. The beach seine is rapidly paying out from the left side of the stern. Note that the first corks and net end are some distance from shore but will rapidly be winched back to shore. (Photo: Richard E. Bailey, Canada Department of Fisheries and Oceans.)

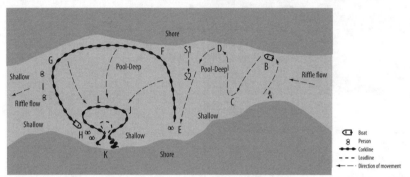

FIGURE 16. — A fast-pursuit variation of the simple arc set where fish are herded, the seine is set from bank to bank, and snorkelers are used to maximize the effectiveness of the set. (Illustration: Andrew Fuller from design by Hahn et al. 2003.)

Procedure: A 61-m beach seine was stacked on the bow of a jet sled, which approached the pool from upstream and held against a bank above the pool, engine off, until signaled. The remaining crew drove to an access point and quietly deployed to locations I, H, and E, crossing at the tailout I if necessary (Figure 16). Alternatively, a second boat could have been used, stopping to disembark crew members above the pool. Two snorkelers deployed by shore to S1 and held quietly against the far bank. One or two crew members with sticks waited at I in the tailout. When everyone was set, a snorkeler passed a signal to the jet sled operator, who started the engine and noisily proceeded to zigzag downstream into the pool (such as the path A–B–C–D–E). When the sled approached E and prepared to hand the end of the seine to a shore crew, the two snorkelers moved to midstream (S2) and thrashed the water with their arms, holding themselves against the current. This kept fish from turning back upstream while the boat was near shore at E. The sled then moved rapidly backward across the river (to F) and backed downstream towards G, while a sled crew facilitated laying the seine off the bow. As soon as the boat reached the far bank, the crew at I began flailing the water with sticks to keep fish from swimming downstream and leaving the pool. The crew at E began to pull the seine end towards K. The two snorkelers monitored the seine and lifted the net over minor snags (see Safety section, page 311). They also monitored the number, position, and behavior of the fish, which often dashed into the net between E and F and sometimes attempted to wiggle under the lead line if it happened to be upstream of the cork line. A snorkeler could move to prevent them from escaping and call for increased lead line retrieval speed.

The sled finished the arc set by backing to H and handing the seine end to the shore crew. The crew at I moved and joined in pursing the seine towards the shore at K, pulling the lead line more rapidly than the cork line. Quick and coordinated action was sometimes needed to prevent the seine from being sucked downstream into the riffle. The snorkelers followed the seine to J and L and lifted the cork line if fish attempted to jump out when in shallow water. When the lead line was fully retrieved, a portion of the seine was kept open (K–J–L) and one crew stationed at J to hold the net against the current and keep the corral from collapsing. This person recorded data as the other crew members processed fish.

Variations: When the pool tailout remained a modestly deep glide, and fish could easily swim downstream, the seine was fully deployed down the far bank (to G or beyond) and the end ropes released. A snorkeler kept the end of the net near shore while the sled moved downstream and then turned and zigzagged noisily back upstream, herding fish into the net. The snorkeler handed the tow line back to the sled crew, and the sled backed across to the other shore. The pursing was completed as described above.

7. Seine method: Double-seine simple arc set

Citations: None known; devised by Hahn, personal communication.

Procedure: Two seines and two fast boats are used, one seine per boat. The second boat drifts or motors quietly to G (similar to Figure 16) and waits until the first boat reaches E. The first boat begins a downstream set. As soon as it reaches the opposite bank and turns downstream, the second boat begins deploying its net across and slightly downstream (left of the area shown in Figure 16). A person holds the shore end and tow line of each seine. When the first boat reaches G, the person on shore hands the tow line of the second seine to this boat crew, and the first boat either (a) pulls each seine end over to the pursing shore, or (b) overlaps the two seines and waits to be pulled to shore by the pursing crew.

Analysis: This fast-pursuit method could be used to maximize the capture of fish for tagging and release, not for CPUE analysis.

Where: Large, medium, or small rivers or tidal channels. Currents may be modest to vigorous. Faster currents mandate good timing and rapid retrieval. The second seine acts like a block net in situations where the downstream channel configuration does not inhibit fish escape. It can be used in strong currents that would not allow a stationary block net to be held in place.

Variations: None.

8. Seine method: Beach-lay elliptical arc set

Citations: Sims and Johnsen 1974; Dawley et al. 1981; Dawley et al. 1986 (see Appendix B).

Procedure: The full seine was stretched onto the beach at water's edge (BD in Figure 17), with the short anchor wing (BN) towards the current, the lead line next to the water, and the seine fixed to shore (to an anchor or log at A). A boat was used to pull the end of the long wing (D) from shore (D to E) and then along the shore (E–F–G–H–I) into the current until the entire seine entered the water (MI). The boat stayed relatively close to, but kept a constant distance from, the shore while pulling and then brought the end to shore at J. The seine was then progressively pulled onto shore, starting with the long wing (JK), lead and cork lines first, while fish were concentrated towards the bunt section (KL). The anchor wing (LM) was brought ashore before the final gathering of the bunt.

Analysis: May be used for fish/set analysis if done consistently. Fish/area or fish/volume could be calculated if the distance from shore is measured and if the maximum water depth is measured; however, there are several difficulties. The length of the set is somewhat longer than twice the length of the seine because the anchor end may shift from B to M. Measuring the distance from shore may require a range finder and may not be constant, depending on the path of the boat. Maximum depth may be difficult for the boat crew to measure during seining. Furthermore, because of the current, there is greater volume actually filtered than estimated from swept area and depth. This method works well for capturing small fish for biological and mark sampling (e.g., relative survival).

Where: Lakes, estuaries, or protected marine shoreline with no to moderate current and small or no waves.

Variations: None.

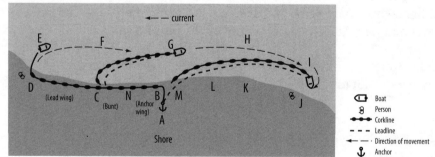

FIGURE 17.—Beach-lay elliptical arc set deployment and retrieval of a beach seine. (Illustration: Andrew Fuller from design by Peter Hahn.)

9. Seine method: Rectangular arc set

Citations: Yates 2001 (see Appendix B).

Procedure: Beach seining at the sites was performed by first setting metal stakes at 12-m (40-ft) intervals along the beach and setting anchored floats 6-m (20-ft) directly offshore. A short waiting period allowed nearby fish to recover from any preparatory disturbance. The seine, one end hooked to the first stake, was deployed from a floating tub waded out 6 m from the waterline to a marked net float held by a second person, turned up current for 12 m parallel to the beach to another marked net float and then returned perpendicular to the beach at the second stake. The up-current end of the net was then dragged along shore back to the original stake, the seine pulled onto the beach, and the trapped fish removed. Four contiguous rectangular sets were made at each sample site.

Analysis: May be used for fish/set, fish/area, or fish/volume analysis (note: length, width, and depth must be measured). This method is probably one of the most consistent and accurate for calculating area and volume. Works best for small fish that are not greatly alarmed by persons wading and that do not escape from the slowly set net.

Where: Lakes, estuaries, or protected marine shoreline with no to moderate current and small or no waves. Possibly the near shore areas of large rivers with slow currents.

Variations: See enclosure net tide set on page 302 (Toft et al. 2004).

10. Seine method: Circle set

Citations: Bayley and Herendeen 2000 (see Appendix B).

Procedure: A seine is set carefully in a circle using canoes, rowboats, or small or large motorized boats. If the area is vegetated, heavier weights must be used on the lead line, and the vegetation may have to be pulled by hand and sorted for entrapped fish. After setting, the net is slowly pursed together and the lead line gathered to allow all remaining fish to be concentrated in a section of the webbing.

Analysis: May be used for fish/set, fish/area, or fish/volume analysis (note: diameter and maximum and minimum depths at opposing points of the perimeter must be measured). If the seine can be set in a nearly perfect circle and the length of seine is known (subtracting the length of overlapped net), then the area is given by πR^2, where $R=C/2\pi$, and C=circumference. The volume is given by $0.5\pi R^2(h_1+h_2)$, where h_1=minimum depth and h_2=maximum depth. Efficiency must be calibrated.

Where: Submerged tide flats and estuaries, floodplains of large rivers, shallow lakes and ponds with no currents.

Variations: Possibly could be set around four poles to assume a rectangular shape. The poles should be set the day before or several hours prior to net deployment.

BEACH SEINING

Block or Trap Sets

11. Seine method: Cable-L trap set

Citations: Farwell et al. 2006 and Richard E. Bailey (personal communication).

Procedure: On each day prior to seining, a taut, strong cable (13-mm or larger nylon or polypropylene rope) (ABC in Figure 18) is suspended across the river at the head end of a deep pool known to contain migrating salmon. Sturdy trees greater than 60 cm diameter at breast height are recommended as anchors, and sections of 2 × 4 lumber should be positioned to prevent girdling the trees (Figure 19). One or two jet sleds are used for setup. The cable must be flagged and high enough to preclude danger to boaters, kayakers, and rafters. Warning signs must be posted upstream. The pool should have a sluice of fast water entering at an angle towards the far bank, such that the near bank has a back eddy that is attractive to salmon and the far half of the river has moderate-to-fast current. The cable should be over or slightly upstream of the point where the back eddy meets the entering fast water.

A short sling with a pulley is attached to the cable by means of a prusik knot (a mountaineering knot typically made with a sling fastened to a line; the loose knot can be slid easily along the line but binds tightly when tension is applied to the free end of the sling [Cox and Fulsaas 2003]) and prepositioned on the cable above where the eddy line goes down the pool (at B). One end of a tender-line is passed from the far bank near C out to, and loosely attached to, the sling at B. The other end is held by a crew member near C. This line later moves downstream with the sled to become ED in Figure 18. A clip-line KB is passed through the pulley, and one end with a carabiner is clipped loosely to the sling or cable at B (Figure 19), and the free end of the clip-line is handed ashore and fastened at K. The sled with the seine stacked on the bow (the seine is folded back and forth with the lead lines towards the stern and cork line towards the bow) is then gently nosed to shore at H. Shore crew members pull part of the seine from the bow (which has a bridle connecting the lead line and cork line to a tow line) and then attach the tow line to a shackle (or carabiner) fastened securely to shore, perhaps at A. A second tender-line is tied or clipped to the sled and the remainder held on shore at H. This tender-line later moves downstream with the sled to become FD in Figure 18. The second sled (without the seine) is beached away from the pool on the near shore.

When ready to deploy the trap-seine, the seine-sled crew (two-person minimum) quietly grasps the cable and pulls the sled bow along the cable towards the middle of the river, letting the seine pay out as they go. The tenderline handlers on each shore (at H and C) can help move the sled across. When the sled reaches B, the sled crew unfastens and then clips the carabiner end of the clip-line to the cork line of the seine. Tension is applied to the clip-line from shore and the free end is refastened at K, thus holding the seine in place from H to B. The sled crew then allows the sled to drift downstream, paying out the remaining length of seine but remaining attached to the end of the seine (ending up at D, with the shore crew moving tender-lines from C to E and from H to F). The tenderline FD is then clipped to the end of the bridle joining the lead line and cork line (but remains attached to the sled, perhaps held by a crew), while tender-line ED remains attached to the bow of the sled. The shore crew at F and E help position the sled and seine end slightly to the fast-water side of the eddy, and keep the seine from collapsing into the eddy. The distance BD may be longer than HB.

All crews now quietly wait for 45 min or more, watching for salmon moving into or within the back eddy. A shore crew member (spotter) may be in a tree or at some vantage point with a two-way radio to relay fish sightings to the sled operator. When sufficient salmon are within the upper part of the back eddy a signal is made; the shore crew at K releases clip-line KB; the sled is released from the net and both tender-lines, and motors to E to pick up the shore crew; the seine end at D is quickly pulled towards shore by tender-line FD; and the lead line at H is quickly retrieved to make sure that it is downstream of the cork line at L (to keep fish from nosing under the lead line and escaping). As the seine is pursed towards shore, the sled motors along the cork line to help free any hang-ups (two crew members lie down and lean over the edge to pull up netting by hand as needed). As soon as further snags are not anticipated, the sled is beached and all hands help tend the seine. Both seine ends, and all the lead line, are pulled onto shore (at M and N), leaving the fish trapped near shore in the remaining pursed net (J). The purse is kept from collapsing while crew members handle and mark the fish (keeping them in the water, using a cradle restraint if appropriate). The processed fish are gently released over the cork line while one crew member records data. (More than 600 chinook salmon have been caught in a single set [Bailey, personal communication]). Generally, only one or two such sets are made per day per site.

Analysis: Qualitative only; used for the capture of fish for marking or for biological samples and inspection for marks.

Where: Small to moderate-size rivers with pools or other areas where fish are known to concentrate. Developed for the capture of adult salmon that are migrating upstream.

Variations: None.

12. Seine method: Block net set

Citations: Wiley and Tsai 1983; Lyons 1986; Parsley et al. 1989; Allen et al. 1992; Bayley and Herendeen 2000 (see Appendix B).

Procedure: Block nets are generally used (a) for efficiency tests on seine nets set within the block net perimeter, (b) to allow some other sampling gear (such as electrofishing and mark–recapture estimates between two block nets in a stream) to be used on a temporarily captive population, or (c) to prevent the escape of fish actively being pursued with another seine (see method 7 above).

Analysis: For efficiency tests, marked fish may be released into the enclosure prior to the seine set for which the efficiency is being calibrated (see previous section of this chapter). The block net is generally pursed carefully to shore and all remaining fish counted after the tested net is set one or more times within its perimeter. The known loss of marked fish is used to estimate the initial population of unmarked fish. The initial population forms the basis for estimating the efficiency of the tested net.

Where: For efficiency tests, any waterbody where currents and waves are sufficiently absent to allow the block net to remain in place. For creating a captive population, generally a stream or small river where currents and debris are modest enough to allow the net to remain anchored to both shores for the duration of the study.

Variations: See methods 13 and 14 below.

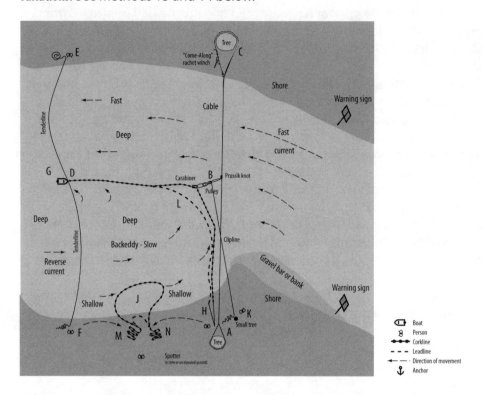

FIGURE 18. — Cable-L trap set with a beach seine to capture adult chinook salmon in the Shuswap River in British Columbia. (Illustration: Andrew Fuller from Richard E. Bailey, Canadian Department of Fisheries and Oceans.)

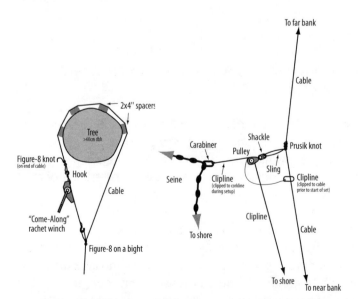

FIGURE 19. — Cable-L trap set details showing (1) how the cable is fixed to sturdy trees on each bank, and (2) how the beach seine is temporarily attached to the middle of the cable, allowing half the seine to stretch downstream. For "Figure-8 on a bight," see Cox and Fulsaas (2003). (Illustration: Andrew Fuller from design by Peter Hahn.)

13. Seine method: Enclosure net tide set

Citations: Toft et al. 2004 (see Appendix B).

Procedure: As described by Toft et al. (2004): Enclosure net sampling "consisted of using a 60 m long, 4 m deep, 0.64 cm mesh net placed around poles to corral a 20 m^2 rectangular section of the shoreline. The poles were installed at low tide the day before net deployment so as to minimize disturbance at time of sampling. The enclosure net was installed at high tide. Fish were removed with either a small pole seine (1.2 m × 9.1 m, 0.64 cm mesh) or dip nets as the tide receded, usually starting at midtide a few hours after net deployment. All fish were removed before low tide."

Analysis: May be used for fish/set, fish/area, or fish/volume analysis (note: depth needs to be measured along outer wall of the seine).

Where: Saltwater beaches with tidal fluctuations that allow the draining of the area enclosed by the net. This could include complex habitat (like boulders) that could not be swept by a seine, although care must be used to account for fish that might remain stranded in micropools under such structures. Probably most efficient for species that are pelagic (e.g., salmon fry) rather than demersal (e.g., sculpins).

Variations: None.

14. Seine method: Channel trap tide set

Citations: Cain and Dean 1976; Levy and Northcote 1982; Yates 2001; SSC 2003 (see Appendix B).

Procedure: A beach seine is pulled across the lower portion of a tidal channel at high tide and anchored to both banks. The seine may be unmodified or have a bag or a bag plus a box or hoop trap. Generally, poles are pounded into the substrate at intervals, and the lead line and cork line are tied to them. Keeping the cork line raised above the water level is important if any target species are likely to jump. The lead line may have to be pushed into the mud along the entire length of the net or very heavy weights should be used. Crabs may cut through the mesh and allow escape. Rotenone may be used for a complete kill but should be administered near low tide and channel areas monitored for stranded fish.

Analysis: Total number and biomass, fish/length of channel, fish/area, and diversity indices. Area can be calculated from aerial photographs taken at high tide. Volume is difficult to calculate due to sinuosity and varying channel dimensions; perhaps flow and cross section area could be monitored at the net site.

Where: Estuaries and marine bays where there are anastomosing channels that drain completely at low tide and have only one outlet.

Variations: Site specific. Trap design can be variable.

Purse Seine Sets

15. Seine method: Purse seine set

Citations: Durkin and Park 1967; Johnson and Sims 1973; Dahm 1980; Healey 1980; Dawley et al. 1981, 1986; Hayes et al. 1996 (see Appendix B).

Procedure: Two boats are used to lay the seine out in a circle, in water too deep for the lead line to reach the bottom. The seine boat passes the end of the net to the skiff, which attaches it to a stanchion. The skiff motors to hold the end of the net nearly stationary until the seine boat completes a circle. The pursing line runs through rings attached to (or hung from) the lead line. When both ends of the pursing line are winched onto the seine boat, the bottom of the net is closed together. Once the bottom is sealed, the cork line and remainder of the net is brought aboard, gradually concentrating the fish into the remaining section of seine (which may have smaller mesh than the wings). When the net is brought aboard and stacked, the lead line should be under the cork line. This reduces the chance of the lead line becoming looped over the top of the cork line when released into the water and assures maximal sinking rate of the bottom of the net.

Analysis: Fish/volume is typically used to report results, but fish/set may also be used. The known depth of net, plus the radius of the set, allows calculation of the approximate volume ($\pi R^2 h$) (approximate because a perfect circle can rarely be accomplished in practice).

Where: Any lentic waterbody deep enough to operate boats, and water depth is generally greater than the reach of the net. Some large rivers are also suitable for purse seining. Marine waters with turbulent currents are not suitable (but may be fished successfully at slack tide stages).

Variations: (See also Circle Set and the double arc single net option of the Simple Arc Set.)

- The end of the net initially released from the seine deck can be attached to a drogue chute, anchored buoy, or sea anchor that holds the net end stationary while the rest of the net is released by the rapidly moving boat. When the boat completes the circle, a boat hook can be used to grab the cork line and the pursing line and bring them aboard for retrieval.
- The seine can be set in shallow water so that the lead line and pursing line touch the bottom. The bottom must be free of snags to allow proper pursing of the net.
- A motorized or towed barge and small skiff can be used to set small purse seines.
- Two equal-sized boats, each carrying half the seine, can be used to set the net in opposite directions.

Measurement details

Sample processing
After the net is brought to shore, the fish can be handled to meet the objectives of the study design (Klemm et al. 1993; Meader et al. 1993; British Columbia Ministry of Environment, Lands and Parks 1997; Lazorchak et al. 1998; Moulton II et al. 2002). Beach seining induces relatively low stress on the fish, and thus mark–recapture techniques can be employed. All standard measurements (e.g., species, length, weight, scales, gut contents, sex) can be gathered. Additionally, if counts are all that is required, it is quick and easy to count and release the fish as the net is brought in.

Preventing transmission of disease and exotic organisms
The cross-watershed transmission of invasive aquatic diseases exotic animals is a serious threat to ecosystems. The U.S. Forest Service has developed an Invasive Species Disinfection Protocol (<www.reo.gov/monitoring/watershed/docs/InvasiveSpeciesProtocolFinal.pdf>) that has been adopted by a wide array of land management agencies. This protocol currently reflects the best known way to prevent the spread of New Zealand mud snails *Potamopyrgus antipodarum*, Port Orford cedar root rot *Phytophthora lateralis*, and sudden oak death syndrome *P. ramorum* in the western United States. The basic techniques include rinsing wading boots and sampling gear in a mild bleach solution and then in boiling water and using a high-pressure sprayer or car wash to clean vehicles prior to traveling to a new watershed. These techniques may prevent the spread of other organisms and diseases. Methods for disinfecting large seines are needed.

Data Handling, Analysis, and Reporting

Data categories
Data collection for each seine set should include the following:
- Time and date of set
- Tidal stage (e.g., ebb, flood, high-tide slack, low-tide slack)
- Water surface area seined (or measurement to allow calculation)
- Length of time the set is held open (large net only)
- Surface and bottom temperature of area seined
- Surface and bottom salinity of area seined (estuarine areas only)
- Maximum depth of area seined
- Average surface water velocity (small net only) using a flow meter
- Substrate class of area seined (small net only; see Appendix C)
- Vegetation type of area seined (small net only; see Appendix C)
- Fish catch records by species
- Subsample of fish lengths and weights (where appropriate, based on objective) (SSC 2003)

Data analysis

Once the surface area sampled has been calculated, data are generally reported as densities (fish/ha). Multiple sets from the same area should be averaged to get the fish number for that area. Multiple sets allow for more rigorous statistical analysis on mean density data and comparisons among sites for various variables. Nobriga et al. (2005) describe deployment and analysis for numerous sites and times in detail. Data can be extrapolated to larger areas if species distribution and habitat conditions are similar. Note that the reported estimate of fish/ha is an index value (unless adjusted for known net efficiency for the specific conditions) that would only be comparable to other results of very similar sampling gear, species composition, size distribution, and habitat conditions.

Effectiveness of seining is limited by gear, species, and habitat sampled. Rozas and Minello (1997) compiled a chart of effectiveness of different sampling gear, listing their advantages and disadvantages. Seines are easy to use, give clean samples, and have a large sample unit area (SUA). The disadvantages are that they can have a low and variable catch efficiency and can be ineffective in vegetation/soft substrate and that SUA can be difficult to define. As with all sampling methods, there is a degree of bias to seining. Mesh size, speed of area encirclement, and method of retrieval all affect the selectivity of the method towards certain species and sizes. It is important to understand these biases when analyzing data and perhaps incorporating other sampling methods in order to capture the entire range of fish assemblage in an area. Introductory insights into analysis of the data are noted under each of the 15 methods

Seining is typically used for six main purposes:

(1) biological sampling,

(2) species presence and diversity,

(3) relative abundance estimation,

(4) absolute abundance estimation via indirect measures,

(5) relative survival estimation, and

(6) absolute abundance estimation via direct measures.

Biological sampling reflects the basic collection and listing of fish species captured and the acquisition of morphological measurements (e.g., length, weight) and other biological samples (e.g., scales, tissues, presence of disease). Species presence and diversity reflects a listing of species captured, with data reported in CPUE, fish/set, fish/area, or fish/volume sampled. When using presence sampling to identify species richness, rare fish distributions, or simple presence/absence of a species at a particular geographical locale, considerations of sampling efficiency should be taken into account. Failure to identify an individual species at a location does not demonstrate that it does not exist there and may be the result of poor sampling efficiency. Detailed habitat descriptions often are reported as part of these sampling efforts, in support of subsequent fish–habitat relationship analysis.

Seining is frequently used for capturing small juvenile salmonids, where a measure of relative abundance or CPUE is needed. The use of standardized nets and deployment methods has provided a means to characterize abundance over time and space, either within or across years. A common assumption that is made when estimating relative abundance is that capture efficiencies are the same for different species and/or for different age-classes of fish. This assumption is unlikely to be true, particularly for species of different sizes and those that use different habitats.

Estimating population abundance with high accuracy and precision requires mark–recapture efforts. Seines allow the selective capture and subsequent release of a wide range of salmonid fish sizes. This characteristic makes seining a useful capture method for many mark–recapture-based salmonid assessments, where marking more fish allows for greater precision of the population estimate. There have been many statistical advances in evaluations of mark–recapture data through the years. Discussing the statistical developments associated with evaluating mark–recapture data is beyond the scope of this protocol. We refer readers to the detailed review and references contained in Schwartz and Seber (1999).

The basic premise of mark–recapture estimates is that the ratio of marked/unmarked fish collected in a sample where M fish are marked is the same as the ratio of marked fish in the total population (M/N). An estimate of abundance from mark–recapture data can therefore be calculated from equation 3:

$$N = MC/R \quad \text{(eq 3)}$$

where N = the population estimate, M = number of fish marked during the mark run(s), C the total number of fish in the recapture sample(s), and R the number of marked fish captured in the recapture sample.

Bailey (1951) and Chapman (1951) presented mathematical corrections to the Petersen estimate when it was recognized that it may be biased when sample sizes are low. Chapman's modification of the Petersen estimate is provided in equation 4:

$$N = \frac{(M=1)(C+1)}{(R+1)} - 1 \quad \text{(eq 4)}$$

Robson and Reiger (1964) suggest that an unbiased estimate of N can be generated from the Chapman modification of the Petersen estimate when one or both of the following conditions are met:

1. The number of marked fish M plus the number of fish captured during the recapture sample C must be greater than or equal to the estimated population N.
2. The number of marked fish M multiplied by the number of fish taken during the recapture sample C must be greater than four times the estimated population N.

Calculations providing approximate 95% confidence intervals about the population estimate N are summarized by Vincent (1971) and can be calculated using equations 5 and 6:

$$Estimate \pm 2\sqrt{Variance} \quad \text{(eq 5)}$$

where

$$\text{Variance} = \frac{(PopulationEstimate)^2 (C - R)}{(C + 1)(R + 2)} \quad \text{(eq 6)}$$

Equations 3–6 allow for hand calculation of population abundance estimates and confidence limits when all assumptions are met during sampling. There are many more complex estimators that can be used to estimate population abundance when assumptions cannot be tested in field settings due to budget or time constraints. Many of these estimators are available through Internet resources and are often free or inexpensive. These resources contain information as well as Internet links to computer software programs for estimating various population and community parameters beyond simple abundance calculations. Many computer programs contain complex procedures that will select appropriate population estimators based on the observed field data. Other programs use iterative calculus techniques to produce maximum likelihood estimates that are the most likely based on observed mark–recapture data.

Personnel Requirements and Training

Roles and responsibilities
The number of crew members required to deploy and retrieve the net will vary depending on the size of seine net being deployed, the force required to recover the net, and method of deployment. A minimum of two persons, up to crew sizes over five, may be required. Furthermore, where currents are strong and/or large nets are used, power winches may be needed to assist in net retrieval.

Boat operator
The boat operator is typically the crew supervisor and is responsible for ensuring that the net is properly loaded and will deploy freely from the boat. The boat operator is also responsible for securing the boat at the end of the set and ensuring all applicable safety regulations are met. An on-board assistant is responsible for throwing the tow line to shore crews as the set is closed, and, once the tow line is under control on shore, releasing it from the tow post. Where sets are made at speed, it is strongly recommended that boat crews wear swift-water helmets to prevent head injuries from rapidly moving cork and lead lines.

Shore crews
Shore crews are responsible for attaching the head end of the net securely to the shore and for net retrieval once the set is closed. Shore crews also clean, repair, and load the net with assistance from and under supervision of the boat crew. All personnel assist with sample processing.

Qualifications
All crew members operating in swift-water environments should have experience in safe swiftwater operations, including safe wading techniques. Boat operators should be experienced in operating the vessel type to be used in riverine environments. At least one crew member, preferably a qualified fish biologist, needs to identify fish species found in the system being sampled. Additional specialized qualifications may be needed, depending on specific information being collected and processed (e.g., fish disease, scale sampling, tagging

techniques, fish preservation for biological sample collection). All snorkelers (and seining crew members) should be vaccinated against tetanus, hepatitis, typhoid fever, and polio (Lazorchak et al. 1998, 2000).

Safety and training

Water safety: Requirements and considerations
As with any field activity, safety is of paramount importance. Snorkelers, divers, seiners, and boat operators should always assess the potential hazards of the site before entering the water (Dolloff et al. 1996). In addition, a health and safety plan should be developed for any surveys of this type in which a risk assessment is conducted, and appropriate countermeasures for each risk are identified and implemented during the survey.

Snorkeling safety considerations
(a) A safety plan should be written in advance.
(b) The person in charge of any snorkeling operation should, at a minimum, have a SCUBA certification. All others participating in the operation should have mastered and demonstrated the basics skills needed to safely conduct the operation. These skills include strong swimming ability; familiarity with working in a wet or dry suit, proper snorkeling technique, the use of dive knives, and the hazards of working in and around nets; and the ability to hold one's breath while manipulating objects underwater. Greater expertise is required for working with seines than is required for simple snorkel surveys to count fish.
(c) All snorkelers must be good swimmers.
(d) Rivers and streams can be dangerous and unpredictable. Snorkelers must be familiar with river dynamics. This knowledge can be achieved through experience with kayaking, canoeing, drift boating, rafting, and/or snorkeling in rivers.
(e) A full wet (or dry) suit, including gloves, booties, and hood, should be worn. These items offer protection from snags and fishing lures that may be entangled on objects. Such suits also provide buoyancy, a positive safety factor. The preferred suit is smooth and without pockets or flaps that can snag. Buoyancy compensators or life vests add bulk, drag, and potential snag points. Knee pads built into the suit are a good addition.
(f) Narrow, smaller fins are best for acceleration and speed in rivers. Avoid large "jetfin" styles for river work. Tape fin buckles if they look like they could snag in webbing.
(g) When working around nets, two sharp knives should be worn. Both should be capable of cutting the type of webbing quickly. Both should be located in a position that allows for quick access and eliminates snag potential (e.g, inside of an arm or leg).
(h) No weight belt is to be used when assisting the seining operation (due to risk of snagging). It may be used when scouting sites for snags or preparing sites for seining. The belt must have a quick release mechanism that is in good working order. Additionally, the weight belt should be less

than that required to gain neutral buoyancy. In other words, the snorkeler should always maintain some positive buoyancy.

(i) Two snorkelers should be suited and in the water if working with the seining operation. Both must remain reasonably close to each other and make frequent visual contact.

(j) A single snorkeler may be used for scouting and mapping but only when accompanied by a jet boat (i.e., no propeller drives) or other craft suitable for reaching and assisting the snorkeler.

(k) All stretches of river must be checked visually prior to snorkeling, and good judgment needs to be used to avoid debris jams, whitewater, or any other perceived hazard (Thurow 1994). It is remarkably easy for a snorkeler to negotiate most rivers, especially at low flows, but caution must always be exercised. High flows during freshets can make a normally benign river dangerous.

(l) Water quality can sometimes be of concern. Recent rain can flush contaminants from roads, parking lots, pastures, and backyards. Industrial outfalls should be noted and downstream areas avoided if at all questionable.

(m) If recreational or commercial boaters are likely to be encountered while seining, additional procedures should be devised and used to warn, reroute, or stop their approach. A dive flag would be useful to warn that there are snorkelers in the water.

(n) Never attach ropes or lines to divers in areas where currents or tidal action are factors.

(o) Hypothermia can be a hazard for snorkelers (although overheating can also occur). Crew members should all be trained in CPR and first aid, with an emphasis on recognizing and treating hypothermia (Dolloff et al. 1996). Swiftwater rescue training would also be useful for crew members working in larger river systems.

Scuba safety considerations

(a) Scuba diving is usually the method of choice when seining sites need to be modified. It offers the advantage of prolonged underwater work and speeds the process greatly. Snorkelers can sometimes install a small number of modification devices, but it is not practical for deeper pools and larger modification operations.

(b) Both certification by an internationally recognized training program (such as PADI, PADI Americas, 30151 Tomas Street, Rancho Santa Margarita, California 92688 USA; www.padi.com/padi/default.aspx) and certification through the employing agency's diving safety program are absolutely required for all divers. Certificates should be current. All dives should be logged. Divers are responsible for the use of safe equipment and following all agency safety procedures. If the agency does not have such a program, then it is in conflict with Occupations Safety and Health Administration regulations in the United States. Development of a diving policy must occur prior to any agency participation in SCUBA diving operations.

(c) A minimum of two divers shall be in the water at all times. One or more snorkelers may assist, especially to direct the placement of modifying devices.

(d) There should be a jet boat (i.e., no propeller drives) or other craft involved in all river dive operations (unless exceptions are granted by the agency's diving safety officer). This craft should be suitable for ferrying, carrying spare tanks, and reaching and assisting the divers. Two boat operators may be preferred, with the pilot focusing on maneuvering and the assistant watching the progress of the bubble streams and handing out tools and devices.

(e) The divers must be familiar with the snag bridging devices and how to install them before entering the water. All equipment, tools, and devices must be staged in advance to avoid delays.

Managing a river seine with snorkelers

A properly prepared and tested seining site should not cause major snagging events, but minor net-stopping hang-ups may still occur. Prior to production seining, inexperienced crew and snorkelers should be trained and should familiarize themselves with the seine, boat, and netting process in the water. A benign site such as a lake, pond, or pool should be used for practice deployments. The snorkelers should use this trial to familiarize themselves with the layout of the net and ensure that all dive gear is free from snags. Additionally, scenarios that test snorkelers' ability for dealing with snags should be rehearsed. They should also practice the procedures for getting in and out of the boat. The boat operator, crew, and snorkelers should review the expected seining procedures. A trial set of the net should be made. Snorkelers should always be upstream or to the outside of the seine until a substantial part of the seine has been pursed and/or is under control in shallow water. Note: The mesh size of a beach seine is sufficiently small, and twine size large, making them safe for working snorkelers. Gill or tangle nets are not safe for snorkelers to be near; however, drift or set gill-net sites can still be prepared in advance by snorkelers and divers.

Freeing a snagged net in moderate to strong currents

The snorkelers must evaluate the strength of the water flow near any snag point. If the divers cannot swim against the flow safely, a vessel must be used to free the snagged net. When lifting part of a seine net that is stuck on any object by currents, it is critical that the diver should never grab the lead line in a manner that would entrap fingers or hands. The snorkelers should also be aware of and try to avoid fishing lures that may be stuck to the net or the snagging object.

A snagged net quickly shows a cork line that points upstream in an inverted V. The lead line will be some distance upstream from the point of the V. After evaluating the cause of the snag, the snorkeler should position him/herself pointing upstream over the snag, while looking for the fold of netting that leads to the snag. After a few deep breaths, the snorkeler free dives down to the webbing, grabs it, and pulls him/herself down towards the snag. He/she should then pull upstream on the webbing and let the positive buoyancy of the exposure suit free the net. If not, on the next dive, the snorkeler should get closer to the lead line, grab the netting again, and, while kicking upstream, let the positive buoyancy of the exposure suit free the net. Another method involves grabbing the leadline on either side of the snag. Again, the snorkeler should swim upstream and let the positive buoyancy of the exposure suit free the net. If that fails, a line attached to a carabiner can be taken down by the snorkeler and attached to the lead line as close to the snag as possible. Once the line is attached, it can be used by snorkelers at the surface to try to free the snag by pulling upstream while swimming. If the current is strong, this can be facilitated by a drift boat or jet boat. In the unlikely event that the net can still not be freed, the set will have to be aborted. The surface support vessel can then be used along with additional rope or a long tender-pole to pull hard upstream and free the snag. If all else fails, the snorkeler can cut the lead line and/or webbing and free the net.

Boating safety
(1) Write a safety plan.
(2) Jet-boat and propeller-boat operators should be thoroughly familiar with the equipment and operation of their craft. They must also be familiar with boat behavior in flowing water, when heavily loaded (with seine and crew), and when towing or pulling objects.
(3) The boat operator must constantly be conscious of where all snorkelers or divers are.
(4) The boat operator must be trained/experienced in swift-water boat operations and rescue.
(5) The crew must be trained in swift-water operations.
(6) All field staff members need to be trained in wilderness first aid.
(7) All field staff members need to have medical clearance for field operations (e.g., routine physical).

Operational Requirements

Workload and field schedule
(1) Seining for adult salmon, if conducting census data, may continue throughout the run of salmon past seining sites.
(2) Seining can be continued all day by making sets at multiple sites or by allowing fish to reenter the seine site prior to making another set.
(3) Night seining is also possible if conditions are safe to do so. Night seining is more frequently used for juvenile salmon.
(4) Seining activity may be reduced as migrations taper off and crews are reassigned to other activities associated with the study.

Equipment needs
(1) Vessel suitable for setting seine. Vessel choices are jet-powered riverboats, propeller-powered riverboats, rafts, and drift boats. Choices are governed by the operating environment, size of net, and species to be seined. Surveying fish that are capable of fleeing rapidly require powered boats.
(2) Fuel and oil for powered craft
(3) Beach seines suitable for the operating environment

(4) Spare ropes

(5) Net repair equipment

(6) Dip nets

(7) Polarized glasses for boat operators

(8) Waders

(9) Life jackets

(10) Rain gear

(11) Swift-water operation helmets for boat crews

(12) Throw bags

(13) Marking, tagging, and sampling supplies (as needed); data sheets; and pencils

(14) Hydraulic winch (optional)

(15) Anti-snag devices (optional)

(16) Long-handled gaff hook

Budget considerations (in U.S. dollars)

(1) Personnel costs (2–5 or more people)

(2) Capital costs for boat ($10,000 to $50,000)

(3) Nets (under $500 to $5,000 or more)

(4) Expendable field equipment (e.g., waders) ($200)

(5) Fuel for boat(s)

(6) Transportation costs to/from project site

Literature Cited

Allen, D. M., S. K. Service, and M. V. Ogburn-Matthews. 1992. Factors influencing the collection efficiency of estuarine fishes. Transactions of the American Fisheries Society 121:234–244.

Backiel, T. 1980. Introduction. In T. Backiel and R. L. Welcomme, editors. Guidelines for sampling fish in inland waters. Food & Agriculture Organization of the United Nations, Rome.

Backiel, T., and R. L. Welcomme, editors. 1980. Guidelines for sampling fish in inland waters. Food & Agriculture Organization of the United Nations, Rome.

Bagenal, T. B. and W. Nellen. 1980. Sampling Eggs, Larvae and Juvenile Fish. Pages 13–36 in T. Backiel and R. L. Welcomme, editors. Guidelines for sampling fish in inland waters. Food & Agriculture Organization of the United Nations, Rome.

Bailey, N. T. J. 1951. On estimating the size of mobile populations from capture-recapture data. Biometrika 38:293-306.

Bayley, P. B., and R. A. Herendeen. 2000. The efficiency of a seine net. Transactions of the American Fisheries Society 129:901–923.

Bax, N. J. 1983. The early marine migration of juvenile chum salmon (*Oncorhynchus keta*) through Hood Canal—its variability and consequences. Ph.D. dissertation. University of Washington, Seattle.

Brandes, P. L., and J. S. McLain. 2001. Juvenile chinook salmon abundance, distribution, and survival in the Sacramento–San Joaquin Estuary. In R. L. Brown, editor. Contributions to the biology of Central Valley salmonids. Fish Bulletin 179(2):39–136.

British Columbia Ministry of Environment, Lands and Parks. 1997. Fish collection methods and standards, Version 4.0. Prepared by the B.C. Ministry of Environment, Lands and Parks, Fish Inventory Unit for the Aquatic Ecosystems Task Force, Resources Inventory Committee, Victoria, British Columbia.

Cain, R. L., and J. M. Dean. 1976. Annual occurrence, abundance and diversity of fish in a South Carolina intertidal creek. Marine Biology 36:369–379.

Chapman, D. G. 1948. A mathematical study of confidence limits on salmon populations calculated from sample tag ratios. International Pacific Salmon Fisheries Comm. Bulletin 2:69–85.

Cox, S. M. and K. Fulsaas, editors. 2003. Mountaineering: The freedom of the hills, 7th edition. The Mountaineers, Seattle.

Craig, J. A., and R. L. Hacker. 1940. The history and development of the fisheries of the Columbia River. Bulletin of the U.S. Bureau of Fisheries 32:133–216.

Dahm, E. 1980. Sampling with active gear. Pages 71–89 in T. Backiel and R. L. Welcomme, editors. Guidelines for sampling fish in inland waters. Food & Agriculture Organization of the United Nations, Rome.

Dawley, E. M., C. W. Sims, R. D. Ledgerwood, D. R. Miller and J. G. Williams. 1981. A study to define the migrational characteristics of chinook and coho salmon in the Columbia River estuary and associated marine waters. National Oceanic and Atmospheric Administration, National Marine Fisheries Service, Northwest Fisheries Service Center, Seattle.

Dawley, E. M., R. D. Ledgerwood, T. H. Blahm, C. W. Sims, J. T. Durkin, R. A. Kirn, A. E. Rankis, G. E. Monan, and F. J. Ossiander. 1986. Migrational characteristics, biological observations, and relative survival of juvenile salmonids entering the Columbia River estuary, 1966–1983. Report to Bonneville Power Administration #DOE39652-1. National Oceanic and Atmospheric Administration, National Marine Fisheries Service, Northwest Fisheries Service Center, Seattle.

Dewey, M. R., L. E. Holland-Bartels, and S. J. Zigler. 1989. Comparison of fish catches with buoyant pop nets and seines in vegetated and nonvegetated habitats. North American Journal of Fisheries Management 9:249–253.

Dolloff, A., J. Kershner, and R. Thurow. 1996. Underwater observation. Pages 533–554 in B. R. Murphy and D. W. Willis, editors. Fisheries techniques, 2nd edition. American Fisheries Society, Bethesda, Maryland.

Duffy, E. J., D. A. Beauchamp, and R. M. Buckley. 2005. Early marine life history of juvenile Pacific salmon in two regions of Puget Sound. Estuaries, Coastal and Shelf Science 64:94–107.

Durkin, J. T., and D. L. Park. 1967. A purse seine for sampling juvenile salmonids. The Progressive Fish-Culturist 29(1):56–59.

Farwell, M. K., R. Diewert, L. W. Kalnin, and R. E. Bailey. 1998. Enumeration of the 1995 Harrison River chinook salmon escapement. Canadian Manuscript Report of Fisheries and Aquatic Sciences 2453.

Farwell, M. K., D. C. Allen, N. D. Trouton, and R. E. Bailey. 2006. Enumeration of the 2000 lower Shuswap River chinook salmon escapement. Department of Fisheries and Oceans, Division of Fisheries Management, Kamloops, British Columbia.

Flotemersch, J. E., and S. M. Cormier. 2001. Comparisons of boating and wading methods used to assess the status of flowing waters. U.S. Environmental Protection Agency, 600/R-00/108, Washington, D.C.

Fresh, K. L., D. Rabin, C. Simenstad, E. O. Salo, K. Garrison, and L. Mateson. 1979. Fish ecology studies in the Nisqually Reach Area of southern Puget Sound, Washington. Fisheries Research Institute, College of Fisheries, University of Washington, Seattle.

Fryer J. K. 2003. Expansion of Hanford Reach tagging project, 2002. Report to Pacific Salmon Commission—Chinook Technical Committee. Columbia River Inter-Tribal Fish Commission, Portland, Oregon.

Hahn, P. K. J., T. Cropp, and Q. Liu. 2004. Assessment of chinook salmon spawning escapement in the Green-Duwamish River, 2002. Washington Department of Fish & Wildlife, Olympia.

Hahn, P. K. J., M. C. Mizell, and T. Cropp. 2003. Assessment of chinook salmon spawning escapement in the Green-Duwamish River, 2000. Washington Department of Fish & Wildlife, Olympia.

Hayes, D. B., C. P. Ferreri, and W. W. Taylor. 1996. Active fish capture methods. Pages 193–220 in B. R. Murphy and D. W. Willis, editors. Fisheries techniques. American Fisheries Society, Bethesda, Maryland.

Healey, M. C. 1980. Utilization of the Nanaimo River estuary by juvenile chinook salmon, *Oncorhynchus tshawytscha*. Fishery Bulletin 77(3):653–668.

Holland-Bartels, L. E., and M. R. Dewey. 1997. The Influence of seine capture efficiency on fish abundance estimates in the upper Mississippi River. Journal of Freshwater Ecology 12(1):101–111.

Johnsen, R. C., and C. W. Sims. 1973. Purse seining for juvenile salmon and trout in the Columbia River estuary. Transactions of the American Fisheries Society 2:341–345.

Kagley, A. N., K. L. Fresh, S. A. Hinton, G. C. Roegner, D. L. Bottom, and E. Casillas. 2005. Habitat use by juvenile salmon in the Columbia River estuary: Columbia River Channel Improvement Project research. Report by National Oceanic and Atmospheric Administration, National Marine Fisheries Service, Northwest Fisheries Science Center, Fish Ecology Division to the U.S. Army Corps of Engineers, Portland District, Oregon.

Klemm, D. J., Q. J. Stober,, and J. M. Lazorchak. 1993. Fish field and laboratory methods for evaluating the biological integrity of surface waters. U.S. Environmental Protection Agency, Office of Research and Development, EPA/600/R-92/111, Cincinnati, Ohio.

Kubecka, J., and M. Bohm. 1991. The fish fauna of the Jordan Reservoir, one of the oldest man-made lakes in central Europe. Journal of Fish Biology 38:935–950.

Lazorchak, J. M., D. J. Klemm, and D. V. Peck, editors. 1998. Environmental Monitoring and Assessment Program—surface waters: field operations and methods for measuring the ecological condition of wadeable streams. U.S. Environmental Protection Agency, National Exposure Protection Laboratory, Cincinnati, Ohio.

Lazorchak, J. M., B. H. Hill, D. K. Averill, D.V. Peck, and D.J. Klemm, editors. 2000. Environmental Monitoring and Assessment Program—surface waters: field operations and methods for measuring the ecological condition of non-wadeable rivers and streams. U.S. Environmental Protection Agency, National Exposure Protection Laboratory, Cincinnati, Ohio.

Levings, C. D., C. D. McAllister, and B. D. Chang. 1986. Differential use of the Campbell River estuary, British Columbia, by wild and hatchery-reared juvenile chinook salmon *(Oncorhynchus tshawytscha)*. Canadian Journal of Fisheries and Aquatic Sciences 43:1386–1397.

Levy, D. A., and T. G. Northcote. 1982. Juvenile salmon residency in a marsh area of the Fraser River estuary. Canadian Journal of Fisheries and Aquatic Sciences 39:270–276.

Lyons, J. 1986. Capture efficiency of a beach seine for seven freshwater fishes in a north-temperate lake. North American Journal of Fisheries Management 6:288–289.

Meador, M. R., T. F. Cuffney, and M. E. Gurtz. 1993. Methods for sampling fish communities as part of the National Water Quality Assessment Program. U.S. Geological Survey open-file report 93–104, Denver.

Miller, B. S., C. A. Simenstad, L. L. Moulton, K. L. Fresh, F. C. Funk, W. A. Karp, and S. F. Borton. 1977. Puget Sound baseline program nearshore fish survey: final report July 1974–June 1977. Fisheries Research Institute, College of Fisheries, University of Washington, Seattle.

Moulton, S. R. II, J. G. Kennen, R. M. Goldstein, and J. A. Hambrook. 2002. Revised protocols for sampling algal, invertebrate, and fish communities as part of the national Water Quality Assessment Program. U.S. Geological Survey open-file report 02-150, Reston, Virginia.

Murphy, B. R., and D. W. Willis, editors. 1996. Fisheries techniques, 2nd edition. American Fisheries Society, Bethesda, Maryland.

Nelson, T. S., G. Ruggerone, H. Kim, R. Schaefer, and M. Boles. 2004. WRIA 9 juvenile salmonid survival studies: juvenile chinook migration, growth and habitat use in the lower Green River, Duwamish River and nearshore of Elliot Bay, 2001–2003. Department of Natural Resources and Parks, Water and Land Resources Division, King County, Seattle.

Nobriga, M. L., F. Feyrer, R. D. Baxter, and M. Chotkowski. 2005. Fish community ecology in an altered river delta: spacial patterns in species composition, life history strategies and biomass. Estuaries 28(5):776–785.

Parsley, M. J., D. E. Palmer, and R. W. Burkhardt. 1989. Variation in capture efficiency of a beach seine for small fishes. North American Journal of Fisheries Management 9:239–244.

Penczak, T., and K. O'Hara. 1983. Catch-effort efficiency using three small seine nets. Fisheries Management 14(2):83–92.

Pierce, C. L., J. B. Rasmussen, and W. C. Leggett. 1990. Sampling littoral fish with a seine: corrections for variable capture efficiency. Canadian Journal of Fisheries and Aquatic Sciences 47:1004–1010.

Rawding, D,, and T. Hillson. 2003. Population estimates for chum salmon spawning in the mainstem Columbia River, 2002. Bonneville Power Administration, project number 2001–05300, Portland, Oregon.

Robson, D. S. and H. A. Regier. 1964. Sample size in Peterson mark–recapture experiments. Transactions of the American Fisheries Society 93(3):215–226.

Rozas, L. P., and T. J. Minello. 1997. Estimating the densities of small fishes and decapod crustaceans in shallow estuarine habitats: a review of sampling design with a focus on gear selection. Estuaries 20(1):199–213.

Schreiner, J. U. 1977. Salmonid outmigration studies in Hood Canal, Washington. M.S. thesis. University of Washington, Seattle.

Schwartz, C. J., and G. A. F. Seber. 1999. A review of estimating animal abundance III. Statistical Science 14:427–456.

Simenstad, C. A., C. D. Tanner, R. M. Thom, and L. L. Conquest. 1991. Estuarine habitat assessment protocol. U.S. Environmental Protection Agency, Region 10, Office of Puget Sound, Seattle.

Sims, C. W., and R. C. Johnsen. 1974. Variable-mesh beach seine for sampling juvenile salmon in Columbia River estuary. Marine Fisheries Review 36(2):23–26.

SCC (Skagit System Cooperative). 2003. Estuarine fish sampling methods. Skagit System Cooperative, La Conner, Washington.

Threinen, C. W. 1956. The success of a seine in the sampling of a largemouth bass population. The Progressive Fish Culturist 18:81–87.

Thurow, R. F. 1994. Underwater methods for study of salmonids in the intermountain West. U.S. Forest Service, Intermountain Research Station, General Technical Report #INT-GTR-307, Ogden, Utah.

Toft, J., C. Simenstad, J. Cordell, and L. Stamatiou. 2004. Fish distribution, abundance, and behavior at nearshore habitats along city of Seattle marine shorelines, with an emphasis on juvenile salmonids. University of Washington, School of Aquatic and Fisheries Sciences, Seattle.

von Brandt, A. 1984. Fish catching methods of the world. 3rd edition. Fishing News Books Ltd., Farnham and Surrey, England

Vincent, R. 1971. River electrofishing and fish population estimates. Progressive Fish Culturist 33(3):163–169.

Weinstein, M. P. and R. W. Davis. 1980. Collection efficiency of seine and rotenone samples from Tidal Creeks, Cape Fear River, North Carolina. Estuaries 3(2): 98–105.

Weisstein, E. W. 2006. Cylindrical wedge. From MathWorld—A Wolfram Web Resource. Available: http://mathworld.wolfram.com/CylindricalWedge.html (February 2007).

Wiley, M. L., and C. Tsai. 1983. The relative efficiencies of electrofishing vs. seines in piedmont streams of Maryland. North American Journal of Fisheries Management 3:243–253.

Yates, S. 2001. Effects of Swinomish Channel jetty and causeway on outmigrating chinook salmon (*Oncorhynchus tshawytscha*) from the Skagit River Washington. M.S. thesis. Western Washington University, Bellingham.

BEACH SEINING

Appendix A: Key to Seine Methods

Marine/estuary/lake
- **Open coast** (with surf)
  - **Sand/gravel/cobble beach**
    - Seining may not be feasible. Consider size of breakers in the range of typical weather and time of year, shape & material of beaches, near-shore depths and currents at all tide stages, type of boats or other net deployment that might be needed. Consult maps and aerial photos. Test with bare leadline for snags.
    - **Large, fast fish**
      - Inhabiting the full surf zone and beyond
      - Simple arc set. Very long, strong nets and specialized motor boats and retrieval systems will be needed (von Brandt 1984)
      - Parallel set
    - **Small, inshore fish**
      - Small nets deployed by wading may be feasible, but rip tide hazards must be considered and overcome or avoided
      - Pole seine
      - Simple arc set
  - **Rocky beach**
    - These could be sinking or floating seines, and night or day sets, depending on the size and species of target fish.
- **Sheltered coast** (without surf (large waves))
  - **Strong currents** (Qualitative sets primarily)
    - Simple arc set (usually down-current, but possibly into current)
    - Perpendicular set, pole seine, pull along beach (down-current) – either as a human powered or boat & human.
    - Beach-lay elliptical arc set (possibly)
  - **Mild to no current**
    - These could be sinking or floating seines, and night or day sets, depending on the size and species of target fish.
    - **Large tidal variation**
      - (Note: both are shallow water. Some sets listed under "Little tidal variation" could be tried here, at specific tide stages.)
      - Lampara seine set
      - Purse seine set
    - **Little tidal variation**
      - **Deep water**
        - (Quantitative if dimension or distance pulled can be measured)
      - **Shallow water**
        - **Little/no vegetation** (All could be quantitative)
          - Simple arc set
          - Half arc, hold open against current, then finish set
          - Rectangular set
          - Parallel set, pull to shore
          - Perpendicular set, pull along beach (pole seine)
          - Perpendicular set, quarter arc
          - Beach-lay elliptical arc set
        - **Abundant vegetation**
          - (Option to remove vegetation after set but prior to pursing, if overall habitat impacts are small. Quantitative if calibrated.)
          - Circle set
          - Simple arc set

River (above tidal influence)
- **Wadeable** (Small, no boat needed)
  - **Small &/or slow fish** (Generally qualitative)
    - Wandering pole seine
    - Simple arc set
  - **Larger & faster fish** (Generally qualitative)
    - Pole seine plus block net seine
    - Double-seine simple arc
- **Non-wadeable** (Medium to large)
  - Note: In turbid water snag detection becomes a problem. Bare leadline or mobile DIDSON SONAR may be the only method of exploring sites underwater. Snag elimination may not be possible (avoid snorkel or SCUBA divers). In clear water (sufficient to see perhaps 1m or more) you can use snorkeling to find, evaluate, then modify sites. Boat(s) needed to conduct seining.
  - **Extensive shallow margins** (Generally qualitative)
    - Beach-lay elliptical arc set
    - Perpendicular set, pull along beach (pole seine)
    - Simple arc set
  - **Occasional shallow margin** (Generally qualitative)
    - Simple arc "pursuit" sets (With variations, see text)
    - Double-seine arc set (Medium river)
    - "Cable-L" trap set (Medium river)
  - **Deep water** (Mostly qualitative, possibly quantitative)
    - Purse seine set

Appendix B: Seine Specifications and Citations

Citation	Target species	Habitat/Location	Net dimensions/construction Length	Depth	Mesh	Cork/leadline	Twine	Crew size	Methods
Threinen 1956	Largemouth bass, other warmwater species	Eutrophic lake (Browns Lake, Wisconsin) USA	609.6m	4.6m	76.2mm & 50.8mm (50:50)	Trailer sticks added to leadline to prevent rolling.	?	?	Simple arc sets, somewhat variable in shape. 3.6 to 11.7 hectares enclosed by each set. Vegetation & soft mud presented difficulties.
Sims & Johnson 1974; Dawley et al. 1981 Dawley et al. 1986	Juvenile Chinook salmon (and coho, steelhead):	Estuary of large river (Columbia River, Washington-Oregon, 1800–6500 m^3/s) USA	95 m (4 panels)	3.6m wings to 5m at bunt	Variable		?	?	Simple arc sets
Yates 2001	Juvenile Chinook (35-100mm); pink & chum salmon	Estuary & ship channel near large river (Skagit River, Washington) USA	24m	3m	3.2mm	?	?	2 ?	Rectangular sets (6x12x6m), four at each site, by wading. (~75 m^2 per set).
Cain & Dean 1976	Misc. intertidal fish such as menhaden, killifish, & mummichog	Tidal channel in estuary (South Carolina) USA	33m with 8m bag	3.3m	6.4mm	Poles held cork line >=0.3m above water. Leadline pushed into mud.	?	2?	Seine set as a block (trap) net across a tidal channel at high tide. Rotenone used to help remove all fish in channel.
Levy & Northcote 1982	Juvenile Chinook, chum & pink salmon	Tidal channels of large river (Fraser River, British Columbia), Canada	("large"), bag with removable trap box	2.4m	6.4mm	Floats every 30 cm; 0.5kg/m (2.0 lb/fathom)	?	2+	Seine set as a block (trap) net across a tidal channel at high tide. Removable trap box allowed sampling periodically during ebb tide. Estimated gear efficiency.
Toft et al. 2004	Juvenile salmon, other fish species, crabs.	Protected marine, near-shore (Puget Sound, Washington). USA	1) 60m 2) 9.1m	4m 1.2m	6.4mm 6.4mm	? ?	? ?	2–3 2	1) "Enclosure net" 20x20x20m set at high tide around poles driven into beach on previous day. 2) Pole seine.
Durkin & Park 1967	Juvenile steelhead, coho & sockeye salmon	Brownlee Reservoir, Snake River, Idaho, USA	182.9m	10.7m	9.5mm wing; 6.3 mm bunt	Corks every 38cm Lead 0.3 kg/m (1.2 lb/fathom)	knotted nylon wing; knotless bunt	4	Purse seine set by motorized raft & 4.3m flat-bottom skiff (28 hp outboard on each).
Fred Goetz, U.S. Army Corps of Engineers, Seattle, Washington	Juvenile Chinook, coho, sockeye & steelhead (smolts).	Inside a lock, or in Lake Washington ("Ballard Locks"), Seattle WA). USA	221m, tapered	9.1 to 3.8m	17.5mm	539g float/0.3m; 1.41kg/m leadline (5.7 lb/fathom). Doubled each end.	"210/15" black bonded nylon	4	Purse seine set from motorized barge, with motor skiff (25 hp outboard on each).
Healey 1980	Juvenile Chinook salmon.	Estuary & tidal channels of small river (Nanaimo River, British Columbia), Canada	1) 90m 2) 18m 3) 216m	7m 3m 18m	? 12mm ?	? ? ?	? ? ?	3+? 2–3? ?	1) Purse seine, hand hauled, set over tide flats at high tide. 2) Beach seine, simple arc set, pulled from beach. 3) "Drum seine" (purse seine used in outer estuary, near-shore areas).
Johnson & Sims 1973; Dawley et al. 1981; Dawley et al. 1986; Dahm 1980	Juvenile steelhead, coho & yearling Chinook salmon	Estuary & off-shore marine (Columbia R. Washington-Oregon, 1800–16500 m^3/s) USA	1) 228.6m 2) 152.4m	10.7m 4.9m	1+2)main 9.5mm; bunt 6.4mm	?(with braided nylon purse line)	1+2) knotted nylon main, knotless nylon bunt	3 ?	Purse seine set by boat ("gillnetter") 8.7m by 2.4m with 260 hp engine) & 6.1m surf dory skiff. Large net for main estuary & ocean, small net for shallower channels.

BEACH SEINING

Citation	Target species	Habitat/Location	Net dimensions/construction					Crew size	Methods
			Length	Depth	Mesh	Cork/leadline	Twine		
Allen et al. 1992 (Efficiency test)	Misc. intertidal fish such as menhaden, killifish, & mummichog	Tidal channel in estuary (South Carolina) USA	15.2m with 8m bag	1.2m	6mm	"Heavily weighted"	?	2+?	Used block nets at each end of isolated tidal pool (22 x 14m, 1m deep). Pole seine repeatedly swept ~90% of pool, then rotenone used.
Bayley & Herendeen 2000 (Efficiency test)	Misc. South American species	Large river floodplain (Amazon River)	1) 25m (40m netting hung in pockets 2) 50m (85m netting)	1) 6m middle to 0.6m at ends 2) ?	1) 5mm 2) 5mm	1) 15cm dia. floats 30cm on-center. 120g lead cylinder 35cm on-center 2) Floats & leads 25 & 50cm on-center	1&2) 0.5mm twine, knotless, blue nylon	4+	1) Three methods: (a) Circle set, like purse seine, (b) simple arc set from beach, (c) like lampara seine. 2) Block set first in all trials. About 25-50% of blocked area was seined. Marked fish also released to measure efficiency of block net.
Holland-Bartels and Dewey 1997 (Efficiency test)	Misc. warmwater species in central North America	Large river (upper Mississippi River)	9.1m	8m	3.2mm	tubular lead weights spaced 30.4cm	?	2+	Perpendicular set within rectangular enclosure, seine ends kept close to sides & shore. Block net enclosures were set around posts & were 15.2x7.6m & 15.2x4.6m in size.
Lyons 1986	7 taxa: mimic shiner perch, logperch, bluntnose minnow, Iowa & Johnny darter, rock bass	Mesotropic clear water lake (Wisconsin)	1) 15.2m with 1.8x1.8x1.8m bag 2) 33m	1) 1.8m 2) 1.8m	1) 6.4mm 2) 6.4mm	1) 7x3.5cm Styrofoam floats 35cm on-center; tubular lead 23.5cm on-center	?	2+?	1) Parallel set, just inside block net with ends kept 0.5m from sides (thus ~92.5% of area seined). 2) Block net around rectangular area, 13.4m long, 5 to 10m wide.
Parsley et al. 1989 (Efficiency test)	Juv.Chinook salmon sunfish, sculpin, pikeminnow, shad, sucker, sandroller...	Impoundment of a large river (John Day Reservoir, Columbia River, Washington) USA	1) 30.5m 2) 92.5m	1) 2.4m 2) 3.1m	1) 6.4mm 2) 6.4mm	1) 61cm spacing for floats & leads. 2) 30.5cm spacing for both.	1&2) knotless nylon	?	1) Perpendicular set along a side of block net, offshore end pulled in ¼ circle to shore=64% of area in #2 2) Block net, square set (30m sides).
Pierce et al. 1990 (Efficiency test)	Perch, shiners, pumpkinseed sunfish, +17 other.	Littoral zone of 10 lakes (southern Quebec), Canada	1) 52m 2) (> 52m)	1) 2.6m 2) ≥2.6m	1) 6mm 2) 6mm	Plastic floats, lead-core bottom line.	1&2	?	1) Simple arc set (~430m²). 2) Block net set 2nd, close to wall of seine, left in place for more seine sets. Marked fish & rotenone used.

Appendix C: Substrate and Vegetation Types (SSC 2003)

Table 1. Definitions of intertidal substrate types modified from Dethier (1990).

Substrate Type	Definition
Bedrock	75% of the surface is covered by bedrock, commonly forming bluffs and headlands.
Boulder	75% of the surface is covered by boulders (>256 mm).
Cobble	75% of the surface is covered by clasts 64 to 256 mm in diameter.
Gravel	75% of the surface is covered by clasts 4 to 64 mm in diameter.
Mixed Coarse	No one size comprises > 75% of surface area. Cobbles and boulders are > 6%.
Fines with Gravel	No one clast size comprises more than 75% of the surface area. Cobbles and boulders make up > 6% of the surface area; coarse sediments combined make up < 55%. Rich with epibenthic fauna.
Sand	More than 75% of the surface area consists of sand 0.06 to 4 mm in diameter.
Mixed Fines	Fine sand, silt, and clay comprise 75% of the surface area, with no one size class being dominant. May contain gravel (<15%). Cobbles and boulders make up < 6%. Walkable.
Mud	Silt and clay comprise 75% of the surface area. Often anaerobic, with high organics content. Tends to pool water on the surface and be un-walkable.
Artificial	Anthropogenic structures replacing natural substrate within the intertidal zone, including boat ramps, jetties, fill, and pilings.

Table 2. Definitions of intertidal vegetation types from Dethier (1990).

Vegetation Type	Definition
Eelgrass	More than 75% of vegetative cover is Zoster marina, Zoster japonica Phyllospadix spp., Ruppia maratima
Brown Algae	More than 75% of vegetative cover is brown algae belonging to taxonomic groups Division Phaeophyta.
Green Algae	More than 75% of vegetative cover is algae belonging to the taxonomic group Division Chlorophyta.
Red Algae	More than 75% of vegetative cover is algae belonging to the taxonomic group Division Rhodophyta.
Mixed Algae	Areas in which red, green or brown algae coexist, no single type occupies more than 75% of vegetated cover.
Kelp	More than 75% of vegetative cover is large brown algae (Order Laminariales).
Salt Marsh	More than 75% of vegetative cover is emergent wetland plants.
Spit-Berm	More than 75% of vegetative cover is plants such as dune grass, gumweed, and yarrow, which generally occur above the highest tides, but still receive salt influence.
Unvegetated	More than 75% of the total surface area is unvegetated.

BEACH SEINING

Snorkel Surveys
Jennifer S. O'Neal

Background and Objectives

Snorkeling is the underwater observation and study of fish in flowing waters. Snorkeling gear is worn by biologists who, individually or in small teams, survey fish abundance, distribution, size, and habitat use while slowly working in (generally) an upstream direction. This technique is most commonly used to survey juvenile salmonid populations but can also be used to assess other species groups. This document discusses the critical elements necessary for the design and implementation of a snorkel survey program to encourage a standardized procedure for the use of underwater techniques to survey fish species in streams. Much of the information in this paper was adapted from Thurow (1994), who provided vital information for use in standardized snorkel surveys.

Snorkel surveys are widely used to monitor fish populations in streams and to estimate both relative and total abundance (Slaney and Martin 1987). Snorkeling can also be used to assess fish distribution, presence/absence surveys, species assemblages (i.e., diversity), some stock characteristics (e.g., length estimation), and habitat use. Each objective will affect how the survey should be conducted. Specific directions for implementing snorkeling to address various objectives are presented in this chapter.

A variety of fish species can be assessed using snorkel surveys; however, salmonids, due to their territorial nature in freshwater and propensity for using habitats with high water clarity, are the group for which snorkel surveys are most frequently conducted. Snorkel surveys have recently been used to assess sculpin diversity (C. Jordan, NOAA Fisheries, personal communication). Snorkel surveys are often selected as the best method for surveying salmonids because they result in relatively little disturbance to the target species, reasonable accuracy, and less cost outlay. Cost optimization methods are discussed at the end of this protocol and explained in detail in Dolloff et al. (1993).

Rationale

Snorkeling is often feasible in places where other methods are not; for example, deep, clear water with low conductivity makes electrofishing prohibitive. Because of the small amount of equipment required for snorkeling, the method can be used in remote locations where it may be difficult to use other methods, such as traps, nets, and electrofishing. Because fish are not handled and disturbance is minimized, snorkeling is especially useful for sampling rare or protected stocks. Snorkel surveys provide an alternative to traditional and more disruptive methods, such as electrofishing and gillnetting (Mueller et al. 2001). Information on individual or group movement, behavior, and habitat associations can also be collected (Mueller et al. 2001). Snorkel surveys can be combined with other methods such as sonar and tools such as geographic information systems (GIS) to generate three-dimensional maps of habitat use by fish species (Mueller et al. 2001). Less time and cost are required for snorkeling than for methods such as mark–recapture or removal, which are often used to estimate abundance (Schill and Griffith 1984; Thurow 1994; Hankin and Reeves 1988;). Snorkelers can observe fish behavior such as spawning, feeding, and resting with minimal disruption.

Objectives

Snorkel surveys can provide quantitative information on the abundance (Schill and Griffith 1984), distribution (Hankin and Reeves 1988), size structure (Griffith 1981), and habitat use (Fausch and White 1981) of salmonids. Snorkel surveys can also be used to provide estimates of salmonid populations within the reaches surveyed (Thurow 1994) and of species distribution within a habitat. In fact, underwater observational techniques are one of the few methods that enable scientists to assess how fish actually use habitats and structural components within habitats, such as boulders and large woody debris. Diversity of species can be assessed using observational surveys for salmonids or transect and quadrant surveys for benthic-dwelling species such as sculpin. With proper training, snorkelers can make relatively precise (within 25 mm) visual estimates of length (Griffith 1981).

Presence/absence

Generally, snorkeling works well in detecting presence/absence of most salmonid species. One exception is bull trout *Salvelinus fontinalis*, which are elusive and difficult to detect using snorkel surveys. Rodgers et al. (2002) provides a detailed discussion on determining presence and detection probabilities for bull trout using three different methods: day snorkeling, night snorkeling, and electrofishing. Snorkeling does not work well when water is turbid or tannic-stained (due to the inability to see the fish). When water temperatures are less than 9°C, fish are generally inactive and therefore not visible for counting during daylight hours, so night snorkeling should be used.

Relative abundance and mark–recapture counts

If a closed population estimate is needed in order to calibrate surveys with another method within a single habitat unit, block nets can be used with snorkel surveys to prevent fish from entering or leaving the unit (Hillman et al. 1992). Hankin and Reeves (1988) list formulas for estimating total fish abundance and calculating confidence limits around the estimates.

If the area to be surveyed is too large for one snorkeler, additional snorkelers can be added to cover the entire channel width. The counts from all snorkelers are then summed for the total count for the reach sampled. This method, called an expansion estimate, assumes that counts are accurate and that snorkelers are not counting the same fish twice (Thurow 1994).

Snorkel surveys can be used with other techniques to estimate abundance using mark–recapture techniques. The use of snorkel surveys for mark–recapture estimates provides a calibration factor for the counting efficiency of snorkel surveys as compared to other methods such as electrofishing and seining. For this technique to be used, the system should be closed off with block nets so that the total population observed is constant. For mark–recapture estimates, fish need to be collected, marked, and released back to the sample area. Collection occurs with electrofishing, seining, or trapping; marking is done using tags or dyes with color coding or patterns; and release is into the sample reach. Snorkel surveys can then be conducted in the same sample reach to count the number of marked fish. The ratio of the number of fish resighted (i.e., those with marks that are counted) can be used as a calibration factor for the effectiveness of a snorkel survey compared to other survey methods. Fish can be marked differently based on size-class and species recorded by observers, which can help identify those species and size-classes that may have been undercounted using a snorkel survey approach.

Habitat Use

Snorkel surveys can be used to observe the direct use of habitat by fish species. This method is especially effective for determining the use of habitat improvement structures placed to benefit rearing juvenile salmonids. Direct use of specific structures can be measured by counting fish within a small radius (1–2 m) of each structure. This method can also be used to determine the relative effectiveness of different types of structures placed in the same stream in providing cover for juvenile fish (O'Neal 2000).

Recommendations for implementation of snorkel surveys

TABLE 1a. — Staff requirements.

Habitat	Staff requirements
Wadable stream	One surveyor (if water clarity allows observation from one bank to the other)
Nonwadable river or stream	Add surveyors as sight distance requires

TABLE 1b. — Water temperature recommendations

Water temperature	Survey timing
Between 10°C and 18°C	Daytime survey
Less than 10°C or greater than 18°C	Night survey (1 h after sunset)

TABLE 1c. — Objective-based recommendations

Objective	Approach
Distribution and average density	One pass, no calibration
Habitat use (baseline)	One pass, no calibration
Species assemblage (diversity)	One pass, no calibration
Relative abundance	One pass with calibration
Total abundance	One pass with controls such as block nets to close the system; mark–recapture

Sampling Design

This protocol addresses snorkel surveys of resident and anadromous salmonids in streams. Before implementing a snorkel survey program, the specific objectives of the study must be identified. Once the research question is clearly defined, the program designer can select the appropriate implementation techniques, equipment, and training methods (Thurow 1994).

Site Selection

The ability to view fish in the water is critical to the survey, so site characteristics must be chosen to facilitate viewing. Reach length to be surveyed is also a concern. A reach is defined as "a section of a stream at least twenty times longer than its average channel width that maintains homogeneous channel morphology, flow, and physical, chemical, and biological characteristics" (Flosi and Reynolds 1994). Within a reach, habitat units are identified as pools and riffles; reaches may or may not contain side channels. As side channels are often used by juvenile salmonids, these areas should be considered for snorkeling if assessing juvenile use of habitats. Generally, a trained snorkeler can survey a maximum of 1.6 km of stream per day, assuming the stream is wadable. In larger rivers, where teams of surveyors are floating downstream, more than 1.6 km may be surveyed. In either case, a reach length with start and end points that can be accessed by surveyors is required. Reach selection for snorkel surveys includes consideration of stream depth and width, velocity, water clarity, and temperature. It is recommended to sample all habitats within a sample reach (not just pools) and report fish observed as fish per square meter of surface area snorkeled.

Several factors can bias results, including the behavior of target fish species and attributes of the physical habitat (e.g., stream size, water clarity, temperature, and cover) (Thurow 1994). Thurow (1994) notes that smaller fish and bottom-dwelling fish that use camouflage are more difficult to count in a snorkel survey. Differences in fish behavior and the amount of cover available may also affect the accuracy of counts (Rodgers et al. 1992; Thurow 1994). Surveyors may misidentify fish, double-count fish, or fail to see all fish. Other limitations of snorkel surveys include the need for estimates of length instead of direct measurements and the inability to weigh or tag fish.

Minimum criteria for depth, temperature, and visibility need to be met for snorkel surveys to be optimal. Surveyors need to be able to submerge a mask to see fish. A minimum recommended water depth for successful surveys is 20 cm. Water temperature influences fish behavior and may bias counts. As temperature falls below 10°C many salmonids will seek cover (Edmundson et al. 1968; Bjornn 1971; Hillman et al. 1992). At water temperatures below 9°C most juvenile salmonids hide during the day and night surveys are likely to be more effective. Hillman et al. (1992) found that above 14°C snorkelers counted about 70% of the juvenile salmonids present; below 14°C they observed less than half of the juvenile fish present. Below 9°C daytime snorkelers observed less than 20% of the juvenile fish present (Dolloff et al. 1993). Water temperature also affects the species composition found within a given habitat.

Visibility or water clarity can severely affect the accuracy of surveys. The minimum recommended visibility for surveys is 1.5 m (P. Roni, NOAA Fisheries, personal communication). Visibility during typical summer base-flow periods is often much greater. Snorkelers can evaluate clarity by using a silhouette of a salmonid with parr marks and spots, as described in Thurow (1994). A surveyor should approach the silhouette until the parr marks are clearly visible and then move away from the silhouette until the marks cannot be distinguished. The average of these two distances provides the water visibility (Thurow 1994). Turbulence may also affect visibility and should be avoided if it affects the surveyor's safety. In small or wadable streams, fish often use turbulence or bubble curtains for cover or have nearby feeding stations along an eddy fence or at the head of a pool. If the surveyor can do so safely, it is recommended to make observations in these areas.

Sample reach selection may also be influenced by the overall sample design. For example, for programs with a probabilistic and spatially balanced sample design, snorkel survey locations may be generated randomly, as in programs such as the U.S. Environmental Protection Agency Environmental Monitoring and Assessment Program (Peck et al. 2003). In these cases snorkelers may be sent to a randomized location and may have to snorkel the most appropriate habitat at that location using some of the guidelines previously recommended.

The selection process for sample units and the number of units that need to be surveyed vary based on study objectives and precision requirements. Sample units used in snorkel surveys range from single habitat units to large sections of a stream or river. Investigators may stratify watersheds into sections and survey units within each section. Streams may also be stratified into habitat units, and abundance can be extrapolated from surveys of a subset of the units (Hankin and Reeves 1988, Thurow 1994). Sampling by habitat type may reduce the variance of the expanded estimate by accounting for the influence of habitat type on fish abundance. Surveys of a single habitat unit or sample reach should be completed within 1 to 2 d to reduce the effect of changes in environmental conditions such as rainfall events (WSRFB 2003).

Dolloff et al. (1993) recommend that 25% of the total habitat units in the sampling universe be sampled to estimate fish populations. For example if 400 total pools are within the sample reach, snorkel surveys should be conducted in every fourth pool, starting with a randomly selected pool from the first four. Surveyors should note when units are unsafe or inaccessible, skip to the next unit of that type, and then return to their original sequence. For stratified samples of habitat units, at least 10 units should be sampled for each habitat type, and 10% of the units sampled should be calibrated using another method, such as electrofishing. Dolloff et al. (1993) provide data from Bull Creek in California. Their findings suggest that when snorkel surveys monitored juvenile steelhead *Oncorhynchus mykiss* abundance once a year within a 2.5 km reach, surveyors detected a 10% change in fish density at a power of 0.8 in 3 years with 17 sample habitats. Adding habitats to increase the sample size to 25 did not significantly improve the power of detection. Decreasing the sample size to 9 required that 5 years of sampling be conducted at each reach to detect a 10% change in fish density at a power of 0.8 (Dolloff et al. 1993).

If sample reaches are to be resurveyed, the reach should be permanently marked using physical objects, such as rebar markers or monuments. Additional methods, including photo points, a reach map, and a global positioning system (GPS) location, should also be used.

For additional insights on sampling design aspects, readers are strongly encouraged to review the applicable sections in the electrofishing protocol by Temple and Pearsons (2006).

Sample Timing and Frequency

Timing of snorkel surveys depends on the objectives of the study and the behavior of the target fish species. If life stage–specific information is desired, timing of the survey must match the use of the surveyed habitat by that life stage. Knowledge of behavior and life history of the target species is essential for effective survey design. Snorkel surveys are most effective during minimal fish migration. The juvenile rearing period of the target species is often the most effective season to obtain data on juvenile populations. The low-flow season is generally selected for summer estimates of population density.

Daytime water visibility is generally best between late morning and early afternoon, when the sun is directly overhead. Cloudy or overcast days are generally better for sampling reaches with significant cover to reduce the dark shadow that may be cast by cover elements. A small waterproof light may be useful to search for fish in dark conditions or shady areas. If criteria for depth, water clarity, and temperature are met, direct sunlight may be less of a critical factor.

Night surveys may be more effective for studying salmonids under certain conditions. During winter and when water temperatures fall below 10°C, night surveys are generally considered to have greater effectiveness. In addition, in any habitats in which temperatures are greater than 18°C, night surveys should be considered (Roni, personal communication). Night surveys in the winter months often have better results than daytime surveys (Campbell and Neuner 1985; Goetz 1997). Surveys should be conducted starting no sooner than 1 h after sunset to allow fish to emerge from hiding. If night survey data are to be compared, the surveys should be conducted during the same moon phase to avoid bias due to the effects of moonlight on fish behavior.

Field/Office Methods

Basic Survey Procedures

For preselected sample reaches, field crews should use GPS units and maps, aerial photos, or landmarks such as road bridges or trail crossings to find the sample reach location (Rodgers 2002). If sample reaches are on private land, landowner permission must be granted before the reach is accessed (Rodgers 2003).

Before conducting a field survey, available information should be collected on species that are likely to be encountered. This effort should include information requests to local, state, or federal resource management agencies that may have fish distribution information or jurisdiction over aquatic species. Where snorkelers expect to find protected species or species listed under the Endangered Species Act, they should avoid fish spawning areas. Snorkelers should not touch or otherwise disturb protected fish while conducting surveys.

The sampling team should record the reach location information, weather conditions, water temperature, and water visibility at the start of the day and several times throughout the day. Each unit sampled should be classified by type and assigned a unique number (Dolloff et al. 1993). Rodgers (2002) recommends ranking turbidity at each reach surveyed. Data can then be segregated for precision based on visibility.

Considerations in field sampling include the size of the water body, which will affect the number of snorkelers needed for the survey and the direction of the survey, and the objective of the study. Surveying in the upstream direction is the most effective technique for small streams where this is feasible (Figure 1). Most salmonids hold position facing the current, so upstream sampling is less likely to cause disturbance (Thurow 1994). Some salmonids such as steelhead tend to hold close to the bottom of the stream and quickly seek cover if they detect a disturbance. Upon entering the water, observers should survey the bottom for any trout that may quickly take cover, as opposed to starting with other species, such as coho salmon *O. kisutch*, that are more resistant to disturbance (Dolloff et al. 1993). Movements should be controlled and sudden movements should be avoided to reduce disturbance to fish. Fish are counted as the snorkeler passes them to reduce double counting. In larger systems, upstream counts may not be possible; in these cases snorkelers may conduct counts while floating downstream and remaining as motionless as possible (although some swimming may be required to maintain lane position relative to other snorkelers).

FIGURE 1. — Snorkel survey conducted in the upstream direction in a small stream.

The number of observers needed to complete the survey is dependent upon water clarity, the size of the stream or river system, and the objectives of the study. Thurow (1994) recommends using enough snorkelers to complete the survey in a single pass. It is recommended that one surveyor be used for streams less than 2 m bank-full width, two surveyors be used for 2–5 m bank-full width, and more than two surveyors should be used in places greater than 5 m bankfull width. During the survey, any impediments to observation, such as deep water, turbidity, or significant cover elements, should be noted (Dolloff et al., 1993).

For smaller streams in which one observer can see the entire channel width, Thurow (1994) recommends proceeding in a zigzag pattern across the channel to see both sides, especially in the margins of the streams where small fish will hold in slow water areas. Other areas that deserve intense scrutiny are eddies behind logs and boulders and underneath logjams and under cut banks (see Figure 2). For these surveys, the surveyor generally proceeds upstream, against the current.

FIGURE 2. — Snorkeler investigating area under logjam.

If two observers are employed (due to stream/river size, water clarity, or habitat complexity), the observers should remain adjacent to one another and move at the same speed. These surveys are generally conducted in the downstream direction with the current, due to the size of the river. Observers should mentally divide the channel into "lanes" in which each observer will look for fish and only count fish that pass within his/her snorkeling lane (Thurow 1994). If two observers are used, each surveyor should be in the middle of the channel looking towards the bank to cover his/her lane. If more than two surveyors are used, middle lanes will have to be designated. Diagrams of these configurations are shown in Thurow (1994). In these situations, snorkelers should frequently check on each other for safety and to maintain position.

If a downstream survey is required, a length of polyvinyl chloride (PVC) pipe can be held between observers to maintain distance, but the distance should never be greater than the water visibility. If rocks, debris, or variable current conditions are present in the channel, it may not be feasible to use PVC pipe, and visual distances between observers should be maintained. If visibility is limited, observers may subsample a portion of the channel and use the estimate to extrapolate an estimate for the entire channel (Thurow 1994).

Measurement Details

Data on fish species and size-class can be collected and recorded on a PVC cuff or diver's slate (see Figure 3) and later transferred to a data sheet or database. Data can also be called out to an onshore observer who records the information on a data sheet or in a digital data collector. After completing the count of fish, observers should measure the surface area of the snorkeled unit. The total length of the unit should be recorded as well as the width at three or more equally spaced intervals. Maximum pool depth should be recorded (Rodgers 2002) if additional habitat surveys are not being conducted. The surface area can be calculated by multiplying reach length by average width or by summing the surface area of individual habitat units, which the crew can measure as it moves upstream. Density of fish is typically expressed as the number of fish per square meter (Thurow 1994). An example of a field form for data collection is offered in Appendix A.

Information on reach location, water temperature, weather conditions, and water clarity should be recorded for each survey (Washington Salmon Recovery Funding Board 2004). Additionally, the staff collecting the data should be identified on the data sheet.

Data Handling, Analysis, and Reporting

Record Keeping

Snorkel surveys require special considerations for record keeping. Fish counts, species identification, and lengths can be recorded directly by a diver or communicated to an assistant on shore. Preformatted data sheets should always be used to ensure that all pertinent information is collected and that it is in a standard format for data entry when the time comes (interpreting nonstandardized field notebooks or PVC cuffs months after the fact is problematic). Waterproof slates or cuffs are likely the most popular method for recording data during a survey. Slates can be made of plastic, Formica board, or Plexiglas, and cuffs are often fashioned from PVC pipe and surgical tubing. Electronic data recording devices are also used but are more costly (Dolloff et al. 1996).

FIGURE 3.—Diving slate for recording data underwater.

Mark–recapture estimates

If sample sizes are sufficient, population estimates are calculated for each size class using Chapman's modification of the Peterson mark–recapture technique (Ricker 1975) as shown in equation 1.

$$N = (M+1)(C+1) / (R+1) \qquad \text{(eq 1)}$$

where M = Number marked
where C = Number captured (observed)
where R = Number recaptured (observed)
where N = Population estimate

Summing the estimates for each size-class results in a total population estimate. Ricker (1975) lists the formulas for calculating confidence intervals around the estimate.

Accuracy and Precision Considerations

Accuracy of snorkel surveys can be estimated by replicating the surveys or using other methods to calibrate the surveys (Northcote and Wilkie 1963; Zubik et al. 1988; Rodgers et al. 1992; Thurow 1994). Replication of snorkel surveys should be done after fish have had time to recover from disturbance but before conditions that would affect the survey results (such as light levels or turbidity) change. Calibration of team member estimates offers a good opportunity to resample the same habitat unit to determine the precision of the survey method (Thurow 1994). Data in Thurow (1994) show close alignment of snorkeling estimates from two different snorkelers at 22 reaches on the South Fork Salmon River.

Since variation among repeated counts is small, snorkel survey data can be readily compared, but the question remains: how accurately does the survey data reflect the true abundance of the sample unit? Rodgers et al. (1992) suggest that snorkel survey data should be calibrated with other methods because accuracy varies by sample reach and sampling conditions. Accuracy of snorkel counts is 90% for juvenile salmon in clusters of fewer than 40 fish, but coho salmon and chinook salmon *O. tshawytscha* were undercounted when in mixed groups of fish species (Hillman et al. 1992). Small steelhead (less than 100 mm) were undercounted by

40% in water less than 15 cm in depth and in areas with abundant cover (Hillman et al. 1992). Of 13 studies reviewed by Thurow (1994), 11 produced snorkel estimates within 70% of other methods; results, however, varied by species, life stage, and sample reach conditions such as water temperature and surveyor experience (Hillman et al. 1992). Dolloff and Owen (1991) found that calibrated snorkel counts were more accurate than electrofishing results for species that maintain position in the water column and are easily seen, such as trout. Other species, such as sculpin and darters, which are cryptically colored or hide in the crevices of the streambed, are more difficult for observers to count during snorkeling (Dolloff et al. 1993).

It is recommended that snorkel surveys be conducted using standard procedures and calibrated periodically with estimates from other methods. Methods for calculating population estimates using calibrated snorkel surveys are detailed in Dolloff et al. (1993).

Personnel Requirements and Training

Safety and training

As with any field activity, safety is of paramount importance. Swift rivers, cold temperatures, poor visibility, contaminants, and other environmental factors may affect divers conducting snorkel surveys. Surveyors should always assess the potential hazards of the reach before entering the water (Dolloff et al., 1996). In addition, a health and safety plan should be developed for any surveys of this type in which a risk assessment is conducted and for which appropriate countermeasures are identified and implemented during the survey.

Several risk issues and recommendations are offered here:

1. Never attach ropes or lines to divers in areas where currents or tidal action are factors.
2. Divers should avoid areas of extreme water velocity and turbulence because appendages and/or equipment may become wedged against rocks or debris.
3. Always enter and leave the study area from low-velocity waters (Dolloff et al. 1996).
4. If turbulent stream reaches are within the survey area, surveys should be completed along the channel margins and the most turbulent areas should be avoided. If a snorkeler becomes caught in strong turbulent flow, the safest position is to float feet first and steer by "back paddling" with the arms. The surveyor should not try to stand in swift water because feet or legs could be caught or entrapped on the bottom and the current can force the surveyor underwater.
5. If debris jams are surveyed, snorkelers should remain alert to ensure that they do not become entangled in debris (Thurow 1994). For safety and counting accuracy, debris jams are best approached from downstream where there is frequently an eddy and the snorkeler can make a slow, controlled, safe approach.

6. Hypothermia is one of the most common hazards for snorkel surveys. Divers in all conditions are at risk, although night surveys and surveys conducted during low temperatures pose a greater risk (Dolloff et al. 1996).

Basic safety in snorkel surveys requires that surveyors are comfortable in water and can effectively use a snorkel and mask. Surveyors should be instructed on how to use and clear their snorkels before beginning any survey. Adequate insulation in the form of a wet or dry suit should be used depending on water temperature and season of survey, and whether the survey is conducted during the day or at night. For night surveys, handheld dive lights should be used for safety and to view fish species. Crew members should all be trained in cardiopulmonary resuscitation (CPR) and first aid, with an emphasis on recognizing and treating hypothermia. Swift-water rescue training would also be useful for crew members working in larger river systems.

Identification and length estimation

Several identification keys are helpful for correct identification of juvenile and adult fish observed during snorkel surveys (Carl et al. 1959; McConnell and Snyder 1972; Martinez 1984; Thurow 1994; Pollard et al. 1997). There are limits on the ability to identify very small fish accurately, but fish as small as 25 mm can often be identified correctly in ideal conditions, so limits to young of the year (Thurow 1994) seem unnecessary (Roni, personal communication).

Training and practice are required to identify fish correctly underwater, to estimate fish sizes accurately, and to complete precise species counts. Annual training should be required for all surveyors. Surveyors should clearly understand the objectives of the study and review a standard protocol before beginning practice sessions. Practice sessions should be conducted at reaches similar to those that will be encountered by snorkelers during the field surveys. Snorkelers should practice identifying, counting, and estimating the size of target species. Fish identification practice can be done using known species in live cages or by having the instructor point to a fish to be identified and estimated by the staff member and then comparing the results with the instructor (Thurow 1994).

It is important to note that objects viewed underwater are magnified 1.3 times. A calibration technique for length estimates, such as marking centimeters on the dive slate or PVC pipe used to record data as a comparison, can be helpful. Snorkelers can also carry a ruler or use a known distance such as index-finger-to-thumb for reference. It is recommended that a measuring tape be fixed to the diver's glove to use as a calibration tool for length measurements (Roni, personal communication) (Swenson et al. 1988). Snorkelers should practice estimating the length of objects or fish of known size before conducting surveys. For example, measured wooden dowels can be used as training aids for learning to estimate length underwater (Roni, personal communication). Blocks of wood or plastic can be cut into fish shapes at varying lengths (50 mm, 100 mm, and 150 mm) and carried with the team to be used for frequent quick calibration and for calibration at varying distances. Training can greatly improve surveyors' accuracy in length estimation. One hour of practice allowed observers to increase their percentage of accurate (+/−25 mm) estimates from 62% to 90% (Griffith 1981).

Surveyors should be familiar with the size of sampling units to be surveyed and the method to be used to estimate fish abundance. The selection of sample units is dependent on the study objectives and the habitat conditions in the stream. Training should provide snorkelers with several opportunities to count the total number of fish in a sample unit; results should be compared across surveyors. Sampling units that contain known numbers and sizes of fish are ideal for training.

Operational Requirements

Operational requirements for snorkel surveys will vary based on the width of the habitat to be surveyed and conditions during the survey. Larger streams will need more than one surveyor, while smaller streams can be surveyed with one observer and an assistant on shore.

Equipment

Daytime surveys (water temperature greater than 9°C).

- Full neoprene wet suit (6.4 mm) or dry suit with knee pads, preferably black or blue or brown
- Wet suit hood
- Neoprene gloves
- Neoprene socks
- Wading boots with felt soles
- Fins (can be useful in large systems when surveys must be conducted downstream)
- Mask (can be worn over contact lenses or ordered with a prescription lens) with side window to increase visibility
- Snorkel
- Extra mask and snorkel for remote reaches
- Data recorders such as a dive slate scroll (Ogden 1977) or a plastic cuff (10-cm PVC pipe in 20-cm lengths will work and can be secured with surgical tubing [Helfman 1983])
- Grease pencil
- Knee and elbow pads (for turbulent or shallow streams)
- Wet suit cement or Aquaseal® for repairing suits
- Thermometer
- Small halogen light
- Canteen
- Food
- First-aid kit
- Mobile phone and/or radio for emergency contact
- Data forms
- Measuring tape
- Flagging

Nighttime surveys or daytime surveys where water temperature is less than 10°C. Everything listed above and the following

- Dry suit (required) (should not have valves if possible)
- Adequate insulation under dry suit (one-piece fleece suits are available at dive shops, or layers of wool or synthetic fabrics can be used)
- Handheld halogen dive lights (a red filter can be helpful for reducing fish disturbance; filters can be made from red Plexiglas) (Thurow 1994).

Budget Requirements

Hankin and Reeves (1988) compared the cost effectiveness of using snorkel surveys calibrated by multiple pass removal electrofishing against electrofishing alone and found that for the same cost, the combination of snorkel surveys and electrofishing was 1.7–3.3 times more accurate than electrofishing alone (Dolloff et al. 1993). They attributed their results to the high cost of electrofishing. Although fish counts by divers may be less accurate than estimates based on depletion electrofishing, snorkelers can move faster and can examine more habitat units in a given time period (Dolloff et al. 1993). Cost optimization methods are explained in detail in Dolloff et al. (1993).

Literature Cited

Bjornn, T. C. 1971. Trout and salmon movements in two Idaho streams as related to temperature, food, streamflow, cover, and population density. Transactions of the American Fisheries Society 11:324–438.

Carl, G. C., W. A. Clemens, and C. C. Lindsey. 1959. The freshwater fishes of British Columbia. Handbook 5. British Columbia Provincial Museum, Department of Education, Victoria.

Campbell, R. F. and J. H. Neuner. 1985. Seasonal and Diurnal shifts in habitat utilized by rainbow trout in western Washington Casade mountain streams. In F. W. Olson, R. G. White and R. H. Hamre (eds), Proc. of the Symp. on Small Hydropower and Fisheries, American Fisheries Society, Bethesda, Maryland. 39-48.

Dolloff, A. C, and M. D. Owen. 1991. Comparison of aquatic habitat survey and fish population estimation techniques for a drainage basin on the Blue Ridge Parkway, Completion Report. U.S. Department of the Interior, National Park Service, Cooperative Agreement CA-5000-3-8007, Blacksburg, Virginia.

Dolloff, C. A., D. G. Hankin, and G. H. Reeves. 1993. Basinwide estimation of habitat and fish populations in streams. U.S. Forest Service, Southeastern Forest Experiment Station, General Technical Report SE-GTR-83, Asheville, North Carolina.

Dolloff, C. A., D. G. Hankin, and G. H. Reeves. 1993. Basinwide estimation of habitat and fish populations in streams. U.S. Forest Service, Southeastern Forest Experiment Station, General Technical Report SE-GTR-83, Asheville, North Carolina.

Dolloff, A., J. Kershner, and R. Thurow. 1996. Underwater observation. Pages 533–554 in B. R. Murphy and D. W. Willis, editors. Fisheries techniques, 2nd edition. American Fisheries Society, Bethesda Maryland.

Edmundson, E., F. H. Everest, and D. W. Chapman. 1968. Permanence of station in juvenile chinook salmon and steelhead trout. Journal of the Fisheries Research Board of Canada 25:1453–1464.

Fausch, K. D., and R. J. White. 1981. Competition between brook trout (Salvelinus fontinalis) and brown trout *(Salmo trutta)* for positions in a Michigan stream. Canadian Journal of Fisheries and Aquatic Sciences. 38:1220–1227.

Flosi, G., and F. L. Reynolds. 1994. California salmonid stream habitat restoration manual. California Department of Fish and Game, Technical Report, Sacramento.

Goetz, F. 1997. Diel behavior of juvenile bull trout and its influence on selection of appropriate sampling techniques. Pages 387–402 in Mackay, W. C., M. K. Brewin, and M. Monita (eds.). Friends of the bull trout conference proceedings. Bull Trout Task Force (Alberta), c/o Trout Unlimited Canada, Calgary.

Griffith, J. S. 1981. Estimation of the age-frequency distribution of stream-dwelling trout by underwater observation. Progressive Fish Culturalist 43:51–53.

Griffith, J. S., D. J. Schill, and R. E. Gresswell. 1984. Underwater observation as a technique for assessing fish abundance in large rivers. Proceedings of the Western Association of Fish and Wildlife Agencies 63:143–149.

Hankin D. G., and G. H. Reeves. 1988. Estimating total fish abundance and total habitat area in small streams based on visual estimation methods. Canadian Journal of Fisheries and Aquatic Sciences. 45:834–844.

Helfman, G. S. 1983. Underwater methods. Pages 349–370 in L. A. Neilson and D. L. Johnson, editors. Fisheries techniques. American Fisheries Society, Bethesda, Maryland.

Hillman, T. W., J. W. Mullan, and J. S. Griffith. 1992. Accuracy of underwater counts of juvenile chinook salmon, coho salmon, and steelhead. North American Journal of Fisheries Management 12:598–603.

Martinez, A. M. 1984. Identification of brook, brown, rainbow, and cutthroat trout larvae. Transactions of the American Fisheries Society 113:252–259.

Mc Connell, R. J., and G. R. Snyder. 1972. Key to field identification of anadromous juvenile salmonids in the Pacific Northwest. NOAA Technical Report NMFS Circular 366.

Mueller, K. W., D. P. Rothaus, and K. L. Fresh. 2001. Underwater methods for sampling the distribution and abundance of smallmouth bass in Lake Washington and Lake Union. Washington Department of Fish and Wildlife, Fish Management Program, Technical Report # FPTO1-17, Olympia.

Northcote, T. G., and D. W. Wilkie. 1963. Underwater census of stream fish populations. Transactions of the American Fisheries Society 92:146–151.

Ogden, J. C. 1977. A scroll apparatus for the recording of notes and observations underwater. Marine Technology Society Journal 11:13–14.

O'Neal, J. S. 2000. Biological evaluation of stream enhancement: a comparison of large woody debris and an engineered alternative. University of Washington, Masters Thesis, Seattle, Washington 103 p.

Peck, D. V., J. M. Lazorchak, and D. J. Klemm. 2003. Environmental Monitoring and Assessment Program: Surface Waters–Western Pilot Study Operations Manual for Wadeable Streams. U.S. Environmental Protection Agency, Corvallis, Oregon.

Pollard, W. R., G. F. Hartman, C. Groot, and P. Edgell. 1997. Field identification of coastal juvenile salmonids. Harbour Publishing, Madeira Park, B.C.

Ricker, W. E. 1975. Computation and Interpretation of Biological Statistics of Fish Populations. Bull. Fish. Res. Board. Can. 191:1–382.

Rodgers, J. 2002. Abundance monitoring of juvenile salmonids in Oregon coastal streams, 2001. The Oregon Plan for Salmon and Watersheds: Report No. OPSW-ODFW-2002-1. Oregon Department of Fish and Wildlife, Corvallis.

Rodgers, J. 2003. Protocols for conducting Oregon Plan surveys of juvenile salmonid in Oregon coastal streams. Oregon Department of Fish and Wildlife, Corvallis.

Rodgers, J. D., M. F. Solazzi, S. L. Johnson, and M. A. Buckman. 1992. Comparison of three techniques to estimate juvenile coho salmon populations in small streams. North American Journal of Fisheries Management 12:79–86.

Schill, D. J., and J. S. Griffith. 1984. Use of underwater observations to estimate cutthroat trout abundance in the Yellowstone River. North American Journal of Fisheries Management 4:479–487.

Slaney, P. A., and A. D. Martin. 1987. Accuracy of underwater census of trout populations in a large stream in British Columbia. North American Journal of Fisheries Management 7:117–122.

Swenson, W. A., W. P. Gobin, and T. D. Simonson. 1988. Calibrated mask bar for underwater measurement of fish. North American Journal of Fisheries Management 8:382–385.

Temple, G. M., and T. N. Pearsons. 2007. Electrofishing: backpack and drift boat. Pages 95–132 in D. H. Johnson, B. M. Shrier, J. S. O'Neal, J. A. Knutzen, X. Augerot, T. A. O'Neil, and T. N. Pearsons. Salmonid field protocols handbook—techniques for assessing status and trends in salmon and trout populations. American Fisheries Society, Bethesda, Maryland.

Thurow, R. F. 1994. Underwater methods for study of salmonids in the Intermountain West. U.S. Forest Service, Intermountain Research Station, General Technical Report INT-GTR-307, Odgen, Utah.

Washington Salmon Recovery Funding Board. 2004. Protocol for monitoring the effectiveness of fish passage projects. Washington Salmon Recovery Funding Board, Olympia.

Zubik, R. J., and J. J. Fraley. 1988. Comparison of snorkel and mark–recapture estimates for trout populations in large streams. North American Journal of Fisheries Management 8:58–62.

Appendix A: Snorkeling Survey Data Sheet

Station
Date
Reach ID
Visit #
Team ID
Reach Name

Time Start End
Snorkeler
Weather

Species	0 to 50 mm	51 to 80	80 to 100	100 to 120	120 to 150	150 -200	200 - 250
Chinook							
Coho							
Steelhead							
Rainbow							
Cutthroat							
Cut-bow							
Sculpin							

Turbidity   Reach Area (m^2)

0 = not snorkelable due to high turbidity or hiding cover
1 = high amount of hiding cover and/ or poor water clarity
2 = moderate hiding cover and/or moderate water clarity
3 = little hiding cover and good water clarity
Comments

Tangle Nets
Charmane E. Ashbrook, Kyong W. Hassel, James F. Dixon, and Annette Hoffmann

Background and Objectives

Background
Tangle net (or tooth net) research first came about in the U.S. Pacific Northwest and Canada as a way to evaluate whether a conservation goal could be achieved for a salmon species or stock of concern. Specifically, researchers wanted to determine the survival rates of nontarget salmon captured and released from tangle nets. This knowledge—whether the survival rate is acceptable—enables fish managers to allow commercial "selective fishing" while protecting weak stocks. Commercial selective fishing is deemed successful if conservation goals are achieved for the species or stock of concern and if a harvest goal is met to make the fishery economically viable.

Work done by Vander Haegen et al. (2002, 2004) compared tangle nets with traditional gill nets and showed that the use of a tangle net, with careful handling techniques, can reduce spring chinook salmon *Oncorhynchus tshawytscha* mortality. The handling techniques included an abbreviated soak time, a shorter net, and meticulous removal of fish from the net. They also included the use of a revival box, which Farrell et al. (2001a) showed could reduce the physiologic stress on coho following their capture in a gill net. Research by Ashbrook et al. (2004a, 2004b) has provided further information for estimating survival of bycatch following their release from tangle nets. Consequently, Columbia River fishery managers have instituted selective tangle net fisheries for upriver spring chinook salmon. In addition to their use in selective fishing studies, applications for tangle nets include capturing fish to apply tags, and for broodstock and biosample collection.

Rationale
Selective capture and subsequent release of nontarget bycatch is possible because the tangle net can efficiently capture salmonids in large rivers and estuaries in short time periods with low immediate mortality rates and relatively low postrelease mortality rates (Vander Haegen et al. 2002 and 2004; Ashbrook et al. 2004). Experienced gill-net fishermen can transition easily to tangle nets, which operate similarly. Tangle nets are visually comparable to gill nets (see Figure 1), and the two gears are fished in the same manner; however, the mesh of the tangle net is smaller than that of a conventional gill net, which results in the fish being caught by the snout or teeth. Ideally, the smaller mesh size increases the chance for fish to continue respiring in the net so they can be released live.

Although different species can be easily sorted, stock-selective fisheries in the U.S. Pacific Northwest and Canada rely on a physical mark—in most cases, adipose fin excision—that allows fishers to distinguish easily between stocks of hatchery-origin fish (which can be retained) and unmarked naturally spawning stocks (which must be released).

TANGLE NETS

FIGURE 1. — A gill net (left) alongside a tangle net.

Objectives

Tangle nets enable practitioners to capture a representative sample of fish to assess survival, to tag, or to collect biological data. Regardless of the goal of tangle net fishing, it is important for fishers to release captured fish in a manner that allows for a high postrelease survival rate. The use of tangle nets also gives scientists a sampling of the percentage of hatchery versus wild-origin fish as they are entering the river system or are in its lower reaches.

Sampling Design

Site selection

Tangle net fishing may be conducted in any area where gill-net fishing is suitable, and in a variety of habitats. Timing depends on the species, life stage, and population(s) that are targeted. In general, the environment is most favorable to tangle net capture and release when water temperatures are relatively cool. Fishing in poorer conditions (e.g., among predators, in waters with relatively high velocity or warm temperatures) is expected to increase postrelease mortality. Water flow and depth as well as snags on the river bottom may affect the use of a tangle net. Adult salmonids may be captured in habitats adjacent to and along migratory routes to spawning grounds. Populations sampled may include visually marked and unmarked populations of the same species as well as other fish species (including salmonids). As fish size varies, a tangle net for one species may act as a gill net for another species, and therefore, practitioners must choose the mesh size carefully.

Sampling frequency

Depending on the purpose of tangle-net fishing and the target species, fishing can occur at or below spawning areas. To capture fish for biological samples or to assess abundance or diversity, one or many sets may be employed, either on one day or throughout a longer period, and this could occur over a wide range of habitats or areas. In most instances, fish should be captured and evaluated in proportion to their abundance. For salmon, this would mean fishing with the same effort throughout the time-frame of their spawning run. The number of sampling sites will vary depending on the question the researcher is trying to answer. For example, to evaluate selective fishing ideally, practitioners should fish in typical fishing areas and during typical fishing seasons to capture the targeted adults. To evaluate the use of a tangle net and estimate survival with jaw tags, approximately 1,000 fish need to be captured per treatment group each year. In our project, the jaw tag recovery rate (which allowed estimation of postrelease survival) ranged from 10% to 20%. If a tag that has greater detection (e.g., a passive integrated transponder [PIT] tag) or recovery can be used, fewer fish will need to be tagged. To calculate survival estimates with the tangle net, sampling over multiple years is most useful because it incorporates year-to-year variation.

Field/Office Methods

Setup

Permits

Permitting requirements may take a few years if a research permit is not already in place. In the event that a research permit has been acquired for an existing study, it is more efficient to add the tangle-net study onto the existing research permit. In the United States, if there is a possibility that species listed under the Endangered Species Act (ESA) may be captured, setup preparations will require obtaining a research take permit under the ESA.

Contracting fishers and setting up nets

Contracting local gill-net fishers who have years of experience fishing for the target species is ideal because these fishers know the most effective manner, location, and times for fishing. Furthermore, in the case of a selective fishing study, the fishers can be asked to mimic the fishery both in manner and location fished. It is recommended that researchers initiate a competitive bid process, which takes into consideration fees and fishing experience. Setting up a contract can take several months. The process includes the following steps: publicize and send out a request for bids, evaluate the bids based on amount as well as the extent to which applicants meet the minimum qualifications for experience and insurance, and contact the fishers and have them send in a signed contract and proof of insurance.

When the fishing contracts are in place, researchers can make arrangements for the fishing sites and meeting locations. Discussions with the fishers will be helpful in choosing the appropriate mesh and net sizes. Finally, regarding net acquisition and preparation, often it is more efficient and less expensive to have the fishers order the nets and hang them on the cork line (hanging the net with floats and weights) than to do this in a separate process.

- For capturing adult spring chinook salmon and releasing those without clipped adipose fins, we recommend using an 11.4-cm (4.5-in), four-strand, 1.5-mm multifilament mesh net that is hung at a ratio of 2:1. The hang ratio describes the length of mesh relative to the length of cork line. This compares to the conventional spring chinook salmon gill net of 20.3-

cm (8-in), monofilament single-strand mesh hung at a ratio of 2:1. We used a net with a total length of 275 m (150 fathoms).

- For capturing coho salmon and releasing unmarked coho salmon and chinook salmon, we recommend a 9-cm (3.5-in), four-strand, 1.5-mm multifilament mesh that is hung at a ratio of 3:1. This compares to the conventional coho salmon gill net of 14.6-cm (5.75-in), monofilament single-strand mesh hung at a ratio of 2:1. We used a net with a total length of 295 m (160 fathoms).

Diver nets, which sink and follow the bottom contours, may be used in addition or as an alternative to floating nets, which remain at the surface. For comparing two nets, the nets may be shackled together to form a complete panel that is 275 m in length. Gear types should be of similar depth. For example, for capturing coho salmon, we used a tangle net that was 74 meshes deep shackled to a gill net that was 45 meshes deep to ensure that both nets fished at the same depth. The depth and color of the nets should be suitable to the area fished.

Preparations for using a control
Preseason preparations can vary if a control is used and depending on the control. If a fish trap that is located within a dam is used (e.g., the adult fish facility in the Bonneville Dam, Columbia River), a request will need to be made to the facility managers. This can entail filling out forms, obtaining signatures, fulfilling safety courses, getting passes and keys for each employee, learning how to operate the trap, and coordinating research schedules with the other research that is taking place at the trap site. If another net type such as a beach or purse seine is used, fishers contracted must be competent with this gear.

If the control fish need to be released near the fish captured in the tangle net, researchers need to consider obtaining an oxygenated fish transport tank, a vehicle to pull the tank, oxygen tanks, and a means of filling the tank with water and hiring additional staff.

Field office
If the fishing location is more than 2 h away from the office, setting up a field station may be necessary; this process takes around 2 d. A field station provides a place to communicate with the headquarters, enter data, and store gear. It can also serve as a place to sleep and make meals if the project duration is too short to rent accommodations or if hotel arrangements are not possible. Among the elements to include are a computer with e-mail capability, a phone, a desk, and office supplies. A dry-erase board can be used to list out the sampling schedule and indicate changes that occur on a daily basis. The field station may also house an area for preparing tagging equipment and drying out rain gear.

Hiring staff
For each boat, one to three researchers will be needed to collect data and tag fish. For studies in which two or more boats fish simultaneously, temporary staff is warranted. Advertising for temporary staff, interviews, hiring, and training can take a few months. Furthermore, each staff member will require safety and first-aid training as well as rain gear.

Tags and biosamples

If fish are to be tagged, the appropriate tag needs to be identified and ordered. Each tagging process has strengths and weaknesses; choosing the tag that best suits the study is an important decision. Some tags, such as telemetry tags, require that receivers (which pick up the signal) be placed along the banks of the river. Use of these tags also necessitates coordinating with other researchers who are using telemetry in the area. If PIT tags are used, detection areas and handheld detectors are required.

Other tags, such as jaw and Floy tags, are less labor-intensive and less costly. Jaw tags, which consist of a piece of copper wire with a colored plastic sheath and individual alphabet-number combination, enable the researcher to connect the recovered tag back to the individual fish. Furthermore, different sheath colors can be used to differentiate study groups or treatment and control groups. If the study depends on tag recovery from recreational fishers, we recommend creating and putting up posters that request recovery information, provide a toll-free telephone number, and offer a lottery-drawn cash prize as incentive.

When collecting biosamples such as tissue, blood, or scales, we recommend developing and rehearsing a quick and efficient method before the study begins. Consider in advance the fact that inclement weather can frustrate tagging and biosampling efforts on the boat.

Database and gear preparation

Prior to the field season, a database and data sheets should be created using a program such as Microsoft Access. The datasheets can be printed onto waterproof paper and multiple copies made. (See the appendices for examples.)

For each vessel used, a sampling tote can be prepared that includes a first-aid kit, a global positioning system (GPS), batteries, headlamps for night fishing, pencils, data sheets, a clipboard, and a thermometer. Each vessel also requires a revival box. For a multiple-year study, the revival box made in the first year can be used during the successive years.

Events sequence

The night before or on the morning of each field day, researchers prepare a sampling tote that includes a first aid kit, GPS unit, thermometer, data sheets, tags and tagging equipment, pencils, and plastic bags. The researchers meet the fisherman at a designated location. One researcher primarily handles the fish while the other primarily collects and records information on the data sheets. If multiple tags are applied—such as a jaw tag, a PIT tag, and a telemetry tag—a total of three researchers may be needed. The fisher drives to the fishing site and deploys the net. As this occurs, the data researcher collects the GPS reading and notes on the data sheet the time the net was deployed, GPS information, date, time, presence or absence of pinnipeds, temperature, first net out (if more than one net is deployed), and weather condition and the fisher's name, boat name, and names of researchers for that day. A large plastic tote is filled with water. Just prior to bringing in the net, the pump that operates the revival box is started.

The fishing vessel is equipped with a hydraulic reel mounted in the bow that deploys and retrieves the net(s). The nets are set by reeling them across the river in a curved pattern and allowing both ends to drift freely. Initially, as they remove fish from the net, fishers are instructed on proper fish handling, particularly to avoid

TANGLE NETS

touching the gill area or holding a fish by its caudal peduncle. When possible, fishers look over the bow as the net is pulled up so that they can lift fish over the roller. If two nets are shackled together, the fisher—if at all feasible—alternates the end of the net that is closest to the shore on subsequent sets so that the fishing effort of each net type is similar.

Depending on the species, fishing location, and safety conditions, fishing may take place during the day or night. In an estuarine area, fishing typically occurs during daylight and at optimal tides for catching fish. Vessels may also be fished simultaneously and proximately to mimic a commercial fishery. In a fishery that captures fish live, the amount of time the tangle nets are left to drift is less than it is in a traditional gill-net fishery. For example, a tangle net fishery deploys nets for a number of minutes where a traditional gill-net fishery deploys nets for a number of hours. During our studies, the time the first cork is deployed until the last cork is pulled in ranged from about 20 min to an hour, depending on the number of fish captured and their condition. If fish are to be released, fish must be carefully handled and revived.

To further ensure fish survival following release, all vessels are to be equipped with a revival box similar to that described by Farrell et al. (2001a). The boxes can be made from 2-cm thick plywood painted black or from stainless steel (see Figure 2) and contain two compartments for holding fish. Each compartment is about 107 cm long, 41 cm high, and 19 cm wide. The compartments of the revival box are wide enough to allow a chinook salmon to fit with its head facing the fresh water flow but narrow enough to prevent the fish from turning around. A 12 V, 240 L/min submersible bilge pump or a 5-cm gas-powered Honda water pump is connected to a discharge hose that supplies fresh water through pipes located at the front and near the bottom of the box. The front panels of the box where the tubes are attached for water flow are constructed to slide vertically. Lifting the panels provides a water slide so that fish can be released in a stream of water into the river. Overflow outlets are located at the opposite end of the revival box.

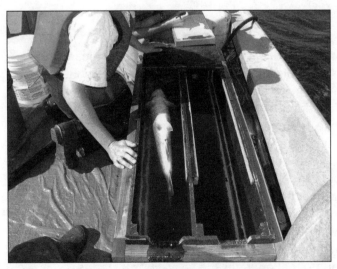

FIGURE 2. — Adult salmon in revival box.

Fish captured by tangle net

As fish are brought in, the fisher extracts them from the net (see Figure 3) and places them into a large plastic tote box (see Figure 4). Data collection then takes place (see pages 348–349). The researcher handling the fish then returns the fish to the water, taking care to release the fish so that it is not recaptured by the net. Any fish that die are checked for a coded wire tag with a detection wand and donated to a public food bank.

Fish captured as a control to the tangle net (i.e., for evaluating survival)

Fish may be collected by a trap or purse or beach seine. They are placed individually into a tank and their condition is evaluated, length and distinguishing marks are recorded, and tags are applied as described above. Depending on the study design, control fish are either returned to the water from where they were captured or transported to a release site near the fishing location. (See Appendices A–B for field forms.)

FIGURE 3. — Deployed tangle net with fisher retrieving net. The fisher occasionally looks over the bow for additional fish.

FIGURE 4. — After the fisher extracts the fish from the tangle net, he places it in a tote of water. The researcher evaluates the condition of the fish while carefully holding it with both hands.

Data collection

This section describes the data collection technique used to evaluate survival (see Appendices A–B for field forms). For other studies, some of these steps may not be pertinent. A series of mark–recapture experiments is used to estimate the survival of adult salmon captured and released from tangle nets. To compare different sized nets, the following elements are examined: (a) immediate survival; (b) catch-per-unit-effort; (c) species composition; and, when possible, (d) long-term survival. To obtain this information two research personnel are on board each vessel during test fishing; one person primarily records data while the other monitors, handles, and tags fish. For each set (one deployment and retrieval of the net), personnel record the time when the first part of the net is placed in and removed from the water, the time the end of the net is brought on board, and the longitude and the latitude for the set (using a handheld GPS unit). If two nets are shackled together, observers also record the time the shackle between the two nets is removed from the water, which net type is put in the water first, and which net type is removed from the water first. Personnel also record the date, skipper's name, boat name, personnel names, set number, weather conditions, water and surface temperatures, presence of pinnipeds, and any other observations pertaining to a particular set. Researchers inform fishers when to start picking up nets to ensure short soak times (the time from when the first cork goes in the water until the last cork is removed from the water).

As they remove fish from the net, fishers are instructed on proper fish handling, particularly to avoid touching the gill area or holding fish by its caudal peduncle. Fishers remove each fish from the net and either place it in a holding tank of fresh water located in the bow or release it overboard, as directed by the researcher. Any unusual observations about fish handling are recorded. For each salmon caught, the data collector notes the net type where it was captured, the type of capture, whether the adipose fin was missing, and the condition of fish at capture.

The researcher who is handling the fish calls out to the data researcher the manner of capture (e.g., tangled by teeth or mouth, gilled [net around the gills], wedged [web around body posterior to the gills], or mouth clamped [net wrapped around mouth, clamping it closed]).

A numerical rating of the condition of the fish is then recorded:

- 1 = Vigorous
- 2 = Vigorous and bleeding as a result of capture
- 3 = Lethargic
- 4 = Lethargic and bleeding as a result of capture
- 5 = No signs of ventilation
- 6 = Dead from being torn into pieces as from pinniped predation

The length and tags or other distinguishing marks, including scratches or wounds from pinnipeds, lamprey marks, descaling, and Floy tags, are then recorded.

If PIT tags are to be applied, a PIT-tag detector is run over the fish at this point. If the fish is less than vigorous (i.e., worse than condition 2), it is placed into the revival box until it recovers to condition 1 or 2 or dies. A tag can be applied or a biosample collected at this point. The person handling the fish measures the fork length and tags the fish. We typically attach a plastic colored jaw tag printed with a number and a color corresponding to the net type where the fish was captured.

The data researcher records whether fish are placed into the recovery box as well as the condition at release or when resuscitation fails and a fish is determined to be dead. Loss of scales, damaged fins, and other visible injuries are also recorded. Nontarget species encountered are counted, and the net mesh size where they are captured is noted. (Another option for every nontarget bycatch encountered is to collect the same information as for the targeted fish; this decision will depend on the goal of the study.) If a fish with a numbered tag is brought on board, its condition at capture, tag number, and whether the fish was caught in the top, middle, or bottom third of the net is recorded. Salmon are released when they are judged to be in good condition. Once dead salmon are scanned for coded wire tags, they are donated to a public food bank.

Control fish counted at dams

At dams, observers record any visible injuries and whether a captured fish is missing its adipose fin. Fish are then transferred to a holding tank with fresh water until they revive in lively condition; at that point, they are either released into a chute and diverted back to the fish ladder to continue migration or are transported to a site near where the tangle net is being fished. Trapping occurs throughout the test fishery to ensure that the same populations of migrating fish are tagged.

Data Handling, Analysis and Reporting

See the appendix for metadata collection procedures (including descriptions of fields and sizes), sample collection information, site description, and quality assurance procedures.

Database design

We created a simple series of flat tables to accommodate our field data—four tables that hold all our field collections and one table that houses our tag recovery information. All the data collected during the release of fish from our fishery are entered into three of the tables (*FinalSetInformation, *Finalsalmonids, and *FinalBycatch). The tables are joined via a three-field key. A separate table (Bonn-Trap) was created to house the control fish releases from the Bonneville Dam. Recovery information was housed in the *FinalTagRecoveryInfo table. This table was connected via a cross-table query to both test fishery and control release groups. Relationships among tables can include date, set number, boat skipper, fish number, and net type.

Data entry

Following field review and verification of field data sheets, the data is entered into the database. Prior to analysis, the data sheets are checked against the database again to ensure accuracy. Data sheets are photocopied and brought to the headquarters each week. The database is backed up weekly on compact discs.

Data summaries

Estimating survival

Survival is estimated as described in Ashbrook et al. (in press). The protocol enables estimation of immediate, postrelease, and total survival. Immediate survival was estimated as the binomial proportion

$$\hat{S}_I = \frac{a}{n} \qquad \text{(eq 1)}$$

where
n = total number of fish captured with the tangle net, and
a = number of fish retrieved from the tangle net that lived.

The sample error for immediate survival was estimated by

$$\sqrt{\hat{V}ar(\hat{S}_I)} = \sqrt{\frac{\hat{S}_I(1-\hat{S}_I)}{n}} \qquad \text{(eq 2)}$$

with confidence intervals estimated by the normal approximation (Zar 1984). Postrelease survival was estimated using the Ricker relative recovery model, where

$$\hat{S}_P = \frac{\left(\dfrac{t}{T}\right)}{\left(\dfrac{c}{C}\right)} \qquad \text{(eq 3)}$$

where
T = number of tangle net fish released,
t = number of tags recovered from tangle net fish,
C = number of control fish released, and
c = number of tags recovered for control fish.

The variance of was estimated by the delta method, where

$$\hat{V}ar(\hat{S}_P) = \hat{S}_P^{\;2} \left[\frac{1}{c} - \frac{1}{C} + \frac{1}{t} - \frac{1}{T} \right] \qquad \text{(eq 4)}$$

Total survival was calculated as the product of the immediate and postrelease survival estimates, where

$$\hat{S}_T = \hat{S}_I \cdot \hat{S}_P \qquad \text{(eq 5)}$$

with approximate variance estimator

$$\hat{V}ar(\hat{S}_T) = \hat{S}_I^{\;2} \cdot \hat{V}ar(\hat{S}_P) + \hat{S}_P^{\;2} \cdot \hat{V}ar(\hat{S}_I) - \hat{V}ar(\hat{S}_I) \cdot \hat{V}ar(\hat{S}_P) \qquad \text{(eq 6)}$$

Summarized tangle net data are reported to managers and researchers in the format shown in Table 1.

TABLE 1. — Format for reporting summary tangle net data to managers and other researchers.

Study year	Study group	Spring chinook adult catch	Immediate adult mortality (%)	Total tagged and released adults	Jaw tag recoveries (n)	Relative long-term adult mortality (%) (95% conf. int.)	Total mortality due to treatment (%)
2004	Tangle net						
	Gill net						
	Controls						
2005	Tangle net						
	Gill net						
	Control						

Report format

Most often the key data analyses can be broken into two categories: immediate and postrelease. The immediate category includes number of target fish captured in the tangle net, their immediate survival, recapture history, visible injuries, species composition or bycatch, temperature, and catch efficiency. The postrelease category includes postrelease and long-term survival, percentage of fish recovered by skipper, and location fished. If a control is used, similar information is collected.

Immediate

The immediate category includes number of fish captured in the tangle net, immediate survival, condition at capture and at release, recaptured fish, catch-per-unit-effort (CPUE), and species composition.

We list the dates fishing began and ended and list the total number of boat days (defined as an individual boat-date combination). We state the total number of captured and recaptured target adults and provide the same information for juvenile fish. Immediate survival is calculated as shown in the "Estimating Survival" section, and the results are shown in a table. The cumulative number of fish captured is shown in a histogram figure. We then list the number of fish that were given tags before release from the tangle net.

TABLE 2. — Immediate survival (%) of adult and juvenile fish captured including recaptures during test fishing in each net type. N is the number of target fish encountered.

Net type	Adults		Jacks	
	Survival (%)	N	Survival (%)	N
Tangle				
Gill				

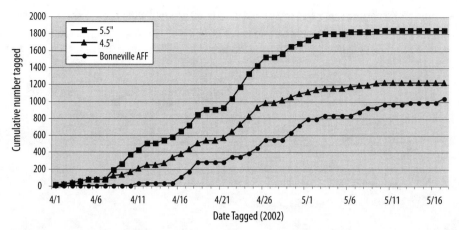

FIGURE 5.—Cumulative number of adult target fish tagged and released during test fishery and at the adult trapping facility with 11.4-cm (4.5-in) and 14.0-cm (5.5-in) mesh nets.

Recaptured fish

The number of recaptured fish are listed and the recaptured percentage is tallied. We also record the number and percent of recaptured fish that survive and are released in good condition.

Condition at capture

We provide a table that lists the number of fish by their initial condition as they are brought aboard the boat; if more than one net is used, a chi-square test is conducted to show if the distribution of the condition of targeted fish in each category is significantly different.

TABLE 3.—Adult fish (including recaptured fish) scored in each condition at capture category that were released (Rel'd) or died for each net

	Condition at capture									
	1		2		3		4		5	
	Lively		Lively, bleeding		Lethargic		Lethargic, bleeding		No visible movement or ventilation	
Net type	Rel'd	Died	Rel'd	Died	Rel'd	Died	Rel'd	Died	Rel'd	Died
Tangle										
Gill										
Total										

Next, we test whether the proportions of fish caught by the different capture types are significantly different between the two net types using a chi-square test.

TABLE 4.—Capture types of adult spring chinook salmon (includes recaptures) that were released (Rel'd) or died.

Capture type	Net type			
	Tangle		Gill	
	Rel'd (%)	Died	Rel'd (%)	Died
Gilled				
Mouth clamped				
Tangled				
Wedged				
Total				

Visible injuries

Visible injuries from net capture or marine mammal wounds are recorded. If more than one net type is used, they can be compared with a chi-square test.

TABLE 5.—Occurrence of visible injuries (%) on fish captured in each net type. The other category contains low occurrences of torn gills, torn operculum, and hook wounds.

Net type	De-scaling (%)	Net marks - body (%)	Net marks - head (%)	Marine mammal wounds (%)	Torn fins (%)	Other (%)
Tangle						
Gill						

Fork length

The range and average fork length for fish captured in the tangle net is provided. The mean fork lengths for fish that died before release and for recovered fish are provided. The soak times of the net for sets with dead fish can be compared to the average soak time for all sets using a *t*-test.

Species composition or bycatch

The number of nontarget species captured in the tangle net can be provided. Often, the actual numbers of nontarget fish may be underreported if this is not the primary goal of the study. We also mention the condition of the bycatch at release.

TABLE 6.—Count of nontarget species in the tangle net caught during the test fishery. ("Other fishes" includes walleye, flounder, carp, bass, and so forth, for which only a small number were encountered.)

Species	Tangle net
Shad	
Northern pike minnow	
Sturgeon	
Sucker	
Other fishes	
Total	

Temperature

The range and average for surface and water temperatures during test fishing are provided. The mean surface temperature for sets including fish mortalities is compared to the average temperatures for all sites and tested with a *t*-test to learn if temperature affects immediate survival.

Catch efficiency

For each day we are able to fish both nets equally, we can compare the catch per hour of targeted fish in each net. For adult-focused research, juvenile animals are omitted from the analysis. The fishing time includes only the time the nets are actually fishing and not the time spent preparing for the next set. Because we record only the time the first cork goes into the water (not when the shackle goes in), we designate the time to set the first net as 3 min in every case. The total fishing time for each net can then be calculated as the time from when the first cork of that net is placed in the water to the time when the last cork of that same net type is removed from the water. We calculate the total number of targeted fish for each net for each set. The total numbers of fish are summed with the set time for each net type by skippers for each day fished. The catch efficiencies of each net type can be compared using a paired t test or, for non parametric data, a Wilcoxon signed rank test.

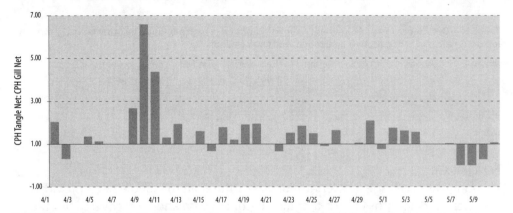

FIGURE 6. — Relative catch of adult spring chinook salmon per hour (CPH) for the tangle net compared to the gill net. Values at 1 indicate equal efficiency, those below 1 indicate when the tangle net was more efficient, and those above 1 indicate when the gill net was more effective. Paired sets were pooled by day across boat skippers.

TABLE 7. — Example of targeted fish per hour (CPH) capture rates during comparable sets for each net type.

Net type	Min. CPH	Max. CPH	Avg. CPH
Tangle	1.6	14.1	6.2
Gill	0.0	22.6	9.2

Postrelease

Postrelease analysis includes how recovery information is collected, the number of fish released from tangle nets, the number of fish recovered following their release from tangle nets, estimation of postrelease survival, comparison of recovery rates by boat skipper and location fished, and fish passage through detection centers.

We evaluate long-term survival of released fish by their returns to hatcheries, sport and treaty harvest, and at spawning ground surveys. Posters are produced that request the following information: date of harvest, location of harvest, tag color, and tag number. They are posted at various locations to target fishers. In addition, hatchery crews and stream surveyors are contacted and asked to return the same information. See Estimating Survival section for a description of how survival is estimated.

We record the number of fish that were given tags and released from the tangle net(s). Tags are recovered throughout sport and commercial fisheries, at hatcheries, and on the spawning grounds (Figures 8 and 9). We record when the first and last tags were recovered and provide a figure that shows the breakdown of jaw tag recoveries by fishery and the size of the estimated return to the natal stream over time. We subsequently report on the number of tags recovered.

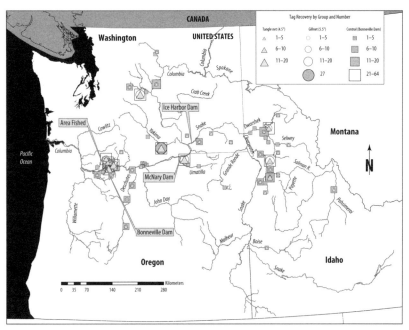

FIGURE 7. — Example of recovery locations of target fish captured and released from tangle nets and from the control fish trap. The area fished denotes the location where the test nets were fished and tagged fish were released from the test nets. The geographic area is the Columbia and Snake rivers.

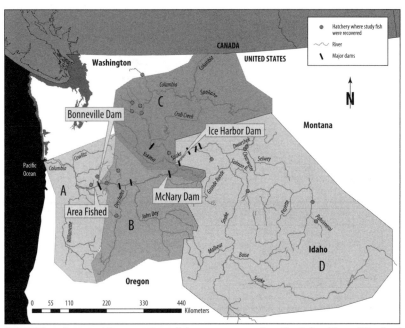

FIGURE 8. — Geographic areas where fish were recovered on the Columbia and Snake rivers: (A) Below Bonneville Dam, (B) between Bonneville and McNary dams, (C) Upper Columbia above McNary Dam, and (D) Snake River above Ice Harbor Dam.

TANGLE NETS

TABLE 8. — Target adult fish released and number recovered by geographic area and chi-square test results.

Both nets		Control	Total	Chi-square test results
				$P > 0.05$
Number of fish released				
Recoveries: below Bonneville Dam				
Recoveries: Bonneville Dam to McNary Dam				
Recoveries: Snake R. above Ice Harbor Dam				
Recoveries: Columbia R. above McNary Dam				
Total				

The Z-statistic as described in Zar (1984) can be used to compare the two proportions and assess whether net size has a significant effect on postrelease survival.

TABLE 9. — Recovery of tag groups from hatcheries, fisheries, and spawning grounds.

Group	Number tagged	Number recovered	Percent recovered	Relative survival rate	95% confidence interval
Control				100%	N/A
Gill net					
Tangle net					

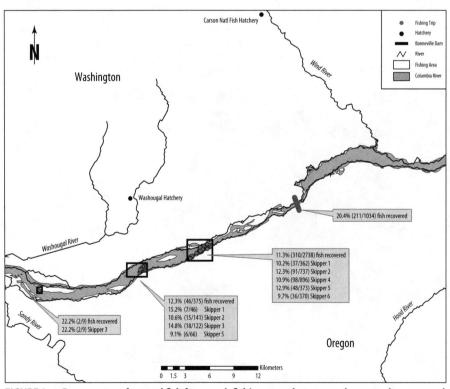

FIGURE 9. — Percentages of tagged fish from each fishing area that were subsequently recovered by boat skipper.

Passage over dams
This evaluation can occur if PIT or telemetry tags are used and if there is suitable placement and a suitable number of PIT detection centers or telemetry receivers. We give the range of passage days and dates for the first and last targeted fish detected at the first and last detection center.

Control fish
To evaluate survival for tangle net captured fish, a control is needed. With the exception of being captured in the tangle net, ideal control fish will have passed through all the same predatory and fishing pressures as the fish caught in the tangle net.

Archival procedures
Hard-copy data are archived in folder files by year. Electronic data are archived on computers, an external hard drive, and CD-ROMs.

Personnel Requirements and Training

Roles and responsibilities
The contracted fisher or fishers are responsible for driving the boat, deploying and retrieving the net, and ensuring that all applicable safety regulations are met. The researchers should not handle any of the boat operations. The contracted fisher is responsible for repairing the net. The fisher extracts the fish from the net and then hands the fish to the researchers. The researchers are responsible for data collection, sample processing, and tagging. These roles and responsibilities should be stipulated in the fisher's contract and made clear to the research staff.

Qualifications
All crew members must be able to handle, lift, and collect samples from fish. For spring chinook salmon, this means being able to lift, hold, tag, and release overboard fish that weigh as much as 30 lbs continuously for each set. Each set is about 45 min. Crew members need to stand for hours at a time, lift water totes and gear onto and off of the vessel, and repetitively tag fish, often in inclement weather. Furthermore, crew members should not be prone to seasickness. All crew members must wear life jackets and should be willing and able to work quickly and efficiently, and have excellent data collection and data entry skills.

Training
All crew members need to be trained in first aid, fish handling, species identification and basic biology, and tag application. Safe handling of the fish is a critically important part of this work: fish should be held with two hands—one hand around the caudal peduncle and the other hand holding the head. Holding the salmon in the tote with its belly up usually makes collecting length measurement and tag application easier.

Operational Requirements

Workload and Field Schedule

The employees needed for tangle-net monitoring include a research scientist, field leader, database lead, and research technicians. One to three research personnel are needed on each boat per boat day. In a recent research project targeting spring chinook or coho salmon, each boat day comprised 6–10 sets. A field leader schedules the research crew, arranges fishing days with the contract fishers, ensures that data are being entered accurately into the database, queries the database weekly, and sends a report to the research scientist in charge of the project. A database lead sets up the database for data entry, queries the database, and makes final correction checks. The database lead may also download portions of the database into a geographic information system and create appropriate maps of fishing and recovery site locations. The research scientist, field leader, and database lead write up final reports. Experimental design and any in-season changes are the responsibility of the research scientist.

The length and timing of the field season depend on the targeted species and stocks. For example, for spring chinook adult salmon returning to the Columbia River, the field season could begin in March and continue through mid-June. Tag recovery would likely occur from March through December. Data entry, data review, and results and report writeup would begin in March and be finalized by the following June.

Tangle-net fishing can occur at any time and in any area where gill-net fishing is suitable. In the case of targeting migrating fish, as fish capture tapers off, the research personnel are assigned to other activities associated with the study.

Equipment Needs

- Tangle net(s) suitable for capturing target fish
- Lead line and floats for the tangle net
- Plastic tote box to hold fish for measurement and sampling
- Dip nets
- Rain gear
- Life jackets
- Marking, tagging, and sampling supplies, data sheets, and pencils
- GPS units
- Tags and tagging equipment
- Computer for data entry and analysis
- Field office

For postrelease survival estimates, a suitable control is needed. This could be an adult trap or another fishing method such as a beach or purse seine. If a purse seine is used, the contracted fisher can provide all required equipment, including a net, skiff, and crew for deploying and retrieving the net. If the control fish need to be transported, a fish hauling tank and vehicle to haul the tank are needed.

Budget Considerations
- Cost to hire contract fishers (fishers supply boat and fuel and show proof of valid insurance)
- Cost for personnel
- Net, cork line, and recovery box costs (net may need to be replaced each year); cost to have net hung onto cork line
- Field equipment (e.g., life jackets, boots, rain gear, GPS units, thermometers, rainproof paper for data collection, computer, software)

Acknowledgments

Geraldine Vander Haegen contributed many of the experimental design and data analysis techniques that are now standard for tangle net research efforts.

Literature Cited

Ashbrook, C. E., K. W. Yi, J. D. Dixon, A. Hoffmann, and G. E. Vander Haegen. 2004. Evaluate live capture selective harvest methods for 2002 study. Annual Report for BPA Contract #2001-007-00. Washington Department of Fish and Wildlife, Olympia.

Ashbrook, C. E., J. R. Skalski, J. D. Dixon, K. W. Yi, and E. A. Schwartz. In press. Estimating bycatch survival in a mark-selective fishery. In J. Nielsen, J. Dodson, K. Friedland, T. Hamon, N. Hughes, J. Musick, and E. Verspoor, editors. Proceedings of the Fourth World Fisheries Congress: Reconciling Fisheries with Conservation. American Fisheries Society, Symposium 49, Bethesda, Maryland.

Farrell, A. P., P. E. Gallaugher, C. Clarke, N. DeLury, H. Kreiberg, W. Parkhouse, and R. Routledge. 2000. Physiological status of coho salmon (*Oncorhynchus kisutch*) captured in commercial non-retention fisheries. Canadian Journal of Fisheries and Aquatic Science 57:1668–1678.

Farrell, A. P., P. E. Gallaugher, J. Fraser, D. Pike, P. Bowering, A. K. M. Hadwin, W. Parkhouse, and R. Routledge. 2001. Successful recovery of the physiological status of coho salmon on board a commercial gill net vessel by means of a newly designed box. Canadian Journal of Fisheries and Aquatic Science 58:1932–1946.

Vander Haegen, G. E., C. E. Ashbrook, K. W. Yi, and J. F. Dixon. 2004. Survival of spring chinook salmon captured and released in a selective commercial fishery using gill nets and tangle nets. Fisheries Bulletin 68:123–133.

Vander Haegen, G. E., J. F. Dixon, K. W. Yi, and C. E. Ashbrook. 2002. Commercial selective harvest of coho salmon and chinook salmon on the Willapa River using tangle nets and gill nets. Final report, IAC Contract #01-1018N. Washington Department of Fish and Wildlife, Olympia.

Vander Haegen, G. E., K. W. Yi, C. E. Ashbrook, E. W. White, and L. L. LeClair. 2002. Evaluate live capture selective harvest methods for 2001 study. Final Report for BPA Contract #2001-007-00. Washington Department of Fish and Wildlife, Olympia.

Zar, J. H. 1984. Biostatistical analysis. Prentice-Hall, Inc., New Jersey.

Appendix A: Tangle Nets Data Sheet

Fish #	Jaw Tag Color (B or R)	Jaw Tag No.	Fork L (cm)	Adipose (H or W)	Syringe No.	Pit Tag #	Sex (U or J)	NMH	NMB	D S	SW	H K	T O	T G	S S	T F	Comments

Entered and Checked by: _____ (Initials and date)

Appendix B: Tangle Nets Data Sheet

Idaho Cooperative Fish and Wildlife Research Unit & WDFW
Adult Tag Form: Lower Columbia- 2003

Date _____
Boat _____ Skipper _____
Sampler _____
Recorder _____
LAT: __ __.__ __ __ N
LON: ___ __.__ __ __ W

GPS unit: _____
Waypoint: _____

(circle one)
net IN water first: 4.25 4.5
net OUT of water first: 4.25 4.5

Set # ____ of ____
Time 1st float In:
Time 1st float Out:
Time Shackle Out:
Time Last Float Out:

Fish #	Sp.	Mesh Size	Capt. Type	Cond. at Capture	Sex: M/F/U/J	Adipose (H or W)	Fork L (cm)	Jaw Tag Color	Jaw Tag #	Syringe	Pit Tag #	Recovery Box Y/N	Sthd Tag Time	Release Cond.	SS	NMH	NMB	DS	SW	HK (Visible In)

Capture Types: T=Tangled, G=Gilled, MC=Mouth Clamp, W=Wedged
Condition Codes: 1=Vigorous, not bleeding, 2=Vigorous, bleeding, 3=Lethargic, not bleeding, 4=Lethargic, bleeding, 5=No movement or ventilation

Entered and Checked by: _____
(Initials and date)

Date entered into: _____
(file name)

TANGLE NETS

Tower Counts
Carol Ann Woody

Background and Objectives

Rationale
Counting towers provide an accurate, low-cost, low-maintenance, low-technology, and easily mobilized escapement estimation program compared to other methods (e.g., weirs, hydroacoustics, mark–recapture, and aerial surveys) (Thompson 1962; Siebel 1967; Cousens et al. 1982; Symons and Waldichuk 1984; Anderson 2000; Alaska Department of Fish and Game 2003). Counting tower data has been found to be consistent with that of digital video counts (Edwards 2005). Counting towers do not interfere with natural fish migration patterns, nor are fish handled or stressed; however, their use is generally limited to clear rivers that meet specific site selection criteria.

The data provided by counting tower sampling allow fishery managers to

- determine reproductive population size,
- estimate total return (escapement + catch) and its uncertainty,
- evaluate population productivity and trends,
- set harvest rates,
- determine spawning escapement goals, and
- forecast future returns (Alaska Department of Fish and Game 1974–2000 and 1975–2004).

The number of spawning fish is determined by subtracting subsistence, sport-caught fish, and prespawn mortality from the total estimated escapement.

The methods outlined in this protocol for tower counts can be used to provide reasonable estimates (± 6%–10%) of reproductive salmon population size and run timing in clear rivers.

Objective
Tower counts enable practitioners to systematically sample a selected salmon population to estimate reproductive population size and determine run timing.

Background
Counting towers provide an elevated vantage point for visually sampling Pacific salmon spawning migrations. Aluminum scaffolding is typically used (see Figure 1), but biologists are creative and employ tower surrogates, such as tall trees, bridges, dams (see Figure 2), or high river banks to accomplish their sampling. Since the 1950s, counting towers have played a central role in Pacific salmon management in Alaska and to a lesser extent in Canada and Washington (Rietze 1957; Cousens et al. 1982; Anderson 2000; Kohler and Knuepfer 2002; Fair 2004). Towers are used on both single- and multispecies systems (see Table 1) and on small to large (10–130+ m in width) clear water rivers.

TABLE 1. — Sample of counting tower projects for estimating Pacific salmon escapement in Alaska. (ADF&G = Alaska Department of Fish and Game; reference list is not comprehensive but will assist research efforts.)

Location	River	Species	Years in operation	References
Bristol Bay	Egegik	O. nerka	1959–present	ADF&G 1974–2004; Anderson 2000
	Igushik	O. nerka	1958–present	"
	Kvichak	O. nerka	1955–present	"
	Alagnak[1] (branch)	O. nerka	1957–1976 2002–present	"
	Newhalen[1]	O. nerka	1980–1984 2000–present	Poe and Rogers 1984; Woody 2004
	Tazimina[1]	O. nerka	2000–2003	Woody 2004
	Naknek	O. nerka	1958–present	ADF&G 1974–2004; Anderson 2000
	Nuyakuk	O. nerka	1959–1988 1995–present	"
	Togiak	O. nerka	"	"
	Ugashik	O. nerka	1957–present	"
	Wood	O. nerka	1956–present	"
Norton Sound	Eldorado	O. kisutch, keta, gorbuscha		
	Kwiniuk[2]	O. kisutch, keta, gorbuscha, tshawytshaw	1965–present	Hamazaki 2003
	Niukluk	O. kisutch, keta, gorbuscha		"
	Nome	O. kisutch, keta, gorbuscha		"
	Pilgrim	O. kisutch, keta, gorbuscha		"
	Snake	O. kisutch, keta, gorbuscha	1960–1973	"
Lower Tanana River	Chena	O. tshawytscha, O. keta	2002–present	
	Salcha	O. tshawytscha, O. keta	2002–present	Tanana Chiefs Conference[3]
	Goodpaster	O. tshawytscha	2004–present	"
	Gulkana River	O. nerka, O. tshawytscha	2002–present	Taras and Sarafin 2005

[1] Rivers within the Kvichak River watershed.

[2] Not all species were monitored in all years; see review by Hamazaki (2003) for details.

[3] Go to <www.tananachiefs.org/natural/fisheries.html> for contact information.

FIGURE 1. — Example of a counting tower used on the Newhalen River, Bristol Bay, Alaska. This tower is constructed of lightweight aluminum scaffold and is stabilized with guy wires. Here, Libby Baney conducts a 10-minute systematic count.

FIGURE 2. — Example of a counting tower "surrogate" used to estimate escapement on the Chena River, Alaska. Biologists conduct counts from the closest piling. Note contrasting substrate panels that improves visibility of migrating fish. (Photo by the Alaska Department of Fish and Game.)

History

Until development of counting towers in the 1950s, estimates of the number of salmon that "escaped" the commercial fishery to spawn (escapement) eluded Alaskan fishery managers. Weirs proved too expensive and difficult to maintain, and they caused excessive delays to salmon returning to their spawning grounds. Estimation methods such as mark–recapture and other indices (e.g., aerial estimates) were expensive, imprecise, and inconsistent (Eicher 1953; Bevan 1960; Thompson 1962; Seibel 1967; Symons and Waldichuk 1984; Cousens et al. 1982).

In 1951 a young fish biologist named Charles Walker reported that he was able to count migrating sockeye salmon *Oncorhynchus nerka* from a high riverbank at the outlet of Lake Alegnegik, on the Wood River in Alaska (Thompson 1962). Rietze (1957) later detailed their migration behavior:

> Fish closely followed the contour of each bank of the river in water about three to six feet deep and rarely more than thirty feet from the shore ... migrations occurred in a narrow band of about four to ten fish swimming abreast and appearing in a steady stream. The right bank carried ... the greater number of fish, but sporadically, greater numbers appeared to follow the left bank. There appeared ... little, if any, crossing from bank to bank ...

W. F. Thompson, then a director of fisheries research in Bristol Bay, realized that such behavior (see Figure 3) might allow abundance estimates through systematic sampling (Cochran 1977). He proposed this new salmon escapement estimation technique in 1953, conducting a pilot study to test this approach by counting fish from high towers. The pilot study was auspicious, and a series of studies ensued to verify the method's accuracy and to optimize sampling protocols (Fisheries Research Institute 1955; Thompson 1962).

FIGURE 3.—Examples of sockeye salmon migration pattern at low (a) and high (b) density. Sockeye salmon generally migrate in a band within 10 m of the shore, making them an ideal candidate for systematic sampling from counting towers.

Accuracy of tower counts was first examined by comparing tower estimates to weir counts (Rietze 1957; Spangler and Rietze 1958); it was assumed that weirs provided total abundance. Rietze (1957) described how, in just a few days, four counting towers were erected on each bank of the Egegik River, both above and below a 230-m picket weir, which took 3 weeks to install. Researchers divided each hour into two systematic 15-min counts, followed by a 15-min break to reduce fatigue and possible error. The sum of 24-h counts was expanded by two for the daily abundance estimate (Rietze 1957). Researchers then defined relative error between tower and weir counts as

$$\frac{\text{tower estimate} - \text{weir count}}{\text{weir count}}$$

Rietze's (1957) estimated relative error was about −7.4% (see Figure 4); however, when he dropped the 2 d it took to build one tower (see Figure 4; 16–17 July) from the comparison, the relative error declined to −1.6%. Tower counts below the weir failed to provide a reliable abundance estimate because natural migration was delayed, causing salmon to mill and rendering accurate counting impossible (see Figure 5).

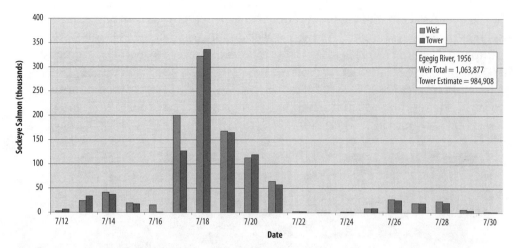

FIGURE 4. — Comparison of daily weir and systematic tower counts (30-min. counts/hour) for sockeye salmon escapement, Egegik River, Alaska, 1956. The relative error (i.e., [tower-weir]/weir) between methods was −7.4% (data from Reitz 1957).

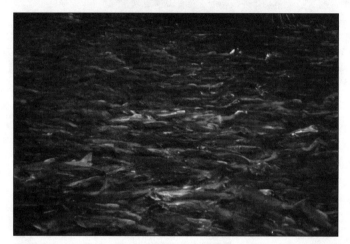

FIGURE 5. — Sockeye salmon backed up behind a weir (Bear Creek, Tustumena Lake, Alaska).

In 1957, towers placed above a weir showed relative error of tower counts to be +12.9% (see Figure 6) (Spangler and Rietze 1958), but biologists noted the weir had not been "fish tight" on at least 6 d (see Table 2), meaning that fish dug under or found a hole in the weir and passed uncounted. Relative error between methods was likely lower than was reported by Spangler and Rietze (1958). These initial studies indicated that compared to weirs, systematic tower counts were both (a) relatively accurate and (b) did not interfere with fish migration.

TABLE 2. — Summary of daily sockeye salmon escapement data from weir counts and counting tower estimates, Naknek River, Alaska, 1957 (data from Spangler and Reitz 1957). Asterisks indicate days when weir was not "fish tight" (i.e., fish passed through uncounted). Note the relatively large differences occurring at the peak of the run, 8–10 July.

Date	Weir estimate	Tower estimate	Tower − Weir	Tower − Weir ÷ Weir × 100 *relative error*
Jun 29*	6,375	8,040	1,665	26.1
Jun 30	7,401	7,560	159	2.1
Jul 1	7,437	7,420	−17	−0.2
Jul 2	1,380	728	−652	−47.2

TOWER COUNTS

Date	Weir estimate	Tower estimate	Tower − Weir	Tower − Weir ÷ Weir × 100 *relative error*
Jul 3	9,831	7,604	−2,227	−22.6
Jul 4	3,704	4,336	632	17.1
Jul 5	6,350	5,284	−1,066	−16.8
Jul 6	13,777	12,896	−881	−6.4
Jul 7*	2,662	3,188	526	19.8
Jul 8	93,592	145,204	51,612	55.1
Jul 9	75,436	100,408	24,972	33.1
Jul 10	149,787	136,072	−13,715	−9.2
Jul 11	58,415	61,256	2,841	4.9
Jul 12	32,052	35,104	3,052	9.5
Jul 13*	17,072	20,708	3,636	21.3
Jul 14	21,079	16,600	−4,479	−21.2
Jul 15	19,659	22,032	2,373	12.1
Jul 16	5,943	9,324	3,381	56.9
Jul 17	16,563	16,788	225	1.4
Jul 18	22,411	28,268	5,857	26.1
Jul 19	8,916	9,160	244	2.7
Jul 20	6,203	7,544	1,341	21.6
Jul 21*	10,862	10,764	−98	−0.9
Jul 22*	8,693	9,352	659	7.6
Jul 23	4,242	5,208	966	22.8
Jul 24	6,228	5,536	−692	−11.1
Jul 25	4,887	5,072	185	3.8
Jul 26	2,536	2,712	176	6.9
Jul 27	1,633	1,880	247	15.1
Jul 28	856	1,152	296	34.6
Jul 29*	2,647	2,304	−343	−13.0
Jul 30	1,336	1,620	284	21.2
Jul 30	1,036	1,000	−36	−3.5
totals	631,001	712,124	−81,123	12.9

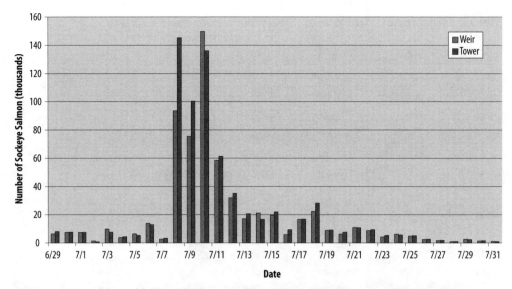

FIGURE 6. — Comparison of daily weir and systematic tower counts (15-min. counts/hr), Naknek River, Alaska, 1957. Relative error (i.e., [tower-weir]/weir) between methods was +12.9%; however, error may have been lower as the weir was not "fish tight" on 29 June, and 7, 13, 21, 22, and 29 July and fish passed through the weir undetected. The large discrepancy in estimates on 8 July suggests that the weir was not fish tight on that date either.

Because tower crews have other duties, such as seining salmon to collect biological samples (e.g., age, sex, length, and tissue for genetic analysis) from the upriver migrating population and because it is difficult for counters to maintain focus for long intervals, researchers sought to reduce sampling intervals without increasing relative error. Becker (1962) examined how counting interval length and sample frequency affected relative error of escapement estimates. Four systematic samples of 10, 20, 30, 40, and 60 min were taken from a continuous 48-h count at a frequency of 1–4 h. Short counts (<40 min) that were conducted every 1–2 h generally ranged within ± 6% of the actual count, whereas a wider range of error was observed for counts taken every 3–4 h (see Figure 7). Because error was not greatly reduced through longer sample intervals and because prior studies indicated relatively low relative error compared to weirs (Rietze 1957; Spangler and Rietze 1958; Becker 1962), the nonrandom systematic 10–20-min sample counts per hour, 24 h a day, were widely adopted. Interestingly, psychologists later conducted attention span studies on students and showed that they could only focus an average of 15–20 min before their attention lapsed (Johnstone and Percival 1976), providing further support for short counting intervals.

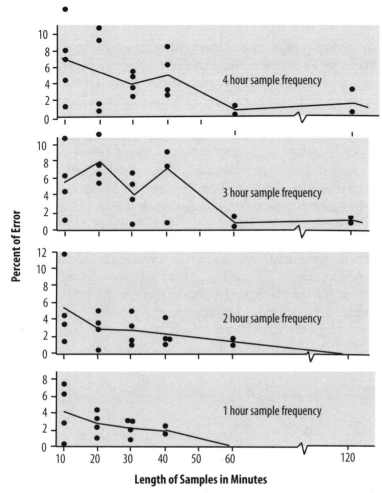

FIGURE 7.—Relationship between counting interval length, sample frequency, and relative error. Dots denote relative errors from different systematic samples from a 48-hour period; solid lines denote the mean relative error (i.e., [estimate − true total]/true total) for a given sampling interval duration. (Graph from Becker [1962].)

Siebel (1967) reevaluated systematic counting protocols for eight rivers in Alaska and found that relative errors ranged from −34.9% to +21.8% but were equally divided between over and underestimates, indicating a lack of bias. Mean relative error was 0.9%, insignificant at the 95% confidence level, with a reported 95% confidence interval of (−7.1%, 8.9%). He recommended sample count intervals be increased to 20 min if migration occurred in less than a week, if migration was highly concentrated, or if short-period escapement estimates were needed for calibrating aerial surveys.

Sources of error

Counting towers do not provide error-free estimates of escapement. The primary factors that affect accuracy and precision of counts are

- observer variability,
- aspects of migration,
- weather conditions, and
- systematic sampling method—nonreplicated versus replicated.

Observer variability

Variability among counting tower personnel in their ability to record data, detect and count fish, or identify species may introduce error in escapement estimates. Becker (1962) examined such error by conducting a series of 32 paired 5-min counts; one observer participated in all 32 counts while three rotated. The total difference between observers ranged from −5.3% to +3.5%; total counts differed by 1% implying that observer error was unbiased and therefore tended to cancel out (Becker 1962). A similar study used paired tower counts with both an inexperienced and an experienced observer over a range of conditions. Each paired test ($n = 3$) consisted of eighteen 10-min counts; six daylight, six crepuscular, and six night counts. Percent errors ranged from −1.8 to +1.3 and resulted in a combined total error of +0.4% (117 fish difference out of a total of 29,000 fish; see summary in Anderson 2000). These studies indicated that observer bias under a variety of conditions was random; when added together, overestimates (+) and underestimates (−) of fish passing the towers tended to cancel out. Observer bias should not be ignored; project leaders can reduce such bias by conducting paired counts with inexperienced personnel until they demonstrate count and species identification proficiency. Computerized training programs that teach estimation techniques are available; go to <www.wildlifecounts.com> for more information.

Aspects of migration

Within a given river system, species generally vary enough in the following traits to allow counters to distinguish among them: size, coloration, migration timing, and/or behavior (Groot and Margolis 1991). For example, in the Kvichak River in Alaska, sockeye salmon and chinook salmon *O. tshawytscha* may migrate by towers at a similar time, but sockeye salmon are much smaller—and hence easy to distinguish with little training. Even the smallest salmon (pink salmon *O. gorbuscha*) are relatively large compared to other fishes, weighing about 2–6 kg and easy to see from towers. Generally, not all species migrate at the exact same time, although in some rivers there can be considerable overlap, making tower counts alone infeasible (Dunmall 2004); in such cases, use of weirs or video monitoring should be considered. In many Alaskan systems, chinook salmon migrate first, followed by chum salmon *O. keta* and pink salmon, then sockeye salmon, and finally coho salmon *O. kisutch*. Many tower systems, such as those in Bristol Bay, oversee rivers that are dominated by sockeye salmon, and other species are relatively rare.

An extreme but rare example of the potential range in daily salmon escapements is from the Kvichak River in Alaska when, in 1980, a strike by fishermen led to a sockeye salmon escapement of 22.5 million fish (Anderson 2000). Daily escapement estimates ranged from 0 to 1.8 million salmon, with an estimated 0–150,000 fish passing the counting towers during every 10-min counting interval. In this situation, observers visually divided migrating bands of salmon into tens, hundreds, and even thousands, and tallied observations accordingly. The impact of this type of error has not been studied; however, Becker (1962) found a slight positive correlation between number of migrants and observer bias with greater variation observed when number of migrants equaled or exceeded 700 fish per 10-min interval. Examination of the data in figures 4 and 6 imply greater observer bias at high migration densities, but further research is clearly needed.

Weather conditions

Glare, overcast skies, high winds, rain, and turbidity all reduce visibility and affect count accuracy. While this source of error has not been quantified, it can be reduced through

1. careful site selection that reduces glare and wind in the counting region (in Figure 1, note wind direction and how the counting region in front of the tower is not turbulent);
2. use of polarized glasses to reduce glare and improve cloudy-day visibility.
3. use of riffle dampeners just upstream of the counting area, which can help reduce surface turbulence in the counting region (these structures are usually floating wood or logs in a V-shape); and
4. use of lighter substrates or panels (figures 2 and 7), which can help in spotting salmon.

Turbidity and associated decline in water clarity is usually uncontrollable, whether due to storm runoff or glacial water intrusion. There is little that can be done with regard to storm runoff, and fortunately the impact is temporary. Most projects use a form of count interpolation to account for missed sample intervals (see section on count interpolation). Glacial water intrusion is a different story. Determine if and when glacial water intrudes at the selected site relative to fish migration; then assess whether it will prevent accurate counts. If glacial water makes tower counts prohibitive, consider using hydroacoustic estimation techniques.

Systematic sampling method (nonreplicated versus replicated)

The sampling design selected can affect both variance of the total escapement estimate and bias in estimates of that variance. See the Sampling Design section on pages 371–374 for guidance.

Sampling design

Site selection

Generally, one tower is installed on each river bank, although up to four have been used on divided channels. During site selection you specifically select noncomplex reaches (e.g., no pools, no woody debris, level bottom) as you must have a clear view of the river in front of the tower, and fish must continually move upstream. The following list will help guide site selection.

1. Ensure that upstream migration of adult fish is in an observable pattern; it may be feasible to divert fish to an observable region with a partial weir (see Figure 8) or bright substrate panels.
2. Avoid sites where fish mill, spawn, or move downstream.
3. Ensure that you have generally clear water during the migration period.
4. Ensure that you have a constrained channel (i.e., not braided).
5. Ensure that the area is relatively protected from glare and prevailing wind; some projects employ alternative counting sites when specific weather

conditions prevail.

6. Ensure that you have relatively laminar flow in the counting region throughout the migration period; because river flow changes throughout the season, it is important to examine flow patterns over a range of discharge to ensure that the counting region remains relatively free from turbulence.

7. Ensure that water depths are ~0.5–3 m where fish travel; again, let fish migration pattern and observability be your guide.

8. Ensure that bottom substrate contrasts with passing fish (see Figure 3) or allows installation of panels or other materials to achieve such contrast (see figures 2, 8, and 9).

9. Situate tower sites (ideally) directly across from each other; if fish do not cross from bank to bank in intervening river passage, tower sites may be somewhat staggered.

10. Install floodlights either above or across the entire river or on the shore near towers on night counts (see figures 1, 8, and 9); light system selection will depend on salmon behavior (e.g., are they migrating near shore or are they distributed across the entire river width?).

FIGURE 8.—Example of a diversion weir used in a counting tower project to guide fish into observable range. (Photo courtesy of the Alaska Dept. of Fish and Game.)

FIGURE 9. — Counting tower, contrast panels, and a riverwide suspended lighting system on the Chatanilka River, Alaska. (Photo courtesy of the Alaska Dept. of Fish and Game.)

Systematic sampling designs are the standard for using counting towers to estimate escapement of Pacific salmon in Alaska; however, there are many possible variants of systematic sampling (Reynolds et al., in press). All provide unbiased estimates of total escapement but differ in the variation of the total escapement estimate and the bias associated with estimates of that variation.

Nonreplicated systematic sample design

The most common sampling design for counting towers in Alaska is nonreplicated systematic sampling of 10 min per hour (see Table 4). Before the initiation of sampling, a random number is drawn from the range 1 to 6; each number represents a 10-minute count interval over a 1-h period. Counts are then made at the selected interval of each hour for the rest of the season. For example, say the first counter of the season randomly selects interval 2. She takes her first count at 12:10, her second at 13:10, her third at 14:10, and so forth. This design does not allow for unbiased estimation of the variance associated with the total escapement estimate. For estimating salmon escapement using nonreplicated systematic sampling designs, Reynolds et al. (in press) show that the best variance estimator is the V5 estimator of Wolter (1984), defined on page 376. Calculations are discussed in the statistical analysis section on pages 376–377.

TABLE 4. — Systematic sampling designs commonly used in estimating total sockeye salmon escapement from tower counts (see Becker 1962, Anderson 2000, and analysis by Reynolds et al. in press) (j = replicate index).

Design		Daily mean escapement, $\bar{y}$	Expansion[1]	Possible samples[2]
Nonreplicated systematic	20 min / 2 hr	$\sum_{i=1}^{n} y_i / n$	$6 \times 24 \times N$	6
	10 min / 1 hr	$\sum_{i=1}^{n} y_i / n$	$6 \times 24 \times N$	6

Design		Daily mean escapement, $\bar{y}$	Expansion[1]	Possible samples[2]
Replicated systematic	4 @ 10 min / 4 hr	$\sum_{j=1}^{4}\left(\sum_{i=1}^{n} y_{ij}/h\right)/4$	$24 \times 24 \times N$	10,626
	2 @ 10 min / 2 hr	$\sum_{j=1}^{2}\left(\sum_{i=1}^{n} y_{ij}/h\right)/2$	$12 \times 24 \times N$	66

NOTE: Total annual escapement is estimated by expanding the daily mean escapement: $\hat{Y}$ = (Expansion) $\times \bar{y}$.

[1] Units/hr $\times$ hrs/day $\times$ days

[2] Number of possible samples given a sampling period of N consecutive days.

Replicated Systematic Sample Design

Replicated systematic sampling designs consist of multiple, independently selected nonreplicated systematic samples (Reynolds et al., in press). A replicated systematic sample of 2–10 min/2 h systematic samples is created by having the first counter of the year randomly draw two numbers ranging from 1 to 12; each number represents a 10-min count interval over a 2-h period. Counts are then made at the selected intervals for the rest of the season to generate two independent (replicated) systematic samples. For example, say the first counter of the season randomly selects intervals 2 and 10. She takes her first count at 12:10, her second at 13:30, her third at 14:10, her fifth at 15:30, and so forth. At the end of the season the counts from interval 2 (12:10, 14:10, etc.) and the counts from interval 10 (13:30, 15:30, etc.) are analyzed separately as nonreplicated systematic samples, each providing an estimate of total escapement. These estimates are analyzed for an overall escapement estimate and associated variance. This design provides unbiased estimates of both total escapement and its variance (Reynolds et al. in press). Calculations are discussed in the Statistical Analysis section on pages 378–380.

Nonreplicated versus replicated systematic sampling designs

A recent comparison of systematic sampling designs for counting towers for large (22 million) and small (2 million) salmon escapements demonstrated that nonreplicated systematic sampling produced estimates of the variance associated with the total escapement estimate that were biased high even with the best variance estimator, while replicated systematic sampling provided unbiased estimates (Reynolds et al., in press). Furthermore, the simulation study showed a 25% reduction in the average estimated variance, averaged across years of simulated high and low escapements, using a replicated design of four replicated systematic samples of 10 min/4 h compared with the standard nonreplicated 10 min/h design.

Field/Office Methods

Setup

Preseason tasks

1. The tower site should be selected in advance of the anticipated project to maximize efficiency and accuracy as well as to provide continuity across years. What seems like an ideal tower site relative to abiotic factors (e.g., water depth, substrate) may not be ideal relative to salmon behavior. The most important factor in selecting a site is that salmon pass the selected counting tower site in an observable pattern. A pilot season to check migration and flow patterns at peak and postpeak migration is advised.

2. The selected site must be able to support a 3–7-m tower, stabilization cables, and a field camp for at least three people. The field camp should be located above flood lines.

3. Evaluate your field site relative to your planned power source. For example, if you plan to use solar cells, make sure you are able to capture sufficient sunlight for your seasonal power needs.

4. Permits (as applicable): Obtain landowner (e.g., state, tribal, federal, private) permits well in advance of field season. Deadlines, requisites, and fees vary. Collecting age, sex, and length data may require a fish handling permit.

5. Prior to going into the field, order, assemble, and test critical counting tower and camp gear, such as scaffolds and anchors, solar panels, lights, camp stove, and water purifier. Bring extra parts and leave enough time to obtain any missing parts.

6. Advertise available positions and recruit personnel.

Events Sequence

Project leaders usually have a known time frame within which the salmon run of interest will occur. Events sequence varies, depending on what data are needed. For total escapement counts over the duration of the salmon migration, our field crews (consisting of 3 to 4 people each) arrive at the main office 1–2 weeks in advance of project mobilization to undergo safety training, get supplies, and pack. Crews and gear generally reach field sites via plane charter and/or boat. Once on the site, towers and light systems typically take a day or two to set up; counts generally begin on the second day. The first day of counts is from 08:00 to 17:00 hours. If no salmon are observed, the next shift begins the next day at 08:00. Once fish begin moving by the towers, crew members work 8-h shifts, 24 h a day. Tower counts stop once the fish stop migrating or when the daily estimated fish numbers drop below 1% of the total run size.

Data Handling, Analysis, and Reporting

Measurement details

Data collection
Data collection at most towers is relatively simple: Observers count fish as they move upstream past the tower. When fish densities are high, counters may use one click on their hand tally to indicate tens, hundreds, and (rarely) thousands of passing salmon. Downstream movement is tracked on a separate hand tally and subtracted from the daily escapement total prior to expansion. Again, downstream movement can be minimized or avoided by careful site selection.

Observers each keep a log of hourly counts and observations (e.g., date, initials, interval time, species counts for left and right banks, visibility comments [see Table 3]) in a waterproof field notebook; data are entered onto standard forms after each shift. Some projects are more complex, requiring observers to keep track of both upstream (+) and downstream (–) movement of fish and/or more than one species. Daily totals are called in to main offices each day, entered into an Excel spreadsheet, graphed, and distributed as needed to cooperators and stakeholders.

TABLE 3.—Water clarity rankings used at salmon counting towers by the Alaska Department of Fish and Game.

Rank	Description	Salmon Viewing	Water condition
1	Excellent	All passing fish observable	No turbidity or glare; all routes of passage observable
2	Good	All passing fish observable	Minimal to very low levels of turbidity or glare; all routes of passage observable
3	Fair	All passing fish observable	Low to moderate levels of turbidity or glare; all routes of passage observable
4	Poor	Some passing fish may be missed	Moderate to high turbidity or glare; some likely routes of passage obscured
5	Unobservable	Passing fish not observable	High turbidity or glare; all routes of passage obscured

Original data sheets are transferred to the main office regularly and are reviewed by the project leader. Any discrepancies between reported and original data are verified with observers to ensure accuracy.

Although it happens relatively rarely, high discharge, turbidity, or lack of access to the tower can be a problem. When it does happen, estimate the missed count with linear interpolation between counts prior to and after the event. If only one bank count was missed, estimate the missing count based on either the bank-to-bank relationship or by the time relationship for that bank, depending on which relationship is stronger.

Statistical analysis

Nonreplicated systematic samples: total escapement $\hat{Y}$
Total annual escapement is estimated by expanding the daily mean escapement:

$$\hat{Y}_{Nonrep} = (expansion) \times \bar{y} = (expansion) \times \sum_{i=1}^{n} y_j / n \quad \text{(eq 1)}$$

where y_j is the count from the jth observation period, there were n total observations, and the *expansion* factor is (# of possible intervals per sample period) × (# of sample periods in a days) × (# of days in season) (Reynolds et al., in press). For example, for a 10-min/h systematic sample, the expansion factor is 6 × 24 × N, where there were N days in the season.

Nonreplicated systematic samples: variance of total escapement
The V5 estimator by Wolter (1984) is recommended for estimating variance of the total escapement estimate. It reduced uncertainty by 38–95% compared to other common variance estimators in a simulation study of systematic sampling for tower counts (Reynolds et al., in press). The estimator is based on sequential differences among observations:

$$V(\hat{Y}_{Nonrep}) = (1-f)(1/n) \sum_{j=5}^{n} c_j^2 / (3.5(n-4)) \qquad (eq\ 2)$$

where

$$c_j = y_j/2 - y_{j-1} + y_{j-2} - y_{j-3} + y_{j-4}/2$$

where y_j is the count from the jth observation period, there were n total observations, and f is the proportion of the possible observations that were actually collected, f = 1 / (# of possible intervals per sample period).

Nonreplicated systematic samples: confidence intervals
Reynolds et al. (in press) examined four confidence interval estimators for both nonreplicated and replicated systematic samples and found that they were effectively identical in terms of both mean width and coverage. Here, the familiar normal interval is recommended:

$$\hat{Y}_{Nonrep} +/- 1.96 \sqrt{\hat{V}(\hat{Y}_{Nonrep})} \qquad (eq\ 3)$$

assuming
$$\hat{Y}_{Nonrep} \sim Normal(\hat{Y}_{Nonrep}, \hat{V}(\hat{Y}_{Nonrep}))$$

Replicated systematic samples: total escapement
Following the example earlier, assume the design provided two replicates, each a nonreplicated systematic sample of 10 min/2 h. One replicate observed the second interval of each period, the other the tenth interval. At the end of the season the counts from interval 2 (12:10, 14:10, etc.) and the counts from interval 10 (13:30, 15:30, etc.) are analyzed separately as nonreplicated systematic samples using the equations above, each providing an estimate of total escapement. The overall estimate of total escapement is the mean of the replicate estimates

$$\hat{Y}_{Nonrep} = (\hat{Y}_{Nonrep\ 1} + \hat{Y}_{Nonrep\ 2})/2 \qquad (eq\ 4)$$

Replicated systematic samples: variance of total escapement
The variance of the total escapement is estimated directly from the replicate total escapement estimates using the usual sample variance formula:

$$V(\hat{Y}_{Rep}) = \frac{1}{k-1}\sum_{i=1}^{k}(\hat{Y}_{Nonrep_i} - \hat{Y}_{Rep})^2 \qquad \text{(eq 5)}$$

where k is the number of replicates ($k = 2$ in the example).

Replicated systematic samples: confidence intervals
The normal interval is recommended for confidence intervals, as it was for nonreplicated systematic samples.

Database design
An excellent example of database design used for salmon escapement is available from the Alaska Department of Fish and Game (ADF&G), Commercial Fishery Division for its Bristol Bay Management Area.

Data entry procedures
All data are field checked and verified by the project leader. In the main office original data (from observer forms) are entered into the database and proofed up to three times for accuracy. Proofreaders initial and date raw forms to indicate task completion. Data are backed up nightly onto the network by system administrators and individual project leaders.

Data summaries
Report formats vary with the agency conducting the escapement estimate. Our projects follow standard scientific reporting outlines (i.e., abstract or executive summary, introduction, methods, results, conclusions) with the addition of any recommendations and problems encountered (Woody 2004). Commercial fishery managers of the ADF&G include tables of daily escapement from all towers in its district as part of a much larger management report (ADF&G 1974–2000).

Archival procedures
Hard data are copied and archived with cooperating agencies; raw forms are stored by year in the main office of the collecting agency with a computer backup of the final database for that year. Electronic data are generally archived on the network hard drive, on the project leaders' hard drive, and on a DVD or similar storage device.

Personnel Requirements and Training

Tower crews generally consist of 3–4 technicians who conduct a daily 8-h shift and a project leader who ensures that crews are properly trained and that data are collected and recorded neatly and accurately.

Responsibilities
Project leaders are responsible for

1. selecting an appropriate counting tower site and conducting a pilot study;
2. obtaining appropriate permissions for camp establishment from landowners;
3. obtaining all appropriate state and federal sampling permits;

4. determining run timing of target species and mobilizing crew in a timely manner;
5. ensuring that towers are correctly installed and secured;
6. installing the electrical system and determining proper light angle and rheostat settings;
7. training field crews by
   a. ensuring that crews have all necessary safety training (e.g., first aid, CPR),
   b. training crews to be consistent and accurate in time and manner of counts (missed or late counts are undesirable and complicate error estimates),
   c. ensuring that crew members understand how the project fits into regional fishery management plans,
   d. conducting independent paired counts with inexperienced staff members for a minimum of 24–48 h over a range of conditions and fish densities, and
   e. ensuring that paired counts with experienced crew members should be made for 20 h over a range of fish densities.
8. periodically visiting each field crew to review accuracy of data collection, recording, and expansion practices;
9. assisting with counts during run peaks to assure quality control and accuracy; and
10. writing the annual report.

Technicians are responsible for
1. setting up and maintaining field camp,
2. learning to identify and enumerate the species of interest,
3. ensuring that the counts made during their 8-hour shift are conducted on time,
4. recording hourly counts in both field notebooks and on daily data sheets,
5. entering data into a computer and graphing results, and
6. calling in daily totals to fishery managers if necessary.

Qualifications

Compared to weirs, mark–recapture , and hydroacoustic projects, counting tower projects are relatively simple to implement. Project leaders should understand systematic sampling protocols, be able to install necessary electrical system (e.g., solar, turbine, generator), and have excellent troubleshooting and supervisory skills.

Field crew observers should be reliable, able to identify the species of interest, make simple mathematical calculations, safely pilot watercraft (if necessary), climb towers, and assist in field camp logistics and maintenance. Observers can generally develop needed skills on site. Because many counting tower projects are conducted in remote areas and require frequent boat travel, crew members should receive appropriate safety training.

It is recommended that field crews receive and pass tests for first aid, CPR, watercraft safety, bear behavior and safety, firearms safety, and proper lifting techniques. Having at least one crew member with watercraft troubleshooting skills is a valuable addition. Many counting tower projects successfully employ volunteers, entry-level technicians, retirees, teachers on summer break, and student interns. The best way to train a new field crew in systematic sampling protocols, fish identification, and data collection is to train them on site. Pairing experienced personnel with new crew members to ensure accurate and consistent data collection and entry is critical.

Budget Considerations

Estimated costs of a remote site counting tower system

Generally, there is a positive correlation between project cost and remoteness of the site. If observers must cross the river to make counts, two dependable boats are necessary; project leaders can determine which boat and motor combination is best suited to river conditions. The following estimates were valid in 2005. To save money on food shipments to remote locations, it may be more economical to shop for nonperishables well in advance and then mail or barge them to the nearest accessible town for later transport to the field site.

TABLE 4.— Cost estimate for establishing a remote counting tower field site, based on 2006 estimates obtained over the Web. Salary, permits, food, fuel, and transportation costs are not included, due to wide variation in cost among potential sites.

Item	Quantity	Cost per unit (USD)
Aluminum scaffolding and anchoring systems	1–4	$4,000–8,000
Solar panels, inverters, and charger accessories	2—one for each bank	$2,000–4,000
12 V deep-cycle marine batteries	4	$100–500
Lighting system for river and camp	Varies by project design	$100–2000
Field gear (e.g., tents, sleeping bags, stove, water system)	Per person	$1,000

Operational Requirements

Equipment needs

1. Enough counting towers to allow complete observation of the migration corridors. Aluminum scaffolding is most often used; however, high bluffs, trees, dams, and bridges can be surrogates.
2. Electronic timers with audible alarms to delineate shifts.
3. Polarized sunglasses; dark pairs for sunny days and lighter pairs for darker days.
4. Waterproof field notebooks and pencils for each observer to record his/her counts.
5. Hand tally counters.
6. Light system: night counts are integral to obtaining good abundance estimates at towers. Fish will avoid bright lights, but if the entire river is illuminated as allowed by the design illustrated in Figure 8, fish have no

choice but to pass through the beam. Rheostat controlled light beams (see Figure 1) may be used when it is not feasible or desirable to illuminate the entire river width. They can be used to either illuminate the fish passage zone or to divert fish closer to shore, where they are more visible.

7. If the site is remote, two boats are recommended for safety; if one is disabled, the other can be employed; fuel and oil for field season, tool kits, and troubleshooting guides are also essential.

8. If there is no electricity to the site, solar panels, generators, or turbine power can be used as energy sources. (We use solar panels to charge four deep-cycle marine batteries from which we run tower and office lights.)

9. If it is difficult to distinguish fish from the background, a contrasting panel or lighter substrate is necessary. A color similar to the bottom will provide contrast but not frighten fish. Materials used to improve contrast range from plastic sheets to metal panels anchored to the bottom. Experimentation will reveal the best substrate for your site.

Acknowledgements

Special thanks to Larry DuBois, Tracy Lignan, Gary Todd, Matt Evenson, Lowell Fair, Pat Poe, and the many Bristol Bay fishery biologists of the Alaska Department of Fish and Game who assisted with this project.

Literature Cited

Alaska Department of Fish and Game. 2003. Bristol Bay salmon fishery. Alaska Department of Fish and Game, Division of Commercial, Fisheries, Regional Information Report No. 2A01-10, Anchorage.

Alaska Department of Fish and Game. 1975–2004. Annual Bristol Bay salmon forecast. Alaska Department of Fish and Game, Division of Commercial Fisheries, Informational Leaflet Numbers 164, 167, 169, 171, 173, 177, 183, 190, 197, 209, 229, 244, 247, 253, 255, and 259; Bristol Bay Data Report Numbers 85-1, 85-13, 86-9, 87-1, 87-5, and 88-5; and Regional Information Report Numbers 2K88-13, 2K90-01, 2A92-12, 2A93-01, 2A94-04, 2A94-28, 2A95-17, 2A96-32, 2A98-02, 2A99-03, and 2A00-38, Anchorage.

Alaska Department of Fish and Game. 1974–2000. Annual "Bristol Bay salmon catch and escapement data compilations," Alaska Department of Fish and Game, Division of Commercial Fisheries, Technical Data Report Numbers 24, 40, 43, 47, 88, 94, 128, 129, 175, and 191 and Technical Fishery Report Numbers 89-06, 89-07, 90-14, 91-15, 92-17, and 94-16, Anchorage.

Anderson, C. J. 2000. Counting tower projects in the Bristol Bay area, 1955–1999. Alaska Department of Fish and Game, Division of Commercial Fisheries, Regional Information Report 2A00-08, Anchorage.

Becker, C. D. 1962. Estimating red salmon escapements by sample counts from observation towers. U.S. Fish and Wildlife Service Fishery Bulletin 61(192):355–369.

Bevan, D. E. 1960. Variability in the aerial counts of spawning salmon. University of Washington, College of Fisheries, Contribution No. 61. Seattle.

Cochran, W. G. 1977. Sampling techniques. John Wiley & Sons, New York.

Cousens, N. B. F., G. A. Thomas, C. G. Swann, and M. C. Healey. 1982. A review of salmon escapement estimation techniques. Canadian Technical Report of Fisheries and Aquatic Science 1108.

Dunmall, K. 2004. The Snake River, Eldorado River and Pilgrim River Salmon Escapement Enumeration and Sampling Project Summary Report, 2003. A report prepared for the Norton Sound Salmon Research and Restoration Steering Committee. Kawerak, Inc. Nome, Alaska. 53 p.

Eicher, G. J. 1953. Aerial census of salmon in Alaska. Journal of Wildlife Management 4:521–528.

Edwards, M. R. 2005. Comparison of counting tower estimates and digital video counts of coho salmon escapement in the Ugashik Lakes. U.S. Fish and Wildlife Service, Alaska Fisheries Technical Report No. 81, King Salmon Fish and Wildlife Field Office, King Salmon, Alaska.

Fair, L. 2004. Critical elements of Kvichak River sockeye salmon management. Alaska Fishery Research Bulletin 10(2):95–103.

Fisheries Research Institute. 1955. How shall we defend the concept of sustained yield conservation? Pacific Fisherman 53(6):18–20.

Groot, G., and L. Margolis. 1991. Pacific salmon life histories. University of British Columbia Press, Vancouver.

Hamazaki, T. 2003. Kwiniuk River salmon counting tower project review and tower estimation. Alaska Department of Fish and Game, Division of Commercial Fisheries, Regional Information Report No. 3A03-20, Anchorage.

Johnstone, A. H., and F. Percival. 1976. Attention breaks in lectures. Education in Chemistry 13:49–50.

Kohler, T., and G. Knuepfer. 2002. Kwiniuk River salmon counting tower project, 2001. Alaska Department of Fish and Game, Commercial Fisheries Division, Regional Information Report No. 3A02-12, Anchorage.

Poe, P. H., and D. E. Rogers. 1984. 1984 Newhalen River adult salmon enumeration program. University of Washington, Fisheries Research Institute, FRI-UW-8415, Seattle.

Reynolds, J. H., C. A. Woody, N. Gove, and L. Fair. In press. Efficiently estimating salmon escapement uncertainty using systematically sampled data. In C. A. Woody, editor. Sockeye salmon evolution, ecology, and management. American Fisheries Society, Symposium 54, Bethesda, Maryland.

Rietze, H. L. 1957. Western Alaska salmon investigations; field report on the evaluation of towers for counting migrating red salmon in Bristol Bay, 1956. Mimeo report to the U.S. Department of the Interior, U.S. Fish and Wildlife Service, Bureau of Commercial Fisheries, Juneau, Alaska.

Seibel, M. C. 1967. The use of expanded ten-minute counts as estimates of hourly salmon migration past the counting towers in Alaskan rivers. Alaska Department of Fish and Game, Division of Commercial Fisheries, Informational Leaflet 101, Juneau.

Spangler, P. J., and H. L. Rietze. 1958. Field report on the evaluation of towers for counting migrating red salmon in Bristol Bay, 1957. Mimeo Report to the U.S. Department of the Interior, U.S. Fish and Wildlife Service, Bureau of Commercial Fisheries, Juneau, Alaska.

Symons, P. E. K., and M. Waldichuck, editors. 1984. Proceedings of the workshop on stream indexing for salmon escapement estimation, West Vancouver, B.C. Canadian Technical Report of Fisheries and Aquatic Science 1326.

Taras, B. D., and D. R. Sarafin. 2005. Chinook salmon escapement in the Gulkana River, 2002. Alaska Department of Fish and Game, Fishery Data Series No. 05-02, Anchorage.

Thompson, W. F. 1962. The research program of the Fisheries Research Institute in Bristol Bay, 1945–1958. In T. S. Y. Koo, editor. Studies of Alaskan red salmon. University of Washington Press, Seattle.

Wolter, K. M. 1984. An investigation of some estimators of variance for systematic sampling. Journal of the American Statistical Association 79(388):781–790.

Woody, C. A. 2004. Population monitoring of Lake Clark and Tazimina River sockeye salmon, Kvichak River watershed, Bristol Bay, Alaska, 2000-2003. Final Report for Study 01-095 to the U.S. Fish and Wildlife Service, Office of Subsistence Management, Anchorage, Alaska.

Weirs

Christian E. Zimmerman and Laura M. Zabkar

Background and Objectives

Weirs—which function as porous barriers built across stream—have long been used to capture migrating fish in flowing waters. For example, the Netsilik peoples of northern Canada used V-shaped weirs constructed of river rocks gathered on-site to capture migrating Arctic char *Salvelinus alpinus* (Balikci 1970). Similarly, fences constructed of stakes and a latticework of willow branches or staves were used by Native Americans to capture migrating salmon in streams along the West Coast of North America (Stewart 1994). In modern times, weirs have also been used in terminal fisheries and to capture brood fish for use in fish culture. Weirs have been used to gather data on age structure, condition, sex ratio, spawning escapement, abundance, and migratory patterns of fish in streams.

One of the critical elements of fisheries management and stock assessment of salmonids is a count of adult fish returning to spawn. Weirs are frequently used to capture or count fish to determine status and trends of populations or direct in-season management of fisheries; generally, weirs are the standard against which other techniques are measured.

To evaluate fishery management actions, the number of fish escaping to spawn is often compared to river-specific target spawning requirements (O'Connell and Dempson 1995). A critical factor in these analyses is the determination of total run size (O'Connell 2003). O'Connell compared methods of run-size estimation against absolute counts from a rigid weir and concluded that, given the uncertainty of estimators, the absolute counts obtained at the weir wer significantly better than modeled estimates, which deviated as much as 50–60% from actual counts. The use of weirs is generally restricted to streams and small rivers because of construction expense, formation of navigation barriers, and the tendency of weirs to clog with debris, which can cause flooding and collapse of the structure (Hubert 1996). When feasible, however, weirs are generally regarded as the most accurate technique available to quantify escapement as the result is supposedly an absolute count (Cousens et al. 1982). Weirs also provide the opportunity to capture fish for observation and sampling of biological characteristics and tissues; they may also serve as recapture sites for basin-wide, mark–recapture population estimates. Temporary weirs are useful in monitoring wild populations of salmonids as well as for capturing broodstock for artificial propagation.

FIGURE 1. — A temporary weir constructed of metal tripods, stringers, and galvanized conduit with both an upstream and downstream trap. Ninilchik River, Alaska.

Rationale

Temporary weirs enable field biologists to quantify the escapement of adult salmonids in streams and rivers. In addition to providing absolute counts of fish migrating through the weir, weirs can be used to capture fish, determine sex ratios and species composition, recapture tagged fish, and collect tissue or scale samples. In locations where management relies on escapement goals, weirs can be used to monitor escapement in order to direct in-season management of commercial or subsistence fisheries.

Objectives

This protocol describes the methods used in counting migrating adult salmonids in streams and rivers using weirs and traps. Generally, weirs and traps are temporarily installed across the stream channel and enable monitoring practitioners to estimate or make an absolute count of fish passing that point in the stream.

FIGURE 2. — A resistance board weir featuring a skiff gate to allow boat passage (in the foreground, marked by pylons), a trap to capture fish in middle of weir, and a passage chute (background), where fish are counted as they pass the weir. Pilgrim River, Alaska. (Photo: Karen Dunmall and Tim Kroeker.)

Temporary weirs may be constructed from a range of materials. Rigid weirs generally consist of a fence and support structure; fences may be constructed from netting (Blair 1956; Noltie 1987) or from rigid material such as pipe or metal pickets (Hill and Matter 1991). These weirs are generally easy to dismantle and transport but are sometimes difficult to maintain during high water or in streams with high debris loads. Weirs constructed of screen or wire panels have a tendency to collect debris such as leaves and algae (Clay 1961). Kristofferson et al. (1986) used a weir constructed of polyethylene (Vexar®) and metal t-posts, similar to that described by Noltie (1987), on an Arctic river but found that clogging by algae and debris led to excessive water pressure that eventually caused the weir to collapse. Noltie (1987), however, reported that the same material could easily be cleaned of leaves using a push broom. Because weirs constructed of these materials are relatively inexpensive, they are probably best used for short-term studies in small shallow streams; practitioners choosing these materials need to take into account the time needed to clean these weirs of debris.

Weirs constructed of metal pickets (which are frequently made of aluminum rods or galvanized conduit) are more resistant to buildup of algae, leaves, and other fine material. In some designs, the pickets can be removed for easy cleaning and to reduce pressure from high flows. The length of the conduit will depend on the depth of the stream and should be long enough that salmon should not be expected to jump over the weir. Anderson and McDonald (1978), Kristofferson et al. (1986), and Hill and Matter (1991) describe construction details for such rigid weirs, which generally consist of structures that support panels of pickets. Supports usually consist of tripods constructed of pipe or wood and support stingers that hold the pickets (see Figure 3). By angling the upstream face of the weir at 120° relative to the stream bottom, the water flows slightly up the pickets before passing through; this movement creates a greater area over which water pressure may dissipate (Anderson and McDonald 1978). To further dissipate water pressure, the weir can be constructed so that the wings of the weir terminate in a 90° angle entering the trap box (see Figure 4). This arrangement allows more water to pass through the weir for a given stream width and guides fish into the trap. Mullins et al. (1991) describe a two-way trap that can be constructed in the apex of two weirs that allows for sampling of both upstream and downstream migrants.

FIGURE 3.—Construction details of tripods and weir panels made of metal stringers and pickets of galvanized electrical conduit.

Rigid weirs work best in rivers that have minimal variation in water flow and depth; these conditions will help avoid, to the greatest extent possible, frequency of washout of the trap and/or weir by increased flows or seasonal freshets. Lake outlets, therefore, are particularly suited to the placement of rigid weirs (Clay 1961). Rigid weirs are also susceptible to damage by large floating debris such as logs or ice. Resistance board weirs were designed to accommodate fluctuation in flow and debris and to allow for inclusion of easy-to-use boat passes (Tobin 1994; Stewart 2002) (see figures 2 and 4). Although not impervious to washout, resistance board weirs can be used in rivers that experience debris-laden high water periods (Tobin 1994). During high water, the resistance board weir will temporarily submerge when pressure created by water and debris loading reaches a point that would typically wash a rigid weir downstream (Tobin 1994). This flexibility requires less maintenance and also reduces the frequency of these occasions when fish cannot be counted.

Resistance board weirs consist of three main components: panels made of capped polyvinyl chloride (PVC) pipe, a rail anchored to the substrate that attaches the panels to the river bottom, and a trap box or chute where fish are captured or counted. Detailed construction and installation manuals for resistance board weirs are available in Tobin (1994) and Stewart (2002, 2003). In summary, a rail is installed across the stream. This rail is anchored to the substrate using steel rod and cables attached to duck-bill anchors placed upstream of the weir. The rail anchors and aligns the cable to which panels are attached. The weir panels are constructed of schedule 40 PVC electrical conduit 6.1 m in length. Electrical conduit is used rather than PVC water pipe because it resists breakdown caused by ultraviolet light. Panels consist of multiple pipes supported by 1.2 m-long stringers that are spaced evenly lengthwise along the panels to provide rigidity to the flexible PVC pipe. Pipe spacing is determined by the desired distance between pipes and is adjusted accordingly based on the size of each target species. A resistance board is attached at the downstream end of the panel to deflect water flow downward, which causes lift and holds the downstream end of the panel above the surface of the water (see Figure 5). Panels are attached to one another and span the width of the stream. At either end of the weir, a short section of fixed weir (similar to the rigid weirs described above) seals the end of the floating weir at either bank (Figure 4). Tobin (1994) and Stewart (2003) describe how to incorporate a skiff gate that allows upstream and downstream passage of boats without opening the weir.

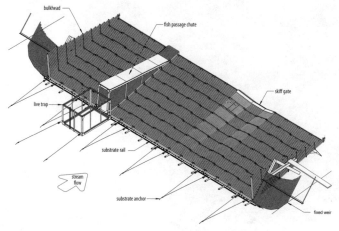

FIGURE 4.— Resistance board weir and major components. (Diagram: Rob Stewart, Alaska Dept. of Fish and Game.)

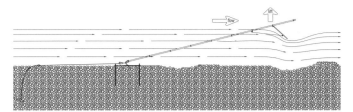

FIGURE 5. — Side view of a resistance board weir panel. The resistance board deflects water at the end of the panel and creates lift to counteract the downward pressure of flow. (Diagram: Rob Stewart, Alaska Dept. of Fish and Game.)

Both rigid and resistance board weirs use fish passage chutes and trap boxes to pass or capture fish. Fish passage chutes allow fish to swim though an opening in the weir and can either attach to a live box for trapping fish or include a counting station where fish are identified and counted. As fish pass through the counting chute they are tallied by observers. In some cases, to minimize personnel costs, automated counters that utilize video technology or resistivity counters are used to quantify fish passing through the fish counting chute (see Figure 6). Trap boxes are used to capture fish for direct examination or for sampling tissues, length, weight, and sex. After having been counted and examined, the fish is passed upstream of the weir. Trap-box designs are presented by Kristofferson et al. (1986), Whelan et al. (1989), and Mullins et al. (1991). In most designs fish enter the trap through a V-shaped passageway (termed a fyke), which inhibits fish from passing back downstream. After fish enter the trap, they either pass through the front gate or counting chute or are trapped for further examination. The trap should be big enough to hold expected numbers of fish comfortably. During the trapping session, the front gate (upstream) is closed and the back gate (downstream) is open to allow fish to enter the trap box. Once the desired number of fish enter the trap, the downstream gate is closed and fish are sampled and released upstream of the weir.

FIGURE 6. — Video monitoring chute, Nikolai Creek, Alaska. Fish passing through the passage chute swim through a video chamber, where they are filmed. Fish can be counted by viewing the video tape in the office.

Sampling design

When creating a sampling design for a weir project, it is important to evaluate the purpose and objectives of each project. In most cases, weirs are used to capture or count all fish passing that point in a stream. For example, if the goal of the project is to determine escapement to a particular river, it is critical that the weir be placed downstream of all spawning habitats and that the weir is operated through the entire migration. In the case of monitoring a stream with small numbers of migrating fish, all fish can be trapped, counted, and passed over the weir without continuous (24-h) counting through the chute. Rather, fish can be examined, counted, and passed at period intervals through the day. A counting chute allows for monitoring larger runs of fish and the counting of each species as they pass uninterrupted through the weir. Counting chutes are fitted with a light-colored floor to facilitate the identification of individual fish; counts are done visually by personnel stationed at the chute, video, or other automated counters. Personnel count fish as long as the counting chute is open to fish passage, which, depending on fish passage rates, may be up to 24 h per day.

Site Selection

Site selection includes two principle considerations. First, the weir must be located at a site that allows sampling and counting of fish to address the objectives of the study. For example, if the objective of the study is to determine escapement of upstream migrating salmonids, the weir must be located downstream of the lowest point of spawning habitat. Second, the location must be conducive to weir construction and maintenance through the range of water flows expected during the operation period.

Site selection is an important consideration in determining the success of weir construction and maintenance. When selecting possible weir sites, substrate, river flow, depth, and width, and timing of high water events should all be considered. Suitable sites for both rigid weirs and resistance board weirs are characterized by wide and shallow areas of stable substrate consisting of gravel or small cobble (Clay 1961). Stable substrates of pebbles or small cobbles allow the weir structure to lie flat on the surface of the substrate and also facilitate secure anchoring using pins or duck-bill anchors. If larger boulders are present, they may impede the structure, act as an obstacle, or create gaps that allow fish to get through the weir. At high water, stream energy is equally distributed across the stream in straight reaches of laminar flow, making it easier to maintain the weir through a range of flows. Water depths less than 1 m during normal flows and water velocity that is slow to moderate are preferred for both rigid and resistance board weirs. If the water is too deep and swift to allow an adult person to wade comfortably at normal flow, the site is not suitable for safe weir operations. At the same time, the weir site should have enough current (especially at passage chutes or traps) to ensure efficient fish passage and attraction flows. Stream width may vary, but practitioners should note that wider locations will require more material in weir construction. Near-vertical stream banks are easier to seal against fish passage but should not contain undercut channels because they are difficult to seal.

Another consideration when choosing weir location and design is the position of a trap or traps on the weir. The recommended number and location of traps depends on the size and morphology of the channel at the weir site. Since fish typically travel in the thalweg (the deepest portions of the river channel), the sampling area or trap should be located in this area. If the river has more than one thalweg or channel, it is recommended that the weir have more than one sampling location for fish to travel through. Trap placement on the weir should account for minimum expected depth. If the trap box is located in a site that is too shallow or too deep, it may be difficult to access the trap and handle fish in the trap, and may lead to excessive stress on fish in the trap. If the system is prone to flooding, it is common to have an alternate sampling location for times when normal operations at the main site are not possible. For example, many systems in Alaska experience flood stages during the rainy months of the summer; by installing a second trap in a different location on a given river, practitioners may avoid sampling delays during flood stages.

Sampling frequency and replication
Weirs are usually used to acquire an absolute count of migrating fish; therefore, sampling considerations need to ensure that weir operations begin before fish migration and continue through the end of migration. In the event that logistic considerations or environmental constraints interrupt weir operation, counts will be incomplete. An important consideration in the development of a fish weir project is replication. Depending on the objectives and goals of the study, many years of data may be needed. Practitioners examining run timing of Pacific salmon will need to monitor the entire run over several years to conduct trends analysis. For example, Korman and Higgins (1997) examined the needed replication of escapement estimates to adequately determine the response of salmon populations to habitat alteration, and they determined that posttreatment monitoring needed to be longer than 10 years.

Stratified sampling designs are a common method for estimating total run abundance and determining overall age, sex, and length composition. In most cases the operational period is stratified into weeks, with escapement determined for each week; and total escapement is simply the sum of weekly escapement. In locations where total fish runs are small, each fish may be handled to gather age, sex, and length data. When run size is too great to allow for sampling of all fish passing the weir, a stratified sampling design is used to estimate sex, age, and length distribution of fish passing the weir. This may be achieved by collecting scales and length from every nth fish of a species passed through the trap. The number of fish sampled needs to be determined prior to the season and in consultation with a statistician. On weirs that incorporate both a counting chute and trap, fish can be sampled in the trap for stratified periods of time throughout the run. This technique, also known as a pulse sampling design, is conducted over a 1–3-d time period, followed by a period without sampling (trapping and scale collection/length determination). In most cases, a target sample size for each species is sought for each sampling stratum (e.g., week). These samples are calculated as a subset of the entire escapement and expanded to characterize the age, sex, and length composition of the total annual escapement (see Data Handling, Analysis, and Reporting, page 394, for details).

Field/Office Methods

Setup

Before trapping can begin, all weir components need to be purchased, assembled, and constructed. For larger weirs, fabrication can take several weeks and requires use of a workshop with necessary tools and staff expertise. Commercially produced traps are available but can be very expensive. After construction the weir components need to be shipped to the weir location. For remote locations, this may involve the use of helicopters or other means of transportation. Once on-site, the weir can be installed in the stream. In Alaska weirs are frequently installed in early spring when water flow is low. Final panels or trap boxes are installed when trapping begins later in the season. In remote locations a camp must be constructed at the weir site to provide sleeping and eating quarters for all personnel required to maintain the weir. Once the crew is on-site, personnel in the office will need to facilitate grocery shipments and safety of the field crew.

Safety and operational training is an important step in preparing for a field season. Safety training should include first aid appropriate for the location, as remote field locations will require higher proficiency in dealing with injuries while waiting for transportation or medical aid. This training is frequently referred to as wilderness first-aid. Training in operation around water is required, and boat training is needed if the weir will be accessed by boat. Similarly, if helicopters or fixed-wing aircraft are used to access the weir site, training may be needed for proper conduct around aircraft. In remote locations in Alaska, training regarding bear encounters and gun handling is needed. Most agencies have established safety programs and requirements, and weir project planners should check with agency safety personnel when starting a weir project to determine what safety training will be needed and where such training can be acquired. Operational training is also important, and all weir personnel should receive training in weir operation, construction, and project implementation (as well as safety training). Personnel who fully understand the protocol and objectives of the project will be in a better position to make day-to-day decisions concerning project implementation.

Events sequence

Although weir projects may have a field season of a few weeks to months, preparation, analysis, reporting, and maintenance usually require year-round activity. Field crews need to be hired early enough to allow for preseason deployment and training. Creating a preseason task checklist is recommended. Appendix A details the preseason checklist used by the U.S. Fish and Wildlife Service on the Kwethluk River weir project. It includes all tasks that must be completed to initiate the Kwethluk River weir project, including scheduling, hiring, training, shipping, travel, and crew gear distribution. This checklist can be modified for use by other projects.

Another pertinent item to include when preparing for field season is an equipment checklist or inventory list. This should include all the equipment needed to complete the project with associated quantities, quality (i.e., used or new), and storage location. If the field project is located on private, state, or federal land, a land-use agreement or lease will be necessary to occupy and use the land,

and permits may be required. Scientific collecting permits may also be required from state fish and game agencies. Before beginning any project, local fish and wildlife managers should be consulted concerning permit requirements. When preparing for the field season it is important to create a timeline that is associated with the preseason checklist. The timeline should outline deadlines for the preparation, installation, operations, and takeout process.

Measurement details and sample processing

During weir operation, a range of data needs to be collected. These data include the number of fish passing the weir (usually recorded on an hourly basis), length, weight, and scale samples collected, water temperature, and flow (stage height or discharge). Data forms printed on weatherproof paper or notebooks should be developed to organize data gathering and ensure that all data are collected at the appropriate time. All data forms should be easy to understand, with proper headings and space provided to include all the data necessary. Fish passage should be monitored continuously, and fish passage should not be hindered. When operating a weir, it is important to ensure that the weir does not delay fish migration. Fleming and Reynolds (1991) found that Arctic grayling *Thymallus arcticus* delayed at a weir did not migrate as far as control fish and suggested that such delays could cause fish to spawn in suboptimal locations. Fish collected in traps should be sampled as quickly as possible to minimize holding stress and migration delay.

There are two different approaches for weir operation: one utilizes the trap to capture fish and release manually, and the other utilizes the trap as a counting station allowing fish to pass uninterrupted. Typically, the size of the escapement or objectives of the study will determine which technique is suitable for each system. The first method involves trapping fish as they pass through the weir, sampling them for genetics, age, sex, or length information, and releasing them by hand above the weir. This method works well on smaller systems. For larger systems, trap operation may be round-the-clock, and the design would include a counting chute that allows fish to pass uninterrupted. An observer or a video camera would then count each fish as it passes upstream through the weir. In this method, a counting session commences as the counting chute is open, and fish are identified and counted as they pass through the weir. Fish can also be examined by simply allowing the counting chute to remain closed and the downstream gate of the trap to remain open, essentially allowing fish to move into the trap, where they are retained until examined and then released upstream of the weir.

Sample processing may take place in-season or, in most cases, will occur postseason. In-season samples, such as daily escapement estimates, scale samples, or genetic samples, may need to be transported for immediate consideration by fishery managers. In this case, it is important that sampling procedures include a detailed component for quality control. If this element occurs in the field, proper steps should be taken to ensure accurate data collection and proper crew training; one individual needs to be responsible for the oversight and handling of samples.

Data Handling, Analysis, and Reporting

To determine total escapement past the weir, the total number of fish passing the weir each day is summed. In the case of weirs that incorporate a fish counting chute that is open to passage at all times, if fish are only counted during a portion of the day, then the count is expanded to estimate the full day's passage (assuming that passage rates are constant throughout the day). For full days missed because of high water or other events that prevent fish counting, linear interpolation of the counts before and after the missed day(s) can be used to estimate passage for that time period. In cases where fish are passed through a fish trap or where all fish are counted (24-h per day), the resulting number is the absolute number of fish passing the weir.

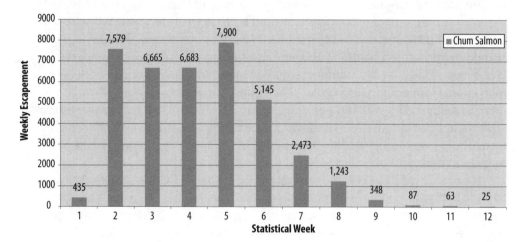

FIGURE 7. — Weekly passage (escapement) of chum salmon *Oncorhynchus keta* at the Kwethluk River weir, 2004 (data from Roettiger et al. [2005]).

When analyzing data from weirs, the sampling period is usually stratified into weekly periods (see Figure 7). Within each stratum (or week), the proportion of fish of a given sex or age (p_{ij}) is calculated as

$$\hat{p}_{ij} = \frac{n_{ij}}{n_j} \quad \text{(eq 1)}$$

where n_{ij} is the number of fish of sex i or age i sampled in week j, and n_j is the total number of fish sampled in week j. Sex or age composition (p_i) for the total run of a species of sex i or age i is calculated as

$$\hat{p}_{ij} = \sum \left(\frac{N_j}{N} \right) \hat{p}_{ij} \quad \text{(eq 2)}$$

where N_j equals the total number of fish of a given age or sex passing through the weir during week j, and N is the total number of fish of a given species during the run.

Personnel and Operational Requirements

A successful weir project requires a dedicated and professional staff. Typically, projects will be staffed by a crew leader who is a fishery biologist. The crew leader is responsible for the day-to-day operation of the weir, including determination of crew schedules and daily tasks, quality control of data, and overall project performance. Crew members are responsible for following crew leader's instructions of the and ensuring that tasks are completed in a safe manner that is consistent with project objectives. When possible, project personnel should be hired locally from surrounding communities. Locally hired personnel are likely to have personal experience in the area; such practices are an important step in establishing community support for the project.

The number of personnel required to operate a weir depends on the project. Crew members should always work in pairs to ensure safety. If few fish are encountered, a two- or three-person crew may be sufficient to sample fish and maintain the weir. Projects intended to count large numbers of salmon around the clock may require up to eight people.

Equipment

- Weir and traps
- Boats (if needed to reach weir site)
- Dip nets for collecting fish from trap
- Scales and measuring boards
- Sample containers or cards for scale and genetic samples
- Floodlights (for night work)
- Tools and equipment for maintaining or repairing weir
- Camp equipment, including tents and cooking gear
- Safety equipment (fire extinguishers, personal flotation devices, medical kits, etc.)
- Radio, satellite phone, or other means of communication
- Brooms or rakes for cleaning front of weir

Budget Considerations

Estimated costs

Generally there is a positive correlation between project cost and remoteness of the site. Approximate amounts for two differing kinds of weirs (a picket weir and a floating weir) are shown in the following budget breakdown; costs are in U.S. dollars as of 2005.

Item	Quantity	Cost per weir (USD)
Picket weir with one trap	30 m length	$65,000
Floating weir with one trap	60 m length	$100,000
Lighting system for river and camp	Varies by project design	$100–2,000
Field gear for remote site (e.g., tents, sleeping bags, stove, water system)	Per person	$1,000

Literature Cited

Anderson, T. C., and B. P. MacDonald. 1978. A portable weir for counting migrating fishes in rivers. Canada Fisheries and Marine Service Technical Report 733.

Balikci, A. 1970. The Nesilik Eskimo. Natural History Press. New York. 276 p.

Blair, A. A. 1956. Counting fence of netting. Transactions of the American Fisheries Society. 86:199–207.

Clay, C. H. 1961. Fences (or weirs) and barrier dams. Pages157–183 in C. H. Clay, editor. Design of fishways and other fish facilities. Queen's Printer, Ottawa.

Cousens, N. B. F., G. A. Thomas, C. G. Swann, and M. C. Healey. 1982. A review of salmon escapement techniques. Canadian Technical Report of Fisheries and Aquatic Sciences 1108.

Fleming, D. F., and J. B. Reynolds. 1991. Effects of spawning-run delay on spawning migration of Arctic grayling. Pages 299–305 in J. Colt and R. J. White, editors. Fisheries bioengineering symposium. American Fisheries Society, Symposium 10, Bethesda, Maryland.

Hill, J. P., and W. J. Matter. 1991. A low-cost weir for salmon and steelhead. The Progressive Fish-Culturist. 53:255–258.

Hubert, W. A. 1996. Passive capture techniques. Pages 157–181 in B. R. Murphy and D. W. Willis, editors. Fisheries techniques, 2nd edition. American Fisheries Society, Bethesda, Maryland.

Korman, J., and P. Higgins. 1997. Utility of escapement time series data for monitoring the response of salmon populations to habitat alteration. Canadian Journal of Fisheries and Aquatic Sciences. 54:2058-2067.

Kristofferson, A. H., D. K. McGowan, and W. J. Ward. 1986. Fish weirs for the commercial harvest of searun Arctic charr in the Northwest Territories. Canadian Industry Report of Fisheries and Aquatic Sciences 174.

Mullins, C. C., P. L. Caines, D. Caines, and J. L. Peppar. 1991. A two-compartment fish trap for simultaneously counting downstream and upstream migrants in small rivers. North American Journal of Fisheries Management 11:358–363.

Noltie, D. B. 1987. Fencing material and methods for effectively trapping and holding pink salmon. North American Journal of Fisheries Management 7:159–161.

O'Connell, M. F. 2003. Uncertainty about estimating total returns of Atlantic salmon, *Salmo salar* to the Gander River, Newfoundland, Canada, evaluated using a fish counting fence. Fisheries Management and Ecology 10:23–29.

O'Connell, M. F., and J. B. Dempson. 1995. Target spawning requirements for Atlantic salmon, *Salmo salar* L., in Newfoundland rivers. Fisheries Management and Ecology 2:161–170.

Roettiger, T. G., F. Harris, and K. C. Harper. 2005. Abundance and run timing of adult Pacific salmon in the Kwethluk River, Yukon Delta National Wildlife Refuge, Alaska, 2004. U.S. Fish and Wildlife Service, Kenai Fish and Wildlife Field Office, Alaska Fisheries Data Series Number 2005-7, Kenai, Alaska.

Stewart, H. 1994. Indian fishing: early methods on the Northwest coast. University of Washington Press, Seattle.

Stewart, R. 2002. Resistance board weir panel construction manual. Alaska Department of Fish and Game, Division of Commercial Fisheries, Arctic–Yukon–Kuskokwim Region, Regional Information Report No. 3A02-21, Anchorage.

Stewart, R. 2003. Techniques for installing a resistance board fish weir. Alaska Department of Fish and Game, Division of Commercial Fisheries, Arctic-Yukon-Kuskokwim Region, Regional Information Report No. 3A02-21, Anchorage.

Tobin, J. H. 1994. Construction and performance of a portable resistance board weir for counting migrating adult salmon in rivers. U.S. Fish and Wildlife Service, Kenai Fishery Resource Office, Alaska Fisheries Technical Report Number 22, Kenai, Alaska.

Whelan, W. G., M. F. O'Connell, and R. N. Hefford. 1989. Improved trap design for counting migrating fish in rivers. North American Journal of Fisheries Management 9:245–248.

Appendix A: Pre-season checklist used to prepare the Kwethluk River weir project, U.S. Fish and Wildlife Service, Kenai, Alaska\

- ☐ Determine Field Schedule _____
- ☐ Set-up Contract Funding _____
- ☐ Build New & Replacement Parts _____
- ☐ Hire Local Crew (see below) _____
- ☐ Hire USFWS Crew (see below) _____
- ☐ Coordinate Short-Term Volunteers (see below) _____
- ☐ Train Crew _____
- ☐ Issue gear to Crew _____
- ☐ Crew Schedules to Kwethluk _____
- ☐ Update Supplies According to Inventory List _____
- ☐ Initiate Satellite Phone Account _____
- ☐ Initiate grocery Accounts _____
- ☐ Initiate Fuel Account _____
- ☐ Ship Supplies (Consider HazMat Lead-Time) _____
- ☐ Update Field Procedures _____
- ☐ Update Emergency Procedures & Acquire Signatures _____
- ☐ Brief Crew on Emergency Procedures _____
- ☐ Set-Up Field Computer _____
- ☐ Make Travel Arrangements _____
- ☐ Refuge for Bunkhouse arrangements _____
- ☐ Ship Vehicle, _____
- ☐ Ship Boat _____
- ☐ Vehicle Key _____

Hire Local Crew: Position   Contact #
- ☐ 1. _____   _____   _____
- ☐ 2. _____   _____   _____
- ☐ 3. _____   _____   _____

Hire USFWS Crew:
- ☐ 1. _____   _____   _____
- ☐ 2. _____   _____   _____

Coordinate Short-Term Volunteers:
- ☐ 1. _____   _____   _____
- ☐ 2. _____   _____   _____
- ☐ 3. _____   _____   _____
- ☐ 4. _____   _____   _____

Train Crew: 1st Aid/CPR   Watercraft   Bear/Firearms
- ☐ 1. _____   _____   _____   _____
- ☐ 2. _____   _____   _____   _____
- ☐ 3. _____   _____   _____   _____
- ☐ 4. _____   _____   _____   _____
- ☐ 5. _____   _____   _____   _____
- ☐ 6. _____   _____   _____   _____
- ☐ 7. _____   _____   _____   _____

Issue Gear to Crew: Hip Waders   Chest Waders   Rain Gear   Sleeping Bag   Other
- ☐ 1. _____   ☐   ☐   ☐   ☐   _____
- ☐ 2. _____   ☐   ☐   ☐   ☐   _____
- ☐ 3. _____   ☐   ☐   ☐   ☐   _____
- ☐ 4. _____   ☐   ☐   ☐   ☐   _____
- ☐ 5. _____   ☐   ☐   ☐   ☐   _____
- ☐ 6. _____   ☐   ☐   ☐   ☐   _____

Aerial Counts
Edgar L. Jones III, Steve Heinl, and Keith Pahlke

Background
Aerial counts of salmon are essential tools in Pacific salmon *Oncorhynchus spp.* management. In Alaska the first recorded aerial count of salmon was made by C. M. Hatton of the U.S. Bureau of Fisheries in the Lake Clark district of Bristol Bay in 1930. As fisheries management progressed, so did the need to cover more streams in shorter periods of time, inspiring the first systematic use of aerial surveys in Alaska by Agent Fred O. Lucas of the Bureau of Fisheries in 1937 (Eicher 1953).

The aerial survey technique is best suited for broad, shallow, clear-water systems with limited overhanging vegetation, undercut banks, and canopy cover. Aerial counts are severely compromised in glacial or turbid waters and in excessively deep water such that fish are beyond the range of visibility (Cousens et al. 1982). Species such as steelhead *O. mykiss* and coho *O. kisutch* salmon can be difficult to survey as these fish are often cryptic in coloration and have the behavior of seeking cover, even during spawning, making them less visible.

The visibility of spawning salmon to observers depends on many factors such as water quality, fish concealment, stream dimensions, and density of fish, among others (Bevan 1961). The ability of the observer to count fish accurately has been the main topic of many aerial survey studies (Bevan 1961; Neilson and Geen 1981; Cousens et al. 1982; Labelle 1994; Symons and Waldichuk 1984; Dangel and Jones 1988; Jones et. al 1998). Furthermore, biased counts of salmon abundance and associated measurement error have been seen to produce seriously biased estimates of optimum harvest rate and escapement in stock-recruitment analysis (Walters 1981; Walters and Ludwig 1981). An interesting phenomenon is that the accuracy and precision of observer counts decreases as abundance increases, and simple linear corrections for bias are not as appropriate as using allometric forms with multiplicative error structure in light of changing magnitudes of fish. In short, humans are overly conservative and tend to underestimate versus overestimate when counting objects (Jones et al. 1998; Clark 1992; Dangel and Jones 1988; Daum et al. 1992; Evensen 1992; Rogers 1984; Shardlow et al. 1987; Skaugstad 1992).

Efforts should be made to minimize the influence of extraneous variables such as weather, water quality, aircraft type, and pilot performance, and observers should minimize the impacts of these variables to the best of their abilities. The density of fish may also be an important variable. Eicher (1953), in work performed in Bristol Bay, said that the accuracy of observer counts might be inversely proportional to the density of salmon. Often, salmon can be seen packed into very tight schools, and in one study on coho salmon, fish were much easier to count once they were disturbed and disbursed, in principle lowering the school density (Irvine et al. 1992). In essence, increasing the density of salmon has much the same effect as increasing the number of undercut banks, water glare and turbidity, and canopy cover (Jones et. al 1998). Prior knowledge of the stream is beneficial with regard to accuracy when performing aerial counts. One study showed that observers familiar with the stream consistently produced more accurate estimates when compared to observers not familiar with the stream (ADFG 1964).

Rationale

Reliable methods for estimating escapements are of critical importance to fisheries management agencies. Such information is vital in forecasting production in subsequent years as well as in measuring the relative success of management charged with achieving adequate escapements over time. Aerial counts are a common method used to index escapement, given the large number of streams around the Pacific Rim (and elsewhere) that produce salmon. Often, these counts can be quite crude, providing little more than an index of escapement from year to year (Neilson and Geen 1981). Specifically, aerial counts are valuable not so much as estimators of the actual magnitude of salmon to each and every stream surveyed; rather, observer counts are useful as general indicators of what is taking place and how it compares within a year and to prior years. Long time series are essential, and the value of observer data increases with the length of the time series of data (Symons and Waldichuk 1984).

Objectives

This protocol describes methods used to achieve estimates of salmon escapement using two primary types of aircraft commonly used in aerial observer counts. Since many factors can introduce bias in observer counts, this paper will detail some key points to follow when performing aerial counts in an effort to produce consistent measures of salmon abundance over time. Two long-term programs in southeastern Alaska that estimate escapement utilizing observer counts from fixed-wing aircraft and helicopter for pink salmon *O. gorbuscha* and chinook salmon *O. tshawytscha*, respectively, are described.

The first and foremost objective when making an aerial count is to try and make the most accurate count possible. The second objective, and probably of equal importance to the first, is to be consistent. An observer who consistently counts at a certain rate produces a better index to abundance than does an observer who is inconsistent. After all, the consistent observer can be modeled for counting rate whereas the inconsistent observer is virtually impossible to model. For example, one aerial observer who did surveys in southeastern Alaska typically counted pink salmon in units of 100 (e.g., every click on the tally whacker equaled 100 fish). This is atypical for most pink salmon aerial observers, who normally count in units of 1,000; however, upon examination, the observer who counted in units of 100 was shown to be the most consistent observer in the group, a characteristic that is vital to creating a reliable index of abundance over time.

Fixed-wing aircraft

Today fixed-wing aircraft—specifically Supercubs—are most commonly used when performing aerial counts of salmon escapement using fixed-wing aircraft. The observer sits directly behind the pilot, allowing viewing access in either direction. Often, aircraft are flown at speeds around 100 km/h and heights of 30 m, and counts are made in an upstream direction. Not knowing the results of the first count, observers will sometimes turn and count the same stretch of stream in a downstream fashion and later compare results for consistency. This has the advantage of familiarizing the observer with the stream conditions and provides a different viewing angle that may eliminate glare or other factors encountered in the first survey. Sometimes, pilots also make counts, but care should be taken to ensure that pilots are most concerned with keeping an open view of the river channel at all times.

Helicopter survey

Although more expensive than using fixed-wing aircraft, helicopters are often deployed for species of conservation concern or when fixed-wing counts are not practical or safe (e.g., in dense canopy or canyons). These aircraft provide slower, more maneuverable counting platforms that can increase accuracy and precision. Jet Ranger or Hughes 500 aircraft are commonly used, and observers sit on the left side of pilots. Normally, the door is removed and observers view the streams at heights of 15 to 70 m. Counts are normally made in only one direction to cut down on fuel costs, and pilots typically are solely concerned with keeping the aircraft level and over the viewing area.

Fixed-wing aircraft: Pink salmon in southeastern Alaska

In southeastern Alaska, the methods used to monitor pink salmon escapements and calculate annual indices of spawning abundance were described by Hofmeister (1998), Van Alen (2000), and Zadina et al. (2004). Using the current method, biologists annually estimate the peak pink salmon abundance in 718 pink salmon index streams (selected from more than 2,500 known pink salmon spawning streams in the region). This assessment is made via aerial surveys, conducted at intervals during most of the migration period. Most pink salmon stocks in southeast Alaska do not show persistent trends of odd- or even-year dominance, and for simplicity, escapement indices of both brood lines are combined (Van Alen 2000; Zadina et al. 2004).

Individual observers track absolute abundance within the streams, but each observer tends to count at his/her own rate, or "bias" (Bue et al. 1998; Dangel and Jones 1988; Jones et al. 1998). In 1995, raw stream survey counts were modified in an attempt to standardize as much observer bias as possible—not by removing bias but rather by adjusting all observer counts within each of the four Alaska Department of Fish and Game (ADF&G) management areas to the same bias level (Hofmeister 1998; Van Alen 2000). The index used only stream surveys conducted by key personnel or "major observers"—individuals who had flown more than 100 surveys per year in more than 4 years. Each major observer's counts in a given management area was converted to the counting rate of the area management biologist, whose conversion rate was set at 1.0. These observations were statistically adjusted so the estimates of the number of fish were comparable among observers within the same management area (Hofmeister 1998). The largest count for the year was then retained for each stream in the survey and termed the "peak-adjusted count" for each stream. The index for each stock group was made up of the peak-adjusted counts, which were summed over this standard set of index streams for a particular area.

If a particular index stream was missing escapement counts for any given year, an algorithm (McLachlan and Krishnan 1997) was used to interpolate the missing value. Interpolations were based on the assumption that the expected count for a given year was equal to the sum of all counts for a given stream, divided by the sum of all counts over all years for all the streams in the unit of interest (i.e., row total times column total divided by grand total). (The unit of interest is the stock group, and interpolations for missing values were made at the stock group level.)

This method is based on an assumed multiplicative relation between yearly count and unit count with no interaction.

This method of assessing escapement does not actually provide an estimate of the total escapement of pink salmon for all of southeast Alaska. In the past, ADFG has multiplied the escapement indices by 2.5 to approximate the total escapement. For example, we found the statement "An expansion factor of 2.5 was applied to the escapement index to convert the index to an estimate of total escapement" (Hofmeister and Blick 1991) and similar statements in published material. The 2.5 multiplier was originally intended to convert peak escapement counts to an estimate of what was actually present at the time of the survey (Dangle and Jones 1988; Jones et al. 1998; Hofmeister 1990).

Another important factor to consider in relating total run size to index series of escapement is the relationship between the total fish that spawn and die and the number of fish that are present in the creek at the time of the "peak observation" (Bue et al. 1998). This factor has not been well studied for systems in southeast Alaska (Zadina et al. 2004). The 718 streams in the current index represent only about one-third of the region's 2,500+ pink salmon streams. Thus, the 2.5 multiplier does not take into account fish that were not present at the time of the survey, nor does it take into account streams that were not surveyed.

Finally, the majority of aerial surveys, particularly those conducted prior to about 1970, were conducted to monitor in-season development of salmon escapements for management purposes, not to estimate total escapements (Jones and Dangel 1981; Van Alen 2000). There is no simple way to convert the current index series to an estimate of total escapement in southeast Alaska. Moreover, escapement indices are clearly less than total escapements (Hofmeister 1990; Van Alen 2000; Zadina et al. 2004).

Helicopter surveys: chinook salmon in Southeastern Alaska

There are 34 river systems with populations of wild chinook salmon in southeastern Alaska. Three transboundary rivers—the Taku, Stikine, and Alsek—are classed as major producers, each with potential production (harvest + escapement) greater than 10,000 fish (Kissner 1974). There are nine rivers that are classed as medium producers, each with production of 1,500 to 10,000 fish. The remaining 22 rivers are minor producers, with production of less than 1,500 fish. Small numbers of chinook salmon occur in other streams of the region but are not included in the above list because successful spawning has not been documented. Chinook salmon are counted via aerial surveys or at weirs each year in all three major producing systems, six of the medium producers, and one minor producer. Abundance in the Chilkat River is estimated only by a mark–recapture program. These index systems, along with that used in the Chilkat River, are believed to account for about 90% of the total chinook salmon escapement in southeast Alaska and transboundary rivers (Pahlke 1998).

Pahlke (1997) provides detailed descriptions of the escapement goals and their origins. Escapement goals have been revised when sufficient new information warrants. Most of the revised escapement goals have been developed with spawner–recruit analysis, as ranges of optimum escapement rather than a single point estimate. Spawner–recruit analysis requires not only a long series of escapement estimates but also annual age and sex-specific estimates of escapement (McPherson and Carlile 1997).

Spawning chinook salmon are counted at 26 designated index areas in nine of the systems; total escapement in the other two systems are estimated by complete counts of chinook salmon at the Situk River weir and by annual mark–recapture estimates on the Chilkat River. Counts are made during aerial or foot surveys during periods of peak spawning or at weirs. Peak spawning times—defined as the period when the largest number of adult chinook salmon actively spawn in a particular stream or river—are well documented from surveys of these index areas conducted since 1976 (Kissner 1982; Pahlke 1997). The proportion of fish in prespawning, spawning, and postspawning condition is used to judge whether the survey timing is correct to encompass peak spawning. Index areas are surveyed at least twice unless turbid water or unsafe conditions preclude the second survey. Survey conditions on each index survey are rated as poor, normal, or excellent for that particular index area, and are coded as to whether that survey is potentially useful for indexing or estimating escapement. Factors that affect the rating include water level, clarity, light conditions, and weather.

Only large chinook salmon—typically age 3-, 4-, and 5-year, and greater than or equal to 660 mm mideye-to-fork length (MEF)—are counted during aerial or foot surveys. No attempt is made to record an accurate count of small (typically age 1- and 2-year) chinook salmon less than 660 mm MEF (Mecum 1990). These small chinook salmon (also called jacks) are early maturing, precocious males that are considered to be surplus to spawning escapement needs. Under most conditions they are distinct from their older counterparts because of their short, compact bodies and lighter color. They are, however, difficult to distinguish from other smaller species such as pink salmon and sockeye salmon *O. nerka* salmon. In some systems, it may be difficult to avoid counting age 2-year fish that are larger than 660 mm MEF.

During aerial surveys, pilots are directed to fly the helicopter 15–70 m above the river at speeds of 6–16 kph. The helicopter door on the side of the observer is removed, and the helicopter is flown sideways while observations of spawning chinook salmon are made from the open space. Foot surveys are conducted by at least two people walking in the creek bed or on the riverbank.

Weather, distances, run timing, and other factors can make it difficult for a single surveyor to complete all the index surveys annually under normal or excellent conditions. Thus, alternate surveyors are selected to conduct the counts when the primary surveyor is unavailable. New surveyors also take on primary responsibilities at infrequent intervals. Since between-observer variability and bias can be significant (Jones et al. 1998), new surveyors must be trained and calibrated against the primary surveyor to provide consistency and continuity in the data. Alternate observers accompany the primary observer on regularly scheduled surveys to learn survey methods and counting techniques (on training flights). Each alternate observer also accompanies the primary observer on additional regularly scheduled surveys to independently count chinook salmon (on calibration flights). Each calibration flight consists of two passes over the index area, so that the two observers, in turn, sit in the preferred location in the helicopter during one pass along the river. Count results are not shared during the calibration surveys but are shared and discussed following the completion of the second pass of each flight. Calibration data is collected annually for several years. The relationship between observer escapement counts will be determined from accumulated data and applied to counts as appropriate.

Several chinook salmon index areas are routinely surveyed by more than one method. For example, Andrew Creek, a small tributary in the lower Stikine River in Alaska, is surveyed from airplanes, from helicopters, and by foot. The various surveys are conducted as close as possible to each other to promote comparison and calibration of the different methods.

Estimates of total escapement are needed to model total production, exploitation rates, and other population parameters. Since indices are only a partial count of spawning abundance, observer counts from index areas are increased by an expansion factor to estimate escapement. The expansion factor is an estimate of the proportion of the total escapement counted in a river system during the peak spawning period. Expansion factors are based on comparisons with weir counts, mark–recapture estimates, and spawning distribution studies. They vary among rivers according to how complete the coverage of spawning areas is and the difficulties encountered in observing spawners, such as overhanging vegetation, turbid water conditions, presence of other salmon species (e.g., pink salmon and chum salmon $O.\ keta$), and protraction of run timing. In southeastern Alaska, chinook salmon expansion factors range from 1.5 for the King Salmon River to 5.2 for the Taku River.

Survey expansions are not necessary for those streams in which weirs or other estimation programs are used to count all migrating chinook salmon. In southeastern Alaska, estimates of total escapement are obtained from a weir on the Situk River and by use of mark–recapture on the Chilkat River. Still, observer counts of spawning abundance are regularly conducted in these systems because managers rely on counts and observation for more than just escapement objectives.

Finally, to estimate the total southeastern Alaska regional escapement, estimates from the 11 index systems are expanded to account for the unsurveyed systems. The total estimated escapement in the index areas represents approximately 90% of the region total (Pahlke 1998).

Expansion factors for individual rivers have been revised based on results from experiments to estimate total escapement and spawning distribution. For example, estimated total escapement and radio-tracking distribution data were used to revise tributary expansion factors for the Taku and Unuk rivers (Pahlke and Bernard 1996; Pahlke et al. 1996; McPherson et al. 1998). Mark–recapture studies to estimate spawning abundance on the Unuk River in 1994 (Pahlke et al. 1996) and on the Chickamin River in 1995 and 1996 were used to revise expansion factors for those two rivers in 1996; results were also applied to the nearby Blossom and Keta rivers. More mark–recapture studies were conducted on all four rivers and the expansion factors for the Behm Canal systems were revised again in 2002 (McPherson et al. 2003). On Andrew Creek, a weir was operated for 4 years (1979, 1981, 1982, and 1984), during which time index counts were also conducted, establishing a new expansion factor for that system in 1995. Also in 1997, 10 years (1983–1992) of matched weir and index counts were used to revise the expansion factor for the King Salmon River (McPherson and Clark 2001). The expansion factors for the Taku River were revised in 1996 and again in 1999 based on the results of mark–recapture studies (Pahlke and Bernard 1996; McPherson et al. 2000).

These studies have improved estimates of total escapement in southeastern Alaska and have shown, in most cases, that the surveyed index areas provide reasonably accurate trends in escapements; however, Johnson et al. (1992) demonstrated that expansion factors used before 1991 on the Chilkat River system were highly inaccurate, because the index areas received less than 5% of the escapement. Consequently, since 1991, escapement to the Chilkat River has been estimated annually by mark–recapture experiments (Ericksen and McPherson 2004). Studies on the Taku, Stikine, Alsek, Unuk, Chickamin, Blossom, Keta, and King Salmon rivers, as well as on Andrew Creek, have shown that the index expansion factors used on those systems were much more accurate than those used on the Chilkat (PSC 1991).

Sampling Design

Site selection is vital when choosing a suitable location for conducting observer counts. Areas surveyed should be representative of the population of concern and readily accessible and visible from the air. During periods of low abundance, the optimal spawning habitat will be that area containing salmon (assuming that salmon will seek out optimal spawning habitat). At higher levels of abundance, salmon may choose to spawn in less suitable habitats due to any number of reasons, and pinpointing the optimal habitat may be problematic. Ideally, the optimal spawning habitat will be contained in the area surveyed along with examples of less optimal habitat so that trends in abundance are captured entirely from year to year.

Multiple counts should be made annually in each area so that all components of the run are captured, especially the peak. Many programs use peak counts for index purposes, but it is well understood that these counts do not represent the total escapement due to variability in run timing and stream life. At best, observers will get an index (an unknown portion) of the number of adult salmon returning to spawn, even if corrected for the changing population. In practice, it is more convenient to estimate the peak versus the average and assume that stream life is consistent among years; the peak count is a useful index (Bevan 1961). Some programs often go one step further by expanding indices by some factor to gain an estimate of total escapement; yet this in itself may introduce error.

More reliable estimates of escapement can be obtained through use of area-under-the-curve (AUC) methodologies (English et al. 1992). These methods rely heavily on multiple counts performed within a year and, when coupled with an estimate of stream life, provide the information necessary to estimate total escapement (Cousens et al. 1982).

Personnel Requirements and Training

The experience of the personnel performing the counts is vital in any program. Pilot experience is also very important because the observer and pilot must work as a team to produce dependable estimates. Fatigue can play a big role in the accuracy of counts, yet studies have shown that utilizing one observer consistently from year to year is the best means available for providing an accurate index over that time. Knowing this, observers should keep the amount of survey time at a reasonable level; yet at the same time maintain adequate site coverage (Cousens et al. 1982).

Surveys performed during adverse weather conditions can produce entirely dissimilar results from surveys performed during ideal weather conditions. Surveys can be impacted by an array of adverse conditions that can delay surveys for weeks or more, resulting in the majority of the run being missed. Nevertheless, effort should be made to perform surveys during optimal conditions whenever and wherever possible and to maintain consistency from year to year. Precision (consistency) of escapement estimates is more important than accuracy for defining long-term stock-recruitment relationships (Symons and Waldichuk 1984).

Recommendations for Aerial Surveys

Aerial surveys should be performed with the understanding that the information gathered is first and foremost useful as an index—and only with significant study can observer counts be expanded further to estimates of total escapement. Information gathered during surveys should be clearly labeled as Aerial Survey Index Information gathered through observer counts so that it will not be misinterpreted as actual numbers. Surveys should be performed each year by a single observer, and when possible, other observers should perform overlapping surveys to gain information regarding observer variability, with the eventual change in survey personnel in subsequent years. It should be understood that any factor relating one observer to the next may vary from stream to stream and from year to year (Bevan 1961).

Along with safety, maintaining consistency during surveys is of the utmost priority and concern. Counting units, aircraft and pilots used, and most certainly areas surveyed should be consistent from year to year to maximize the utility of any information gathered.

Literature Cited

ADFG (Alaska Department of Fish and Game). 1964. Studies to determine optimum escapement of pink and chum salmon in Alaska. Alaska Department of Fish and Game, Final Summary Report, Contract 14-17-007-22, Juneau.

Bevan, D. E. 1961. Variability in aerial counts of spawning salmon. Journal of the Fisheries Research Board of Canada 18(3):337–348.

Bue, B. G., S. M. Fried, S. Sharr, D. G. Sharp, J. A. Wilcock, and H. J. Geiger. 1998. Estimating salmon escapement using area-under-the-curve, aerial observer efficiency, and stream-life estimates. Pages 240–250 in D. W. Welch, D. E. Eggers, K. Wakabayaski, and V. I. Karpenko, editors. Assessment and status of Pacific Rim salmonid stocks. North Pacific Anadromous Fish Commission Bulletin 1.

Clark, R. A. 1992. Abundance and age-sex-size composition of chum salmon escapements in the Chena and Salcha rivers, 1992. Alaska Department of Fish and Game, Fisheries Data Series No. 93-13, Anchorage.

Cousens, N. B., G. A. Thomas, S. G. Swann, and M. C. Healey. 1982. A review of salmon escapement estimation techniques. Canadian Technical Report of Fisheries and Aquatic Sciences 1108.

Dangel, J. R., and J. D. Jones. 1988. Southeast Alaska pink salmon total escapement and stream life studies. Alaska Department of Fish and Game, Division of Commercial Fisheries, Regional Information Report No. 1J88-24, Juneau.

Daum, D. W., R. C. Simmons, and K. D. Troyer. 1992. Sonar enumeration of fall chum salmon on the Chandalar River, 1986–1990. Alaska Fisheries Technical Report 16.

English, K. K., R. C. Bocking, and J. R. Irvine. 1992. A robust procedure for estimating salmon escapement based on the area-under-the-curve method. Canadian Journal of Fisheries and Aquatic Sciences 49:1982–1989.

Eicher, G. J., Jr. 1953. Aerial methods of assessing red salmon populations in western Alaska. Journal of Wildlife Management 17(4):521–528.

Ericksen, R. P., and S. A. McPherson. 2004. Optimal production of chinook salmon from the Chilkat River. Alaska Department of Fish and Game, Division of Sport Fish, Fishery Manuscript No. 04-01, Anchorage.

Evenson, M. J. 1992. Abundance, egg production, and age-sex-length composition of the chinook salmon escapement in the Chena River, 1992. Alaska Department of Fish and Game, Fisheries Data Series 93-6, Anchorage.

Hofmeister, K. 1990. Southeast Alaska pink and chum salmon investigations, 1989-1990. Final report for the period July 1, 1989 to June 30, 1990. Alaska Department of Fish and Game, Division of Commercial Fisheries, Regional Information Report 1J90-35, Juneau.

Hofmeister, K. 1998. Standardization of aerial salmon escapement counts made by several observers in southeast Alaska. Pages 117–125 in Proceedings of the Northeast Pacific Pink and Chum Salmon Workshop, 26-28 February 1997, Parksville, British Columbia, Department of Fisheries and Oceans, 3225 Stephenson Point Road, Nanaimo, B.C., V9T 1K3.

Hofmeister, K., and J. Blick. 1991. Pages 39–41 in H. Geiger and H. Savikko, editors. Preliminary forecasts and projections for 1991 Alaska salmon fisheries and summary of the 1990 season. Alaska Department of Fish and Game, Division of Commercial Fisheries, Regional Information Report No. 5J91-01, Juneau.

Irvine, J. R., R. C. Bocking, K. K. English, and M. Labelle. 1992. Estimating coho salmon *(Oncorhynchus kisutch)* spawning escapements by conducting visual surveys in areas selected using stratified random and stratified index sampling designs. Canadian Journal of Fisheries and Aquatic Sciences 49:1972–1981.

Johnson, R. E., R. P. Marshall, and S. T. Elliott. 1992. Chilkat River chinook salmon studies, 1991. Alaska Department of Fish and Game, Division of Sport Fish, Fishery Data Series No. 92-49, Anchorage.

Jones, D., and J. Dangel. 1981. Southeastern Alaska 1980 brood year pink *(Oncorhynchus gorbuscha)* and chum salmon *(O. keta)* escapement surveys and pre-emergent fry program. Alaska Department of Fish and Game, Division of Commercial Fisheries, Technical Data Report No. 66, Juneau.

Jones, E. L., III, T. J. Quinn, II, and B. W. Van Alen. 1998. Observer accuracy and precision in aerial and foot survey counts of pink salmon in a southeast Alaska stream. North American Journal of Fisheries Management 18:832–846.

Labelle, M. 1994. A likelihood method for estimating Pacific salmon escapement based on fence counts and mark–recapture data. Canadian Journal of Fisheries and Aquatic Sciences 51:552–566.

Kissner, P. D., Jr. 1974. A study of chinook salmon in southeast Alaska. Alaska Department of Fish and Game, Annual report 1973–1974, Project F-9-7, 16 (AFS-41), Anchorage.

Kissner, P. D., Jr. 1982. A study of chinook salmon in southeast Alaska. Alaska Department of Fish and Game, Annual report 1981–1982, Project F-9-14, 24 (AFS-41), Anchorage.

McLachlan, G. J., and T. Krishnan. 1997. The EM algorithm and extensions. John Wiley and Sons, New York.

McPherson, S. A., D. R. Bernard, M. S. Kelley, P. A. Milligan, and P. Timpany. 1998. Spawning abundance of chinook salmon in the Taku River in 1997. Alaska Department of Fish and Game, Division of Sport Fish, Fishery Data Series No. 98-41, Anchorage.

McPherson, S. A., D. R. Bernard, and J. H. Clark. 2000. Optimal production of chinook salmon from the Taku River. Alaska Department of Fish and Game, Division of Sport Fish, Fisheries Manuscript No. 00-2, Anchorage.

McPherson, S. A., and J. Carlile. 1997. Spawner-recruit analysis of Behm Canal chinook salmon stocks. Alaska Department of Fish and Game, Commercial Fisheries Division, Regional Information Report 1J97-06, Juneau.

McPherson, S. A., and J. H. Clark. 2001. Biological escapement goal for King Salmon River chinook salmon. Alaska Department of Fish and Game, Commercial Fisheries Division, Regional Information Report No. 1J01-40, Juneau.

McPherson, S. A., D. R. Bernard, J. H. Clark, K. A. Pahlke, E. Jones, J. Der Hovanisian, J. Weller, and R. Ericksen. 2003. Stock status and escapement goals for chinook salmon stocks in southeast Alaska. Department of Fish and Game, Division of Sport Fish, Special Publication No. 03-01, Anchorage.

Mecum, R. D. 1990. Escapement of chinook salmon in southeast Alaska and trans-boundary rivers in 1989. Alaska Department of Fish and Game, Fishery Data Series No. 90-52, Anchorage.

Neilson, J. D., and G. H. Geen. 1981. Enumeration of spawning salmon from spawner residence time and aerial counts. Transactions of the American Fisheries Society 110:554–556.

PSC (Pacific Salmon Commission). 1991. Escapement goals for chinook salmon in the Alsek, Taku, and Stikine rivers. Transboundary River Technical Report, TCTR (91)-4, Vancouver.

Pahlke, K. A. 1997. Escapements of chinook salmon in southeast Alaska and transboundary rivers in 1996. Alaska Department of Fish and Game, Division of Sport Fish, Fishery Data Series No.97-33, Anchorage.

Pahlke, K. A. 1998. Escapements of chinook salmon in southeast Alaska and transboundary rivers in 1997. Alaska Department of Fish and Game, Division of Sport Fish, Fishery Data Series No.98-33, Anchorage.

Pahlke, K. A., and D. R. Bernard. 1996. Abundance of the chinook salmon escapement in the Taku River, 1989 to 1990. Alaska Fishery Research Bulletin 3(1):8–19.

Pahlke, K. A., S. A. McPherson, and R. P. Marshall. 1996. Chinook salmon research on the Unuk River, 1994. Alaska Department of Fish and Game, Division of Sport Fish, Fishery Data Series No. 96-14, Anchorage.

Rogers, D. E. 1984. Aerial survey estimates of Bristol Bay sockeye salmon escapement in Symons , P. E. and M. Waldichuk. Proceedings of the workshop on stream indexing for salmon escapement estimation. Canadian Technical Report of Fisheries and Aquatic Sciences 1326:197–208.

Shardlow, T., Hilborn, R., and D. Lightly. 1987. Components analysis of instream escapement methods for Pacific salmon (Oncorhynchus spp.). Canadian Journal of Fisheries and Aquatic Sciences 44:1031–1037.

Skaugstad, C. 1992. Abundance, egg production, and age-sex-length composition of the chinook salmon escapement in the Salcha River, 1992. Alaska Department of Fish and Game, Fisheries Data Series 93-23, Anchorage.

Symons, P. E., and M. Waldichuk. 1984. Proceedings of the workshop on stream indexing for salmon escapement estimation. Canadian Technical Report of Fisheries and Aquatic Sciences 1326.

Van Alen, B. W. 2000. Status and stewardship of salmon stocks in southeast Alaska. Pages 161–194 in E. E Knudsen, C. R. Steward, D. D. McDonald, J. E. Williams, and D. W. Reiser, editors. Sustainable fisheries management: Pacific salmon. CRC Press, Boca Raton. Florida.

Walters, C. J. 1981. Optimum escapements in the face of alternative recruitment hypotheses. Canadian Journal of Fisheries and Aquatic Sciences 38:678–689.

Walters, C. J., and D. Ludwig. 1981. Effects of measurement errors on the assessment of stock-recruitment relationships. Canadian Journal of Fisheries and Aquatic Sciences 38:704–710.

Zadina, T. P., S. C. Heinl, A. J. McGregor, and H. J. Geiger. 2004. Pink salmon stock status and escapement goals in southeast Alaska and Yakutat. In H. J. Geiger and S. McPherson, editors. Stock status and escapement goals for salmon stocks in southeast Alaska. Alaska Department of Fish and Game, Divisions of Sport and Commercial Fisheries, Special Publication No. 04-02, Anchorage.

AERIAL COUNTS

Fyke Nets (in Lentic Habitats and Estuaries)
Jennifer S. O'Neal

Background and Objectives

Fyke netting is a passive capture method used for sampling juvenile salmon and steelhead *Oncorhynchus mykiss* that use lentic habitats and estuary areas and, in some cases, stream habitats. Fyke nets are large hoop nets with wings (and/or a lead) that are attached to the first frame and act as funnels to direct swimming fish into the trap (see Figure 1). The second and third frames each hold funnel throats, which prevent fish from escaping as they enter each section. The opposite end of the net may be tied with a slip cord to facilitate fish removal. These nets are typically used in shallow water (where the first hoop is less than 1 m under the water's surface), although some lake studies have used fyke nets where the water was as deep as 10 m over the first frame. This deep-set approach has resulted in comparable data to shallower sets, except for 0-age fish where the deeper sets had lower catch values.

FIGURE 1.—Diagram of a fyke net (from Dumont and Sundstrom 1961).

The net is set so that the leads intercept moving fish. When the fish try to get around the lead, they swim into the enclosure. Leads and wings are held in place by poles or anchors. Modified fyke nets have rectangular frames to enhance their stability. The square or rectangular frames prevent the net from rolling on the bottom substrate (Hubert 1996). Fyke nets are suspended between buoyant and weighted lines much like a gill net.

History

Fyke nets have their origins in salmon wing nets and have been used in river fisheries for hundreds of years. According to Kustaa Vilkuna, a Finnish academic, large fyke nets were used in Finnish sea regions to catch herring, whitefish, and salmon. The size of the catch determined the mesh size of the netting used. The first version of this fyke net was used in Finland before anyone registered the invention of a new gear. From that area, it was adopted in the Vaasa archipelago to be used as a herring trap in the 1860s. On the coasts of Sweden, the gear was first called "finnryssja" meaning "the Finnish trap net."

With fyke nets, as with many other fish collection nets, the size of the mesh used is dependent on the intended composition of the catch. Large fyke nets with mesh size of 13 mm (0.5 in.) tend to capture larger fish, since they cannot detect the bigger mesh very well, whereas fyke nets with net mesh of only 10 mm (0.38 in.) are better at capturing smaller fish.

Fyke nets have been used to assess populations of salmon and steelhead juveniles in the Pacific Northwest and other regions. Fresh (2000) used fyke nets to capture juvenile chinook salmon *O. tshawytscha* in the Green River in Washington to assess migration patterns, growth, and habitat use. Gallagher (2000) used fyke nets to monitor downstream migration for steelhead in the Noyo River in California. The objectives of this latter study were to assess abundance, size, age, survival, migration timing, and distribution.

Rationale

Assessing salmon and steelhead survival at specific life stages is critical to effective management of populations and evolutionarily significant units (ESUs). An ESU is defined as a population that (1) is substantially reproductively isolated from conspecific populations, and (2) represents an important component in the evolutionary legacy of the species (Johnson et al. 1994; McElhany et al. 2000). Identification of lifestage-specific survival and the factors limiting that survival will allow scientists and managers to better address these factors or "threats" to species recovery. NOAA Fisheries has determined that estimates of juvenile and adult abundance for listed ESUs are a critical component in the recovery of these ESUs. Assessment of juvenile abundance in areas that are turbid or in which substrate or obstructions do not allow for active capture of fish can be accomplished using fyke nets. This gear type avoids many of the issues that arise when visibility is impaired or net snagging is a problem. The use of fyke nets is presented here as an option for population assessment when other methods are not suitable for use.

Fyke netting is a useful method for sampling fish that use lentic habitats and estuarine areas. It is commonly used to monitor the yearly changes in fish species abundances in sites where seining cannot be used alone or in combination with other methods for a mark–recapture study. If the habitat has large and uneven substrate, significant woody debris, or other obstructions, seining may not be possible, and fyke nets may provide a viable alternative. Fyke nets tend to be the most useful in capturing cover-seeking mobile species and migratory species that follow the shorelines, and have been used to a sample juvenile salmon in estuary habitat in the Skagit River in Washington (E. Beamer, Skagit River System Cooperative, personal communication). Fyke nets induce less stress on captured fish than do entanglement gears (Hopkins and Cech 1992), and most captured fish can be released unharmed. Fyke nets are widely used in the assessment of fisheries stocks because of the low mortality of fish and aspects of their species and size selectivity. Trap mortality for steelhead caught using fyke nets in the Noyo River in California was less than 1% (Gallagher 2000).

Objectives

This protocol describes methods used to capture fish using fyke nets. Objectives that could be addressed using this method include the following:

- determining relative abundance and yearly changes in abundance of juvenile fish populations in lentic and estuarine environments;
- determining population characteristics such as size, age, growth, migration timing, and distribution; and
- determining diversity of juvenile fish species in a lentic or estuarine habitat or determining the ratio of hatchery versus wild fish in these habitats.

Sampling Design

When describing the use of fyke nets or other passive sampling gears, the aspects of gear selectivity and efficiency must be addressed (Kraft and Johnson 1992). Selectivity can include a bias for species, sizes, and sexes of fish. Efficiency of a gear refers to the amount of effort expended to capture target organisms. A quantitative understanding of gear selectivity is needed to interpret the data, but little such information is available for most sampling devices. Variables that affect capture efficiency include season, water temperature, time of day or night, water level fluctuation, turbidity, and currents. Changes in animal behavior lead to variability in data collected among species and age-groups because animal capture with passive gear is a function of animal movement.

Standardizing the use of fyke nets for a specific objective in a specific habitat for juvenile salmonids will be helpful in interpreting the data from different studies and for calibration of fyke nets with other capture methods. Currently, habitats where fyke nets are most often used are lentic habitats such as lakes, off-channel habitats, and estuarine areas. Estuarine use has been particularly important for the assessment of juvenile salmon use, especially where seining is not feasible due to large substrate (K. Fresh, NOAA Fisheries, personal communication). These habitats are critical to the survival of salmon species such as chinook, which require significant growth in estuarine habitat before venturing into the open ocean. Sample design will be dependant upon project objectives, but minimally, three sets with the fyke net should be used to assess variability of sampling.

Other applications of fyke nets include capturing juvenile steelhead in streams or for off-channel habitat use by juvenile salmon species such as chinook salmon and coho salmon *O. kisutch*, where those habitats have significant obstacles that prevent effective seining. Coho salmon were observed to have a higher probability of capture as compared to steelhead when using a fyke net in the Noyo River in California. This difference may be due to stricter life cycle timing for coho, which spend one year in freshwater, versus steelhead, which have a more flexible freshwater residence time (Gallagher 2000).

Fyke nets can be used to collect data on relative abundance as well as indices of change in stock abundance. Combining fyke nets with gill nets and electrofishing can be used to assess species assemblages in lake habitats (Bonar et. al 2000). Mark–recapture sampling with fyke nets has been used to assess steelhead abundance, size, growth, age, migration timing, and distribution (Gallagher 2000).

Field/Office Methods

Pre-field Activities

Field staff should obtain standardized fyke nets for sampling. A set of multiple fyke nets of similar dimensions is effective for lake sampling, where the number of nets used is dependent on the size of the lake and the study objectives. For sampling juvenile salmonids, a typical fyke net is approximately 12 m long and consists of two rectangular steel frames, 90 cm wide by 75 cm high, and four steel hoops, all covered by 7-mm delta stretch mesh nylon netting. An 8-m long by 1.25-m deep leader net made of 7-mm delta stretch nylon netting is attached to a center bar of the first rectangular frame (net mouth). The second rectangular frame has two 10-cm-wide by 70-cm-high openings, one on each side of the frame's center bar. The four hoops follow the second frame. The throats, 10 cm in diameter, are located between the second and third hoops. The net ends in a bag with a 20.4-cm opening at the end, which is tied shut while the net is fishing.

Modified fyke nets are widely used to sample lakes and reservoirs. These nets have 1–2 rectangular frames to prevent the net from rolling on the bottom. These smaller fyke nets are 10 m long (including the lead) with one rectangular frame followed by two aluminum hoops. The aluminum frame is 98 cm wide by 82 cm tall, and is constructed of 2.5-cm tubing with an additional vertical bar. The hoops are 60 cm in diameter and constructed of 5-mm-diameter aluminum rods. The single net funnel is between the first and second hoops and is 20 cm in diameter. The lead is 8 m long, 1.25 m deep, and constructed from 7-mm delta stretch mesh.

Other pre-field tasks include

- obtaining a map of the survey site before sampling. Use the map to measure the shoreline perimeter;
- determining what species and life stages are of greatest interest to sample;
- determining how to stratify habitat based on where the density of the species of interest or species diversity would be highest based on life history;
- designating strata locations on the map based on predicted level of species diversity and distribution, or on fish density or habitat;
- selecting needed sample size (see Appendix A); and
- allocating sampling effort based on nonuniform probability allocation, if the degree of difference in species diversity, distribution, or catch-per-unit-effort (CPUE) by habitat is known, or proportionally allocate effort based on habitat distribution.

Field Activities

Each fyke net is set in shallow water perpendicular to shore such that the net mouth is covered by about 1 m of water when possible (Fletcher et al. 1993; Hubert 1996) (see Figure 2). When the net is properly set, the lead is perpendicular to shore (vertical and not twisted), the mouth of the net is upright and facing shore, and all the hoops are upright. Fyke nets should be set in the evening or late afternoon and retrieved the next morning. All nets should be checked and emptied 12–24 h after setting (Klemm et al. 1993). Record set time, pickup time, and location of the net on a map or global positioning system (GPS).

If the bottom is soft and the water is shallow, the fyke nets are suspended by placing floats at the apex of each hoop and on top of the opening frames. This is done to prevent the nets from sinking into the soft sediments at the bottom of the lake.

When the net is set from a small boat, it is placed on the bow with the pot on the bottom and the lead on the top. The end of the lead is staked or anchored and played out as the boat moves in reverse. When the lead is fully extended, the pot is put overboard and staked or anchored into position. Fish are then removed by lifting only the pot into the boat and placing the fish in a live well.

Fyke nets may also be deployed away from shore in pairs with a single lead between them (Hubert 1996). This type of set is generally made parallel to the shore along the outer edge of vegetation or along shallow off-shore reefs.

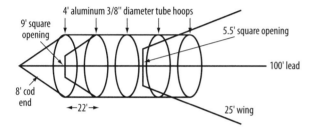

FIGURE 2. — Setting a fyke net in a lake. (Diagram: Andrew Fuller from Regents of the University of Minnesota 2003).

Measurement Details

Nets should be checked and emptied 12–24 h after being set (Klemm et al. 1993). When the net is pulled in, the hoops and frames are gathered together and lifted into the boat. The net is positioned over a livewell with the net mouth upward. Frames are lifted one at a time, and any fish present are shaken down into the next chamber until all of the fish are in the bag, which is then emptied into a livewell (see Figure 3). Each fish is then identified to species, measured to the nearest millimeter, and released back into the water. Age can also be determined by removing a scale for later analysis or by using a size/age relationship.

The field team should record the time that the net was set, the time it was pulled in, the total fishing time, the number of nets used, and the location of each set on the map.

Sample Processing

1. Measure each specimen to the nearest millimeter, identify to species, and weigh to the nearest gram.
2. Field data recording should be standardized and should include the following:
   a. Habitat type
   b. Sampling date
   c. Gear type
   d. Net location (shore orientation, depth, placement time, collection intervals, universal transverse mercator or latitude/longitude coordinates)

e. Hours fished
   f. Species collected
   g. Weight for each species
   h. Length for each species
   i. Names of personnel involved in sampling

Data Handling, Analysis, and Reporting

Data collection by passive sampling can be used to determine relative abundance, which is expressed as number collected per 24 h and weight (kg) collected per 24 h (Ohio EPA 1989, as cited in Klemm et al. 1993). Other data that may be collected using fyke nets would be species assemblage or diversity data or distribution data (which generally does not require significant analysis). If fyke nets are used as part of a mark–recapture study, additional analysis will need to be done depending on the other method that was used in the study.

CPUE data is one type of data that can be used as an index for population density. One critical assumption is that the CPUE results are proportional to stock density (Hubert 1996). The true density of the species is still unknown, and the proportionality constant that relates CPUE to true density is also unknown. But as long as this constant is not expected to change, differences in CPUE should reflect changes in the species abundance. This method can be used for relative abundance, but total abundance estimates would not have a high level of accuracy.

Variability in fish behavior can alter the accuracy of CPUE data and reduce its utility even for relative abundance. Standardized gear, methods, and sample designs must be used for estimates to be comparable. Time of year and placement of nets also affect comparability of data.

Additionally, CPUE data are generally not normally distributed. At low and moderate densities, the distribution of the number of fish captured with fyke nets will not have a normal distribution. At low densities, the distribution approximates a declining logarithmic function (Hubert 1996). At moderate densities, the distributions are skewed. Only at very high densities, when the target species are caught very often is the distribution approximately normal; hence, descriptive statistics designed for normally distributed data cannot be used with CPUE data. Additionally, no single transformation can be used to address the variability in distribution that is seen as the fish density changes.

Nonparametric statistics offer equivalents to most of the procedures that require an assumption of normal distribution. A more appropriate descriptor for CPUE data than the mean is the median or 50th percentile for the density. The frequency of zero catches can also be used to report an index of fish density but cannot be transformed into abundance.

Personnel Requirements and Training

Responsibilities and Staff Requirements
The net should be set with two persons whenever possible, especially when deploying from a boat. During boat use, one person deploys the net while the

other operates the vessel. This reduces the chances of net entanglement and ensures that the net will be deployed properly. If only one person is available, initial preparation of the net is critical. During net retrieval, two persons are needed: one who pulls the net on board and another who removes fish and ensures a successful and careful transfer to the livewell. One person should be responsible to record weights, lengths, and other data on each fish.

Qualifications
The person using the fyke net needs to have been trained by an experienced field biologist who should have a degree in biology or 1 year of experience in sampling fish in the geographic area where the sampling is to occur.

Training
Training should be provided on the job and/or through videos and demonstrations prior to the season. On-the-job training should be provided by an experienced field biologist. All personnel on project must have training in fish identification.

Operational Requirements

Workload and field schedule
The field schedule for setting and retrieving fyke nets is seasonal—ideally when the fish are active and before there is a lot of recreational lake activity; however, as noted above, the sampling can occur at any time, depending upon the objectives of the study and the needs of the monitoring.

The collapsible nature of this trap is very popular because it allows biologists to carry many more traps per outing. Safety during deployment from a boat is also increased due to the lower space requirements for each net.

Equipment needs
- Fyke net(s)
- Small motorized boat (optional, depending on habitat)
- Livewell
- Field forms
- Hip waders, boots, and rain gear
- Meter board
- Electronic scale
- Dip nets
- Net tubs/buckets
- GPS unit
- Materials for taking any biological samples (e.g., scales)
- Fish species identification guides
- Communications gear (e.g., cell phone, two-way radio, satellite phone)

Budget considerations

With careful handling and placement, a fyke traps can be expected to last for several seasons. General guidelines for a single fyke net survey are as follows:

Equipment	Cost/time
Time for two biologists to set the net, retrieve the net, and process the catch	3 h
Travel time	Site dependent
Preparation time	4 h
Training	1 h
Lab work	2 h
Data analysis	2–8 h

A season of fyke netting would take far more effort than a single sampling event. In Gallagher's study (2000), a crew of two checked six fyke nets daily for 4 months, and recommended a longer sampling season. Their effort required 13,440 person-hours for field data and 550 person-hours for data analysis and reporting; a database was created to manage the data.

Literature Cited

Bonar, S. A., B. D. Bolding, and M. Divens. 2000. Standard fish sampling guidelines for Washington state ponds and lakes. Washington Department of Fish and Wildlife Report, Olympia.

Dumont, W. H., and G. T. Sundstrom. 1961. Commercial fishing gear of the United States. Fish and Wildlife Circular 109.

Hubert, W. A. 1996. Passive capture techniques. Pages 157–192 in B. R. Murphy and D. W. Willis, editors. Fisheries techniques, 2nd edition. American Fisheries Society, Bethesda, Maryland.

Fletcher, D., S. A. Bonar, B. Bolding, A. Bradbury, and S. Zeylmaker. 1993. Analyzing warmwater fish populations in Washington State. Washington State Department of Fish and Wildlife Report, Warmwater Survey Manual, Olympia.

Fresh, K. 2000. Juvenile chinook migration, growth, and habitat use in the Lower Green River, Duamish River, and nearshore of Elliot Bay. Prepared for King County Department of Natural Resources and Parks, Water and Land Resources Division, Seattle.

Gallagher, S. P. 2000. Results of the 2000 steelhead (Oncorhynchus mykiss) fyke trapping and stream resident population estimations and predictions for the Noyo River, California with comparison to some historical information. California Department of Fish and Game Steelhead Research and Monitoring Program, Fort Bragg.

Hopkins, T. E., and J. J. Cech, Jr. 1992. Physiological effects of capturing striped bass in gill nets and fyke traps. Transactions of the American Fisheries Society 121:819–822.

Johnson, O. W., R. S. Waples, T. C. Wainwright, K. G. Neely, F. W. Waknitz, and L. T. Parker. 1994. Status review for Oregon's Umpqua River sea-run cutthroat trout. U.S. Department of Commerce, NOAA Technical Memorandum NMFS-VWFSC-15, Seattle.

Klemm, D. J., Q. J. Stober, and J. M. Lazorchak. 1993. Fish field and laboratory methods for evaluating the biological integrity of surface waters. U.S. Environmental Protection Agency, EPA/600/R-92/111, Cincinnati, Ohio.

Kraft, C. E., and B. L. Johnson. 1992. Fyke-net and gill-net size selectivity for yellow perch in Green Bay, Lake Michigan. North American Journal of Fisheries Management 12:230–236.

McElhany, P., M. H. Ruckelshaus, M. J. Ford, T. C. Wainwright, and E. P. Bjorkstedt. 2000. Viable salmonid populations and the Recovery of Evolutionarily Significant Units. U.S. Department of Commerce, NOAA Technical Memorandum NMFS-NWFSC-42, Seattle.

Appendix A: Using sequential sampling or previous year's data to calculate CPUE sample size during a survey

To determine appropriate sample size for the survey, first reach a decision about survey objectives. Is the survey purpose to get a point estimate of value or to measure change? What degree of confidence is required in the results (e.g., 70%)? If change is to be measured, what degree of change should be detected? Then select a sample size for fyke netting that will be appropriate to meet these goals. The best method to calculate CPUE sample sizes so they will be tailored to individual lakes is to use previous estimates of variance that are available from the specific lake, taken at the same time of year. These estimates can be obtained either through sequential sampling or through previous year's sampling.

A.1. Calculating a sample size to estimate CPUE within certain bounds

If the biologist wants to measure CPUE within certain bounds, use the following equation to calculate needed sample sizes (Willis 1998; Cochran 1977).

$$n = \frac{(t^2)(s^2)}{[(a)(x)]^2}$$

where

n = sample size required
t = t value from a t-table at n-1 degrees of freedom for a desired sample size (1.96 for 95% confidence)
s^2 = variance
x = mean CPUE
a = precision desired in describing the mean expresses as a proportion

Simply plug in values obtained from last year's survey or while the survey is in progress to calculate how many samples are needed to get the precision required.

(Bonar 2000)

Appendix B: Example measurements of fish captured in Fyke Net

Data from Klemm et al. (1993); reflecting total catch, length frequency by 1/2 inch group, percent of total catch in lake, and mean length; measurements are in English units, as shown in original document.

½ inch group	Bluegill	Pumpkinseed	Rockbass	Smallmouth Bass	Black Crappie	Walleye	Yellow Perch	Largemouth Bass	Brook Trout	White Sucker	Golden Shiner	Common Shiner
2.0–2.4												
2.5–2.9	1											
3.0–3.4	1		1	1	1		1					1
3.5–3.9	1	2	1	1			1				2	
4.0–4.4	2	6	6	2								
4.5–4.9	4	5	11	1								
5.0–5.4		4	25	2						1		
5.5–5.9	1	7	19	1								
6.0–6.4	4	12	12									
6.5–6.9	48	42	10	4							1	
7.0–7.4	91	49	8	8		1					1	
7.5–7.9	71	22	5	1		1						
8.0–8.4	26	1	7									
8.5–8.9							1				1	
9.0–9.4				1					1			
9.5–9.9				1								
10.0–10.4				4	4							
10.5–10.9				2	4							
11.0–11.4				2	1							
11.5–11.9				1								
12.0–12.4				4	2							
12.5–12.9				7								
13.0–13.4				3	1					2		
13.5–13.9				2						1		
14.0–14.4				1						2		
14.5–14.9										9		
15.9–15.4										5		
15.5–15.9						1		1		5		
16.9–16.4										7		
16.5–16.9			1							3		
17.0–17.4										7		
17.5–17.9										9		
18.0–18.4										6		
18.5–18.9										4		
19.0–19.4										2		
19.5–19.9										4		
20.0–20.4										3		
20.5–20.9										3		
21.0–21.4										3		
21.5–21.9												
22.0–22.4												
22.5–22.9												
23.0–23.4												
23.5–23.9												
24.0–24.4												
24.5–24.9												
25.0–25.4												
25.5–25.9												
26.0–26.4												
26.5–26.9												
27.0–27.4												
27.5–27.9						1						
28.0–28.4												
28.5–28.9												
Unmeasured	31											
Total	275	150	106	50	10	7	3	1	1	75	5	1
% of Total	40.1%	21.9%	15.5%	7.3%	1.5%	1.0%	0.4%	0.1%	0.1%	11.1%	0.7%	0.1%
Mean length (inches)	7/2	6.7	5.9	9.6	9.9	13.7	5.3	15.7	9	17	5.9	3

Appendix C: Example Fyke Net Survey

summary from Klemm et al. (1993); note: measurements remain in English units as seen in original publication

Summary of GLIFWC Summer 2003 Fyke Net Survey
Lake: Kentuck County: Vilas
Dates of Survey: 6/23 – 6/27/03 Objective: General Fish Survey
Sampling Method: Fyke Net
Crew: Ed White, Phil Doepke, Henry Mieloszyk, Michael Preul

Sampling Information:
Data Summarized: Phil Doepke
Water Temperature (F) 72–76
Number of Nets: 7, 3/8" bar mesh 6, 3' × 6' frame; 1, 4' × 6' frame; 1,1" bar mesh 5' × 6' frame
Net Locations: 8   Number of Net Nights: 32   Number of Lifts: 32

Catch Data Summary:

Fish Summary	Total Number Caught	Length Range (inches)	Modal Size (0.5" group)	Catch/Unit (# fish per net night)
Bluegill	275	2.6–8.4	7.0	8.6
Pumpkinseed	150	3.9–8.0	7.0	4.7
Rockbass	106	3.4–8.4	5.0	3.3
Smallmouth Bass	50	3.4–16.8	7.0	1.6
Black Crappie	10	3.2–11.2	10.0, 10.5	0.3
Walleye	7	7.4–27.6	12.0	0.2
Yellow Perch	3	3.3–8.8		0.1
Largemouth Bass	1	15.7		0.03
Brook Trout	1	9.0		0.03
White Sucker	76	5.0–21.4	14.5, 17.5	2.4
Golden Shiner	5	3.5–8.5		0.2
Common Shiner	1	3.0		0.03

Muskellunge were observed during an electro-fishing run on the night of June 24, 2002, however, none were caught in the fyke nets of this survey.

Comments
1) Weather: Clear to overcast skies, winds strong and gusty, occasional thunderstorms during week.
2) Only the first 25 fish of each species were measured for each net lift.
3) Bluegill and pumpkinseed sunfish were not guarding nests, females of these species had extended abdomens, likely full of eggs, rockbass were guarding nests. Bass were guarding brooding schools.
4) Fishing pressure: There were between 6 and 10 boats fishing Kentuck Lake at any time while the survey crew was present.
5) Yellow perch (4–10") were commonly observed during an electro-fishing surveys during the same week, but not readily caught in the fyke nets.

6) Two SCUBA divers spent 20 minutes searching in front of the east end boat landing in 10 to 20 feet of water; the search revealed no evidence of zebra mussels.

7) Conversations with anglers:

   One pair of anglers mentioned catching and releasing a 45" muskellunge. They reported a catch rate of approximately 12 walleye per hour, and the walleye were averaging about 13 inches.

   The campground hosts mentioned that fishing for panfish was slower this year than in the past. They reported catching roughly 15 fish (perch and bluegill) during a typical evening outing.

FYKE NETS (IN LENTIC HABITATS AND ESTUARIES)

Variable Mesh Gill Nets (in Lakes)
Bruce Crawford

Background and Objectives

Background
Along areas of the Pacific Northwest coast, gill nets were traditionally constructed of a coarse fiber twine made from willow bark (Coffing 1991) and other materials, such as seal skin (as reported in 1844 by Zagoskin [Michael 1967]) and moose or caribou sinew (Oswalt 1980; Stokes 1985). Linen twine was used for making gill nets beginning in the 1920s (Coffing 1991). Gill nets were used both for set net and drift net fishing. In the 1960s, nets made from synthetic fibers such as nylon came into wider use. Most nets were 50 m or less in length until the 1980s. Nets are generally 50–70 m long, with mesh size varying depending on the salmon species targeted (Charnley 1984).

Variable mesh gill nets have been used for fish population evaluation for about a century. The efficiency with which gill nets capture fish and the versatile use of these nets in lakes and streams have made them a common tool for fishery evaluation (Hamley 1975).

This supplemental technique addresses the use of gill nets targeting salmonids in the Pacific Northwest but can be used for other species as well. The chapter draws extensively from the following papers: Bernabo (1986); Baklwill and Combs (1994); Bonar et al. (2000); and Klemm et al. (1993). Additional insights into use of gill nets can be found in Hubert (1996).

Rationale
Variable mesh gill nets are appropriate for sampling when fish mortality is not a limiting factor. Gill nets normally kill a high percentage of fish due to the trapping mechanism of the net around the gills. Careful net tending can reduce but not eliminate the mortality percentage. The use of variable size mesh panels in the gill net allows capture of fish of different sizes. As such, this method can be used to collect data on population abundance, stock characteristics, population distribution, and species richness. Gill nets are not species-selective, and as a result, it can be expected that as many or more nontarget species will be captured as target species. In addition, small aquatic mammals and birds will also occasionally become entangled in the mesh and drown.

Objectives
- Determine relative abundance of lake or stream populations by measuring the catch-per-unit-effort (CPUE).
- Determine total abundance of lake populations by measuring the recapture rate of marked fish.
- Determine the length, sex, phenotypes, and genotypes of fish by collecting a representative catch of each sample.
- Determine the species composition and relative biomass of a lake or a stream.

Sampling Design

Trend information based on results of gill-net sampling will only be as reliable at the reproducibility of the sampling technique for each monitored site. Location of nets, orientation along the bottom in relation to the shoreline, diel time of placement and collection, and season of placement must be standardized for each site. Because lake sampling programs will be site specific, standardization must be within a given lake and not between lakes. Each lake has a unique morphometry, and net placement must be carefully considered according to lake characteristics and target species.

The types of data acquired from gill nets include fish age, growth, relative weight, and proportional stock density calculations. Also, estimates derived from gill nets are typically given in CPUE or abundance within restricted habitat zones such as nearshore areas or coves (Dauble and Gray 1980; King et al. 1981; Johnson et al. 1988; Rider et al. 1994). CPUE methods assume that the calculated index is proportional to total population size, allowing trends through time to be detected. Unfortunately, violating this assumption is easy, but detecting the violation is not (Hillborn and Walters 1992). Given this situation, suggestions for estimating population abundance in deeper-water habitats must be tentative. Alternatively, one can use active capture gear or define a series of equally spaced transects over the entire water surface from which to sample randomly or systematically (Thompson et al. 1998).

Borgstrøm (1992) assessed the effect of population density on gill-net catchability of Brown trout *Salmo trutta* in four Norwegian high-mountain lakes. Catchability was found to be inversely related to the number of fish present; brown trout populations with low densities were more vulnerable to gill nets than high density populations; gill-net catches as an estimator of population density were biased.

While there are many ways to utilize gill nets, two examples of gill-net use in lakes are offered here.

McLellan (2001) used electrofishing and gill nets to sample resident fish in eastern Washington reservoirs and streams. Of the taxa involved, four were salmonids (cutthroat trout *O. clarki*, rainbow trout *O. mykiss*, brown trout, and lake trout *Salmo namaycush*). A total of 10 horizontal experimental monofilament sinking gill nets (2.4 × 61.0 m; four 15.2-m panels with square mesh sizes 1.3, 2.5, 3.8, and 5.1 cm) were set at randomly selected shoreline sites per season. Two horizontal gill nets were set in reaches 1, 3, and 4, and four nets were set in reach 2. The nets were set perpendicular to the shore, with the smallest mesh size closest to shore. A total of eight monofilament vertical gill nets were set per season, four in the pelagic zones of both reaches 1 and 2, except during the spring, when flows were too high and the verticals were not set in the forebay. The nets (2.4 × 29.9 m), one of each mesh size (1.3, 2.5, 3.8, and 5.1 cm), were set in the upper 29.9 m of the water column at randomly selected pelagic locations. During the summer, two additional horizontal nets were set in the pelagic zone of the forebay, one at the surface and one at the bottom (61 m). Data collected from the pelagic horizontal gill nets were not used in the relative abundance or CPUE calculations; however, the data were included in age, growth, relative weight, and proportional stock density calculations. Gill nets in reaches 2, 3, and 4 were set at dusk and retrieved within 4 h. The gill nets set in reach 1 were set in the early morning (~02:00 hours) and retrieved within 4 h.

Since 1981, the Center for Limnology at the University of Wisconsin–Madison, with support from the U.S. National Science Foundation, has been administering the Northern Temperate Lakes Long-Term Ecological Research (LTER) program. The center is focusing its attention on eight deep-water lakes in Wisconsin for monitoring. Vertical gill nets are used to monitor yearly changes in the abundance of pelagic fish species (<http://lter.limnology.wisc.edu/fishproto.html>). Researchers sample the deep basins of these lakes with seven nets, each a different mesh size, hung vertically from foam rollers and chained together in a line. Each net is 4 m wide and 33 m long. From 1981 through 1990, the nets were multifilament mesh, in stretched mesh sizes of 19, 25, 32, 38, 51, 64, and 89 mm. In 1991 the multifilament nets were replaced with monofilament nets of the same sizes. One side of the net is marked in meters from top to bottom. Stretcher bars have been installed at 5 m intervals from the bottom to keep the net as rectangular as possible when deployed. The bottom end is weighted with a lead pipe to quicken the placement of the net and to maintain the position of the net on the bottom.

Gill nets are set at the deepest point of all long-term ecological research lakes except Crystal Bog, Trout Bog, and Fish Lake. The nets are set for two consecutive 24-h sets. The nets are set in a straight line, each connected to the next and anchored at each end of the line. Once the nets are in position, they are unrolled until the bottom end reaches the bottom and then tied off to prevent further unrolling. The nets are pulled by placing each net onto a pair of brackets attached to the side of the boat and by rolling the net back onto its float; the fish are picked out as the net is brought up and placed in tubs according to depth. The fish are processed when the net is completely rolled up and before it is redeployed.

Field Methods

Setup

Boats
The investigator should review the size and type of waterbody where the gill nets will be employed. Since gill nets are dangerous to work with and cannot normally be effectively set by personnel on foot, an effective boat, rubber raft, or canoe should be used. For work in remote lakes where transportation is restricted to foot travel, an inflatable rubber raft is the most effective method for setting gill nets. Where helicopters are available, a small skiff or canoe can be used. In lowland areas, a variety of boats are available depending on road access to the waterbody and the size and type of waterbody.

Nets
Recommended lake gill net specifications are as follows:
1. Length: 15–48 m (50–150 ft).
2. Depth: 2–2.5 m (6–8 ft).
3. Each net includes a proportional panel of 1.25, 1.90, 3.54, and 3.80 cm (i.e., 0.5, 0.75, 1.0, and 1.5 in) mesh. This mesh is capable of capturing fish as small as 7–8 cm total length.

4. The gill net is designed with a braided lead line of 7 g/m (0.3 lbs/fathom).
5. Floats are attached such that the lead line lies along the bottom and the floats suspend the net in the water column. Float line must be braided nylon with corks 15–20 cm (6–8 in) apart of a size to make the net either sink or float.
6. Nets are normally constructed of double-knotted monofilament and hung on a 2:1 basis (i.e., twice as much web as lead/cork line). Monofilament is nearly invisible under water and highly entangling, and it is nearly maintenance free. Its disadvantage is that it is more dangerous to handle, and if the net is lost, it continues to fish for years thereafter.
7. All nets must have nylon gables (side panels) of approximately 18 kg test.

For some operations the net may be allowed to float on the surface of the lake. In this case, the floats would be replaced with larger, more buoyant floats capable of suspending the net and the lead line with fish (Balkwill and Combs 1994). Gill nets should be clearly labeled with the researcher's name and contact phone number. In urban areas, the net may cause concern with the public, and special arrangements may need to be made with a local landowner or others to arrange for access and to prevent vandalism to the equipment.

Other equipment

The sampler should plan to bring measuring boards, scale envelopes, buckets, global positioning system (GPS) unit, weighing scales, clipboards, waterproof forms, and other equipment if genetic information is also being collected. Proper collecting permits may need to be obtained depending upon species collected, jurisdictions, and other factors.

Events Sequence

Setting the net

1. Set the net along the bottom in shallow waters not exceeding 5–7 m in depth to capture a representative sample of the total fish stock when sampled at night.
2. Nets should be placed perpendicular to the shoreline in shallow water or at a 45° angle in deep water, with the small end of the mesh nearest the shoreline.
3. The deep end of the net should have a line with a float attached to it to aid in retrieving the net if it becomes snagged. The location selected should be free of sunken logs, jagged rocks, pipes, and other objects that can snag the net and keep it from being retrieved.
4. Nets should be set at dusk (one hour before sunset) and retrieved at dawn (1 h after sunrise).
5. The net is coiled in the bow of the boat with the lead line on one side and the float line on the other side. The small mesh end is tied to the shore or to a log or an anchor near the shore, and the boat is moved out towards deep water. The net is allowed to pay out over the bow. The person paying out the net should be vigilant to keep the net from snagging on the

vessel. When the net has been fully deployed, the net should be stretched as tightly as possible before being released. (Note: It is very important that the person deploying the net ensure that all buttons, zippers, and other apparel that could be entangled in the net are secured.)

6. Nets can be set with or without bait. Baiting is effective for many purposes and can be accomplished by dispensing the contents of a can of tuna fish along the length of the net.

7. Net sets may need to be modified when testing for presence of bass and other nonsalmonid species. Trout and salmon tend to swim forward when encountering the net and then quickly become entangled. Bass will tend to back up when encountering the net and swim perpendicular to an obstacle to avoid it. When sampling waters where bass and other non-salmonid species are present, at least a few of the sites should be set with one net perpendicular to the shore and another perpendicular to the first net to increase the probability of capturing the bass avoiding one of the nets.

8. The following morning, the net is retrieved by the sampler paddling out to the float and bringing up the deep end of the net first. This minimizes snagging and allows the sampler to work the net towards shore. If the net is pulled from shore, the net has a higher chance of snagging on the bottom obstructions as it slides along the bottom.

9. Repeat steps 1–7 for each sample site.

Setting nets on ice-covered lakes

1. Gill nets can be fished through the ice when necessary. This can be accomplished by first determining the net location during the summer when bottom contours and obstacles can be assessed.

2. During the winter, the sampler must locate the net site and then mark out the length of the net on the ice perpendicular to shore as before.

3. Use an ice saw or chain saw to cut two parallel lines in the ice the distance of the net set. Remove blocks of ice and clear the hole of debris.

4. Lower the net into the hole as described in steps 3–5 under Setting the net.

5. The next morning, the hole may have refrozen and will need to be cleared of ice either with a saw or an axe.

6. In subfreezing conditions, the net should be pulled quickly from the hole and spread out on the ice as straight as possible. This will allow the net to be picked after it freezes.

Number of nets

Following is a rough guideline for the number of nets to use:

Lake size (ha)	Number of nets
< 4	1
4–10	2
10–20	3
20–40	4
Each additional 40 ha	Add 1 net

If the initial sampling effort yields few or no fish, the sampling stations should be moved and the sampling effort repeated. A careful description of the sampling location is important in order to find and duplicate the same location the following sampling period.

Sample Processing

The following steps should be used when processing fish caught in the gill net:

1. Fish should be carefully removed from the gill net. For most of the fish species, it will require untangling the gill opercles from the net mesh. Some species, such as catfish *Siluriformes* tend to spin once they are entangled and will require a lot of work to release.

2. Measure to the nearest millimeter, identify to species, weigh to the nearest gram, and take scale samples from the left side (just posterior to and below the dorsal fin and above the lateral line). If genetic samples are needed, take samples per the protocol being employed for DNA or electrophoresis. Depending on the purpose and need, stomach samples may be taken, most commonly via gastric lavage, and internal organ, examined for parasites, gender, and maturity.

Field data recording should be standardized and should include the following:

1. lake
2. sampling date
3. gear type
4. net location (shore orientation, depth, placement time, collection interval)
5. hours fished
6. species
7. weight (gm)
8. total length (cm)
9. scale number
10. parasites observed
11. deformities observed
12. wounds observed
13. universal transverse mercator or latitude/longitude coordinates
14. names of survey personnel

Other physical measurements such as temperature, pH, and visibility may also be taken. These factors often affect fish activity and net visibility and efficiency and should be tracked.

Personnel Requirements and Training

Responsibilities

The net should be set with two people, whenever possible, with one person deploying the net and one person propelling the vessel. This reduces the chance that the net will become entangled, and it helps ensure that it will be deployed properly. If only one person is available, the way the person initially prepares the net is crucial to successful deployment. In turn, during net retrieval, two persons are ideal: one person controls the vessel against water and wind conditions while the other person slowly brings the net on board and either picks the net as it is brought on or brings in the entire net and later picks the fish out of the net on shore under more stable conditions.

The samplers can determine who will record and who will weigh, measure, and conduct other examinations of the fish when they are being processed.

Qualifications

The person using a gill net should have been properly trained by an experienced field biologist and should have a degree in biology or one year of experience in sampling fish in the geographic area where the sampling is to occur. The use of volunteers should be carefully evaluated due to the danger involved and potential adverse reactions with the public.

Training

Training should either be provided through videos and demonstrations under cover prior to the season or through on-the-job training by accompanying an experienced field biologist.

Operational Requirements

Field Schedule

The field schedule for setting gill nets is normally during the spring when fish have become active and before there is a lot of recreational lake activity; however, as noted above, the sampling can occur at any time, including winter, depending upon the objectives of the study and the needs of the monitoring.

Equipment List

Use this list to help in developing a budget estimate.

Item	Comments
4 m aluminum rowboat with oars	
15 hp outboard motor	
25-L gas tank and other motor repair items	
One-person raft with kayak paddle	Used for remote applications
Net tubs or buckets	
Meter board	
Anchor and anchor lines	
Dissecting kit	
10% formalin or 70% ethanol	
Screw-top vials	
Scale envelopes	
Collecting permits	
Secchi disk	Measures transparency of water

Item	Comments
GPS unit	Location of nets
Life jackets	
Sharp knives	To cut loose an entangled person or net; for fish samples
First-aid kit	
Ice saw or chain saw	For under-ice sampling
Axe	For under-ice sampling
Net labels	
50-m variable mesh gill nets	Number as needed
Clipboard	
Sample forms	
Thermometer	
Wire clippers	Used for catfish spine removal
Cell phone, two-way radio, or satellite phone	Communications

Personnel Budget

The following guidelines can be used to estimate time budget for personnel:

Activity/item	Cost
Staff time*	3 h
Travel time	Variable
Preparation time	4 h
Training	1 h
Lab workup	2 h
Data analysis, report writing	8 h

* two biologists to set the net and to retrieve and process the catch

Literature Cited

Balkwill, J. A., and D. M. V. Combs. 1994. Lake survey procedure manual for British Columbia. BC Environment, Lands and Parks Report, Victoria.

Bernabo, C., and B. Hood. 1986. Protocols for establishing current physical, chemical and biological conditions in remote alpine and subalpine ecosystems. U.S. Forest Service, Fort Collins, Colorado.

Bonar, S. A., B. D. Bolding, and M. Divens. 1993. Standard fish sampling guidelines for Washington State ponds and lakes. Washington Department of Fish and Wildlife Report, Olympia.

Borgstrøm, R. 1992. Effect of population density on gillnet catchability in four allopatric populations of Brown trout (*Salmon trutta*). Canadian Journal of Fisheries and Aquatic Sciences 49:1539–1545.

Charnley, S. 1984. Human ecology of two central Kuskokwim communities: Chuathbaluk and Sleetmute. Alaska Department of Fish and Game, Division of Subsistence, Technical Paper No. 81, Juneau. Available: www.subsistence.adfg.state.ak.us/TechPap/tp081.pdf (accessed August 2006).

Coffing, M. W. 1991. Kwethluk subsistence: contemporary land use patterns, wild resource harvest and use, and the subsistence economy of a lower Kuskokwim River area community. Alaska Department of Fish and Game, Division of Subsistence, Technical Paper No. 157, Juneau. Available: www.subsistence.adfg.state.ak.us/TechPap/tp157.pdf (accessed August 2006).

Dauble, D. D., and R. H. Gray. 1980. Comparison of small seine and backpack electro-shocker to evaluate nearshore fish populations in rivers. Progressive Fish-Culturist 42:93–95.

Hamley, J. H. 1975. Review of gillnet selectivity. Journal of the Fisheries Research Board of Canada. 32: 1943–1969.

Hillborn R., C. J. Walters. 1992. Quantitative fisheries stock assessment: choice, dynamics and uncertainty. Chapman and Hall, New York. 570 p.

Hubert, W. A. 1996. Passive capture techniques. Pages 157–192 in B. R. Murphy and D. W. Willis, editors. Fisheries techniques, 2nd edition. American Fisheries Society, Bethesda, Maryland.

Johnson, B. M., R. A. Stein, and R. F. Carline. 1988. Use of a quadrat rotenone technique and bioenergetics modeling to evaluation prey availability to stocked piscivores. Transactions of the American Fisheries Society 117:127–141.

King, T. A., J. C. Williams, W. D. Davies, and W. L. Shelton. 1981. Fixed versus random sampling of fishes in a large reservoir. Transactions of the American Fisheries Society 110:563–568.

Klemm, D. J., Q. J. Stober, and J. M. Lazorchak. 1993. Fish field and laboratory methods for evaluating the biological integrity of surface waters. U.S. Environmental Protection Agency, Office of Research and Development, Report EPA/600/R-92/111, Cincinnati, Ohio.

McLellan, J. G. 2001. 2000 WDFW Annual Report for the Project Resident Fish Stock Status above Chief Joseph and Grand Coulee Dams. Part I. Baseline assessment of Boundary Reservoir, Pend Oreille River, and its tributaries. Pages 18–173 in N. Lockwood, J. McLellan, and B. Crossley. Resident fish stock status above Chief Joseph and Grand Coulee dams. 2000 Annual Report. Report to Bonneville Power Administration, BPA Report DOE/BP-00004619-1, Portland, Oregon.

Michael, H. N., editor. 1967. Lieutenant Zagoskin's travels in Russian America, 1842-1844. The first ethnographic and geographic Investigations in the Yukon and Kuskokwim valleys of Alaska. Anthropology of the north: translations from Russian sources No. 7. University of Toronto Press, Toronto.

Oswalt, W. H. 1980. Kolmakovskiy redoubt: the ethnoarchaeology of a Russian fort in Alaska, Monumenta Archaeologica Volume 8. University of California, Institute of Archaeology, Los Angeles.

Rider, S. J., M. J. Maciena, and D. R. Lowery. 1994. Comparisons of cove rotenone and electrofishing catch-depletion estimates to determine abundance of age-0 largemouth bass in unvegetated and vegetated area. Journal of Freshwater Ecology 9:19–27.

Stokes, J. 1985. Natural resource utilization of four upper Kuskokwim communities. Alaska Department of Fish and Game, Division of Subsistence, Technical Paper No. 86, Juneau.

Thompson, W. L., G. C. White, and C. Gowan. 1998. Monitoring vertebrate populations. Academic Press, New York.

VARIABLE MESH GILL NETS (IN LAKES)

Foot-based Visual Surveys for Spawning Salmon

Bruce Crawford, Thaddeus R. Mosey, and David H. Johnson

(Note to readers: this supplemental protocol complements two full protocols—Carcass Counts, page 59, and Redd Counts, page 197.)

Background and Objectives

Background

Adult salmonids return from the sea to spawn in their natal streams. Since adult spawner abundance is, for many populations, the principle measure of abundance and enables estimation of potential egg deposition, it is crucial to be able to obtain accurate estimates of the number of spawners to monitor abundance, relative to escapement goals or other management objectives. As salmon and steelhead *Oncorhynchus mykiss* enter rivers or streams over an extended period of time, and since there is continuous mortality (for salmon) and/or emigration (for steelhead), it is very difficult to obtain an estimate of the total number of spawners. Spawner counts, therefore, ideally should be replicated weekly throughout the spawning period to estimate total escapement. More frequent surveys will provide a more accurate estimate of spawner abundance.

Streamflow and turbidity, among other factors, influence the accuracy of spawner counts. Given the spawn timing of salmon and steelhead, counts must often be conducted under suboptimal conditions; however, streams in many regions are relatively clear, and thus, it is possible to count spawning salmon directly as they dig their egg nests (redds) and actively mate. Counts of spawning fish are usually done on foot or from a boat (although boat-based surveys are not discussed in this chapter). Spawning counts are often done on steelhead and chinook salmon *O. tshawytscha*, chum salmon *O. keta*, and coho salmon *O. kisutch*.

In addition to supporting spawning escapement estimates, live fish counts describe spawning timing, particularly when redd and carcass counts are also available. Comparatively high live fish and low redd and carcass numbers may indicate that spawning is just beginning. The opposite (low live fish and high redd and carcass numbers) can indicate that spawning in that stream is nearly complete. Material in this protocol has been extensively drawn from Heindl (1989). This method is mainly used where road networks are highly developed (for access) in southern British Columbia and Japan.

Rationale

Counting live salmon while walking along a stream is one of the oldest and simplest methods of estimating spawner abundance. Foot surveys are an effective method of counting spawning salmon because the fish are large enough to observe directly, and when done correctly, surveyors can achieve measurable accuracy for certain species of salmon and for steelhead (Waldichuk 1984; Irvine et al. 1992; Nickelson 1998). Furthermore, by combining the estimated number of spawning fish and total redd counts, one can determine the number of carcasses needing to be recovered (sampled) to evaluate key demographic aspects of hatchery and supplementation programs. Labor costs associated with visual spawner index counts have been found to be less than those associated with mark–recapture or fish counting weir techniques (Irvine et al. 1992); however, live fish are wary, mobile, and easily camouflaged and, in some situations relatively nocturnal in their spawning habits, thus reducing the accuracy of visual direct counting methods.

Visual spawner surveys may be the basis for estimating total escapement in an entire basin or for simply monitoring spawner abundance in one or more index reaches. Surveys may seek only to estimate peak spawner abundance or to describe total spawning timing. Reaches for survey and survey frequency must, therefore, be selected carefully to achieve the desired objective.

Objective

The objective of visual spawner surveys is to determine total or relative abundance of river or stream populations by directly counting the numbers of spawning fish on their redds.

Sampling Design

Site selection

Site selection for foot surveys of spawners is based on timing of returning fish, water clarity, and access to banks with a view of the stream. Foot surveys need to be conducted where the clarity of water is high enough to distinguish spawning fish from the substrate. Timing of spawning can affect potential for high water clarity. For example, spring spawners such as steelhead may be spawning at higher water levels when water is more turbulent and has higher turbidity. Foot surveys conducted for steelhead should be timed for when water clarity is expected to be the greatest. Salmon that return to spawn in the fall are generally spawning when water clarity is higher than that for steelhead. Surveys are frequently conducted under suboptimal visibility, so a visibility factor must be subjectively estimated by the surveyor and applied to counts. This can create a source of bias in visual counts. Additional insights into behavioral aspects of salmon spawning activity can be found in Perrin and Irvine (1990), Barlaup et al. (1994), Fleming (1996), Webb and Hawkins (1989), and Webb et al. (1990).

Access to banks with a clear view of the stream is critical for effective implementation of the foot survey protocol. If the stream is in a steep V-notch and surveyors cannot safely get close to the stream to view spawning activity, then a foot survey is not an effective method. Similarly, if the survey is to be conducted over a considerable length of stream, thick vegetation may impede surveyors' progress and reduce the efficiency of the survey. Surveyors should not need to walk in the stream to conduct the survey as this action would disturb the spawning fish.

To improve the accuracy and precision of escapement estimates obtained using visual surveys, a variety of sampling approaches can be used to identify portions of a stream to survey for fish. The easiest approach may be to use a simple random sampling design. With this design, each section of a stream has the equal probability of being selected; however, if sampling units can be divided into homogeneous groups, then stratified random sampling is generally preferable to simple random sampling, since stratification usually results in a smaller variance than that given by comparable simple random sampling (Cochran

1977). A stratified random sampling design to estimate escapement is achieved by stratifying the watershed and then randomly selecting sample sites within each stratum. Alternatively, sample units within each stratum can be selected on the basis of easy access by surveyors and/or on the basis of expected or actual locations of fish. This latter type of sampling has been termed stratified index sampling (Johnston et al. 1987; Bocking et al. 1988). As visual surveys are typically conducted on foot, surveyors should evaluate a few representative stream reaches (e.g., 250 m long each) prior to the start of spawning surveys to determine suitability for surveys based on water depth and flows.

In their 3-year project with coho salmon in British Columbia, Irvine et al. (1992) examined stratified random and stratified index sampling designs for estimating the numbers of spawning fish. The approach was evaluated in two streams with widely varying escapements; estimates of adult fish using the stratified index design were always similar to estimates obtained through an independent mark–recapture program. The stratified random design underestimated fish numbers in every case but one. Because the distribution of fish was aggregated, with random sampling there was a higher probability of sampling low fish abundance areas than high-abundance areas. Numbers of jack coho were underestimated with each sampling design, probably because the surveyors overestimated the efficiency of seeing these fish.

Sampling frequency and replication

Surveys are typically conducted on 6 (Irvine et al. 1992) to 10 (T. Mosey, Chelan County Public Utilities District, personal observation) stream reaches per crew per day. Foot counts are generally conducted every 7–10 d throughout the spawning season to observe fish on their redds. Replication of foot surveys is necessary to count the total or relative number of spawners over the spawning season. For each target species, the expected spawning season should be identified, and surveys should ideally be scheduled to start before the first spawner enters the stream and continue until after the last spawner completes the spawning process. Counts that are conducted more frequently will have a higher accuracy than those that are conducted less frequently. Estimates of residence time, as determined from redd life, vary from several days to 21 d. Irvine et al. (1992) found that the residence time of coho in their British Columbia study area was between 13 and 17 d (with one exception of 8 d). It is important to note that residence timing varies among different reaches, annually for a given population, and between rivers. A general recommendation is to conduct spawner surveys weekly.

Field/Office Methods

Setup

Prior to arrival

Streams should be divided into survey reaches prior to sampling. This can be done either through a randomized sampling design or through previous surveys that identify where the target salmonid is known to spawn. Before a survey is undertaken, the surveyors must be well aware of what reaches they are to survey, the target species, when the spawning season begins and ends, what the time interval for the survey is, what data are needed, and what equipment is required. The project leader must ensure that each surveyor or team carries all necessary equipment.

Events Sequence

Survey Description

Surveys should begin at the earliest anticipated beginning of spawning in order to reflect the overall spawning interval. Live fish may be encountered practically anywhere in the stream. Prior to spawning, fish may concentrate in deep pools (e.g., chinook) or hide in the foam and boil of the riffles (e.g., steelhead). They also may hide under cut banks and in logjams. As spawning progresses fish are more often observed on or near the spawning riffles.

Live fish survey

The following are tasks to perform upon arrival at the survey reach

(1) Turn on the global positioning system (GPS) device and after gaining satellite contact, record the coordinates (latitude–longitude or universal transverse mercators [utm]) of the beginning of the survey reach.

(2) Begin at the downstream terminus of the spawning reach and proceed slowly upstream. Fish are less likely to observe you approaching from downstream and, therefore, are less likely to swim to cover before they can be counted.

(3) Wearing polarized glasses will help reduce surface glare and allow the fish to be more easily observed; wearing dark clothing/rain gear will help the sampler blend into the streamside forest vegetation.

(4) Fish that detect your movement will usually dart downstream but may dart upstream into a nearby pool; the sampler should wait a few seconds to determine if the fish will return to the redd to avoid double counting farther upstream.

(5) Record the live fish observed as you move upstream; look carefully for auxiliary males, especially jacks that may be positioned near the actively spawning pair or the redd.

(6) It is not unusual to find live spawned-out fish in slack water below spawning riffles. These live "carcasses" should not be included in the live fish count.

(7) Walk all stream channels (side channels and backwater pools).

(8) At the end of the reach, verify your position with the GPS device and record the coordinates.

(9) Record survey data in the survey field book or on appropriate forms and transfer to a database for data management.

(10) Repeat steps 1–9 for each reach sampled and for each day sampled.

(11) Survey visits should be conducted every 7–10 d; wait until the end or termination of each survey visit to record weather, flow, and visibility for each reach surveyed.

Data Handling, Analysis, and Reporting

Data from spawner surveys should be entered into a database for data management. A combined data set for redd counts and/or carcass counts is often used; however, separate counts for live spawners should be maintained in the database. Table 1 provides metadata for variables that should be collected during the spawner surveys.

TABLE 1.—Database parameters for spawner count data.

Description	Metric	Format	Comments
Species	Text	Text	Note the species sampled.
Live	#	XXXX.	Number of live fish observed.
River	Text	Text	Record the stream name and section being surveyed.
Reach length	Meters	XXXX.X	Record the total distance of the survey reach.
Days sampled	#	XXXX.	Record the total number of days sampled.
Reach	Text	Text	Record the reach description and reach code.
Date	Date		Record the date of the survey.
Samplers	Text		Record the last name of the samplers.
Start latitude	D, M, S	XX,XX,XX	Record the latitude of the beginning survey point.
Start longitude	D, M, S	XX,XX,XX	Record the longitude of the beginning survey point.
End latitude	D, M, S	XX,XX,XX	Record the latitude of the end survey point.
End longitude	D, M, S	XX,XX,XX	Record the longitude of the end survey point.
Temp	°C	XX.X	Record the stream temperature at the time of the survey.
Time	Time	Time	Record the time the survey began.
Conditions	Text	Text	Record the water conditions and weather at the time of the survey.
Other	Text	Text	Describe other samples taken and remarks.

Data analysis procedures will vary with the objectives of the survey. Generally, however, data are evaluated with respect to the abundance or relative abundance of spawners throughout the spawning season. Relative abundance could be as simple as replicated spawner counts at the same site over several spawning seasons to compare the size of the counts across years. Using these data to estimate population abundance can only be done if sample sites are representative of that population both spatially and temporally. Other methods, such as area-under-the-curve (AUC) (English et al. 1992), may be used to extrapolate the total number of spawners from the survey data.

Personnel Requirements and Training

Responsibilities
A crew of two surveyors should be used. They can split up part of the survey area if needed, but two persons should be employed for safety reasons.

Qualifications
The person conducting a spawning survey should have been properly trained by an experienced field biologist and should have a degree in biology or 1 year of

experience in sampling fish in the geographic area where sampling is to occur. Volunteers can be used when carefully trained and evaluated.

Training

Crews should be trained in the classroom first with illustrations of counting techniques, equipment needed, process for location of survey reaches, and so forth. Training should include at least one survey (a paired count), with individual crew members matched up with an experienced instructor who can assist the student in locating redds and in detecting and identifying spawners as they hold in the stream.

Where spawner stream reaches have been randomly selected, a pilot survey should be conducted during the summer to verify accessibility of survey segments and collect other physical information.

Operational Requirements

Field schedule

The field schedule for spawner surveys is generally in the fall or winter for salmon species and in the spring for steelhead. The schedule is determined by the spawning season for the target species of the survey.

Equipment list

Equipment

The foot surveyor should be equipped with hip boots or waders and dark clothing or rain gear to minimize disturbance to spawners, polarized sunglasses, appropriate survey forms, and a handheld GPS device. Additional equipment for boat/canoe surveys (as needed for access to some stream/river sections) are life vests, dry bags, a survival kit (containing matches, food, whistle, emergency blanket, etc.), helmets, and a throw line. (See equipment list below.)

Item	Comments
Data forms and backing	E.g., clipboard or digital data device
Pencils	
GPS handheld device	Used to record survey latitude and longitude or UTMs
Cell phone and/or two-way radios	Useful for emergencies and for maintaining contact with other crews
Satellite phone and personal locator beacon	For emergency situations; carried by crew members in areas where cell phone or two-way radio communication is restricted due to topography or limited coverage network
Polarized glasses	
Wading gear (hip boots/chest waders)	Not recommended if rafting
Metric measuring tape	
Water thermometer	
First-aid kit	
Colored flagging or biodegradable spray paint	For marking multiple survey reaches
Indelible markers	For marking flagging

Item	Comments
Day pack	For carrying supplies, lunch, etc.
Extra clothing	Jacket, rain gear, hat
Rubber raft	If floating a river
Air pump	If floating a river
Spare oar or paddle	If floating a river
Spare oarlock	If floating a river
Tool kit	Pliers, crescent wrench, wire, heavy tape, bolts, nuts, washers, etc.
Rope	If floating a river
Waterproof bag	If floating a river
Flotation vest	If floating a river

Budget

The following guidelines can be used to calculate the budget:

Activity/Item	Cost
Equipment	
Staff time*	3 h
Travel time	(dependent on survey site)
Preparation time	1 h
Training	1 h
Lab workup	2 h
Data analysis	8 h

* for two biologists to walk the stream reach

Literature Cited

Barlaup, B. T., H. Lura, H. Saegrov, and R. C. Sundt. 1994. Inter- and intra-specific variability in female salmonid spawning behavior. Canadian Journal of Zoology 72:636–642.

Bocking, R. C., J. R. Irvine, K. K. English, and M. Labelle. 1988. Evaluation of random and indexing sampling designs for estimating coho salmon (Onchorhynchus kisutch) escapement to three Vancouver Island streams. Canadian Technical Report on Fisheries and Aquatic Science 1639.

Cochran, W. C. 1977. Sampling techniques. 3rd edition. John Wiley & Sons, New York.

English, K. K., R. C. Bocking, and J. R. Irvine. 1992. A robust procedure for estimating salmon escapement based on the area-under-the-curve method. Canadian Journal of Fisheries and Aquatic Science 49:1982–1989.

Fleming, I. A. 1996. Reproductive strategies of Atlantic salmon: ecology and evolution. Reviews in Fish Biology and Fisheries 6:379–416.

Heindl, A. L. 1989. Columbia River chinook salmon stock monitoring project for stocks originating above Bonneville Dam. Field operations guide. Columbia River Intertribal Fish Commission, Technical Report 87-2 (Revised), Portland, Oregon.

Irvine, J. R., R. C. Bocking, K. K. English, and M. Labelle. 1992. Estimating coho salmon (Onchorhynchus kisutch) spawning escapements by conducting visual surveys in areas selected using stratified random and stratified index sampling designs. Canadian Journal of Fisheries and Aquatic Science 49:1972–1981.

Johnston, N. T., I. R. Irvine, and C. J. Perrin. 1987. Instream indexing of coho salmon (*Oncorhynchus kisutch*) escapement in French Creek, British Columbia. Canadian Technical Report on Fisheries and Aquatic Science 1573.

Nickelson, T. E. 1998. ODFW coastal salmonid population and habitat monitoring program. Oregon Department of Fish and Wildlife, Corvallis.

Perrin, C. J., and J. R. Irvine. 1990. A review of survey life estimates as they apply to the area-under-the-curve method for estimating the spawning escapement of Pacific salmon. Canadian Technical Report Fisheries and Aquatic Science 1733.

Waldichuk, M. 1984. Chairman's summary. Pages Ix–xv in P. E. K. Sumons and M. Waldichuk, editors. Proceedings of the workshop on stream indexing for salmon escapement estimation. Canadian Technical Report on Fisheries and Aquatic Science 1326.

Webb, J. H., and A. D. Hawkins. 1989. The movements and spawning behavior of adult salmon in the Girnock Burn, a tributary of the Aberdeenshire Dee, 1986. Scottish Fisheries Research Report 40.

Webb, J. H., J. L. Bagliniere, G. Maisse, and A. Nihouarn. 1990. Migratory and reproductive behavior of female adult Atlantic salmon in a spawning stream. Journal of Fish Biology 36:511–520.

Video Methodology
Jennifer S. O'Neal

Background

Video equipment has been used to assess aspects of migrating salmonids since about 1990. Efforts to assess abundance of salmonid species more accurately have intensified due to declines in many stocks. Additionally, management of stocks requires the accurate assessment of abundance and the diversity of species in freshwater systems. In assessing adult salmon moving upstream, video surveys are well suited as an alternative to manual counting at fishways and dams (Hatch et al. 1994; Hiebert et al. 2000). Manual counts require significant staff time during the migration season, and video surveys have been tested as an alternative approach to reduce this staff time requirement.

Video surveys have been successfully applied in remote areas where tower counts or aerial surveys have generally been conducted (see Figure 1). Aerial surveys have been used in Alaska to assess salmon populations over large areas using relatively little staff time, but variable conditions and observer variability limit the accuracy of these counts (Otis and Dickson 2002). Data from aerial surveys are often a rough estimate and therefore may not be suitable for measuring abundance, productivity, or recovery of salmon populations; this information should be considered an index rather than a true measure of adult escapement (Hetrick et al. 2004).

Counts at weirs are used as another way to track both upstream migration of adult salmon as well as downstream migration of smolts. While weir counts provide the most accurate assessment of migrating fish populations through a freshwater system, their operation requires daily monitoring and maintenance. Additionally, weirs can be overtopped by high flows or damaged by debris. In an attempt to address issues of cost and the ability to provide a permanent record of fish counted, video equipment has been used to count and establish species composition for a variety of salmonid species.

Video techniques have also been used to assess the movement (including leaping ability) of juvenile coho salmon *Oncorhynchus kisutch* at culvert outfalls (Pearson et al. 2005). This work is particularly relevant as very large numbers of perched culverts are blocking upstream fish passage to thousands of kilometers of stream habitat.

Deepwater spawning of fall chinook salmon *O. tshawytscha* downstream of the Bonneville Dam on the Columbia River was assessed using video-based boat surveys (Mueller 2005). A total of 293 redds were detected in a 14.6-ha area; an expanded redd count based on the percentage video coverage of the area was 3,198 redds. Chinook redds were found 1.07–7.6 m deep in predominately medium-cobble substrate.

A particularly thorough and practical review of video and acoustic camera technologies for studying fish under ice has come from Mueller et al. (2006). These researchers have examined fish species presence, abundance, size, and behavior under ice cover in northern and arctic lakes, rivers, and streams.

Remote-controlled, underwater video techniques have recently been used to review the specific reproductive behavior of fishes in the subfamily Salmoninae (Esteve 2005). Esteve used underwater video recordings taken in wild and in seminatural channels together with literature references to review how *Oncorhynchus*, *Salmo*, and *Salvelinus* behave during reproduction (e.g., phases related to female nest selection, construction, and completion, and male strategies and tactics).

Lighting has been a significant issue in videography efforts, and important insights into the use of visible versus infrared light have been explored (e.g., Hiebert et al. 2000; Mueller et al. 2006). There have been several changes in the equipment used in video surveys as technology improves. Initial work was conducted with black-and-white recorders and storage was on 160-min videotapes. Recent use of digital video technology has allowed images to be stored on DVDs and hard drives. Digital technology has increased the general quality of the video image, but storage requirements, increased expense for the equipment, software compatibility, and practical aspects of demanding field conditions may require additional resources to implement correctly.

An additional application of video survey has been in the monitoring of juvenile fish in nearshore marine and kelp *Macrocystis* sp. areas (Brady and Schwartz 2005). This technique requires the use of a diver to operate mobile cameras along transects. This methodology is still under development, and interested readers are directed to Brady and Schwartz (2005).

FIGURE 1.—Video camera set up on tower in an Alaskan river (note contrasting panels on river bottom).

Rationale

Video equipment can be used to create high-quality records of fish passage using relatively inexpensive equipment and adequate lighting (Hatch et al. 1994). These data are collected at less cost than direct observation or weirs. Video also creates a permanent record of the observations, which can be reanalyzed if necessary to allow for specimen identification or development of confidence bounds on escapement estimates (Hatch et al. 1994). Staff fatigue can be reduced by spacing out the time used for interpretation or by using technology to help interpret images (Irvine et al. 1991). Recent users of video technology have been working on computer image processing automation (Hatch et al. 1994, 1998) and on recording triggered by motion detection to reduce the number of blank images on tape. Video offers a permanent record of the number of fish and often the species and size that pass an area, and it can be used to compare differences between years and site conditions as well (Hiebert et al. 2000). A permanent record is especially helpful when time constraints do not allow for immediate data analysis (Irvine et al. 1991). Additionally, video can be used to assess populations without any handling of the fish species, reducing potential stress on fish. Finally, video surveys have little impact on the environment and do not cause fish mortality.

Video surveys can be used at existing facilities such as dams and fishways but can also be applied in remote locations (see Figure 1). Systems that are lightweight, transportable, and run under their own power (solar, wind, or hydrogenerated power) have been developed for remote application (Otis and Dickson 2002).

Time savings from video surveys vary by application. Hatch et al. (1994) found that 3 d of migration could be reviewed in 6 h, or 2 h per d. Otis and Dickson (2002) determined total fish passage in an average of 38 min per d of video at one sample site and 16 min per d at another. An additional 29 min and 40 min per d of video were required to determine species composition at each site, respectively, using a subsample of the first 15 min of every h (Otis and Dickson 2002). Efficiency rates of 3.6% and 2.1% were calculated for these two sites, reducing the amount of staff time needed for results by 96.4% and 97.9%. Hatch et al. (1994) found an overall time savings of 92% using video surveys versus visual observation.

Objectives

The primary objectives of using videography with salmonids include

(1) determining, or estimating, total numbers of moving or migrating fish at a sampling site;

(2) determining, or estimating, species composition of moving or migrating fish at a sampling site;

(3) determining, or estimating, the numbers of hatchery versus wild fish by close examination of body markers or fin clips;

(4) characterizing the body sizes of fish moving or migrating past a sampling site;

(5) assessing spawner and redd density, distribution, and habitat characteristics (e.g., deepwater environments); and

(6) characterizing behavioral aspects of fish (e.g., mating, habitat selection, hesitation in migration).

The first five objectives are used by fisheries scientists and managers concerned with population status and trend information. While objective 6 is mainly used by behaviorists for a variety of reasons (including in the lab), video is also used in fish ladders to determine how well fish pass certain structures.

Video technology can be used to provide counts of fish entering a system or passing a certain point in the system (Irvine et al. 1991; Hatch et al. 1994; Hiebert et al. 2000; Otis and Dickson 2002; Hetrick et al. 2004; Mueller et al. 2006). Video recording has been used to assess passage of chinook salmon, coho salmon (Hiebert et al. 2000), sockeye salmon *O. nerka*, pink salmon *O. gorbuscha*, and chum salmon *O. keta* adults (Otis and Dickson 2002), as well as counting and measuring of coho smolts (Irvine et al. 1991). Video-recording technology was implemented to reduce staff requirements and fatigue and to provide a permanent record of fish passage that could be analyzed or further reviewed. This technology has been applied mostly in areas with existing dams or passage facilities in the Pacific Northwest (Irvine et al. 1991; Hatch et al. 1994; Hiebert et al. 2000) but also in remote sites in Alaska without dams or counting facilities (Otis and Dickson 2002; Hetrick et al. 2004). Dolly Varden *Salvelinus malmo* have also been detected using this technology, but they were not the target species; their migratory movement along the margins of rivers when other species such as sockeye are densely occupying the main channel may make video a lesser choice for assessing Dolly Varden.

Additional lighting is required for night surveys or for accurate species identification (e.g., Hiebert et al. 2000; Mueller et al. 2006). Hiebert et al. (2000) report that more than half of the chinook salmon counted during their study moved upstream at night; however, other studies (Otis and Dickson 2002) have found a low percentage (<1.5%) of the population that moved at night.

Determining the species composition of fish passing a certain point is another common objective of using video technology (Hatch et al. 1994; Hiebert et al 2000; Hetrick et al. 2004; Otis and Dickson 2002; Pearson et al. 2005; Mueller et al. 2006). Additional image quality is required to determine the species of fish from recorded or digital images. Achieving this objective may require additional above-water or underwater cameras, and the lens of the camera generally needs to be within 2–2.5 m of the fish to identify the fish to determine species (Hetrick et al. 2004). Otis and Dickson (2002) note that the survey is simplified in areas without multiple species returning to spawn at the same time.

A third objective identified in video surveys is the ability to distinguish hatchery fish from wild fish based on the presence of tags or fin clips (Hiebert et al 2000). Hatcheries are playing a role in the recovery of listed stocks, and the use of video to determine the ratio of hatchery-to-wild fish within a system is important for harvest and habitat management.

A fourth objective identified in video surveys is the ability to measure fish that pass through the camera view (Irvine et al. 1991). This objective is mostly suited for fixed facilities with narrow passage chambers, where the distance of the fish from the camera can be tightly controlled and a measurement bar in the viewing screen can be used for scale. Automated measurement by computer was implemented by Irvine et al. (1991); however, correction by an observer was still required. Additional developments in this technology are anticipated.

A fifth objective is in assessing spawning redd density, distribution, and habitat characteristics, especially in deepwater areas (>1 m) (Mueller 2005), which are generally not counted in surface-based (e.g., boat, aerial) surveys. Here, boat-based video transects are typically made within a determined sample area.

A sixth objective is the characterization of the behavioral aspects of fish (e.g., mating) (Esteve 2005). Here, cameras are mounted underwater, with videography used to capture detailed movements and behaviors of reproducing salmonids.

Sampling Design

Site Selection

Video recording has often been used in places with existing counting facilities such as dams with fishways or spawning grounds. Surveys can be cost effective at remote sites or areas with multiple sites. Other work has constructed camera arrays in remote areas to reduce the need for staff to be constantly present at remote locations. Site selection requirements at remote sites are complex. If solar energy is used for power, the site needs to have adequate sun exposure and plenty of power storage capacity. Stream and river channels best suited for remote video surveys include the following characteristics (Otis and Dickson 2002):

- Relatively narrow (<30 m)
- Shallow depth (<1.5 m)
- Smooth, even bed of cobble substrate
- Smooth current flow (>6 m^3/s)
- Good source for wind, solar, or hydropower
- Good water clarity
- Low surface turbulence

Power requirements from Otis and Dickson (2002) were 3.5 A/h without lighting for night surveying. A single 75-W (4.3-A) solar panel can provide this amount of power in full sunlight, but more panels are needed for overcast days and in areas with high canyon walls (Otis and Dickson 2002).

Otis and Dickson (2002) found that sites influenced by tidal conditions with pink and chum runs were not the best sites for video surveys because large numbers of spawners were likely to congregate in the area, making fish counting difficult.

Field and Office Methods

Field Setup and Operation

Video recording systems vary in their setup, but all systems require one or more cameras (analog or digital; digital is recommended) and one or more recording and storage devices (e.g., a VCR with tapes, a computer with DVDs or a large hard drive attached). These elements of the system require protective housings if they are to be operated outside an enclosed facility. Additionally, review of the recorded data requires a playback device (often the same as the recording device), the ability to see footage from multiple cameras at the same time if necessary, and a high-quality monitor. Finally, the system has to have an adequate power source from the facility or use other means of power generation at remote sites.

Cameras used by video survey crews vary by objective, site requirements, and budget. Since the requirements for remote sites are the most challenging, we report on the equipment used at a remote study site that can be integrated with digital technology. Hetrick et al. (2004) utilized three Toshiba IK-64WDA 24-V day/night cameras for above-water filming. The cameras were housed in outdoor enclosures, set for "super day/night" exposure (which allows the camera to activate a slow shutter speed automatically), and switched to black-and-white in low light situations. Power needs for this camera were less than 5 W per camera (Hetrick et al. 2004).

Settings for camera recording can affect both the image quality and camera efficiency. Cameras can generally be set in several time-lapse recording modes. If using more than one camera and a multiplexer, Otis and Dickson (2002) recommend using 72-h normal mode on an analog camera to balance tape duration, the ability to track and individual fish, and reviewing efficiency. If only one camera is used, the 120-h normal mode allows a 160-min cassette to last 8 d at one image per sec. Camera lenses can greatly affect glare, and an auto-iris lens was found to allow a camera to adjust to varying light conditions (Otis and Dickson 2002). Hetrick et al. (2004) recommend using polarizing filters on all lenses and mounted their above-water cameras 6 m above the water surface. The lenses and filters they used were Pelco F1.6/5-40 mm variable focus, auto-iris lenses with 40.5 mm-Tiffen circular polarizing filters. They pointed the lens towards directional north to minimize glare. Lenses for underwater cameras likely need to be cleaned once per week, and camera enclosures should contain desiccant to reduce moisture (Hetrick et al. 2004).

For digital recording, a digital video recorder (Digicorder 2000 Deluxe digital video recorder, Alpha System Laboratory, or Gyyr four-channel Digital Video Management System [DVMS 400, recommended for higher resolution]) (see Table 1) was used by Hetrick et al. (2004) for both recording and as a four-channel multiplexer to view images from multiple cameras at once. Otis and Dickson (2002) used video cassette recorders for analog recording and viewing but were not satisfied with the multiplexer because it slowed the reviewing process. Hetrick at el. (2004) captured footage at four to five frames per second, at one of four resolutions (720×486, 352×480, 352×240, and 176×120). These files were compressed and recorded to a hard drive. The date and time was recorded on each frame as well as the title of the camera from which the footage came. Video files were then loaded onto DVDs using a firewire DVD-RAM drive (Institute of Electrical and Electronics Engineers, IEEE 1394; see Table 1). Storage requirements reported by Hetrick et al. (2004) consisted of 1.2 gigabytes/h recording at 5 frames per second for one camera with a compression ratio of 10:1 and 4.7 gigabytes/h for four cameras at the same settings.

Lighting techniques are critical for high image quality and the ability to determine the species of fish passing the camera. Surveyors need to balance the image quality against lighting that may affect the passage of fish over the viewing substrate (Hiebert et al. 2000). Infrared light has been tested to determine if it would present lesser disturbance to migrating fish (Hiebert et al. 2000). Visible light caused some hesitation in coho and chinook (<9 min) (but not for steelhead *O. mykiss*), particularly at night as compared to infrared light. Chinook salmon in this study at Prosser Dam on the Yakima River, Washington, also migrated more at night (70%) under infrared light than under visible light (49%) as compared to migration during daylight. Infrared lights operating at wavelengths longer than 800 nm can be useful for identifying fish in low light or during the nighttime; most fish species are unaffected by this infrared range because it falls beyond their spectral range (Lythgoe 1988); however, infrared light dissipates quickly in water and does not result in high image quality. Visible light was superior for determining species and hatchery origin of fish. Hetrick et al. (2004) found that a single high-beam auto headlight mounted on a tower was the most effective lighting system for detecting coho salmon during night video surveys. The need for image quality should be weighed against the potential for slight delays in fish passage when selecting a lighting system.

Additional work has been done in lighting for video surveys. In a study of coho smolts at an enumeration fence on the Keogh River in British Columbia, Canada, Irvine et al. (1991) used a black plastic shroud around the lighting and filming setup to eliminate stray reflections. For low light situations, a super low-lux black-and-white camera with infrared lights could be used (Otis and Dickson 2002).

Transmission of data from the camera to a recording device may be accomplished wirelessly if the camera used is digital. Hetrick et al. (2004) used a wireless FM video transmitter (see Table 1) to transmit the video images collected up to 210 m to a computer.

Additional setup is required if the work is to be conducted at a remote site. If an enclosed facility is used, the equipment is generally mounted in an existing viewing chamber. At remote sites, a substrate needs to be installed to make fish visible as they pass the recording site; the main function of this material is to provide a contrasting background between the channel bottom and the fish passing in front of the camera. Hetrick et al. (2004) used high-density polyethylene panels to provide contrast with fish and found that they were easy to keep clean. This material was selected as it is durable under a wide range of temperatures and does not degrade underwater or in sunlight, and it has a smooth surface. The panels measured 1.2 m × 1.5 m × 3.2 mm and were anchored to the stream bottom using 9.5 mm galvanized cable, set perpendicular to the flow and 19.1-mm S-hooks. Additional connections between the panels should be made, and Hetrick et al. (2004) recommend 50–80-kg tensile strength white cable ties. Duck-bill anchors (size 68) were used to anchor the cable and the recommended spacing on these is 4 m or less (Hetrick et al. 2004). Otis and Dickson (2002) used a light-green substrate material of 0.32-cm mesh seine material, which was also attached to the bottom of the channel.

Constricting the path through which fish pass has been useful in reducing the area where filming needs to be concentrated. Hetrick et al. (2004) placed substrate panels to form a "V" in the channel and found that many fish followed the edge of the panel to the small break in the "V" before crossing the viewing area. The "V" location was monitored using an underwater camera that could capture close-up images of fish to determine species. Limiting the passage space for fish can also reduce the risk of fish moving back through the viewing area a second time and the swarming of fish, which result in overestimation (if the same fish is counted twice) or underestimation (if the viewer cannot see all the fish in a large group).

Towers for cameras are also required for remote site use. These can be constructed or commercially purchased, but need to be sturdy enough to prevent shaking during weather events. Placement of the towers, with respect to viewing area, angle towards the water surface, and direction is critical to reduce glare. The ability to adjust the camera position without disturbing the placement is also critical when setting up the optimum viewing system based on sight conditions. Excessive glare can ruin the image quality of the recording. Additionally, Hetrick et al. (2004) found that a viewing area greater than 10 m wide restricts the viewer's ability to count large numbers of fish, even with very high image quality from digital cameras. Another consideration in camera placement is the use of underwater cameras. It is often necessary to capture an image of the fish in a side view to determine species and especially to determine hatchery origin. Side views of fish are generally captured from below the water surface and in a remote site would require additional anchoring. Hetrick et al. (2004) mounted underwater cameras on steel fence posts pounded into the streambed at the thalweg.

Power generation can be accomplished in several ways, depending on the site conditions and power requirements. A solar-powered system can be used if adequate sunlight is deemed available. Hetrick et al. (2004) used two sets of four Siemens SR90 90 W solar panels wired in parallel using 10-gauge electrical wire in polyvinyl chloride pipe. Each solar array had a breaker box leading to a Trace C-30A, 30 A/h charge control regulator connected via 8-gauge wire. The charge controller was connected to a battery bank of eight 100 A/h 12-V sealed absorbed glass mat deep-cycle batteries, wired in parallel using 2/0 gauge battery interconnects for 800 A/h storage capacity (Hetrick et al. 2004). (If freezing temperatures occur and batteries are not transported by aircraft, flooded lead acid batteries are preferable because they perform in cold temperatures and weigh less for the same storage capacity [Hetrick et al. 2004]). The 12-V direct current from the batteries was converted to 120-V alternating current using a Prosine 1000-W converter (see Table 1). This system produced between 10–11 V in overcast conditions and 19–21 V during clear skies (Hetrick et al. 2004). Additional power was needed (360 A/h per day) by Hetrick et al. (2004), and they used a gasoline generator (Honda EU3000) at night to power lights.

Computerized recording and counting system overview at Rock Island Dam, Columbia River

Gene Colburn and Jeff Bailey (Station 3 Media, Spokane, WA, unpublished report)

In the past, counting of fish species at fish ladders was performed by posting fish counters 24 h a day at a viewing window. The counts were captured using board-mounted mechanical clickers, one clicker for each species. The counts were manually recorded by pencil on paper forms and graphs. As technology evolved, cameras and VCRs were placed into service at each of the viewing windows, thus relieving the need for fish counters to staff the viewing windows 24 h a day. Our objectives were to overcome some of the limitations of conventional VCR recording (e.g., tape and machine longevity, slow seek to data, grainy picture viewing). Easier methods of recording fish counts, utilizing conventional database management techniques, were also desired. Anything we could do to improve the situation, including image enhancement, lighting, and overall resolution, made the process of species identification easier for the fish counters. Because the viewing windows were preexisting, ideal camera placement and lighting needs placed physical limitations on the retrofit. We recognized three aspects that we could implement to make the task easier: camera placement, lighting, and recording techniques.

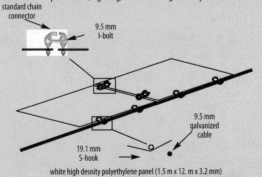

FIGURE 2. — Typical installation design for panels to enhance fish viewing.

The cameras were easy. We installed Sony XC-390 cameras with 3-mm-wide angle-fixed lenses, allowing the cameras to be placed inches away from the viewing window. These are three chip cameras, which had very good overall specifications at the time of installation. Newer high-density (HD) cameras with a wider depth of field could also help here. Lighting was and still is a critical part of the process. Because of the preexisting viewing windows and the cost to redesign, these windows were nearly impossible to light correctly. Cameras do best when there is good conventional three-point lighting (i.e. key, fill, and back lighting). We tried everything, but lighting through the viewing window glass (providing key and fill lighting) resulted in reflected light spots in the camera viewing area. We installed mercury vapor lights over the water, where fluorescent lights had been installed, and the new lighting added top and back kinds of lighting. While the mercury vapor lights did not provide adequate lighting, they were an improvement. We experimented with quartz and light-emitting diode lighting close to the glass window, but we still did not achieve the lighting effect that we needed. The best that we have been able to come up with is to provide light as far back from the viewing window as allowed with a Fresnell lens stage light. This provided a through-the-glass, general flat lighting to the window. Unfortunately, there was still a small reflected light that was seen by the camera. Ideally these windows should be totally revamped. Installing lighting fixtures in the water is difficult to implement. The fixtures could potentially affect fish passage or may be prone to collecting debris; however, lights on the fish passage side of the viewing window would be best. Cameras are also very sensitive to color temperature and ideally look best with either daylight (6,500°kelvin[K]) or Quartz (3,200°K). On the other hand, lighting in water looks best to the eye with 10,000°K lighting used in aquarium types of lighting (metal halide). A small amount of ultraviolet (UV) light would help minimize algae growth and allow the pigments of a fish's skin to fluoresce, but we have not figured how to implement UV lighting in the existing ladders.

It was important to record time lapse data at three to seven frames per second for 24 h or more on any computer platform with sufficient resolution. Because of the spot sizes on smaller species and varying water conditions, recognition of species was extremely difficult if compression algorithms were used in the recording format. We finally settled on some video boards used for broadcast and three-dimensional

Office Methods for Footage Review

A multiplexer can be used to view images from multiple cameras at once. Playback speeds for analog cameras are step-wise and not very flexible (Otis and Dickson 2002). Much greater flexibility is found in digital footage (Hetrick et al. 2004). Each footage day can be divided into three periods (dawn to 1159 hr, 1200 to 1759, and 1800 to dusk) (Otis and Dickson 2002). Reviewers should count the total number of fish swimming upstream and keep track of the amount of time required to review each section of footage. The time for dawn and dusk should be recorded for each footage day as well as the playback speed used to review the footage.

Motion detection algorithms were used by Hetrick et al. (2004) with some success. This approach was designed to help eliminate long sections of blank footage from having to be reviewed. The algorithms undercounted coho salmon during a test by 27%. Motion detection ability could be improved by using wider substrate panels, but potential effects on fish passage over wider panels would need to be evaluated (Hetrick et al. 2004). Underwater cameras used in the fall often triggered false responses from the motion detector when leaves and debris floated in front of the camera. Use of this system during earlier migrations in the summer may reduce the number of false triggers if there are fewer leaves and less debris in the water (Hetrick et al. 2004). Motion detection is most difficult during crepuscular periods, when light conditions are changing rapidly (Hetrick et al. 2004). Hetrick et al. (2004) recommend constant recording during these periods

animation applications. They are Digital Processing Systems (DPS) Perception boards that allow 10-bit recording of uncompressed video data. They are well suited for varying frame rate recordings, with excellent playback and jog-shuttle control. The DPS graphical interface is much like a VCR which the fish counters were already accustomed to using. Using server board computer configurations, we could install two DPS boards into each computer, allowing the fish counter to view the previous day's recordings while still recording the current day. Each video board has its own Small Computer System Interface-embedded controller that allows for the amount of Raid Array storage required for this kind of data. The embedded controller also freed up the central processing unit, allowing for a reliable multitasking environment.

A custom software application was developed to allow users to track fish passage easily and analyze the empirical data. The software consisted of two tiers. The first tier handled the collecting of data by date/time and fish species. A simple user interface was developed that both mimicked the previous mechanical counting and provided more granularity and accuracy than was previously possible. At the end of each day of counting, the fish counters would "push" the passage data to the corporate office for analysis. The second tier of the software system provided a means to aggregate the passage data providing summary results as well as point-in-time analysis. Using popular spreadsheet functionality, biologists could see all data that had been collected. With a push of a button, users were now able to instantly update summary data on a Web site. Additional data could be exported and made available through a file transfer protocol site.

Once a reliable computer configuration was developed, the next task was a facility implementation that would allow the fish counters at Rock Island Dam to operate these systems easily. The accompanying block diagram of the system (see Figure 1) shows the basic signal flow to each computer. All equipment is rack-mounted and powered by a power-conditioning ininterruptible power supply system. The basic signal path from each camera is fed to a time-and-date video insertion generator and then sent to distribution amplifiers to allow for additional recording by VCRs for back-up (in case of a computer crash). The signals are then passed on to a video routing switcher, allowing any computer to record any camera position. Three cameras and four computer work stations are installed at the dam. The video router provides routing as needed for overflow counts or any failure in the overall system. A second matrix router is installed for switching keyboard, mouse, and computer monitor to each fish counter's desk. This was provided to allow the counters to share the workload on busier count days. The video output from each computer's two internal video boards is switched internally to the video overlay, so that each monitor station sees the actual playback video on his/her respective liquid crystal display computer monitor. The count board overlay and jog shuttle overlay appear on the same screen, making for a good logical screen layout. Included in the signal path is the video from the previous day, which is sent to a DVD recorder for archival purposes.

Ergonomics plays a key role in installation. The counters sit and watch these screens 8 h a day and are continually using their mouse to input data and jog-shuttle; therefore, doing everything possible to make this more ergonomically safe is important. The improvements included proper monitor desk height, padded mouse pads, and good overhead lighting.

There are a number of considerations for future improvements for this system. For example, cameras that can operate better in low light with better resolution, while maintaining an entire digital path from camera to computer. We are currently using existing analog fiber paths and will not upgrade until HD cameras and recording techniques are available. Additionally, any modifications to the fish passage area at the viewing window would need a redesign and laboratory testing before implementation. The benchmark is to make the computer overlay (i.e., video resolution) look as good as if a fish counter were still counting fish (in real time) at the viewing window.

versus the use of motion detector triggers. High wind and heavy rain can also cause significant false triggering of motion detection (Hetrick et al. 2004). Counts can be collected from live video coverage, but they will need to be corrected for miscounts (Hetrick et al. 2004). This can be accomplished by playing the footage in slow motion, freezing frames, and replaying footage with high complexity of movements. The DVMS400 allows for detection of movement of individual fish that may pass back and forth in front of an underwater camera (Hetrick et al. 2004).

To determine species composition of the total return, days with high passage (>2%) were reviewed again (Otis and Dickson 2002). For the first 15 min of each hour of footage, the number of each species observed was recorded (Otis and Dickson 2002). This process was also applied to days with low passage to account for changes in species composition during the run. Subsampling for species composition may miss species that are represented in very low numbers in the run.

Data Handling, Analysis, and Reporting

Accuracy

In general, video surveys have been evaluated and found to perform well when compared to other types of counting approaches. Hatch et al. (1994) found that counts from video were within 4% of those made by experienced observers. Additionally, the authors found that variability between video film reviewers was low (Hatch et al. 1994). No significant differences were detected between five reviewers for any single species counted or for the total number of fish counted.

Videography was compared with weir and aerial survey counts (Otis and Dickson 2002). Otis and Dickson (2002) found that video surveys detected 87–93% of the fish counted at a weir located upstream of the video sampling site. However when fish density was low (fewer than 2,000 fish daily), reviewers tended to overestimate numbers of fish from video counts, and when fish density was high (6,000–10,000 fish daily), the reviewers tended to underestimate the numbers of fish. Results of aerial surveys estimated only 19% and 31% of the fish passage detected at a weir from two specific sampling dates, while video surveys detected 73% and 69% of the passage on these same dates, respectively (Otis and Dickson 2002).

Accuracy and the time required for analysis are affected by several factors, including fish density, turbidity, equipment failure, or need for maintenance, and the presence of very similar species (Hatch et al. 1994). Many of these factors cannot be controlled, but downtime in equipment can be reduced by using multiple cameras and ensuring an adequate power source for cameras and lighting.

Accuracy of fish counts can be complicated when more than one fish swims past the video recording device. Irvine et al. (1991) tested the ability of a computer to count and measure coho smolts swimming through an acrylic tunnel and found that fish overlapped when swimming and caused the computer estimates to be low when compared to direct observations of recordings. When an observer was able to accept or reject the computer counts and measurements, these counts were much more accurate. Hatch et al. (1994) report counting up to 4,500 fish per day at the Tumwater Dam on the Wenatchee River, Washington, and the authors felt that more fish could have easily been counted. Hetrick et al. (2004) used digital footage and replayed footage when migration was more complex. Accuracy using this method improved counts by 7%.

Recent work using digital video by the Chelan County Public Utility District in Washington has integrated transmission of fish counting data from passage ladder viewing windows at a large hydroelectric dam (see Computerized recording and counting system overview at Rock Island Dam, Columbia River, page 447, for an overview of the development of this system). Figure 3 reflects the architecture of this video-to-display system (T. Mosey, Chelan County Public Utility District, personal communication).

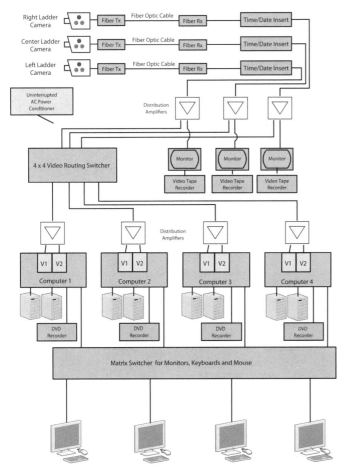

FIGURE 3.—Diagram of fish-counting video and computerization system at Rock Island Dam in Washington.

Personnel Requirements

Setup of a video system will require staff members to be familiar with fish surveys as well as video technology and recording systems. The numbers of staff members required will depend on the type of site in which the surveys will be conducted. More staff members and equipment will be required to survey a remote site.

Operational Requirements

Equipment and Budget

Table 1 provides a list of equipment used by Hetrick et al. (2004). This list provides examples of the types of equipment that would be needed at a remote site. Actual costs may vary.

TABLE 1.—Types and cost for video equipment used for salmon assessments at remote sites (after Hetrick et al. 2004). Prices are shown in USD.

Item Description	Model Number	Manufacture	QT	Unit cost	Total cost
High-resolution video camera	PC33C	Supercircuits	2	$170	$340
Underwater camera	PC81UW	Supercircuits	2	$450	$900
Outdoor camera enclosure	ENCOD	Supercircuits	2	$50	$100
Camera-mounting bracket	MB4	Supercircuits	2	$20	$40
2.5–6-mm variable focal lens	CML2-6MMZ	Supercircuits	2	$130	$260
Digital video recorder	Digicorder 2000 Deluxe	ASL	1	$2,800	$2,800
Underwater lights	SCL VL	Pond Solutions	8	$80	$640
100-ampere-hour sealed AGM battery	PS-100	Power Sonic	8	$147	$1,176
90-W solar panel	SR 90	Siemens	8	$419	$3,352
100-W power inverter	1000/GFCI	STP Prosine	1	$637	$637
Controllers, beakers, fused disconnects	NA	Big Frog Mountain	1	$1,000	$1,000
Solar panel mounting bracket	DP-RGM-SR100-T	DPW	2	$174	$348
HDPE plastic 4′ × 10′ × 1/8″ (white)	NA	NA	15	$58	$870
Cable, duck-bills, connectors	NA	NA		$500	$500
DVD RAM drive (firewire)	QPS525	QPS	1	$710	$710
DVD Discs (9.4 gigabytes)	DRM 94F	Maxell	10	$50	$500
15″ liquid crystal display monitor	ViewPanel VG151	ViewSonic	1	$800	$800
Video transmitter/receiver	CV-191	Nutex Com.	1	$110	$110
Digital video recorder	DVMS 400	Gyyr	1	$5,212	$5,212
Color video cameras	IK-64WDA	Toshiba	4	$534	$2,135
Outdoor enclosure	EH-3512	Pelco	2	$75	$150
Variable-focus auto-iris lenses	13VD5-40	Pelco	4	$151	$604
Circular polarizing filters	NA	Tiffen	3	$30	$90
Color underwater camera	Model 10	Applied Microvideo	1	$475	$475
Power supply box (24 V)	WCSI-4	Pelco	1	$122	$122
Color monitor (17″)	SyncMaster 171MP	Samsung	1	$1,000	$1,000
Portable monitor (5.6″)		Everfocus	1	$288	$288

Item Description	Model Number	Manufacture	QT	Unit cost	Total cost
Charge control regulator (30 ampere-hour)	C-30A	Trace	1	$100	$100
75 ampere-hour charger	NA	DSL	1	$375	$375
Power generator	EU300is	Honda	1	$1,600	$1,600
High-beam auto headlights		AC Delco	1	$100	$100
4 m-tall tripod tower	NA	Cabelas	2	$350	$700

Infrared lights used: Six infrared modules (3 diode wide × 72 diode long rows along the base and sides of the viewing chamber), 600 W total and five 60-W illuminators installed in the floor, another 300 W at 880-nm bandwidth

Camera: Two Supercircuits PC33C high-resolution 12-V color video cameras and two Supercircuits PC81UW 12-V auto-iris, autofocus color underwater cameras.

Lens: 5–20-mm variable-focus zoom lenses with polarizing filters

Budget for Staff

Staff time for video surveying consists of the time to set up the recording and storage system, view and interpret the images, and analyze the information collected from the images. Otis and Dickson (2002) report requiring 25 and 50 h of review time for processing footage from two video sites as compared to 1,120 h that would be required for weir operation at those sites.

Recommendations

The ability to detect passage of fish accurately is critical if video is to be widely used. Video techniques have been tested and compared to visual counts and to weirs constructed upstream of the cameras. Different lighting techniques, as well as various combinations of cameras and recording speeds and techniques, have been tested and evaluated. Different levels of video image quality have been obtained using improvements in video technology, but these improvements have to be balanced against the cost of new equipment. Processing time for the recorded material has been decreased using image recognition software to eliminate sections of tape or digital images without fish presence. These efforts have provided an estimate of the accuracy of video equipment as well as recommendations for improvements to the equipment and the recording process.

- Hetrick et al. (2004) recommend that the field of view of each above-water camera not exceed 10 m.
- Fish movement should be restricted to reduce the number of cameras needed for the survey using a partial floating weir (Hetrick et al. 2004).
- Supplemental lighting is needed to identify species at night, if this is a critical objective (Hetrick et al. 2004).
- Improvements are needed in motion detection algorithms in the form of better image stabilization, additional lighting, and low-level camera heaters to reduce moisture buildup (Hetrick et al. 2004).
- Turbidity of a stream or river should be assessed during the entire planned sampling season to determine if video is an appropriate technique to

use. Turbidity levels above 80 nephelometric turbidity unit resulted in significant image quality degradation (Hetrick et al. 2004).

- Otis and Dickson (2002) recommend the use of real-time microwave or satellite transmission of digital images back to a central office location. This improvement would solve some equipment and maintenance issues, reduce cost, and increase the speed in which data could be analyzed. Hetrick et al. (2004) note that advanced transmitters can send a line-of-sight signal up to 65 km. Using these transmitters could reduce the power requirements at remote sites.

Literature Cited

Brady, B. C., and J. D. M. Schwartz. 2005. Video methodology for surveying nearshore fishes at CRANE sites: with the option of underwater geo-referencing. California Department of Fish and Game, Marine Region, Santa Barbara.

Esteve, M. 2005. Observations of spawning behaviour in Salmoninae: *Salmo, Oncorhynchus* and *Salvelinus*. Reviews in Fish Biology and Fisheries 15:1–21.

Hatch, D. R., M. Schwartzberg, and P. R. Mundy. 1994. Estimation of Pacific salmon escapement with a time-lapse video recording technique. North American Journal of Fisheries Management 14:626–635.

Hatch, D. R., J. K. Fryer, M. Schwartzberg, D. R. Pederson, A. Wand. 1998. A Computerized Editing System for Video Monitoring of Fish Passage. North American Journal of Fisheries Management 18(3):694–699.

Hetrick, N. J., K. M. Simms, M. P. Plumb, and J. P. Larson. 2004. Feasibility of using video technology to estimate salmon escapement in the Ongivinuk River, a clear-water tributary of the Togiak River. U.S. Fish and Wildlife Service, King Salmon Fish and Wildlife Office, King Salmon, Alaska.

Hiebert, S., L. A. Helfrich, D. L. Weigmann, and C. Liston. 2000. Anadromous salmonid passage and video image quality under infrared and visible light at Prosser Dam, Yakima River, Washington. North American Journal of Fisheries Management 20:827–832.

Irvine, J. R., B. R. Ward, P. A. Teti, and N. B. F. Cousens. 1991. Evaluation of a method to count and measure live salmonids in the field with a video camera and computer. North American Journal of Fisheries Management 11:20–26.

Lythgoe, J. N. 1988. Light and vision in the aquatic environment. Pages 57–82 in J. Atema, R. R. Fay, A. N. Popper, and W. N. Tavolga, editors. Sensory biology of aquatic animals. Springer-Verlag, New York.

Mueller, R. P. 2005. Deepwater spawning of fall chinook salmon *(Oncorhynchus tshawytscha)* near Ives and Pierce Island of the Columbia River. 2004–2005 Annual Report, Project No. 199900301. Bonneville Power Administration, DOE/BP-00000652-28, Portland, Oregon.

Mueller, R. P., R. S. Brown, H. Hop, and L. Moulton. 2006. Video and acoustic camera techniques for studying fish under ice: a review and comparison. Reviews in Fish Biology and Fisheries 16:213–226.

Otis, E. O., and M. Dickson. 2002. Improved salmon escapement enumeration using remote video and time-lapse recording technology. Alaska Department of Fish and Game, Division of Commercial Fisheries. Restoration Project 00366 Final Report, Homer.

Pearson, W. H., R. P. Mueller, S. L. Sargeant, and C. W. May. 2005. Evaluation of juvenile salmon leaping ability and behavior at the experimental culvert test bed. Final Report to the Washington State Department of Transportation, Battelle Pacific Northwest Division, WSDOT Agreement No. GCA2677, Richland.

Glossary

abundance. The number of individuals in a stock or a population.

abundance index. Information obtained from samples or observations and used as a measure of the number of fish that make up a stock.

acclimate. The adaptation of an organism to environmental changes.

acclimation pond. Concrete or earthen pond or a temporary structure used for rearing and imprinting juvenile fish in the water of a particular stream before their release into that stream.

acoustic survey. Sonar equipment used to count fish. Sound waves are sent out from a moving research vessel or fixed riverside location, strike fish and are reflected back.

active gear. Active gear moved through the water either by machinery or human power and includes beach seine nets, pole seine nets, purse seine nets, trawl nets, electrofishing (shocking), boat shocking and all types of angling.

adaptive management. A systematic process for continually improving management policies and practices by learning from the outcomes of operational programs. Its most effective form-"active" adaptive management-employs management programs that are designed to experimentally compare selected policies or practices, by implementing management actions explicitly designed to generate information useful for evaluating alternative hypotheses about the system being managed.

adfluvial. Possessing a life history trait of migrating between lakes or rivers and streams.

adipose fin. Modified rayless posterior dorsal fin in some fishes. a small, tapered, fleshy lobe with no rays, located on the dorsal surface between the dorsal and caudal fins.

affluent (stream). A stream or river that flows into a larger one; a tributary.

age. The number of years of life completed, here indicated by an arabic numeral, followed by a plus sign if there is any possibility of ambiguity (age 5, age 5+).

age composition. Proportion of individuals of different ages in a stock or in the catches.

age-class. A group of individuals of a certain species that have the same age.

alevin. Newly hatched salmon when still attached to the yolk sac. The life stage of a salmonid between hatching from the egg and emergence from the stream gravels as a fry. The alevin stage is characterized by the presence of a yolk sac, which provides nutrition while the alevin develops in the protected gravel riverbed.

alluvial. Originating from the transport and deposition of sediment by running water.

anadromous. Fish that hatch and rear in fresh water, migrate to the ocean (salt water) to grow and mature, and migrate back to fresh water to spawn and reproduce. Highly specialized to endure the changes in salinity.

anal fin. Unpaired fin located on the ventral surface posterior to the anus.

annual total mortality rate. The number of fish which die during a year (or season), divided by the initial number. Also called "actual mortality rate" or "coefficient of mortality".

annulus. A mark or ring that forms annually on the otoliths, scales, and other bones of fish, that correspond to the annual period of slow growth that fish go through. Annuli are used by fish managers to determine age and growth of fish.

GLOSSARY

anthroprogenic. Human induced.

aquaculture. The controlled cultivation and harvest of aquatic plants or animals (e.g., edible marine algae, clams, oysters, mussels, and salmon). Also known as marine farming.

aquatic ecosystem. The natural systems of interacting aquatic life within the biological and physical aquatic environment.

area frame. A sampling frame that is designated by geographical boundaries within which the sampling unites are defined as subareas.

artificial propagation. Any assistance provided by human technology to animal reproduction. In the context of Pacific salmonids, this assistance includes (but is not necessarily limited to) spawning and/or rearing in hatcheries, captive broodstock projects, or the use of remote site incubators.

attributes. any living or nonliving feature or process of the environment that can be measured or estimated and that provide insights into the state of the ecosystem. The term Indicator is reserved for a subset of attributes that is particularly information-rich in the sense that their values are somehow indicative of the quality, health, or integrity of the larger ecological system to which they belong (Noon 2003). See Indicator.

beach seining. A fishing method where a net and a length of rope are laid out from and back to the shore and retrieved by hauling on to the shore.

benthic. bottom-dwelling; living on the substrate or sea bed

biodiversity. The variety and abundance of species, their genetic composition, and the natural communities, ecosystem, and landscapes in which they occur.

biological diversity. The variety of species in a community, sometimes expressed by various quantitative measures which reflect not only the total number of species present but also the degree of domination of the system by a small number of species. Includes genetic diversity (within species), species diversity (within ecosystems) and ecosystem diversity. Diversity indices measure the richness (the number and relative numeric abundance) of species in a system, and the connections between them but are indifferent to species substitution, which may, however, reflect ecosystem stress (such as those due to high fishing intensity).

biological significance. An important finding from a biological point of view that may or may not pass a test of statistical significance.

biomass. Total weight of all individuals in a stock or a population.

blue listed species. Any species considered sensitive or vulnerable (in British Columbia). These are indigenous (native) species that are not immediately threatened but are particularly at risk for reasons including low declining numbers, a restricted distribution or occurrence at the fringe of their global range.

brackish water. A mixture of freshwater and seawater.

brailer bags. Huge bags used to carefully lift the salmon catch from fishing boats to the dock.

branchial plume. respiratory structure or external gills, usually located on the dorsal side toward the posterior

branchiostegal ray. One group of dermal bones/rays that close the branchial or gill cavity under the head.

brood stock. Adult fish used to propagate the subsequent generation of hatchery fish.

GLOSSARY

button-up fry. A salmonid fry that has not completely absorbed its yolk sac and has emerged from its spawning gravel.

by-catch. These are the other fish species, birds and marine mammals that fishers may catch while targeting a specific species.

cascade. A series of small steep drops increasing the velocity of the stream.

cast netting (also **throw netting**). A method of fishing where a circular net weighted around the edges is thrown over fish in the shallows.

catadromous. Refers to fishes that migrate from fresh water to salt water to spawn or reproduce such as the American eel.

catch-per-unit-effort. (CPUE) The catch of fish, in numbers or in weight, taken by a defined unit of fishing effort. Also called; catch per effort, fishing success, availability. Ex number of trout per hour, kg of fish per surface area of water seined.

catchment. The area which supplies water by surface and subsurface flow from precipitation to a given point in the drainage system.

caudal fin. The tail fin.

caudal peduncle. The tapering portion of a fish's body between the posterior edge of the anal fin base and the base of the caudal fin.

channelized. A portion of a river channel that has been enlarged or deepened, and often has armored banks.

chinook salmon. An anadromous salmonid of the genus *Oncorhynchus* and species *tshawytscha*. Also known as king, spring, or blackmouth salmon.

chum salmon. An anadromous salmonid of the genus *Oncorhynchus* and species *keta*. Also known as dog salmon, because of the large canine teeth they develop during spawning.

coastal juveniles. Juvenile salmonids inhabiting the waters of the continental shelf.

coded-wire tag (CWT). A small (0.25mm diameter × 1 mm length) wire etched with a distinctive binary code and implanted in the snout of s salmon or steelhead, which, when retrieved, allows for the identification of the origin of the fish bearing the tag. A specialized tag reader detects tags. Codes are applied to batches of fish to also allow identification of stocking date and location. This data is important to assess the impact of a stocking program.

coho salmon. An anadromous salmonid of the genus *Oncorhynchus* and species *kisutch*. Also known as silver or hooknose salmon.

co-location. Sampling of the same physical units in multiple monitoring protocols

compensatory mortality. Mortality is compensatory when the mortality rate (i.e., proportion of population affected) decreases as the population size decreases. This is in contrast to depensatory mortality, were the rate increases as the size of the population decreases.

conceptual Models. purposeful representations of reality that provide a mental picture of how something works to communicate that explanation to others.

conspecific. Individuals of the same species.

critical stock. A stock of fish experiencing production levels that are so low that permanent damage to the stock is likely or has already occurred.

GLOSSARY

critical stream flow period (CSFP). The period of lowest stream flow that the juvenile fish will encounter during the main growing season (occurs throughout most of western North America between August and October). This period represents the stream conditions most likely to limit fish production.

cubic feet per second. A measurement of stream flow; noted as ft^3/s.

cycloid scales. Smooth, flat, round scales that have concentric lines called circuli, found on trout, herring, and other fish.

declining fishery. The state a fishery is said to be in when a fish stock is overfished.

delta. An alluvial landform, typically triangular in shape, composed of sediment at a river mouth that is shaped by river discharge, sediment load, tidal energy, land subsidence, and sea-level changes.

deme. Reproductive or breeding unit (spawning site) comprised of individuals who are likely to breed with each other (i.e., well mixed). A single population may include more than one deme and demes may be partially isolated from one another. Their partial isolation may or may not be persistent over generations. There will always be at least as many demes as populations.

depensatory mortality. Mortality is depensatory when its rate (i.e., proportion of population affected) increases as the size of the population decreases. This is in contrast to compensatory mortality where the mortality rate decreases as the population size decreases.

depressed stock. A stock of fish whose production is below expected levels based on available habitat and natural variations in survival levels, but above the level where permanent damage to the stock is likely.

descaling. A condition in which a fish has lost a certain percentage of scales.

detritus. Organic fragments of plant or animal matter.

dimorphism. Occurring in two distinct forms, as in sexual dimorphism, describing physical differences in the sexual forms of an organism.

dorsal fin. The fin located on the back of fishes, and in front of the adipose fin, if it is present.

drainage basin. A watershed is defined by the stream that drains it. It is the surface area that collects and discharges runoff through a given point on a stream.

drift netting. A fishing method used for catching pelagic fish, the vessel remains tied to one end of the net to stop it drifting too far. Fish swim into the net and are caught behind the gills.

driver. The major external driving forces that have large-scale influences on natural systems. Drivers can be natural forces or anthropogenic.

echo sounders. An instrument that sends out an acoustic pulse in water and measures distances in terms of the time for the echo of the pulse to return.

ecological integrity. a concept that expresses the degree to which the physical, chemical, and biological components (including composition, structure, and process) of an ecosystem and their relationships are present, functioning, and capable of self-renewal. Ecological integrity implies the presence of appropriate species, populations and communities and the occurrence of ecological processes at appropriate rates and scales as well as the environmental conditions that support these taxa and processes.

ecosystem. Defined as, "a spatially explicit unit of the Earth that includes all of the organisms, along with all components of the abiotic environment within its boundaries" (Likens 1992). A community of plants, animals and other organisms that interact with each other and with the physical environment. Rainforests, deserts, coral reefs and grasslands are examples of ecosystems.

ecosystem drivers. Major external driving forces such as climate, fire cycles, biological invasions, hydrologic cycles, and natural disturbance events (e.g., earthquakes, droughts, floods) that have large scale influences on natural systems.

ecosystem management. The process of land-use decision making and land-management practice that takes into account the full suite of organisms and processes that characterize and comprise the ecosystem. It is based on the best understanding currently available as to how the ecosystem works. Ecosystem management includes a primary goal to sustain ecosystem structure and function, a recognition that ecosystems are spatially and temporally dynamic, and acceptance of the dictum that ecosystem function depends on ecosystem structure and diversity. The whole-system focus of ecosystem management implies coordinated land-use decisions.

egg-to-smolt survival. The numerical difference between the number of fertilized eggs produced by a groups of fish and the number of smolts resulting from those eggs.

embeddedness. The degree to which dirt is mixed in with spawning gravel.

emergence. The process during which fry leave their gravel spawning nest and enter the water column.

emigration. Referring to the movement of organisms out of an area. See immigration and migrating.

endemic. Confined to a given or defined geographic area.

enhancement. The application of biological and technical knowledge and capabilities to increase the productivity of fish stocks. It may be achieved by altering habitat attributes (e.g., habitat restoration) or by using fish culture techniques (e.g., hatcheries, production spawning channels).

escapement. The quantity of sexually mature adult salmon (typically measured by number or biomass) that successfully pass through a fishery to reach the spawning grounds. This amount reflects losses resulting from harvest, and does not reflect natural mortality, typically partitioned between enroute and pre-spawning mortality. Thus, escaped fish do not necessarily spawn successfully. (Note: "escapement" is a harvest-centered term; we encourage the development of other terminology, as we do not believe the natural migratory path of salmonids denotes an escape from capture.)

escapement goal. A predetermined biologically derived number of salmonids that are not harvested and will be the parent spawners for a wild or hatchery stock of fish.

estuary. An area at the mouth of a river or stream where salt water and fresh water meet.

euryhaline. Having a wide tolerance to salinity.

eutrophication. To add nutrients to.

evolutionarily significant unit (ESU). The U.S. National Oceanic and Atmospheric Administration (NOAA) definition of a distinct population segment that is smallest biological unit that will be considered to be a "species" under the U.S. Endangered Species Act. A population will be is considered to be an ESU if (1) it is substantially reproductively isolated from other conspecific population units, and (2) it represents an important component in the evolutionary legacy of the species.

exploitation rate. The proportion of a population at the beginning of a given time period that is caught during that time period (usually expressed on a yearly basis). For example, if 720,000 fish were caught during the year from a population of 1 million fish alive at the beginning of the year, the annual exploitation rate would be 0.72.

extinct stock. A stock of fish that is no longer present in its original range, or as a distinct stock elsewhere. Individuals of the same species may be observed in very low numbers, consistent with straying from other stocks.

GLOSSARY

extirpation. The elimination of a species from a particular area.

eyed egg. A fish egg containing an embryo that has developed enough so the eyes are visible through the egg membrane.

fecundity. The number of eggs in the ovaries that are mature or will mature. A common measure of reproductive potential.

fingerling. Juvenile salmonids up to nine months of age and generally 5–10 cm in total length (also called parr). This term is typically used to refer to hatchery juveniles.

fishery. The process of attempting to catch fish, which then may be retained or released.

fishing mortality. Death caused by fishing. Mathematical symbol: F

fish stock. A group of individuals of the same species which are living together in the same area and can intermingle and interbreed freely. Different stocks of the same species, e.g., chinook, can be genetically different.

fishway. A device made up of a series of stepped pools, similar to a staircase, that enables adult fish to migrate up the river past dams.

fitness. The relative ability of an individual (or population) to survive and reproduce in a given environment. Floodplain. The part of a river valley composed of unconsolidated river deposits that periodically floods. Sediment is deposited on the floodplain during floods and through the lateral migration of the river channel across the floodplain. The 100-year floodplain refers to that area of a river valley that is inundated during a large-magnitude flood occurring, on average, once every one hundred years.

fluvial. Migrating between main rivers and tributaries. Of or pertaining to streams or rivers.

focal resources. resources that, by virtue of their special protection, public appeal, or other management significance, have paramount importance for monitoring regardless of current threats or whether they would be monitored as an indication of ecosystem integrity. Focal resources might include ecological processes such as deposition rates of nitrates and sulfates, or they may be a species that is harvested, endemic, alien, or has protected status.

forage fish. Small fish which breed prolifically and serve as food for predatory fish.

fork length. A fish length measurement from the tip of the nose to the fork of the tail fin.

fry. A stage of development in young salmon or trout reflecting a recently hatched fish that can swim and catch its own food. During this stage the fry is usually less than one year old, has absorbed its yolk sac, is rearing in the stream, and is between the alevin and parr stage of development.

galvanotaxis. Response or reaction to electrical stimulus.

gape. To open the mouth wide. In zoological terms, it means the measurement of the widest possible opening of a mouth.

gear restrictions. These are usually imposed to protect young fish, e.g. mesh size restrictions, net size restriction and restrictions on how a net can be set, or to limit by-catch problems.

genetic diversity. The heritable variation within and between populations of species. In this context, this encompasses all the taxonomic classifications below species that are a result of environmental heterogeneity and reproductive isolation.

genetic stock identification. A method that can be used to characterize populations of organisms based on the genetic profiles of individuals. The GSI process consists of a series of steps: (1) collect selected tissues from a representative sample of individuals from the population(s) under investigation; (2) develop genetic profiles for the individuals in each population by conducting starch-gel electrophoresis and histo-chemical staining using tissue extracts; (3) characterize each population by aggregating the individual genetic profiles and computing allele frequency distributions; and (4) conduct statistical tests using the allele counts characterizing each population to identify significantly different populations.

gillnet. Fishing gear; netting with weights on the bottom and floats at the top used to catch fish. Gillnets can be set at different depths and are anchored to the seabed.

gill netting. Commercial fishing with a net that catches the gills of the fish. Salmon swim partway through the mesh and become entangled. Fish are caught when their maxillary or opercular area is caught in the mesh of the net. Fish may also be entangled by their teeth, spines, girth, or scales as they try to pass through or free themselves from the mesh.

gill. Respiratory organ of many aquatic animals (e.g., crustaceans, fishes, amphibians). A plate-like or filamentous outgrowth richly supplied with blood vessels at which gas exchange between water and blood occurs.

gill cover. The operculum; the flat thin bones that cover each side of the head. See Operculum.

glide. A part of a river containing a smooth flow of water with an unbroken surface.

gradient. The amount of vertical drop a stream experiences over a given distance.

grilse. Salmon less than 22 inches (56cm) Fork Length (FL); typically 2 year-old fish; also called "Jacks".

habitat. An area that supplies food, water, shelter, and space necessary for a particular animal's existence.

handlining. Fishing using a line with usually one baited hook and moving it up and down in a series of short movements. Also called "jigging".

harvest. Fish that are caught and retained in a fishery (consumptive harvest).

harvest rate. The proportion of the available numbers of salmonids that is taken by fisheries in a specific time period.

hatchery fish. A fish that has spent some part of its life-cycle in an artificial environment and whose parents were spawned in an artificial environment.

hatchery stock (population). A stock that depends on spawning, incubation, hatching or rearing in a hatchery or other artificial propagation facility (synonymous with cultured stock).

headwaters. The upper reaches of a stream or stream system.

Holarctic. A zoogeographic region that includes North America, Europe, northern Africa, most of Asia. Fish that are Holarctic in distribution may be expected to occur in most of these regions but not necessarily throughout it.

home range. The area that an animal traverses in the scope of normal activities. This is not to be confused with territory, which is the area an animal defends.

homing. The ability of a salmon or steelhead to correctly identify and return to their natal stream, following maturation at sea.

GLOSSARY

homing rate. Of all the fish from a population that successfully return to spawn, the homing rate is the proportion that return to spawn in the same population in which their parents spawned. See also stray rate and gene flow.

hybridization. The interbreeding of fish from two or more different stocks or species.

imprinting. The physiological and behavioral process by which migratory fish assimilate environmental cues to aid their return to their stream of origin as adults.

incidental harvest. (also incidental catch) The capture and retention of species other than those a fishery is primarily opened to target/take. It can also refer to marked fish of the same species.

incubation. The period of time from egg fertilization until hatching.

independent tributary. A small stream flowing directly into marine waters.

indicators. a subset of monitoring attributes that are particularly information-rich in the sense that their values are somehow indicative of the quality, health, or integrity of the larger ecological system to which they belong (Noon 2003). Indicators are a selected subset of the physical, chemical, and biological elements and processes of natural systems that are selected to represent the overall health or condition of the system.

interorbital. The space between the eyes.

introduced species. The opposite of native species, this species does not occur naturally in an area. For example the Atlantic salmon is an introduced species in BC waters. (Also known as "exotic," "alien," and "non-native" species). Brought to an area by means other than its own dispersal ability.

inventory. An extensive point-in-time survey to determine the presence/absence, location or condition of a biotic or abiotic resource.

iteroparous. Species that reproduce repeatedly during their lifetime.

sack salmon. A young male salmon that matures precociously (earlier than other fish in its age-class); see also Grilse.

juvenile. Fish from one year of age until sexual maturity. A young individual resembling an adult of its kind except in size and reproductive activity.

kelt. Spent or spawned-out salmon up until the time it enters saltwater.

keystone species. A species that has a major influence on a community structure.

kokanee. The freshwater form of the sockeye salmon. Kokanee spend their entire life history in freshwater, and in some lakes are known as silver trout.

kype. The hook at the anterior portion of the snout in many breeding male salmonid species.

lacustrine. Referring to a lake environment.

large woody debris. Logs, limbs, or root wads 10 cm or larger in diameter, delivered to river and stream channels from streamside forests (in the riparian or upslope areas) or from upstream areas. LWD provides streambed stability and habitat complexity. LWD recruitment refers to the process whereby streamside forests supply wood to the stream channel to replenish what is lost by decay or downstream transport.

larva (plural **larvae**). A developmental stage in an animal, the larva hatches from an egg, looks very small and different from the adult form, eats different food, and usually lives in a different environment from the adult. The veliger is the typical larval stage of molluscs.

GLOSSARY

lateral line. Longitudinal line on each side of the body of fishes that marks the position of cutaneous sensory cells of the acoustico-lateralis system concerned with the perception of movement and sound waves in water, the cells on the lateral line being collectively known as the lateral line system.

length frequency. An arrangement of recorded lengths which indicates the number of times each length or length interval occurs.

lentic. Characterizing aquatic communities found in standing water.

life history. The events that make up the life cycle of an animal including migration, spawning, incubation, and rearing. There is typically a diversity of life history patterns both within and between populations. Life history can refer to one such pattern, or collectively refer to a stylized description of the 'typical' life history of a population.

limnetic. The open water of a lake that is above the bottom; usually shallow enough for light to penetrate.

littoral zone. The shallow shoreward region of a water body. It usually has light penetration to the bottom and is often occupied by rooted macrophytes; that region of a lake in which the water is less than 6 m deep.

live box. A container filled with water and often equipped with accessories such as aeration equipment that is used to hold and transport live fish.

local population. Reproductive unit (spawning site) comprised of individuals who are likely to breed with each other (i.e., well mixed). A single population may include more than one deme and demes may be partially isolated from one another. Their partial isolation may or may not be persistent over generations. There will always be at least as many demes as populations.

locally adapted population. A population whose members have genetically based characteristics that increase their fitness in their local environment compared individuals that lack these characteristics.

lotic. Meaning or regarding things in running water.

management unit. A stock or group of stocks which are aggregated for the purposes of achieving a desired spawning escapement objective.

marine protected area (MPA). An MPA is a specific area of the ocean that is regulated to protect both the habitat and the species that inhabit it.

marine subadults. Post-juvenile salmonids moving into offshore waters.

maxillae. The upper jaw, the upper jaw bones. (Maxillaries).

maximum sustainable yield (MSY). The largest average annual catch that can be taken over time without reducing the stock's productive potential under existing environmental conditions. (For species with fluctuating recruitment, the maximum might be obtained by taking fewer fish in some years than in others.) Also called maximum equilibrium catch; maximum sustained yield; or sustainable catch.

measures. specific feature(s) used to quantify an indicator, as specified in a sampling protocol. For example, pH, temperature, dissolved oxygen, and specific conductivity are all measures of water chemistry.

mesh size. Size of the mesh of a net. Different fisheries have different minimum mesh size regulation.

GLOSSARY

metadata. Data about data. Metadata describes the content, quality, condition, and other characteristics of data. It's purpose it to help organize and maintain a organization's internal investment in spatial data, provide information about an organization's data holdings to data catalogues, clearinghouses, and brokerages, and provide information to process and interpret data received through a transfer from an external source.

metamorphosis. The change from larva to adult

metapopulation. A population of sub-populations which are in turn comprised of local populations or demes. Note that individual sub-populations can be extirpated and consequently recolonized from other sub-populations. Stability in a metapopulation is maintained by a balance between rates of sub-population extinction and colonization. This is roughly analagous to a "classical" or "Levins" metapopulation after Richard Levins' (1969) mathematical model describing this scenario.

migrant. Life stage of anadromous and resident fish species which moves from one locale, habitat or system (river or ocean) to another.

migrating. Moving from one area of residence to another.

migration. The seasonal movement of an animal from one area to another.

milt. The sperm of fishes.

mitigation. An action intended to reduce the adverse impact of a specific project or development.

mixed stock. A stock whose individuals originated from commingled native and non-native parents; or a previously native stock that has undergone substantial genetic alteration.

monitoring. Collection and analysis of repeated observations or measurements to evaluate changes in condition and progress toward meeting a management objective (Elzinga et al. 1998). Detection of a change or trend may trigger a management action, or it may generate a new line of inquiry. Monitoring is often done by sampling the same sites over time, and these sites may be a subset of the sites sampled for the initial inventory.

mortality. The number of fish lost or the rate of loss.

narcosis. In electrofishing, a state of arrested activity induced by the use of electrical stimulation.

natal stream. Stream of birth.

native species. A species that occurs naturally in an area (is not introduced).

native stock. An indigenous stock of fish that has not been substantially impacted by genetic interactions with non-native stocks or by other factors, and is still present in all or part of its original range. In limited cases, a native population may also exist outside of its original range (e.g. in a captive broodstock program).

natural fish. A fish that has spent essentially all of its life-cycle in the wild and whose parents spawned in the wild. Synonymous with wild fish and with natural origin recruit (NOR).

natural mortality. Mortality due to natural causes. Mathematical symbol: M.

natural return rate. The number of native, naturally produced fish spawning in on generation divided by the total number of naturally spawning fish (hatchery plus naturally -produced fish) in the previous generation.

naturally spawning populations. Populations of fish that have completed their entire life cycle in the natural environment without human intervention.

nectonic. Describes organisms that are capable of swimming independent of water turbulence.

non-native stock (population). A stock (population) that has become established outside of its original range.

non-target population. Any natural populations that is not intended to be integrated with a particular artificial propagation program.

off-channel area. Any relatively calm portion of a stream outside of the main flow.

opercle. Refers to the largest bone in the operculum.

operculum. The bony (opercular bones) covering of the gill cavity of fishes.

otolith. Structure of the inner ear of fish, made of calcium carbonate. Also called "ear bone" or "ear stone". Otoliths are used to determine the age of fish. annual rings can be observed and counted. Daily increments are visible as well on larval otoliths.

outmigration. The migration of fish down the river system to the ocean.

outplanting. Hatchery reared fish released into streams for rearing and maturing away from the hatchery sites.

overfishing. Harvesting a species in such quantity that it reduces the stock biomass and future catches below desirable levels.

parameter. A "constant" or numerical description of some property of a population (which may be real or imaginary). Cf. statistic.

parr. The developmental life stage of salmon and trout between alevin and smolt, when the young have developed parr marks and are actively feeding in fresh water.

parr marks. Distinctive vertical bars on the sides of young salmonids.

passive gear. Passive gear is usually set and left stationary for a period of time. Passive gear includes: gill nets, enmeshing (trammel) nets, gee minnow traps, trap nets, Fyke nets, set lines, inclined plane traps and rotary screw traps.

passive integrated transponder. Passive Integrated Transponder (PIT) tags are used for identifying individual salmon for monitoring and research purposes. This miniaturized tag consists of an integrated microchip that is programmed to include specific fish information. The tag is inserted into the body cavity of the fish and decoded at selected monitoring sites.

peak flows. Extremely high winter-time flows which can cause excessive streambed scour and damage or destroy salmon eggs incubating in the gravel. Peak flows can become more severe as a result of an increase in impervious surfaces and a reduction of hydrologic maturity, both of which increase the rate of water delivery to stream channels.

pectoral fins. The anterior(front) paired fins, attached to pectoral (shoulder) girdle.

pelagic. Of or pertaining to the open waters of a lake or ocean, especially where the water is greater than 20 m deep; term applied to organisms that occupy the open waters of a lake or ocean.

pelvic fins. The most posteriorly located of the paired fins; ventral, paired fins on the underside of body, representing the hind limbs of land vertebrates.

pink salmon. An anadromous salmonid of the genus *Oncorhynchus* and species *gorbuscha*. Also known as humpy or humpback salmon. The most abundant Pacific salmon with a short life cycle of two years.

plankton. Marine plants and animals that drift with the ocean's current; they are usually very small but some are large. including jellyfish. Base of the marine food chain.

GLOSSARY

pond. A body of water smaller than a lake, often artificially formed.

pool. A relatively deep, still section in a stream.

population. Group of interbreeding salmon that is sufficiently isolated from other populations so that there will be persistent adaptations to the local habitat.

population viability analysis. A statistical analysis that provides an estimate of the probability that a population will become extinct over a specific time frame.

post-orbital hypural length. Another measurement often used for salmon that have undergone morphological changes associated with breeding (MacLellan, 1987). It is the distance from the posterior margin of the eye orbit to the posterior end of the hypural bone (last vertebrate).

post-smolt. Life stage of salmon from the time it departs from the river as a smolt until the end of its first winter at sea.

precautionary approach. Set of agreed cost-effective measures and actions, including future courses of action, which ensures prudent foresight, reduces or avoids risk to the resource, the environment, and the people, to the extent possible, taking explicitly into account existing uncertainties and the potential consequences of being wrong.

precocious. Fish that have matured quickly, or faster than the remaining fish of its age-class.

presence/absence sampling. A method of sampling whereby organisms are noted as present or absent without enumeration.

pre-spawning mortality. Generally refers to non-fishery mortality of adult salmon and steelhead between the time the fish enter the Columbia River and the completion of spawning.

productivity. A measure of a biological system's ability to supply organisms with energy and resources to feed, grow, and survive.

protocols. Detailed study plans that explain how data are to be collected, managed, analyzed and reported and area key component of quality assurance for natural resource monitoring programs (Oakley et al. 2003).

purse seine. Fishing gear. large net is laid in a circle around a school of fish and then the bottom is drawn closed, entrapping the fish.

radio-telemetry. Automatic measurement and transmission of data from remote sources via radio to a receiving station for recording and analysis.

rate of exploitation. The fraction, by number, of the fish in a population at a given time, which is caught and killed by man during the year immediately following . The term may also be applied to separate parts of the stock distinguished by size, sex, etc. Also called; *fishing coefficient .

rearing. Refers to the amount of time that juvenile fish spend feeding in nursery areas of rivers, lakes, streams and estuaries before migration.

reach. (see Stream Reach)

recovery project. Artificial production projects primarily designed to aid in the recovery, conservation or reintroduction of particular natural population(s).

recruitment. Addition of new fish to a defined life history stage by growth from among smaller size categories. Also used to describe the number of fish that existing the non-migratory juvenile population, or number of young fish that will eventually enter the migratory population and return to successfully spawn. Often used in context of management, where the stage is the point where individuals become vulnerable to fishing gear.

GLOSSARY

redd. A nest of fish eggs consisting of gravel, typically formed by digging motion performed by an adult female salmon.

red listed species. Any species that is designated as being threatened or endangered. Endangered species are indigenous (native) species facing imminent extirpation or extinction. Threatened species are likely to become endangered if limiting factors are not reversed.

relative abundance. An estimate of actual or absolute abundance; usually stated as some kind of index.

riffle. A shallow gravel area of a stream that is characterized by increased velocities and gradients, and is the predominate stream area used by salmon for spawning.

riparian. Referring to the transition area between aquatic and terrestrial ecosystems. The riparian zone includes the channel migration zone and the vegetation directly adjacent. to the water body, that influence channel habitat through alteration of microclimate or input of LWD.

river mile. A statute mile measured along the center line of a river. River mile measurements start at the stream mouth (RM 0.0).

riverine. Referring to the entire river network, including tributaries, side channels, sloughs, intermittent streams, etc.

riverine sockeye. Small populations of sockeye salmon that spawn and rear in some rivers systems that have no available lake for rearing.

roe. The mass of eggs within the female fish.

run. A group of fish of the same species that migrate together up a stream to spawn, usually associated with the seasons, e.g., fall, spring, summer, and winter runs. Members of a run may interbreed, and may be genetically distinguishable from other individuals of the same species.

run reconstruction. A post-season accounting of all salmon escaping and harvested from individual stocks or management areas.

salmonid. Fish of the family Salmonidae, including salmon and steelhead, trout, and char.

sample. A proportion or a segment of a fish stock that is removed for study, and is assumed to be representative of the whole. The greater the effort, in terms of both numbers and magnitude of the samples, the greater the confidence that the information obtained is a true reflection of the status of a stock (level of abundance in terms of numbers or weight, age composition, etc.)

sampling design. The sampling design of a scientific survey refers to the statistical techniques and methods adopted for selecting a sample and obtaining estimates of the survey variables from the selected sample

selective fishery. A fishery that allows the release of non-targeted fish stocks/runs, including unmarked fish of the same species.

self-sustaining population. A population of salmonids that exists in sufficient numbers to replace itself through time without supplementation with hatchery fish. It does not necessarily produce surplus fish for harvest.

semelparous. Species that reproduce only once during their lifetime.

SCUBA . Self Contained Underwater Breathing Apparatus

siltation. The process of covering or obstructing with silt.

GLOSSARY

smolt. Refers to the salmonid or trout developmental life stage between parr and adult, when the juvenile is at least one year old and has adapted to the marine environment.

smoltification. Refers to the physiological changes anadromous salmonids and trout undergo in freshwater while migrating toward saltwater that allow them to live in the ocean.

sockeye salmon. An anadromous salmonid of the genus *Oncorhynchus* and species *nerka*. Also known as red or blueback salmon. Spawning adults develop dull, green colored heads with brick red to scarlet bodies. (See Kokanee).

sonar. Equipment used to measure surfaces and density of fish groups under water through the use of transmitted sound waves.

spawn. The act of reproduction of fishes; the mixing of sperm of a male fish and eggs of a female fish.

spawner. Sexually mature individual.

spawning grounds. The areas that a fish stock or species will move to, to spawn.

spawning stock. Mature part of a stock responsible for the reproduction. Strictly speaking, the part of an overall stock having reached sexual maturity and able to spawn. Often conventionally defined as the number of biomass of all individuals beyond the "age at first maturity" of size at first maturity (i.e. beyond the age or size class in which 50% of the individuals are mature.

spawning stock biomass. The total weight of all sexually mature fish in the population. This quantity depends on year class abundance, the exploitation pattern, the rate of growth, fishing and natural mortality rates, the onset of sexual maturity and environmental conditions.

spawning stock biomass-per-recruit. The expected lifetime contribution to the spawning stock biomass for a recruit of a specific age (e.g., per age 2 individual). For a given exploitation pattern, rate of growth, and natural mortality, an expected equilibrium value of SSB/R can be calculated for each level of F. A useful reference point is the level of SSB/R that would be realized if there were no fishing. This is a maximum value for SSB/R, and can be compared to levels of SSB/R generated under different rates of fishing. For example, the maximum SSB/R for Georges Bank haddock is approximately 9 kg for a recruit at age 1.

species. A taxon of the rank of species; in the hierarchy of biological classification the category below genus; the basic unit of biological classification; the lowest principal category of zoological classification.

standard length. The straight distance between the tip of the snout and the base of the caudal fin rays.

standardization. The procedure of maintaining methods and equipment as constant as possible.

stock. Individuals that share a particular migration pattern, specific spawning grounds, and subject to a distinct fishery. A fish stock may be treated as a total or a spawning stock. Total stock refers to both juveniles and adults, either in numbers or by weight, while spawning stock refers to the numbers or weight of individuals which are old enough to reproduce. (This level of classification typically subsumes race, metapopulation, subpopulation, and demes. Used in the context of management.)

stock assessment. The process of collecting and analyzing biological and statistical information to determine the changes in the abundance of fishery stocks in response to fishing, and, to the extent possible, to predict future trends of stock abundance. Stock assessments are based on resource surveys; knowldege of the habitat requirements, life history, and behavior of the species; the use of environmental indices to determine impacts on stocks; and catch statistics. Stock assessments are used as a basis to assess and specify the present and probably future condition of a fishery.

stock origin. The genetic history of a stock.

GLOSSARY

stray. An individual that breeds in a population other than that of its parents.

stray rate. The proportion of a population that consists of strays.

straying. A natural phenomena of adult spawners not returning to their natal stream, but entering and spawning in some other stream.

stream reach. A section of a stream at least 20 times longer than its average channel width that maintains homogeneous channel morphology, flow, and physical, chemical, and biological characteristics

stressors. Physical, chemical, or biological perturbations to a system that are either (a) foreign to that system or (b) natural to the system but applied at an excessive [or deficient] level (Barrett et al. 1976, p192). Stressors cause significant changes in the ecological components, patterns and processes in natural systems. Examples include water withdrawal, pesticide use, timber harvesting, traffic emissions, stream acidification, trampling, poaching, land-use change, and air pollution.

subadult. A developmental life stage when fish exhibit most but not all traits of an adult fish.

sub-population. A well-defined set of interacting individuals that compose a proportion of a larger, interbreeding metapopulation.

subspecies. A population of a species occupying a particular geographic area, or less commonly, a distinct habitat, capable of interbreeding with other populations of the same species.

subtidal zone. Shallow-water areas below mean low water.

subyearling. A developmental life stage of fish that are less than one year old.

supplementation. The use of artificial propagation to maintain or increase natural production while maintaining the long-term fitness of the target population, and keeping the ecological and genetic impacts to non-target populations within specified biological limits.

survival rate. Number of fish alive after a specified time interval, divided by the initial number, usually on a yearly basis.

sustainability. Maintaining a population at levels so that exploitation does not affect its reproductive ability and genetic diversity.

sustainable yield. The number or weight of fish in a stock that can be taken by fishing without reducing the stock biomass from year to year, assuming that environmental conditions remain the same.

sustainable. A sustainable way of life is one in which human needs are met without diminishing the ability of other people, creatures, and future generations to survive.

sustained yield. A method of salmon management that guarantees a consistent supply of the resource.

sympatric. Living in the same place, or a least overlapping in ranges.

tagging. Process where scientists catch fish record their physical characteristics, tag the fish and release them. When fishers catch tagged fish they return the tags (and if possible the fish) with information on the fish and where the fish was caught.

taxonomy. Classification of plants and animals into established groups or categories on the basis of their natural relationships.

GLOSSARY

terminal fisheries. A fishery in a river or near the mouth of a river where returning salmon pass through or congregate near to an prior to spawning, and where stocks are relatively unmixed; occurring close to point where local populations separate.

tetanus. State of a muscle undergoing a continuous fused series of contractions due to electrical stimulation.

thermocline. A well-defined vertical temperature change or boundary often associated with stratification in lakes.

total length. Total length is the distance from the most anterior part of the head to the tip of the longest caudal fin ray when the fin lobes of the tail are pressed. In BC, total length is the measurement most commonly used on fish without forked tails such as burbot and sculpins.

trawl. Fishing gear. cone-shaped net towed in the water by a boat called a "trawler". Bottom trawls are towed along the lake or ocean floor to catch species such as groundfish. Mid-water trawls are towed within the water column.

trawl surveys. Scientists catch fish with a trawl net and record what they catch and note the changes when they fish the same area later.

trawling. Fishing methods where a single vessel, or a pair of vessels, tows a large netting bag (trawl net) behind the vessel.

trend. The directional change over time in a series of data via monitoring. Trends can be measured by examining individual change (change experienced by individual sample units) or by examining net change (change in mean response of all sample units).

tributary. A smaller stream which flows into a larger stream.

trolling. A fishing method where baited hooks or lures are towed behind a vessel

turbidity. Muddiness created by stirring up sediment.

undulating. To move in waves. Referring to the movement of a female fish's tail in a waving motion used to move gravel for the construction of a redd.

upwelling. The movement of nutrient rich waters from the bottom of the ocean to the surface.

ventral. On the lower surface; pertaining to the abdomen or belly.

viable population. A population that maintains its vigor and its potential for evolutionary change.

viable salmonid population. NOAA Fisheries term defined as "an independent population of any Pacific salmonid that has negligible risk of extinction due to threats from demographic variation (random or directional), local environmental variation, and genetic diversity changes (random or directional) over a 100-year time frame."

vital signs. are a subset of physical, chemical, and biological elements and processes of ecosystems that are selected to represent the overall health or condition of resources, known or hypothesized effects of stressors, or elements that have important human values. The elements and processes that are monitored are a subset of the total suite of natural resources that include water, air, geological resources, plants and animals, and the various ecological, biological, and physical processes that act on those resources. Vital signs may occur at any level of organization including landscape, community, population, or genetic level, and may be compositional (referring to the variety of elements in the system), structural (referring to the organization or pattern of the system), or functional (referring to ecological processes).

warmwater fish. A broad classification on non-salmonid fish that generally have at least one spiny ray, have pelvic and pectoral fins located behind the gills, and are usually suited for water that consistently exceeds 70° F.

watershed. An area drained by a particular stream. Includes all the vegetation, manmade structures, groundwater and surface water.

weak stock. Listed in the Integrated System Plan's list of stocks of high or highest concern; listed in the American Fisheries Society report as at high or moderate risk of extinction; or stocks the National Marine Fisheries Service has listed. "Weak stock" is an evolving concept; the Northwest Power and Conservation Council does not purport to establish a fixed definition. Nor does the Council imply that any particular change in management is required because of this definition.

weight-at-age. Average weight of individuals in each age class of a particular stock.

weir. Usually a barrier constructed to catch upstream migrating adult fish.

wild population. Fish that have maintained successful natural reproduction with little or no supplementation from hatcheries.

wild salmon. Salmon produced by natural spawning in fish habitat from parents that were spawned and reared in fish habitat.

within-stock diversity. The overall genetic variability among individuals of a single population or stock.

year class (or **cohort**). Fish in a stock born in the same year. For example, the 1987 class of cod includes all cod born in 1987, which would be age 1 in 1988. Occasionally, a stock produces a very small or very large year class that can be pivotal in determining stock abundance in later years.

yield per recruit. The expected lifetime yield per fish of a specific age (e.g., per age 2 individual). For a given exploitation pattern, rate of growth, and natural mortality, an expected equilibrium value of Y/R can be calculated for each level of F.

yolk. The food part of an egg.

Literature Cited

Bain, M. B, and N. J. Stevenson. 1999. Aquatic Habitat Assessment: Common Methods. American Fisheries Society Bethesda, Maryland.

Barrett, G. W., G. M. Van Dyne, and E. P. Odum. 1976. Stress ecology. BioScience 26:192–194.

Elzinga, C. L., D. W . Salzer, and J. W. Willoughby. 1998. Measuring and monitoring plant populations. BLM Technical Reference 1730-1. BLM/RS/ST-98/005+1730.

Likens, G. 1992. An ecosystem approach. its use and abuse. Excellence in ecology. book 3. Ecology Institute, Oldendorf/Luhe, Germany.

Noon, B. R. 2003. Conceptual issues in monitoring ecological systems. Pages 27–71 in D. E. Busch and J. C. Busch, D. E. and J. C. Trexler, eds. 2002. Monitoring ecosystems: Interdisciplinary approaches for evaluating ecoregional initiatives. Island Press, Washington, D.C.

Oakley, K. L., L. P. Thomas and S. G. Fancy. 2003. Guidelines for long-term monitoring protocols. Wildlife Society Bulletin 31:1000–1002.

GLOSSARY

Index

A
abundance (estimating) 70, 89, 95, 96, 99, 113, 124, 135, 156, 158, 202, 268, 269, 308
acoustic survey. *See* hydroacoustic
anesthetic
    formalin 431
    MS-222 181, 187
Arctic char *Salvelinus alpinus* 134, 385
Atlantic salmon *Salmo Salar* 198, 200

B
brook trout *Salvelinus fontinalis* 104, 200, 422
brown trout *Salmo trutta* 104, 200, 426
bull trout *Salvelinus confluentus* 96, 104

C
capture efficiency 88, 91, 92, 101, 102, 105, 117, 121
carcass 59, 435
carcass count 59, 435, 439
    estimation for capture–recapture 60, 62, 68
    scale sampling 66
cast net 87–94
catch-per-unit-effort (CPUE) 88, 100, 104, 112
community richness. *See* species richness
coded-wire tag (CWT). *See* tags
cutthroat trout *Oncorhynchus clarki* 104, 106, 153, 159

D
data management 69, 162, 225
data elements 27, 32, 40
    metadata 31, 40, 55, 349, 439
        quality assurance 55, 27, 166, 184, 218, 349
        quality control 5, 27, 166, 184, 192, 393, 395
data sharing 38
diel sampling 159, 164, 167, 426
DNA sampling 66–67, 270, 430
Dolly Varden *Salvelinus malma malma* 446

E
egg-to-smolt survival 235
escapement (estimating) 59, 62
estuary (sampling in) 269, 272, 276, 280, 412

F
fin marking. *See* marking
fishway 173, 174, 187, 190
fry 186, 222, 239, 243, 256

G
genetic sampling. *See* DNA sampling
gillnet 135, 325
gill cover. *See* marking operculum
grilse 35

H
hydroacoustic surveying
    shore-based 133–147
    boat-based 138, 157

J
jack salmon. *See* grilse

K
kelt 187

L
lake (sampling in) 153, 425

M
mainstem (sampling in) 173
marking (of salmonids) 68, 106, 110, 118
    fin clipping 109, 111
    operculum punches/notches 63, 68
masu (cherry) salmon *Oncorhynchus masou* 88, 91
measurements (of salmonids)
    fork length 59, 73, 109, 348, 353
    total length 109, 211
    mid-eye fork (MEF) 403
    mid-eye to posterior scale (MEPS) 75
    post-orbital to hypural length (POHL) 59
mesh size 87, 267, 272, 277, 281, 308, 313, 341, 411, 425
monitoring plan 11, 15, 21, 55, 200
mortality (during survey methods) 122, 135, 177, 181, 341, 351

N
net. *See* cast net, seine net
nocturnal (nighttime) sampling 274, 448

O
operculum. *See* marking

P
parr 96, 104, 186, 235, 243
pelagic (sampling in) 159, 166, 276, 292, 305, 426
pond (sampling in) 89, 93, 313
pool (sampling in) 89, 109, 117, 173, 254, 272, 298, 313, 328, 332, 438
presence/absence sampling. *See* sampling design

Q
quality assurance. *See* data management

R
radio-telemetry 222
rainbow trout *Oncohynchus mykiss* 96, 104, 107, 112
removal (depletion) technique
    one-pass 104
    multiple-pass 100
riffle (sampling in) 254, 299, 327, 373, 438

S
sample size 20, 61, 97, 117, 260, 273, 329, 391, 414, 420
sampling precision 61, 62, 96, 112
sampling design
    master sample concept 20
    presence/absence 88, 91, 112, 155, 308, 325, 326
    rotating panel design 20, 197, 201, 206
    sampling error estimation 16–17
    simple random sampling (SRS) 16
    spatially-balanced sampling (generalized random-tesselation stratified design, GRTS) 14, 17, 20, 201, 204
    stratified random 16, 436
    systematic sampling 17, 365, 367, 370, 373–376
SCUBA 203, 282, 312
seine nets
    arc sets
    beach-lay elliptical arc set 300
    circle set 301
    double-seine simple arc set 299
    rectangular arc set 300
    simple arc set 292
    pulled linear sets
    lampara set 291
    parallel set 286
    perpendicular set 288
    perpendicular quarter-arc set 290
    wandering pole seine 290
    trap sets
    block net sets 303
    cable-L trap set 302
    channel trap tide set 305
    enclosure net tide set 305
    purse-seine sets
    purse seine set 306
selective fishery 341
smolt (estimating populations) 173, 176, 184

INDEX

smolt-to-adult survival 235
sonar. *See* hydroacoustics
spawner surveys
    aerial surveys 203, 226, 399–410
    foot-based surveys 203, 214, 403, 435–442
species richness 95
stream reach
    determination of for surveys 29, 113, 202, 206
    mapping 217
subtidal zone (sampling in) 272
survival rate (estimating) 341, 342

T

tags
    coded-wire tags (CWT) 59, 61, 63, 64, 69
    Floy tags 68, 345
    maxillae (jaw) tags 68
    passive integrated transponder (PIT) 176, 343, 345, 357
    telemetry tags 59, 63, 345, 357
taimen *Hucho perryi* 7
terminal fisheries 385
trawl surveys 154
tributary 202, 235, 243, 258

W

weir 236, 244, 252, 367, 368, 370, 374, 385–396